Global Issues in Water Policy

Volume 19

More information about this series at http://www.springer.com/series/8877

Bryan R. Jenkins

Water Management in New Zealand's Canterbury Region

A Sustainability Framework

Springer

Bryan R. Jenkins
Hyde Park, SA, Australia

ISSN 2211-0631 ISSN 2211-0658 (electronic)
Global Issues in Water Policy
ISBN 978-94-024-1212-3 ISBN 978-94-024-1213-0 (eBook)
https://doi.org/10.1007/978-94-024-1213-0

Library of Congress Control Number: 2017957640

Printed on acid-free paper

This Springer imprint is published by Springer Nature
The registered company is Springer Science+Business Media B.V.
The registered company address is: Van Godewijckstraat 30, 3311 GX Dordrecht, The Netherlands

This book is dedicated to the people of Canterbury; may you continue to work together to find a sustainable solution to water management in the region.

Preface

The first motivation for this book was to document the development and implementation to date of the Canterbury Water Management Strategy. Reliance on the processes of the legislative framework under New Zealand's Resource Management Act (RMA) was found to be an inadequate basis for sustainable water resource management in the Canterbury region. The RMA was designed when water was a relatively abundant resource. However, the expansion of irrigation in Canterbury, primarily associated with conversions to dairying, led to sustainability limits being reached for water availability and for cumulative effects on water quality. A new paradigm for water management was needed. A sustainability framework based on nested adaptive systems and collaborative governance underpins this new paradigm.

The second motivation for this book was to document the sustainability framework used in Canterbury. Other regions in New Zealand are experiencing pressures on their water resources. Other parts of the world are also experiencing pressures on their water resources. Water globally is an economic and environmental issue. Water crises have been in the top three of global risks in the last five annual Global Risks Reports of the World Economic Forum. Global freshwater nutrient loads are considered to be beyond the safe operating space for the planet. Global freshwater use is considered to have limited safe operating space remaining which is already largely committed to cover the expected water demands for food production to meet projected population increases. There is a need for a sustainability framework, like the one in this book, that links socio-economic systems with natural resource systems at multiple spatial and time scales.

Dr Jenkins has recently retired from the position of Professor, Strategic Water Management at the University of Canterbury and Lincoln University in Christchurch, New Zealand. Prior to that he was chief executive of the Canterbury Regional Council and was responsible for introducing collaborative governance to water management in Canterbury

The third motivation was to reflect on the Canterbury experience. There has certainly been a paradigm shift in water management in the Canterbury region. However, sustainability analysis in the book demonstrates that there is still more to be done to achieve sustainable management. The book provides insights to the further changes needed.

Hyde Park, SA, Australia Bryan R. Jenkins

Acknowledgements

The development and implementation of the Canterbury Water Management Strategy (CWMS) would not have been possible without the support and dedication of the people and communities of Canterbury that have constructively engaged with the collaborative approach. The fact that this paradigm is still underpinning the approach to water management in the Canterbury region is confirmation of Ostrom's thesis that self-governing communities can be an enduring institutional basis for sustainable management.

The CWMS required significant contributions from the staff of Environment Canterbury in strategy development, community engagement, scientific investigations, plan development and regulation of water. Particular mention needs to be made of the contributions from Ken Taylor and Christina Robb.

There were significant creative contributions from Geoff Henley to the community engagement and strategy development process; Phil Driver, the creator of Open Strategies and its application in Canterbury; Gerald Midgley for advice on strategic decision-making; John Friend, the creator of Strategic Choice and the workshop leader of its use in Canterbury; and Martin Ward, Barry Sadler and Shona Russell, the creators of the sustainability appraisal process and its application in Canterbury.

I wish to thank Environment Canterbury, firstly, for my appointment as chief executive for more than 7 years and the opportunity to introduce collaborative governance to the region, develop the CWMS and commence its implementation; secondly, for funding a professorial fellowship at the University of Canterbury which provided the opportunity to commence the book, further develop the sustainability framework and teach sustainable water management; and, thirdly, for continuing to implement the CWMS.

The documentation of the paradigm would not have been possible without the opportunity to teach a postgraduate course in water resource planning and management. The students in the course over the last 5 years provided the stimulus to refine the explanation of nested adaptive systems and collaborative governance. Also, their discussion of the concepts in class, their application of the concepts in assignments and their feedback throughout the course were invaluable in improving the exposition of the concepts. I would particularly like to thank Vicky Southworth who reviewed a draft of the book and provided many helpful comments.

Thanks also go to Ariel Dinar, the series editor, and Fritz Schmuhl, my Springer editor.

I wish to thank Ruth Spearing and the staff at Christchurch Hospital for their medical treatment that enabled me to recover from a debilitating blood cancer.

Special thanks are due to my wife Janet whose support not only throughout the writing of the book but also through earthquakes, repairs to a damaged house, medical treatments and everything else in life made it possible to complete the book.

Contents

List of Figures

List of Tables

List of Boxes

Boxes provide case studies that show either (a) specific Canterbury examples to illustrate a sustainability issue in more detail (Boxes 2.1, 3.1, 6.1, 6.2, 6.3, 8.2, 9.1, 9.2, 12.1 and 13.1); (b) international examples that give additional insight to the application of the sustainability framework to water issues (Boxes 7.2, 8.1, 9.3, 10.1, 10.2, 10.3, 10.4, 10.5, 11.1 and 14.1); and (c) examples of management interventions adopted in other countries that could be applied in Canterbury (Boxes 7.1, 12.2 and 14.2).

Chapter 1
Introduction

Abstract The book is designed to achieve two major purposes. The first is to describe the developments in water management policy in the Canterbury Region of New Zealand. The strategic approach, the collaborative engagement, and, the nested adaptive systems approach represent a paradigm shift in water management in New Zealand. The second is to delineate the sustainability framework that underpins the Canterbury approach. The framework is based on the concept of developing sustainability strategies to address critical failure pathways. While the focus of the book is on Canterbury, comparative applications of the framework to issues in other parts of New Zealand and international issues are included.

The book can be used in at least two ways. The first is the application of a sustainability framework to the management of water in the Canterbury region. The second is the exposition of a sustainability framework that can be applied to the management of water in a region with the application to Canterbury as an illustrative case study.

The book has an introduction and four parts. The introduction (Chap. 1) sets out the scope of the book. The first part (Chaps. 2, 3, 4 and 5) provides the institutional and theoretical framework. The second part (Chaps. 6, 7, 8, 9 and 10) addresses the potential failure pathways for the management of the water resource system in Canterbury. The third part (Chaps. 11 and 12) describes the sustainability methods and their application to Canterbury water issues. The fourth part (Chaps. 13, 14 and 15) looks at the implications for water management in the region and the further changes needed to achieve sustainable management.

Keywords Water management in Canterbury • Sustainability framework • Failure pathways • Nested adaptive systems • Collaborative governance • Management interventions

1.1 Context

In the late 1980s New Zealand introduced major reforms to its institutional and legal framework for the management of natural resources and the government structures with responsibilities for natural resource management. The cornerstone of the legal

B.R. Jenkins, *Water Management in New Zealand's Canterbury Region*, Global Issues in Water Policy 19, https://doi.org/10.1007/978-94-024-1213-0_1

framework was the Resource Management Act (RMA) which replaced 55 statutes and 19 sets of regulations. The RMA has the stated purpose of sustainable management. At the same time, more than 800 government and quasi-government agencies were rationalised into 3 central government agencies, 12 regional councils and 4 unitary councils, and, 70 district and city councils. Regional and unitary councils which had significant responsibilities for water resource management had boundaries based on catchments. There have been numerous gains under the new arrangements but also there are reservations about whether they can achieve sustainable management of water resources.

New Zealand is comparatively fortunate when it comes to water. It has one of the highest rates in the world of annual renewable water resources per capita of 90,000 m^3/year/person. However, the east coast is relatively dry, especially Canterbury on the east coast of the central South Island and in the rain shadow of the Southern Alps. There has been increasing demand for water with the expansion of irrigated pasture for dairying. Canterbury is reaching the sustainability limits of the current means of water abstraction, and, cumulative effects of land use intensification on water quality and freshwater ecology are major concerns.

New Zealand has the highest growth rate of irrigation among developed countries: a 90% increase from 1990–2 to 2001–3 compared the OECD average of 6%. Although it is only 12% of area of New Zealand, Canterbury has 70% of New Zealand's irrigated land and its consumptive water allocation is nearly 60% of water allocated in the country. The situation in Canterbury provides the largest and most critical test of whether the New Zealand framework under the RMA can achieve its purpose of sustainable management.

The experience in Canterbury has demonstrated the limitations of the New Zealand framework based on the RMA. A different approach, based on collaborative governance (Ostrom 1990) and nested adaptive systems (Gunderson and Holling 2002), has been developed as the Canterbury Water Management Strategy and is in the process of being implemented.

1.2 Scope of the Book

This introductory chapter (Chap. 1) sets out the scope of the book. An outline of the chapters in each of the parts is provided.

In Part I, the water management framework for New Zealand is explained (Chap. 2). It covers the institutional arrangements, legislative provisions, resource management practice and current developments, such as the national Land and Water Forum.[1] Chapter 3 provides a description of water management in Canterbury addressing the history and approach under the RMA; it also covers the development

[1] The book draft was completed in March 2017 with text based on developments in water management up to December 2016. Changes in water management are continuing. Some of the significant developments since December 2016 are incorporated in footnotes to the text.

and implementation of the Canterbury Water Management Strategy. Chapter 4 develops the sustainability framework which underpins the analysis and strategic approach for evaluating water management in Canterbury. The failure pathway component builds on the rich literature on theories of collapse of the Classic Maya civilisation and places the collapse theories in the context of nested adaptive systems based on the concepts of Gunderson and Holling (2002). Drawing on the work of Chapin et al. (2009) sustainability strategies to address failure pathways are placed in the framework of nested adaptive systems. Chapter 5 provides examples of how the framework can be used to address water resource systems in identifying failure pathways and then developing sustainability strategies to address potential causes of system failure. It also describes how Part II addresses failure pathways for water management in Canterbury and how Part III covers sustainability assessments and decision making.

Part II provides the failure pathway analysis of water management in Canterbury. The initial concerns were with water scarcity due to inability of run-of-river surface water abstraction and groundwater withdrawals to meet increased demand; there were also concerns with the cumulative effects of increased land use intensification (Chap. 6). Other biophysical system failure pathways relate to the management of climate change implications which are problematic for Canterbury, and the management of water-related natural hazards of droughts and floods (Chap. 7). Chapter 8 addresses the socio-economic component of nested adaptive systems: this chapter considers governance arrangements, institutional arrangements and individual commitment to sustainable water management. Water-borne disease pathways are discussed in Chap. 9; there are particular issues with drinking water quality in rural Canterbury. There are also concerns at the larger spatial scale at the regional/national level (Chap. 10). With agricultural produce representing 70% of New Zealand's export income, regional and national economics is a significant issue. The implications of social change associated with the introduction of water management infrastructure are also addressed. With increasing international concerns with food security, there is increasing overseas interest in New Zealand farm land and food processing industries.

Part III addresses sustainability methods in relation to sustainability assessments (Chap. 11) and sustainability decision making (Chap. 12) and their application in Canterbury. In relation to sustainability assessments, for failure pathway analysis there is a generic approach of resilience assessments, while for sustainability strategies there is a need for the predictions of management interventions. With respect to decision making, multi-stakeholder processes are needed to complement collaborative governance. Also, different decision-making concepts are needed such as "Strategic Choice" which can accommodate uncertainty, complex systems and incomplete information; and, "Sustainability Appraisal" which considers the achievement of multiple sustainability criteria.

The final Part IV considers the implication of the analysis for water management in Canterbury for further changes in order to achieve sustainability. Chapter 13 summarises the biophysical sustainability limits identified in the failure pathway analyses of Part II and points out issues still to be resolved and solutions to these issues. This chapter also considers concerns relating to sustainable management practices that need to be confronted. Chapter 14 summarises socio-economic issues. It analyses the collaborative governance approach taken in Canterbury and identifies refinements needed. It also proposes changes needed to legislation and institutional arrangements as well as putting forward more appropriate evaluation approaches to water resource decision making. Chapter 15 concludes with the key implications for water management in Canterbury to achieve sustainable resource development.

International and local examples of failure pathways and sustainability strategies with respect to water resource management are provided throughout the book in "boxed" text.

1.2.1 Part I: Institutional and Theoretical Framework

1.2.1.1 Chapter 2: Water Management Framework in New Zealand

In the reforms of the late 1980s not only were new institutional arrangements established for the management of water but also there was a significant change in the role of government with a shift in focus from the planning of activities to regulating their effects. Chapter 2 describes the key institutional arrangements and legislative provisions. Implementation was highly devolved to regional councils with reliance on regional policy statements, regional plans and processes for consenting activities to define rules and conditions to manage adverse environmental effects of development proposals. It is only since 2011 that central government has played an active role in defining national policies and national standards.

Chapter 2 also considers issues arising from implementation of the RMA both in terms of improvements in the management of environmental effects and in relation to the limitations in managing resource scarcity and cumulative effects of resource use. Central government is proposing changes to the operation of the RMA. There are also several inquiries in progress looking at improvements to the operation of the RMA.

1.2.1.2 Chapter 3: Water Management in Canterbury

With 58% of New Zealand's water for consumptive use allocated in Canterbury, water allocation is a significant issue for the region. With an area of 507,000 ha under irrigation, Canterbury has 70% of New Zealand's irrigated land and has land suitable to double that area. Water also creates and sustains Canterbury's world-famous braided rivers, high country and coastal lakes, as well as lowland streams and wetlands. However, with current methods of abstraction (primarily run-of-river offtakes and groundwater bores) reaching sustainability limits for many parts of the region, water allocation has become contentious. In addition, cumulative effects of use are contributing to the declining ecological health of lowland streams and coastal lakes as well as water quality in surface and ground water.

There has been a significant increase in irrigation in Canterbury. It was estimated in 1982 there were about 100,000 ha irrigated area in the region (Dommisse 2005). In 2015, this has increased to 507,000 ha (Brown 2016) – a fivefold increase in those 33 years. In recent years, there has been an 11% growth per annum in consented irrigated area.

Reliance on RMA processes had led to long, drawn-out and acrimonious processes to address water management issues. For Canterbury, it was recognised that there was a need for a paradigm shift in water management incorporating:

- Water allocation and availability which addresses sustainability limits and climate variability;
- Management of cumulative effects of water takes and land use intensification; and
- A shift from effects-based management of individual consents to integrated management based on water management zones.

This recognition has led to the undertaking of the Canterbury Strategic Water Study and then the development of a Canterbury Water Management Strategy. This work has been undertaken in four stages:

- Stage 1: an initial study of water availability issues in Canterbury
- Stage 2: an investigation of potential storage sites
- Stage 3: a multi-stakeholder review of storage options
- Stage 4: the development of an integrated water management strategy.

Stage 1 was a study of water availability issues in Canterbury. This demonstrated that under low flow conditions current peak demand cannot be met by run-of-river and groundwater abstractions. On an annual basis water is available to meet future demand but would require storage. Stage 2 was an investigation of potential storage sites for their hydrological feasibility in terms of their supply reliability and effects on flow regimes. Stage 3 was a multi-stakeholder evaluation of the most prospective storage options. This evaluation identified the storage sites that were worthy of further investigation in terms of their sustainability. The evaluation also highlighted the need to address water quality risk from land use intensification. It identified the potential for integrated solutions which improved efficiency of existing use, minimised storage and enabled restoration of lowland streams through higher flows.

Stage 4 involved stakeholder and community engagement on the development of water management strategies based on Ostrom's collaborative governance concepts (Ostrom 1990). There were also strategic investigations of likely outcomes and a sustainability appraisal of strategic options. This generated a set of fundamental principles that the strategy should satisfy and four strategic options: A – business as usual, B – environmental restoration then infrastructure development, C – reconfiguration of existing consents and infrastructure to improve efficiency, and D – advance storage infrastructure with environmental mitigation. The investigations indicated the need to improve current land use practices if further intensification was not to compromise water quality. They also indicated that there were substantial gains to be made in water efficiency through integrated water management. Regional targets were developed for achieving the key components of an integrated strategy.

A nested collaborative governance structure has been established to develop implementation programmes for the Strategy. This comprises ten Zone Committees with community and rūnanga[2] representation who have been tasked with recommending zone implementation programmes; and a Regional Committee with community, Ngāi Tahu[3] and zone committee members to develop a regional implementation programme. Implementation programmes are now being given statutory backing through a regional plan. Solution packages are also being developed to address priority issues in each zone.

1.2.1.3 Chapter 4: Sustainability Framework

This chapter describes the nature of the collapse of the Classic Maya society. It provides an overview of the theories on the cause of the collapse. The Maya collapse is then considered in the context of models for societal collapse. A framework for analysing the resilience of societal systems to major disturbances is then developed. This framework is based on the concept of nested adaptive systems and principles of ecosystem stewardship.[4]

The Maya Collapse provides a fascinating example of the complexity that has to be considered in dealing with societal crises. With increasing sophistication of analysis and increasing levels of data focused on critical issues, more refined findings have been possible. A key purpose of this chapter is to provide an improved framework for analysing the resilience of societal systems dependent on natural resources. To this end the Maya Collapse is considered in the framework of "nested adaptive cycles" of Gunderson and Holling (Gunderson and Holling 2002). This placement provides a systematic framework for the dynamics of system failure and approaches to system management to address the potential causes of system failure. The framework is based on understanding transformations in human (socio-economic) and natural (biophysical) systems and considers how human and natural processes interact. The framework also considers interactions over different time scales and interactions over different spatial scales.

A categorisation of failure pathways is developed for different types of sustainability issues at different geographical scales. Drawing on the work of Chapin and his colleagues (Chapin et al. 2009), a classification of sustainability strategies for a

---

[2] Māori groupings centred on the whanau (family) and hāpu (sub-tribe) of marae (tribal meeting place) based communities.

[3] Ngāi Tahu is the Māori tribe whose rohe (tribal territory) includes the Canterbury region.

[4] Nested adaptive systems concept refers to a systems approach based on adaptive cycles (i.e. a four-phase cycle of exploitation, accumulation, disturbance and recovery) that are linked and nested (i.e. operate at different spatial or time scales (refer Sects. 4.2.1 and 4.2.3)). Principles of ecosystem stewardship refer to sustainability strategies based on the principles of reducing vulnerability, enhancing adaptive capacity, increasing resilience and enhancing transformability (refer Sect. 4.2.8).

nested socio-economic and biophysical system is described that can address potential failure pathways.

1.2.1.4 Chapter 5: Application of Sustainability Framework

The first three chapters of Part I provide the institutional and theoretical framework for the book. Chapter 5 sets out how these will be applied in the remainder of the book. The main components in structuring a sustainability analysis are the identification of failure pathways and then formulating sustainability strategies to address the failure pathways that threaten the sustainability of the water management system under consideration.

Examples of failure pathway analysis for three issues in the Waimakariri River catchment illustrate the definition of adaptive cycles, the nesting of adaptive cycles and the identification of critical variables for sustainable system management. Chapter 4 provided the general types of failure pathways for societal collapse while Chap. 5 describes failure pathways related to water management in the Canterbury region that are addressed in Part II.

Two examples of the development of sustainability strategies are then described: one is the evolution of the Canterbury Water Management Strategy from a focus on water availability to an integrated water management strategy; the second is the development of a rehabilitation strategy for Te Waihora/Lake Ellesmere based on a resilience analysis for the lake. The chapter then sets out how Parts III and IV address the conduct of sustainability assessments, sustainability decision making in Canterbury, and, the implications of the sustainability analysis in Parts II and III for sustainable water management in Canterbury.

1.2.2 Part II: Failure Pathway Analysis

Failure pathway analysis begins with the maintenance of the biophysical system under consideration, in this case water availability for the sustainability of water-dependent systems and then the cumulative impact of human use of water, i.e. the failure pathway of the impact of socio-economic system on the biophysical system (Chap. 6). Other biophysical system failure pathways relate to regional climate change at a higher spatial scale. There are also failure pathways relating to the linkages of the biophysical system to the socio-economic system (Chap. 7). This is the impact of water-related natural hazards (droughts and floods) on communities.

On the socio-economic side, there is a need for maintenance of higher level governance arrangements, co-governance and co-management arrangements with Māori, the management of infrastructure, the institutional arrangements for achieving compliance with water management requirements, and, the individual level commitment to sustainable management (Chap. 8). Chapter 9 identifies water-related health failure pathways: drinking water, contact recreation, shellfish contamination and toxic algal blooms. The system of Water Safety Plans is a good example of vulnerability assessment for contaminant pathways but a nested system resilience assessment indicates it is inadequate to protect water quality. Regional/national level socio-economic failure pathways are addressed in Chap. 10. These comprise external economic viability, social change due to technology change and interventions by external interests.

1.2.2.1 Chapter 6: Cumulative Effects at the Catchment Scale

Chapter 6 addresses two crucial issues relating to the sustainability limits and cumulative effects of water extraction and use at the catchment scale. The first section of the chapter examines water scarcity as the abstraction demands for human use exceed the biophysical capacity of the water resource system to regularly supply water. Supply/demand comparisons are made for run-of-river abstraction indicating lack of reliable supply. However annual comparisons indicate potential capacity but involve storage. Storage investigations indicate that there are limited opportunities for storages that were environmentally and economically sustainable. Other investigations indicated the potential for greater water availability by more efficient use of currently allocated water. Consideration of failure pathways and a broader range of sustainability approaches led to an integrated water management strategy for Canterbury instead of a storage strategy.

The second section considers the cumulative effects of water use. This comprises the impact on river flows (including flow variability) from abstraction; and, the effects on water quality from land use intensification (especially nutrients, bacterial contamination and sedimentation).

One of the major challenges in water management where there are multiple potential failure pathways is the determination of the significance of different pathways when adverse effects occur. The final section provides the example of teasing out the respective contributions to the decline in the aquatic health of groundwater-fed lowland streams of firstly, climate variability (in particular historically low winter rainfall and hence low aquifer recharge), and secondly, increased groundwater abstraction.

1.2.2.2 Chapter 7: Biophysical System Failure Pathways at the Regional Scale

Chapter 7 examines biophysical pathways at the regional scale due to climate change, and, the impact of water-related natural hazards of droughts and floods. The first section of the chapter reviews the New Zealand emissions profile, which has a significant contribution from the agricultural sector for methane and nitrous oxide, and, the projections for climate change. For Canterbury there are some significant implications. There are projections for increased potential evaporation deficit which will increase irrigation demand. It is also projected that the east coast will be drier in winter leading to reduced aquifer recharge and lower flows in foothill rivers. The west coast is projected to become wetter and warmer in winter leading to reduced snow and increased winter flows for the headwaters of Canterbury's alpine rivers. However with reduced snowmelt there will be reduced summer flows of the alpine rivers during the irrigation season. The implications for flows and reliability of supply are analysed at the catchment scale for the Waimakariri River.

New Zealand's response to climate change is summarised in the second section. The current approach is allowing significant increases in New Zealand's greenhouse gas emissions. The potential to reduce agricultural emissions through mitigation measures, such as nitrification inhibitors, and offsets is presented. However, there are no provisions to reduce agricultural emissions. The Western Australian system for addressing greenhouse gas emissions through environmental impact assessment is presented.

The third section considers the vulnerability of societies to droughts. It considers the capacity of society to respond to such hazards and sustainability approaches to management. The fourth section sets out a resilience approach to flood management and its application to the flood protection of Christchurch from flooding of the Waimakariri River. A contrasting international example, the response in New Orleans in 2005 to Hurricane Katrina, is provided.

1.2.2.3 Chapter 8: Socio-economic Failure Pathways

Chapter 8 addresses the socio-economic failure pathways of nested adaptive systems. The first section considers the higher level arrangements in relation to governance. It covers current developments in Canterbury and New Zealand, theoretical developments in democracy and institutional design for resources at sustainable limits, as well as trends in institutional arrangements in North America, Europe and Australia. Māori involvement in water governance is also addressed. The historical shift in New Zealand is described in terms of Ostrom's three governance models for common pool resources (such as water): the "Leviathan" model of government provision of services; the privatisation model with private sector provision of services and government's role limited to that of regulator; and, the self-governing community model. The change in New Zealand has been from a Leviathan model in the form of the Ministry of Works to the regulatory model of the RMA. Because of the shortcomings of the RMA approach in Canterbury there has been a shift to a collaborative governance approach based on the self-governing community model.

The government role in water infrastructure is considered in the second section. The change of the 1980s saw a dramatic shift from an active role of central government as the developer of irrigation and hydro schemes to a reliance on the private sector for infrastructure investment. However, a strategic review of infrastructure sectors in New Zealand rated the water sector very poorly (New Zealand Government 2011). The shift from water being considered an abundant resource in the 1980s to being a constrained resource means there is a greater need for infrastructure coordination. Also with the cumulative effects of land use intensification there is a greater need for investment in ecosystem health and recreation amenity.

The third section examines institutional arrangements for regulatory compliance. The traditional hierarchical controls have not been fully effective. Alternatives to regulatory models for ensuring integration of water management are discussed. The Dairy and Clean Streams Accord, as a voluntary approach of self-regulation is reviewed. The concept of "audited self-management" to overcome the shortcomings of voluntary approaches is described together with its application to the Te Ngawai catchment. The concept of a regulatory spectrum is presented to both encourage beyond compliant behavior as well as address non-compliance.

The individual commitment is addressed in the fourth section. Two individual behavioural models are described: the Lawler and Porter expectancy theory of motivation (Lawler and Porter 1967), and the Dornbusch and Scott evaluation process model for achieving compliance (Dornbusch and Scott 1975). Water management applications in Canterbury are provided: one for the motivational model underpinning the Living Streams programme in the Pahau catchment, and one for the compliance programme for dairy farms in the region. The need for a water ethic by all water users is advocated.

1.2.2.4 Chapter 9: Water-Related Health Failure Pathways

Water-related disease pathways are identified in the first section of this chapter. The main pathways are drinking water, water contact recreation, shellfish gathering from contaminated water and contact with toxic algal blooms. The critical variables for these pathways are described and the thresholds associated with disease risk that are used in management standards and guidelines are identified.

The management approaches to these issues are described in the second section and then analysed in the third section in the context of the sustainability framework based on nested adaptive systems.

The Water Safety Plan approach developed by WHO (World Health Organisation) has been adopted in New Zealand. The approach represents a good example of vulnerability assessment for the impacts of a water resource hazard (i.e. drinking water contamination) on the socio-economic system. However, this has not led to satisfactory drinking water status in Canterbury, particularly for small scale drinking water supplies.

The management approach to recreational water quality for both faecal contamination and toxic algal blooms is primarily based on providing public warnings to avoid water contact when monitoring indicates that critical variables associated with health risk have been exceeded. The decline in recreational water quality is a public concern.

The approach used for commercial shellfish farms is based on a hazard analysis and critical control points (HACCP) approach. This approach has many similarities to sustainable management using nested adaptive systems. However the risk of shellfish contamination is relatively low in Canterbury.

The analysis of management approaches indicates that for sustainability with respect to health risks associated with water-borne disease the following issues need to be addressed: (1) proactive catchment management to prevent contamination, (2) the organizational scale at which drinking water is addressed, and (3) the affordability of management interventions.

1.2.2.5 Chapter 10: Regional Level Socio-economic Failure Pathways

Chapter 10 addresses socio-economic failure pathways at a broader spatial scale: water's role in regional and national economics based on production for export; the potential for external intrusion associated with the availability of water; and, the implications of water technology on society.

Over 95% of New Zealand's agricultural production is exported. New Zealand is the world's largest exporter of dairy products. In 2014, the dairy industry was New Zealand's biggest export earner at $17 billion/year representing 28% of the country's exports and 40% of the world's dairy export market. This had increased from $5.8 billion in 2003 with the greatest growth in Canterbury – a 40-fold increase from 6 million kg of milk solids in 1984–5 to 248 million kg of milk solids in 2009–10 – all based on irrigation.

The first section of the chapter provides a benefit-cost analysis of increasing irrigated area in Canterbury and then compares the net benefit with estimates of the environmental externalities. For 236,000 ha of additional irrigated area the net economic benefit of $0.8 billion is comparable to the net present value of externalities ($0.6 – $0.9 billion). A national level analysis using dynamic computable general equilibrium modelling demonstrates the sensitivity of the dairy industry to the international price of milk solids. A water footprint analysis using the concept of "virtual water" (i.e. the water used to produce a product) shows that New Zealand is a net exporter of virtual water at 2790 m^3/year/person.

In relation to external intrusion, the value of water and concerns about food security have attracted international interest in foreign investment in New Zealand land and food-producing industries. While this is a controversial issue within New Zealand, foreign intrusion as a potential water management failure pathway is more clearly demonstrated by international examples. A summary of the historical conflict over access to water in the Jordan River Basin is presented. Also, the international foreign investment in farmland, heightened by the 2007/8 food crisis, is also discussed.

This chapter also considers the social effects of changing technology. A Canterbury example is presented of the rejection of a proposed storage in the Waianiwaniwa Valley because of its social implications. For comparison, the social impacts of the Three Gorges dam in China is included. There can also be more subtle social changes such as the changes in social structure in the Deccan Plateau in India with the shift from communal water storages to individual tube wells with electric pumps.

1.2.3 *Part III: Sustainability Methods*

Part III considers sustainability methods. Chapter 11 addresses sustainability assessments. The chapter covers two main types of assessments. The first is resilience (or vulnerability) assessments. The focus of these assessments is to determine what components of the nested adaptive system are at or near system thresholds and the critical variables related to the dynamics of system change. The second is the assessments of sustainability strategies which involve predictions of management interventions with a focus on the system outcomes that are achieved as a result of the interventions. Chapter 12 considers decision making methods for sustainability. Multi-stakeholder approaches to decision making are described. There is a need for approaches for finding solutions to complex problems where there is incomplete information, many interconnecting issues, uncertainty about possible effects, and multiple interests with conflicting objectives. The application of "Strategic Choice" to the Canterbury Water Management Strategy is used as an example. There is also a need for appraisal approaches consistent with the sustainability framework. This involves meeting sustainability thresholds across multiple criteria. The application of "Sustainability Appraisal" to the Canterbury Water Management Strategy is used as an example.

1.2.3.1 Chapter 11: Sustainability Assessments

There are two main components of sustainability assessments. The first is the evaluation of potential failure pathways. This involves defining the failure pathways and identifying the critical variables and their thresholds associated with the failure pathways. Vulnerability assessments (also called resilience assessments) evaluate the ability of a system to adapt to disturbance (resilience) or the risk of system change (vulnerability).

The second component of sustainability assessment is the formulation of sustainability strategies. This involves the identification of potential management interventions to address the vulnerabilities of the system and their contribution to sustainability outcomes.

The first section outlines the sustainability framework. The approach of vulnerability assessment is described using the steps defined by Hewitt (1997) in a failure and recovery sequence as an adaptive cycle. Sustainability strategies identify management interventions that target the critical variables and define the level of intervention to maintain the structure and function of the system.

The main part of the chapter sets out the sustainability assessment of six New Zealand lakes subject to eutrophication. The comparative analysis identifies that the lakes have different failure pathways and therefore different critical variables and different thresholds. The current approaches to lake management are also evaluated as possible sustainability strategies. The socio-economic side shows some common elements reflecting the adoption of collaborative organisational arrangements. However on the biophysical side, the analysis indicates that while the interventions will improve water quality (or at least reduce the rate of degradation), the level of intervention is not sufficient to achieve sustainable water quality outcomes for the lakes. The analysis also identifies the links between Chapin's sustainability strategies and management interventions at each of the four phases of the adaptive cycle.

A sustainability analysis is also provided of the decline of the Aral Sea. The analysis shows the adaptive cycles and failure pathways leading to biophysical degradation but also the need to consider multiple spatial scales and the links between socio-economic and biophysical processes. It shows that economic outcomes at the larger basin level are linked to the degradation of smaller scale systems (both biophysical and socio-economic) of the Aral Sea.

1.2.3.2 Chapter 12: Decision Making for Sustainability

Chapter 12 examines decision making processes for sustainability in a collaborative governance setting. The first section considers multi-stakeholder decision processes. The second section discusses decision making methodologies for complex problems, while the third section describes the evaluation of options using sustainability appraisal. A final section looks at the implications for decision making to achieve sustainability.

Collaborative governance approaches involve multi-stakeholder decision processes. There is not a "one-size-fits-all" formula; rather there is a framework of matters to be considered. This includes the context involving the process design, the linkage to decision making, the identification of issues, stakeholder identification, facilitation back-up and funding. It includes framing of the process in terms of group composition, goals and agenda. Inputs comprise stakeholder preparations, agreed rules and procedures, mechanisms to address power gaps between participants, and, capacity building of stakeholders. Dialogue during the process needs established communication channels, facilitation and rapporteuring leading to decision making and closure of issues. Outputs need to be documented and facilitate the implementation of action plans; the ongoing stakeholder processes need to have an impact on official decision making and continue to relate to non-participating stakeholders and the general public. These principles are reflected in the Living Streams programme and the community engagement process for the Canterbury Water Management Strategy (CWMS). The Living Streams programme was a partnership with communities in degraded tributary catchments to improve water quality.

The decision-making process for the CWMS was framed around the stakeholder and community engagement process. It was designed around stakeholder and community engagement on the development of strategic options, the definition of the strategic options for regional water management, community consultation on their option preferences, strategic investigations of outcomes, and, sustainability appraisal of options. The multi-stakeholder process is described and then more detail on defining strategic options and their appraisal is presented.

Defining strategic options in a multi-stakeholder process is a complex problem where there is incomplete information, many interconnecting issues, uncertainty about possible effects of options and multiple interests with different conflicting objectives. Strategic Choice, which was used in the CWMS, is one of the few decision making methodologies designed to accommodate problems of this type. It is more a method of "problem structuring" rather than "problem solving". It comprises four tasks of shaping, designing, comparing and choosing; where shaping involves identifying the range of problems to be considered and the links to other issues and other decisions that could be affected by possible courses of action; designing involves identifying possible options for each decision area and incompatibilities between options in different decision areas; comparing involves undertaking relative assessments of options and recording their comparative advantages; and choosing involves identifying areas of uncertainty, exploring options to address doubts, and devising action schemes for early implementation. The product is a commitment package for early actions, explorations in response to uncertainty, and arrangements for deferred decisions.

The Strategic Choice approach was applied to the CWMS with a workshop for shaping and designing, sustainability appraisal for comparing, and a community consultation and stakeholder engagement process for choosing, leading to a commitment package set out in a strategic framework document.

Sustainability appraisal has been developed for the evaluation of alternatives in a sustainability framework. It has evolved from combining strategic planning to achieve community outcomes across the four well-beings of sustainability as defined in New Zealand (economic, environmental, social and cultural), with the concept of strategic assessment of policies plans and programmes which has developed from project-level environmental impact assessment. Sustainability appraisal is focused on the assessment of the change in capital assets: natural capital, economic capital, social capital and cultural capital. Criteria are developed for all aspects of capital. Sustainability bottom lines are defined for all criteria (quadruple bottom line) and objective-led top lines across all criteria (quadruple top line). The approach involved developing a strategic option that meet all sustainability bottom lines and approached the objective-led top lines.

Sustainability appraisal was applied to options being developed for the CWMS by a multi-stakeholder group with technical support. Four strategic options from the strategic choice process were evaluated. This demonstrated that "business as usual" under RMA processes was below the sustainability bottom line on nearly all criteria. The environment-led option scored well on environmental criteria but was below the economic bottom line. The storage-led option scored well on economic criteria but was below the environmental bottom line. However the efficiency-led option scored above the bottom line on nearly all criteria. Sustainability appraisal demonstrated that the only possible way to achieve sustainable development was by addressing existing uses of water as well as new use and projects. It also showed that the most economically viable source of additional water was from efficiency gains from existing uses rather than storage; that environmental requirements were best met by improved land use practices by existing and new users; and, that there was no capacity for further development unless cumulative effects of existing use were reduced.

To achieve sustainability it is not enough to have a technical basis for decision making; there is a need to find options that address the differing views of multiple stakeholders. This requires a collaborative process. The Canterbury experience has been that innovative solutions have arisen from effective collaboration. It also needs a multi-dimensional evaluation approach like sustainability appraisal. The sustainability appraisal approach differs from other forms of appraisal methodologies. Comparisons are made with benefit-cost analysis, planning balance sheet and multi-criteria evaluation.

Also, examples are provided of problems with effects-based decision making when sustainability limits of resource availability have been reached or cumulative effects have exceeded acceptable levels. Rather than effects-based approaches focused on reducing adverse effects of proposed new development, there is a need for outcome-based approaches to keep overall resource extraction and impacts of use within sustainability limits. Examples of alternative outcome-based approaches from other jurisdictions are described: one is the South African approach to allocation when water is scarce, and the other is salinity management in the Murray-Darling basin to reduce land use effects on river salinity.

1.2.4 Part IV: Implications for Water Management

The insights gained from failure pathway analysis and developing sustainability strategies to address critical variables leads to the identification of the changes needed to achieve sustainable management of water resources. Part IV looks at the implications for water management in Canterbury. Chapter 13 identifies the biophysical sustainability issues that need to be managed, and the sustainability management approaches to address these limits. Chapter 14 analyses the collaborative governance approach, legislative and institutional changes, the socio-economic issues, and approaches to evaluation that should be adopted. Chapter 15 summarises the main implications for Canterbury water management based on the sustainability analysis.

1.2.4.1 Chapter 13: Biophysical Limits and Sustainable Management

While there has been significant innovation in Canterbury, the analysis of water issues indicated that further change is needed to achieve sustainable management. Strategic thinking has evolved from the initial concern with water availability to a multi-dimensional strategy for integrated management across ten target areas. The initial focus on addressing water availability through storage on alpine rivers has shifted to improved water use efficiency and alternative forms of storage. Because of their adverse consequences, storages on the mainstems of alpine rivers have been prohibited. Alternatives of off-river storage, storage on tributaries and groundwater recharge are now being developed for long-term water availability. Smaller storages for irrigation schemes and on-farm storages are being used to address short-term variability in water availability.

However, management of water use efficiency is still rudimentary. Improvements in management of water use efficiency at both the farm and catchment scale would improve economic productivity in a region where water scarcity is a major constraint on the regional economy. There is also a need for greater attention to infrastructure co-ordination and integration of surface and groundwater management to improve water availability. Furthermore, a strategy to address climate change adaptation and greenhouse gas emissions from land use intensification is required.

The sustainability analysis has also identified the main impacts of water abstraction and land use intensification. Effects on surface water quality from land use intensification have been most noticeable in lowland streams and the lower reaches of foothill streams and alpine rivers including nutrient enrichment, algal blooms, faecal contamination, nitrate toxicity and siltation. Coastal lakes fed by these rivers are subject to eutrophication. There is also recent evidence of degradation of high country lakes. Nitrate and bacterial contamination of groundwater is an issue, particularly with the high dependence on groundwater as a drinking water source in Canterbury. However, the sustainability analysis shows that proposed improvements in land use management even with catchment interventions in zone implementation programmes are not sufficient to achieve sustainable water quality.

In addition, there are effects on abstraction on river flows with lower flows reducing aquatic habitat quantity, fewer freshes increasing algal growth, and fewer floods reducing braided character. Lower flows have increased the frequency and duration of river mouth closure. Groundwater abstraction lowered groundwater levels and reduced flows in groundwater-fed lowland streams. Despite recommendations from the regional council to decline further abstractions RMA processes, allowing an "overall broad judgement" between further resource use and environmental limits, have led to overallocation of groundwater resources.

In relation to sustainable management, the adoption of a nested approach has led to a more refined analysis of water management issues. There have also been innovations in management practices to address sustainability issues. To facilitate the sustainable management of cumulative effects, sustainability analysis involves a greater need for modelling and monitoring compared to effects-based management of projects.

The approach of managing to limits is experiencing problems. There are complications with natural variability, measurement uncertainties, modelling inaccuracies, multiple variables, difficulties in enforcement, contributions from past contamination, lag times in effects, uncertain causes of effects, and options for management interventions. In order to achieve sustainable outcomes, a need is seen for using nested adaptive cycles as the basis for management involving a collaborative approach at multiple spatial scales.

1.2.4.2 Chapter 14: Socio-economic Issues and Collaborative Governance

The chapter addresses socio-economic issues. The evolution of collaborative governance in Canterbury is summarised, and evaluations of its implementation reviewed. The means of funding and implementing the solutions packages of the zone committees is yet to be resolved. Also, the uneven implementation of the CWMS targets in undermining the social contract that led to the willingness of multiple stakeholders to accept the CWMS. This is prompting environmental and recreational stakeholders to disengage from the collaborative process.

The water management issues facing Canterbury have changed since the RMA was enacted. Sustainability limits of water availability and cumulative effects of land use intensification have been reached. The limitations of the RMA in managing resources at sustainability limits are set out. The evolution of the concept of sustainable development and the changing role of government in sustainable development since the framing of the RMA are outlined. This illustrates that the RMA no longer provides an adequate legislative basis to address the sustainable management of water issues in Canterbury. The RMA is designed for government having a regulatory role, however, this does not facilitate the more proactive role now required of government. Water framework legislation is proposed to introduce sustainable management goals and strategies to achieve these goals. This would incorporate evaluations of strategies considering economic, social, cultural and environmental criteria.

1.2.4.3 Chapter 15: Concluding Comments

The paradigm shift to a collaborative governance approach based on nested adaptive systems is bringing improvements to water resource management in Canterbury. Some of the key issues arising from the sustainability analysis indicates that further improvement is needed to achieve sustainable development. There is a need for more proactive measures to address recovery of degraded water systems. Reliance on managing to limits is not sufficient, rather management of the basis of nested adaptive cycles is more appropriate. Better measurement and management of water use efficiency would enhance water availability. Water quality measures have been proposed but these are only a first step; more comprehensive measures are needed. For collaborative governance to remain effective, all CWMS targets need to be advanced and all stakeholders need to be engaged. A programme to address climate change adaptation for water and greenhouse gas reduction for agriculture is required. Affordability is a constraint on the ability to implement further improvements. Funding mechanisms to implement proactive measures are required. While the RMA may provide a regulatory framework, there is a need for water framework legislation and water management strategies to facilitate sustainable development.

References

Brown, P. (2016). *Canterbury detailed irrigated area mapping*. Christchurch: Aqualinc.
Chapin, F., Kofinas, G., & Folke, C. (2009). *Principles of ecosystem stewardship: Resilience-based natural resource management in a changing world*. New York: Springer.
Dommisse, J. (2005). *A review of surface water irrigation schemes in Canterbury. Their development, changes with time & impacts on the groundwater resource*. Christchurch: Environment Canterbury.
Dornbusch, S. M., & Scott, W. R. (1975). *Evaluation and the exercise of authority*. San Francisco: Jossey-Bass.
Gunderson, L. H., & Holling, C. S. (2002). *Panarchy: Understanding transformations in human and natural systems*. Washington, DC: Island Press.
Hewitt, K. (1997). *Regions of risk: A geographical introduction to disasters*. Harlow: Longman.
Lawler, E., & Porter, L. (1967). Antecedent attitudes of effective managerial performance. *Organizational Behavior and Human Performance, 2*, 122–142.
New Zealand Government. (2011). *National Infrastructure Plan 2011*. Wellington: New Zealand Government.
Ostrom, E. (1990). *Governing the commons: The evolution of institutions for collective action*. New York: Cambridge University Press.

Part I
Institutional and Theoretical Framework

Chapter 2
Water Management Framework in New Zealand

Abstract There were major changes to the water management framework for New Zealand that occurred in the late 1980s which brought in new legislation, the Resource Management Act (RMA); new water administration bodies based on catchments, the regional councils; an Environment Court; and, a change in the primary role for government from service provider to water regulator.

The RMA created a national standards and policy role for central government; a regional policy, planning, consenting and regulatory compliance role for regional government; and, a local land use planning role for city and district councils.

The RMA led to improvements in the control of point source discharges. The example of the Christchurch City Sewage Outfall is presented. The Act also facilitated the management of project-specific effects through assessment, consenting, monitoring and enforcement provisions. However there are shortcomings of the RMA in relation to water management, for example, the management of diffuse sources from land use activities and the cumulative effects of multiple activities. In addition, a disadvantage in relation to sustainable management is that the Act is focussed on the adverse environmental effects of activities rather than sustainability outcomes. Also with the ability of the Environment Court to deal with technical matters the process has become legalistic and adversarial.

Current developments at the national level include the establishment of a multi-stakeholder process, the Land and Water Forum, to attempt to achieve an agreed approach at the national level for water management in New Zealand. There are also proposals to modify the RMA to make the system easier to use, to increase its certainty and predictability, and reduce unnecessary duplication and cost. More recently three inquiry processes into the RMA have commenced.

Keywords Resource Management Act • NZ water management institutions • Legislative processes • Land and Water Forum

B.R. Jenkins, *Water Management in New Zealand's Canterbury Region*, Global Issues in Water Policy 19, https://doi.org/10.1007/978-94-024-1213-0_2

2.1 New Zealand Institutional Arrangements

There was a major reform of natural resource management in the late 1980s in New Zealand. The number of local and regional government units was reduced from 625 to 94. The most significant innovation for water resource management under the new arrangements was the creation of regional councils whose geographical boundaries were based on natural river catchments. Regional councils are elected local government bodies that coordinate, and set policy for resource management, including water and soil conservation, and transport. They also have roles in pest management, civil defence, navigation safety, coastal management, hazardous waste and more recently biodiversity management (Wallis and Dollery 2000).

Coincident with these reforms was the introduction of the RMA in 1991. The purpose of the Act is to promote sustainable management. The legislation is "effects-based" and concentrates on the environmental effects of activities rather than the activities themselves. The focus is on environmental effects and leaves the pursuit of economic and social goals to other mechanisms. The legislation incorporates the principles of the Treaty of Waitangi.[1]

The reforms also led to the creation of the Ministry for the Environment. The Ministry of Works and Development which had a major water resource development role in building hydro-generation and irrigation infrastructure was abolished in 1988. There is no natural resources agency in central government in New Zealand. The Ministry for the Environment has responsibilities for National Environmental Standards and National Policy Statements. For 20 years there were no standards or policies relating to water management at the national level. The first National Policy Statement on water was gazetted in 2011.

Regional councils are required to prepare Regional Policy Statements identifying environmental issues and responses of significance for its region. Regional councils also have the authority to prepare Natural Resources Regional Plans which can include water management. Territorial authorities, the next tier of local government are required to prepare District Plans which cover land use and subdivision. Regional councils have the authority to issue resource consents for the taking and use of water, and for discharges to water, and, to monitor compliance with those consents. Territorial authorities have the authority to issue consents for land use.

The reforms also created an Environment Court. The Court has extensive powers not only to consider appeals on resource consent decisions but also on regional policy statements and plans. The Court has the ability to review the technical merit of decisions. This has made resource management in New Zealand a highly legalistic process. It has also led to an adversarial style of decision making.

---

[1] The Treaty of Waitangi is the treaty first signed on 6 February 1840 by representatives of the British Crown and various Māori chiefs. It resulted in British sovereignty over New Zealand and is generally considered the founding document of the nation. There is a preamble and three articles. The first article addresses Crown sovereignty. The second article addresses Māori rights in land and other resources. The third article guarantees Māori the same rights as other British subjects (Waitangi Tribunal 2016).

The RMA combined three key environment management activities:

- Resource allocation and consent processes for the management of the use of certain natural resources from the Water and Soil Conservation Act 1967 and Geothermal Act 1953;
- Land use planning and control of the built environment under the Town and Country Planning Act 1977; and
- Environmental regulation functions of the Clean Air Act 1971 and other acts regulating hazardous material.

However there was also a marked change in the role of government. The previous role of government involved directing economic activity and making trade-offs in the interests of the wise use of resources. As stated by Simon Upton, the Minister responsible for the passage of the Act through Parliament: "the Government moved to underscore the shift in focus from planning for activities to regulating their effects" (Upton 1995).

The RMA was also designed on the premise that people know best what it is that they are after in pursuing their well-being (Upton 1995). Thus the responsibility for defining proposals was left to proponents. The RMA was also designed on the basis of "effects management", i.e. that choices by applicants would be constrained by bottom lines of effects that were not to be exceeded.

This approach may be suited to circumstances where there is an abundance of resources. However when resource use approaches sustainability limits, either in terms of resource availability or in terms of cumulative effects of resource use, then the actions of one user can harm all others. Indeed for a common pool resource (i.e. a resource that is readily accessible and difficult to exclude access to, and, is in limited supply so that resource use by one user diminishes the availability for others), allowing all users to act in their own self-interest leads to degradation of the resource for all users (Hardin 1968).

There have been limitations identified in the practice of water management under the RMA (Jenkins 2009). These include:

- Effects based management rather than sustainability outcomes
- Reliance on regulation alone
- Focus on environmental issues alone
- Development pattern based on applicant-driven proposals
- First-come/First-served allocation rather than merit-based allocation
- Absence of provisions for sustainability limits and cumulative effects
- Reliance on adversarial court-based decision making
- The separation of responsibilities in relation to water and land use.

While "sustainable management" is the purpose of the RMA, the Act provides no elaboration on how decision makers can apply this purpose. Amendments to the RMA since its enactment have been focussed on efficiency of process rather than address the definition of sustainable management. It has been left to the courts to make an interpretation. The position from several court cases is that the application of Section

5[2] involves an "overall broad judgement" of whether a proposal will promote sustainable management of natural and physical resources (Skelton and Memon 2002).

Skelton and Memon argue "the fundamental tensions that underpin the section arise from the challenge of crafting a definition of sustainable management that can enable decision makers (elected councils, the Environment Court, the Minister for the Environment) to reconcile the spectrum of values different groups accord the environment in a plural social setting. Such a definition needs to be sufficiently clear, procedurally fair and focussed on the substantive goal of protecting and improving environmental quality" (Skelton and Memon 2002).

One of the purposes of this book is to provide an operational basis for sustainability.

2.2 RMA Legislative Provisions

The main policy and planning instruments of the New Zealand statutory framework under the RMA are:

- at the national level: National Environmental Standards, National Policy Statements and National Coastal Policy Statements
- at the regional level: Regional Policy Statements, and Regional Plans
- at the city/district level: City and District Plans

While there are three tiers under the legislation, water management has been highly decentralised in practice with the first National Policy Statement on water only being gazetted in 2011. The purpose of national policy statements is to state objectives and policies for matters of national significance (RMA s45). The National Policy Statement for Freshwater Management 2011 (New Zealand Government 2011) is a brief document which defines the intrinsic values for water and its use values. It also contains objectives and policies for water quality, water quantity, integrated management and *tangata whenua*[3] roles and interests. One of the key implications is that it directs regional councils to set quality and quantity limits.

The National Policy Statement for Freshwater Management was amended in 2014 (New Zealand Government 2014). The amendments introduced a National Objectives Framework for regional councils to develop freshwater objectives for all

[2] RMA s5 (1) The purpose of this Act is to promote the sustainable management of natural and physical resources. (2) In this Act, **sustainable management** means managing the use, development, and protection of natural and physical resources in a way, or at a rate, which enables people and communities to provide for their social, economic, and cultural well-being and for their health and safety while—(a) sustaining the potential of natural and physical resources (excluding minerals) to meet the reasonably foreseeable needs of future generations; and (b) safeguarding the life-supporting capacity of air, water, soil, and ecosystems; and (c) avoiding, remedying, or mitigating any adverse effects of activities on the environment.

[3] *Tangata whenua* means 'people of the land' in Māori and refers to the roles and interests of indigenous people.

freshwater management units. The amendments also introduced the requirements for accounting systems for freshwater takes and contaminants.

The purpose of regional policy statements is to provide an overview of the resource management issues of a region, and the policies and methods that will be used to address these issues and achieve the integrated management of natural and physical resources of the region (RMA s59). Canterbury's second generation Regional Policy Statement (Environment Canterbury 2013) was made operational in January 2013 (the first was in place in 1998). In relation to fresh water the following issues were identified for the Canterbury region:

- management of water and exercising stewardship and kaitiakitanga[4] over water;
- adverse effects of activities on fresh water;
- need for high quality fresh water for drinking water supplies, customary uses and other activities;
- the benefits of and demand to abstract and use fresh water for economic well-being and the costs and effects of meeting this demand and realising benefits; and
- inefficient allocation and use of water.

Regional plans are designed to give effect to regional policy statements and state: (1) the objectives for the region; (2) the policies to implement the objectives; and (3) the rules to implement policies (RMA s67). The Canterbury Region Natural Resources Regional Plan was made operative in 2011. The Land and Water Regional Plan has been developed to provide statutory backing to the Canterbury Water Management Strategy and the implementation programmes as they are agreed.

Activities in relation to water can be undertaken if allowed by a rule in a regional plan. Where activities would otherwise contravene provisions in the RMA a resource consent is required. For water management the main activities are taking, using, damming or diverting water (RMA s14) for which a water permit is required (RMA s87(d)), and, discharging contaminants into water or on to land that may contaminate water (RMA s15) for which a discharge permit is required (RMA s87(e)).

One other important legislative provision that was in the Water and Soil Conservation Act 1967 and was retained in the RMA is the Water Conservation Order. The purpose of a Water Conservation Order is to recognise and sustain outstanding or intrinsic values of water bodies and may provide for preservation in its natural state. Characteristics which can be considered to be outstanding are as habitat; fishery; scenic and natural characteristics; scientific and ecological values; recreational, historical, spiritual or cultural purpose; and, tikanga Māori[5]. There are three Water Conservation Orders in Canterbury: One on the Rakaia River and another on the Rangitata River specifying environmental flow requirements, and a third for Te Waihora/Lake Ellesmere in relation to lake levels and openings to ini-

---

[4] *Kaitiakitanga* means the exercise of guardianship by the tangata whenua of an area in accordance with tikanga Māori (Māori custom) in relation to natural and physical resources; and includes the ethic of stewardship.

[5] *Tikanga Māori* means rights, customs, accepted protocol, rule, Māori traditions, lore or law, the correct Māori way.

Table 2.1 Main steps in RMA Schedule 1 process for policies and plans and process for notified consents

Process step	Schedule 1 process for policies and plans	Process for notified resource consents
Notification	Public notification of proposed policies and plans with s32 evaluation report: *Sch1s5(1)*	Public notification for application for activity if adverse effects more than minor: *s95A*. Assessment of Environmental Effects available: *s92 (3B)*
Submissions	Any person can make submission and request to be heard: *Sch1s5(2)(b)*.	Any person can make submission on notified consent and seek to be heard: *s96*.
	Local authority to provide summary of decisions requested in submissions: *Sch1s7(1)*.	Consent authority may commission report or request applicant for information: *s92(2)*.
	Opportunity for submitters to make further submissions: *Sch1s7(1)(c)*.	
Mediation opportunity	Opportunity for mediation of disputes: *Sch1s8AA*.	Option of prehearing: *s99*; and mediation: *s99A*.
Hearing	Hearing into submissions: *Sch1s8B*	Hearing with commissioners: *s100A*.
Council decision	Local authority decision on provisions and matters raised in submissions: *Sch1s10(1)*.	Decision by consent authority: *s104(1)*.
Appeal on decision to Environment Court	Ability for submitters to appeal to Environment Court: *Sch1s14*.	Opportunity for submitters and applicant to appeal decision to the Environment Court: *s120*.
Court decision	Environment Court direction to local authority: *s293(1)*	Environment Court may confirm, amend or cancel decision to which appeal relates: *s290(2)*.
Appeal on points of law	Ability for appellants (applicant or submitters) to appeal Environment Court decision on points of law to the High Court: *s299*.	Ability for appellants (applicant or submitters) to appeal Environment Court decision on points of law to the High Court: *s299*.

Items in italics are reference to sections or schedules of the RMA

tially protect water bird habitat and with recent changes now includes long fin eel passage and Ngāi Tahu[6] values.

In addition to the statutory instruments an important component of the RMA is the decision making process for policies, plans and consents. Table 2.1 sets out the main steps of the process defined in Schedule 1 of the RMA for policies and plans. The table also shows a very similar process for resource consents that are publicly notified: these are applications for activities where the adverse effects are likely to be more than minor.

Of particular note in both processes are:

[6] *Ngāi Tahu* is the Māori tribe whose *rohe* (tribal territory) includes the Canterbury Region.

- the public notification with supporting documentation,
- the ability for any person to submit and be heard,
- the opportunity for mediation prior to formal hearings,
- the public hearing process before a decision can be made,
- the authority's decision being able to appealed to the Environment Court,
- the ability of the Environment Court to make technical determinations and revise (or confirm) authority decisions, and
- the only avenue of appeal beyond the Environment Court is on points of law to the High Court.

2.3 RMA Practice

The introduction of the RMA has led to environmental improvements in a number of areas, in particular, the management of point source discharges. Box 2.1 sets out the example of the Christchurch City Sewage Outfall (URS 2004). The RMA process for notified resource consents led to the removal of the discharge from the Avon Heathcote Ihutai Estuary even though the City's original proposal was to retain an estuary discharge because of lower cost. The process enabled the public, who had become submitters to the process, to have the ability to influence the decision. The example also shows the role of the Environment Court in the technical and not just legal, aspects of the process.

The RMA facilitated the management of project-specific effects, not just for major projects like wastewater outfalls, but also smaller scale activities such as water abstraction from groundwater bores; e.g. by limiting pumping rates to restrict the drawdown impacts on adjacent groundwater bores. Such limits could be incorporated into resource consent conditions. The RMA also introduced powers to enable regional councils to monitor and enforce these consent conditions.

Furthermore, the RMA facilitated the setting of environmental flow restrictions on resource consents for run-of-river abstractions. There are 162 environmental flow monitoring points established in Canterbury (Environment Canterbury 2011). When river flows dropped to certain specified levels, the ability to abstract from the river was either partially or fully restricted so that withdrawals did not further reduce flow.

However, there are shortcomings of the RMA in relation to water management.

While point source discharges can be readily addressed by the RMA, the issue of diffuse sources has been problematic, particularly in relation to contamination from land use. Regional councils can exercise controls on discharging contaminants onto land that may contaminate water. However, with land use decisions allocated to district and city councils, there is a separation of responsibilities for land-based sources of contamination. Furthermore, framing consent conditions that are readily enforceable is difficult for contamination from agricultural land use when the level of contamination is related to day-to-day management of farm operations.

While project-specific effects can be addressed by the RMA, the Act is not designed to deal with cumulative effects of multiple activities.[7] This is both in terms of allocating resources, like water, close to their sustainability limits, and, in terms of the impacts of resource use from many sources, such as land use intensification on multiple farms in a catchment.

With allocation under the RMA based on "first-come first-served" and decisions on water takes based on "acceptable effects", issues such as resource productivity are not considered in the context of the RMA[8]. This is not a concern when a resource is abundant: indeed "first-come first-served" is a very efficient and equitable way of administratively managing resource allocation applications when the resource is abundant. However, in dry regions like Canterbury[9], water availability is at sustainability limits and is the constraint on further irrigation as the means of increasing agricultural productivity. There is an opportunity cost to the regional economy for inefficient water use or low productivity use of water.

With the RMA designed for government's role to be the regulator of effects rather than the planner of activities, water management projects are dependent on proposals by applicants. This is not conducive to compatible proposals particularly when there is competition for scarce water. The RMA is designed for case-by-case consideration of projects on their merits. It is not designed for strategic approaches to integrated water management.

Furthermore, "effects-based management" does not equate to "outcomes-based management". Sustainable development is related to achieving community outcomes across multiple well-beings. In New Zealand four well-beings are normally considered: economic, social, cultural and environmental well-being.

With the pivotal role of the Environment Court in policies, plans and major projects, decision making under the RMA is a legal process. Cases for and against issues and projects are argued by lawyers. RMA practice has become highly adversarial.

Sir Geoffrey Palmer, the original architect of the RMA, is quoted as saying: "The provisions of the RMA relating to water haven't actually worked our very well. They need revision. In many ways, New Zealand had a lot of water, we thought, when the act was designed, but now we realise it is a scarce and valuable resource. The boom in dairying has certainly exacerbated that" (The Press 9 June 2008).

For Canterbury there is a need for a paradigm shift in water management. The approach requires:

- water allocation and availability which addresses sustainability limits and climate variability,

[7] As Gunningham notes: "the RMA does not protect the environment by invoking the precautionary principle and the court's narrow interpretation of cumulative 'environmental effects' has further limited the reach of the Act" (Gunningham 2008).

[8] As stated in the Court of Appeal decision in the case of Central Plains v Ngāi Tahu Properties Limited (19 March 2008): "This case concerns competing resource consent applications to take water from the limited free volume available from the Waimakariri River. The RMA says nothing specific about the priority of competing claims to take from a natural resource."

[9] Refer Sect. 3.1 for the water situation in Canterbury.

- management of cumulative effects of water takes and land use intensification, and
- a shift from effects management of individual consents to integrated management based on water management zones.

This book describes the development of the Canterbury Water Management Strategy based on nested adaptive systems approach of Gunderson and Holling (Gunderson and Holling 2002) and the collaborative governance approach of Ostrom (Ostrom 1990) designed to bring about such a paradigm shift.

Box 2.1 Christchurch City Sewage Outfall: An Example of the RMA Schedule 1 Process in Action

The sewage treatment plant for Christchurch City at Bromley was built prior to the enactment of the RMA. The City held an existing consent under the Water and Soil Conservation Act 1967. The discharge was to the Avon Heathcote Ihutai Estuary but there were no water quality standards or monitoring requirements associated with the consent. The Bromley discharge represented 90% of the nitrogen load and 98% of the phosphorus load to the estuary, and, was a major source of algal blooms in an area of high ecological and recreational value.

Under the RMA there was a 10 year grace period for existing consents which meant the Bromley outfall consent expired in October 2001 and a new consent application was required by 31 March 2001. Christchurch City Council undertook a capacity study in 1996 to review treatment upgrades to meet RMA requirements and obtain consent.

A Community Working Party was formed in 1996 involving 15 members of the public from different interest groups. The focus of the consultation was on treatment and disposal options. An Issues and Options Report was released in 1998. From 9 options, two feasible solutions were identified: (1) treatment modifications and estuary disposal, and, (2) treatment modifications and ocean disposal.

The Working Party recommendation in May 2000, after community consultation, was for treatment modifications and ocean disposal. However, because of cost considerations the City Council resolution in December 2000 was for treatment and estuary disposal. Resource consents were then lodged with the Regional Council in March 2001 for estuary disposal.

At the consent hearing the Hearing Commissioners' recommendation was not to grant the consent. The Regional Council released its decision in September 2001 not to grant the consent for discharge to the estuary. The City Council appealed this decision to the Environment Court. Two expert panels, one on ecology and one on public health, advised the removal of the discharge from the estuary and discharge by ocean outfall.

The City Council re-evaluated its decision after the Environment Court ruled that the wastewater should be disposed of through ocean outfall. The City Council negotiated with the registered interested parties to the estuary consent appeal for continuation to discharge to the estuary on conditions to progress an ocean outfall. A short-term consent was granted by the Environment Court to continue estuary discharge on the condition that an ocean outfall was operational by September 2009. After delays associated with some tragic construction accidents the outfall was operational in March 2010.

2.4 Current Developments

There are two contrasting approaches occurring in Central Government. One is the development of water resource policy. The direction for policy initiatives comes from the multi-stakeholder group – the Land and Water Forum. The Forum has provided four reports of recommendations. Recent Central Government policy initiatives have been based on these recommendations. The second relates to governance in local government and to changes in the RMA where it is Ministerial direction that is leading the change. These different approaches are reflected in current government proposals to amend the RMA. There are also two 'blue skies' initiatives to review the appropriate approaches in New Zealand for planning and resource management.

2.4.1 Land and Water Forum

The Land and Water Forum was established in the belief that stakeholders needed to engage directly with each other if a way forward was to be found for water management in New Zealand. The Forum was formed initially with 58 organisations as plenary members with a "small group" of 21 major stakeholders with six active government observers as the main developers of the Forum's recommendations.

The Government asked the Forum to recommend potential reforms for New Zealand's fresh water management. The main areas put forward by the Forum in their first report (Land and Water Forum 2010) were:

- Setting limits for water quality and flows in line with needs, values and objectives of communities;
- Achieving the limits and targets through good management practices, audited self-management, regulatory approaches, price-based measures and investment in clean-up of contaminated water bodies;
- Developing principles of allocation to address scarcity in catchments at or near over-allocation;
- Establishing a framework to enable more flexible transfer of water permits;

- Improving rural water infrastructure;
- Improving water governance through establishing a National Land and Water Commission on a co-governance basis with iwi[10], reporting to a core group of Ministers, promulgating a National Policy Statement, and, improving regional governance by the addition of government appointees;
- Ensuring scientific data is consistently collected, archived and publicly available;
- Investigating the way urban water management systems are organised; and
- Providing greater national direction in flood management.

In a second report (Land and Water Forum 2012a), the Land and Water Forum was asked by Government to make recommendations on setting objectives for water bodies, improving decision making processes and managing within limits.

The main recommendations of the Forum in relation to objective setting were:

- Government should through a national instrument direct regional councils to give effect to national objectives at the catchment scale; and
- Regional councils should be required to set resource use limits for the taking of water and discharge of contaminants as rules in regional plans to give effect to the objectives for all water bodies.

There was recognition that some specific objectives can be national whereas others will vary spatially because each water body is different. It was also recognised that to control cumulative effects limits must be binding.

The Second Report of the Forum also recommends the use of collaborative processes for setting freshwater objectives and limits at the regional level. The Forum states that: "Collaborative processes engage communities in a dialogue about their values and interests and make them responsible for resolving them" (p29).[11]

In addition the Second Report recognises the need for "plan agility". Monitoring, new information or improved analytical methods may indicate the need to adjust limits. Collaborative processes for implementation and the need to manage the transition to a limits based system are also recommended.

The proposals in the Third Report (Land and Water Forum 2012b) present tools and approaches required to manage fresh water to meet limits and achieve freshwater objectives, and, to realise the potential of New Zealand's freshwater economy. The general approach was one of providing guidance through a national objectives framework by central government and catchment-specific planning by regional councils using collaborative approaches.

A framework for water quality management was recommended. This involved catchment-based limits consistent with a national framework, and, the mix of methods and tools to achieve the limits to be incorporated in regional plans. This was to

---

[10] *Iwi* means Māori community or people.

[11] While noting that the Forum has pioneered the use of collaborative processes at the national level, it also acknowledges their use "in a number of regions of New Zealand in various forms for some years now, not least through the pioneering work of the Canterbury Water Management Strategy" (p29).

be followed by an implementation programme that was monitored and reviewed. Also, a water allocation model was recommended to define catchment-based water availability limits within a national framework. The model introduced different approaches to "first-come first-served" as total allocation in the catchment reaches a scarcity threshold (a proportion of the catchment allocation limit). The approaches required accounting for all takes and all contaminant sources in the catchment.

Its fourth report (Land and Water Forum 2015), was in response to a Ministerial request for advice on maximising economic benefits while managing within water quantity and water quality limits, identifying mechanisms to manage the transition from the current regime to effectively manage within limits, exploring iwi rights and interests in freshwater, and recommending regulatory requirements for stock exclusion from streams.

In relation to managing within water quantity and quality limits some of the key recommendations included: (a) encouraging the efficient use of water through transfers to highest-value uses and requirements for water use efficiency; (b) facilitating the development of infrastructure and catchment scale mitigations to increase the amount and reliability of water, and, the assimilative capacity available for economic use; and (c) reducing over-allocation through implementing good management practices, requiring technically efficient water use, targeting critical source areas, developing additional infrastructure, administrative reductions ("haircuts"), and land use controls. The report recognised the importance of the resolution of iwi rights and interests in fresh water but noted the responsibility for reaching agreement on how to do this rests with the Crown and iwi. In relation to stock exclusion, a national framework was recommended that excludes dairy cattle, beef cattle, deer and pigs from waterways on the plains and lowland hills over time while councils would also be able to require stock exclusion in other terrain that were areas of ecological significance or critical source areas of contaminants.

2.4.2 Central Government Initiatives Based on the Land and Water Forum Recommendations

Central Government announced three major initiatives in relation to water management in 2011. These were the:

- National Policy Statement for Freshwater Management: this recognised water as nationally significant and directs regional councils to set water quality and quantity limits;
- Fresh Start for Fresh Water Clean-Up Fund: this allocated $15 m over 2 years to restore waterways affected by historical pollution;
- Irrigation Acceleration Fund: this allocated $35 m over 5 years to unlock economic growth potential through water infrastructure.[12]

[12] The Irrigation Acceleration Fund is designed to support the development of irrigation infrastructure proposals to the 'investment-ready' prospectus stage. In Canterbury (as at 28 July 2014), there

These initiatives were consistent with recommendations from the first report of the Land and Water Forum. The key implication of the National Policy Statement is the requirement by 2030 that regional councils have set freshwater objectives that reflect national and local values. Regional councils are in the process of giving effect to this requirement through regional plans. For Canterbury the requirements have been included as part of the Canterbury Land & Water Regional Plan (Environment Canterbury 2015).

Central Government has also established the Crown Irrigation Investments Ltd. (CIIL) for direct capital investment in regional-scale irrigation schemes. CIIL has provided \$6.5 m of subordinated debt finance for up to 5 years at sub-commercial interest rates to Central Plains Water Ltd. to enable construction of the headrace being built during Stage 1 to support later stages of the scheme. CIIL has also underwritten the expansion of North Otago Irrigation Company and encouraged directors to seek and utilise alternative sources of capital. Furthermore for Stage 2 of Barrhill Chertsey Irrigation Scheme, a clear offer from CIIL led to increased commitment from irrigators and encouraged the scheme's bankers to finance the whole project (Crown Irrigation Investment Ltd. 2015).

Based on Land and Water Forum recommendations, the Government made a series of additions to the National Policy Statement for Fresh Water (Ministry for the Environment 2013). These are to:

- Require regional councils to account for all water takes and contaminant discharges;
- Include a National Objectives Framework to support and guide the setting of freshwater objectives in regional plans;
- Provide explicit recognition of tangata whenua values for freshwater;
- Establish ecosystem health and human health (secondary contact) as compulsory national values in regional plans;
- Introduce 'bottom lines' for ecosystem and human health that apply everywhere;
- Include restricted grounds for exceptions to 'bottom lines'; and,
- Provide an approach for monitoring progress towards freshwater objectives.

Further changes in freshwater management have been foreshadowed (Ministry for the Environment 2016). These follow the fourth report of the Land and Water Forum and discussions with the Iwi Leaders Group. One set of changes are related to amending the National Policy Statement to incorporate the Macroinvertebrate Community Index as a mandatory monitoring method, to allow exceptions in water

had been contributions to the Hurunui Water Project (\$2.385 m) for geotechnical investigations and preliminary design work, Central Plains Water (\$5.300 m) for design completion of the head race and piped distribution system, and prefeasibility contributions for the Ashley River (\$195,000) and Opuha and Rangitata South integration (\$277,550), Lower Waitaki upgrade study (\$133,865), Haka Valley design study (\$194,930), a demand study (\$641,329) and feasibility and design study (\$7.044 m) for Hunter Downs, and, an upgrade from open race to piped distribution for Ashburton Lyndhurst Irrigation (\$742,281) (Ministry for Primary Industries 2014). More recently there have been grants to design Stage 2 of the Central Plains Scheme (\$6.64 m), to pilot aquifer recharge in the Hinds Scheme (\$312,000), for the Sheffield Water Scheme (\$900,000), and a further \$520,000 to refine the scheme layout for the Hurunui Water Project.

quality limits in catchments with significant infrastructure, and, to include lakes and lagoons with intermittent connection to the sea. Another change is to progressively introduce regulations for stock exclusion from streams. Changes are also proposed with respect to iwi rights and interests including Māori principles for water management (Te Mana o te Wai – the quality and vitality of water), and improve iwi participation in freshwater governance and management. There are also changes proposed relating to the economic use of water such as achieving greater water use efficiency, adoption of good management practice, addressing over-allocation, and facilitating water transfer to more efficient uses.[13]

2.4.3 Central Government Initiatives in Governance of Local Government

Central Government has initiated two major changes to local government arrangements with the prospect of further restructuring in local government. One major change was the amalgamation of the regional council and seven city and district councils into one unitary council – the Auckland Council. In 2009 a series of bills were brought to parliament to create the new council to replace the region's existing eight councils. The new Council was formed in November 2010.

In March 2010, Central Government replaced the 14 elected councillors of the Canterbury Regional Council with seven appointed commissioners. This was based on a review report concerned with the significant scale of the water management task in Canterbury and their perception of the capability of the organisation to perform that task. The Minister for the Environment stated that: "Canterbury is strategically important with it holding more than half of the country's irrigation water and hydro storage. Government leadership is needed to address Canterbury's lack of a proper allocation plan, increasing problems with water quality and failure to progress opportunities for storage" (The Press 30 March 2010). Special legislation was passed providing the commissioners with additional powers, in particular, the ability to impose a moratorium on consent applications, and, removal of the ability for appeals to the Environment Court on policies and plans.

2.4.4 Resource Management Act: Proposals for Change

There are currently three reviews in train in relation to the RMA.[14] One is the Resource Legislation Amendment Bill (Ministry for the Environment 2015), a second is the Productivity Commission's Better Urban Planning Review (New Zealand

[13] Changes to the National Policy Statement for Freshwater Management came into effect on 7 September 2017.

[14] At the time this book was going to press a fourth review of the RMA has been announced by the Environmental Defence Society.

Productivity Commission 2015), and third is Local Government New Zealand's Blue Sky discussion about the New Zealand resource management system (Local Government New Zealand 2015). The primary focus of these reviews is the process of urban planning (rather than water resource management). This is based on concerns about planning issues in Auckland.

The Amendment Bill has provisions for greater Ministerial direction to centralise the control of planning through the introduction of planning templates, provisions for more regulatory powers of central government, and requirements for local government to ensure sufficient land availability for residential and business development. In contrast to this increased centralisation, it also introduces an alternative decision-making process based on collaborative planning processes to enable greater public engagement consistent with the concept recommended by the Land and Water Forum but the approach does not follow the Forum's recommendations.

The main purpose of the Productivity Commission's inquiry is "to review New Zealand's urban planning system and to identify, from first principles, the most appropriate system for allocating land through this system to support desirable social, economic, environmental and cultural outcomes" (New Zealand Productivity Commission 2015). The LGNZ review asks the question: "What should a 'fit for purpose' resource management regime which works for communities, businesses, regional economies and New Zealand's environment look like?" (Local Government New Zealand 2015). Both inquiries are considering fundamental change to the approach to resource management in New Zealand.

The LGNZ review highlights that "objectives, values and world views can change with time and context and New Zealand in 2015 is a different place to New Zealand in the late 1980s and 1990s when the core elements of the resource management system were designed". The review identifies three contextual issues that have changed and that are relevant to the context of this book: (1) increasing resource scarcity and competition for access; (2) a changing society (e.g. urbanisation, aging population, increasing role of Māori, and attraction to immigrants); and (3) an increasingly dynamic context (e.g. climate change, and new technologies). In terms of what New Zealanders want, LGNZ concludes they want "a resource management system that allows communities to participate in developing locally-tailored frameworks – these would determine the level where decisions are made and what processes are used for reconciling different views and aspirations – but at the same time avoids unproductive litigation on technical matters and prevents council from having to 're-invent the wheel' to deal with common issues" (Local Government New Zealand 2015).

Also of particular relevance to the context of this book is one of the Productivity Commission discussion papers that explores what complexity theory can tell us about urban planning (Cranford 2016). The paper highlights issues such as the search for efficiency does not address the risk of systemic breakdown; and, that government is one instance of a collective action mechanism and policy should allow for others, particularly that enable bottom-up collective action.

References

Cranford, R. (2016). *What can complexity theory tell us about urban planning?* Wellington: New Zealand Productivity Commission.

Crown Irrigation Investment Ltd. (2015). *Annual report for the year ended 30 June 2015*. Wellington: Crown Irrigation Investment Ltd.

Environment Canterbury. (2011). *Natural Resources Regional Plan*. Christchurch: Environment Canterbury.

Environment Canterbury. (2013). *Canterbury Regional Policy Statement 2013*. Christchurch: Environment Canterbury.

Environment Canterbury. (2015). *Canterbury Land and Water Regional Plan December 2015*. Christchurch: Environment Canterbury.

Gunderson, L. H., & Holling, C. S. (2002). *Panarchy: Understanding transformations in human and natural systems*. Washington, DC: Island Press.

Gunningham, N. (2008). *Innovative governance and regulatory design: Managing water resources*. Canberra: Australian National University.

Hardin, G. (1968). The tragedy of the commons. *Science, 162*, 1243–1248.

Jenkins, B. R. (2009, February 26). *Are regional councils past their use-by date?* Paper presented at the Environmental Defence Society Conference – Reform in Paradise: Threat or Opportunity?, Auckland.

Land and Water Forum. (2010). *Report of the Land and Water Forum: A fresh start to fresh water*.

Land and Water Forum. (2012a). *Second report of the Land and Water Forum: Setting limits for water quality and quantity, and freshwater policy-and plan-making through collaboration*.

Land and Water Forum. (2012b). *Third report of the Land and Water Forum: Managing water quality and allocating water*.

Land and Water Forum. (2015). *Fourth report of the Land and Water Forum*. Wellington: Land and Water Trust.

Local Government New Zealand. (2015). *A 'blue skies' discussion about New Zealand's resource management system*. Wellington: Local Government New Zealand.

Ministry for Primary Industries. (2014). *Irrigation Acceleration Fund – summary of project progress updated 28 July 2014*. Ministry for Primary Industries Wellington.

Ministry for the Environment. (2013). *Proposed amendments to the National Policy Statement for Freshwater 2011: A discussion document*. Wellington: Ministry for the Environment.

Ministry for the Environment. (2015). *The second phase of Resource Management Act Reform*. Wellington: Ministry for the Environment.

Ministry for the Environment. (2016). *Next steps for fresh water: Consultation document*. Wellington: Ministry for the Environment.

New Zealand Government. (2011). *National Policy Statement: Freshwater Management 2011*. Wellington: Ministry for the Environment.

New Zealand Government. (2014). *National Policy Statement for Freshwater Management 2014*. Wellington: New Zealand Government.

New Zealand Productivity Commission. (2015). *Better urban planning*. Wellington: New Zealand Productivity Commission.

Ostrom, E. (1990). *Governing the commons: The evolution of institutions for collective action*. New York: Cambridge University Press.

Skelton, P., & Memon, A. (2002). Adopting sustainability as an overarching environmental policy: A review of section 5 of the RMA. *Resource Management Law Journal, X*(1), 1–10.

Upton, S. (1995). Purpose and principle in the Resource Management Act. *Waikato Law Review, 3*, 17–55.

URS. (2004). *Assessment of environmental effects Christchurch city council ocean outfall pipeline*. Christchurch: Christchurch City Council.

Waitangi Tribunal. (2016). *Meaning of the treaty*. http://www.waitangitribunal.govt.nz/treaty-of-waitangi/meaning-of-the-treaty/. Accessed 6 July 2016.

Wallis, J., & Dollery, B. (2000). *Local government reform in New Zealand*. Working Paper Series in Economics. Armadale: University of New England.

Chapter 3
Water Management in Canterbury

Abstract Water is significant to the Canterbury region both in terms of high and on-going irrigation demand, and, the implications of water extraction and use on the sustainability of its rivers, aquifers and aquatic ecology. There was a need for a paradigm shift in the approach to water management as the RMA processes were found to be inadequate to manage resources at sustainability limits. Strategic studies evolved from a technical investigation of future demand with a focus on additional water storage to a collaborative process for the development on an integrated water management strategy – the Canterbury Water Management Strategy. The implementation of the strategy is still in progress. The strategy is based on ten target areas identified by the community engagement process as the main uses and benefits of water in the Canterbury region. A collaborative governance framework has been established for developing implementation programmes for ten water management zones and for the region.

The strategy identified improvements in the efficiency of use of water already allocated was more effective in improving water availability than a reliance on increased storage alone. New forms of storage with less adverse impacts were introduced, such as managed aquifer recharge and off-river storage. Reductions of nutrients from land use intensification by improving management practices has been a priority. Solutions packages for addressing the degradation of vulnerable lakes and rivers have been prepared. Biodiversity enhancement projects have been initiated. Greater involvement of Māori has led to the incorporation of kaitiakitanga (resource stewardship) into implementation programmes. Emerging issues are the need for improved integration of surface water and groundwater interactions, increased use of modelling of water and financial outcomes to guide decision making, and, equity in allocation among existing users as well as between existing and future users.

Keywords Paradigm shift in water management • Canterbury water management strategy • Collaborative governance framework • Water quality management • Implementation programmes

B.R. Jenkins, *Water Management in New Zealand's Canterbury Region*, Global Issues in Water Policy 19, https://doi.org/10.1007/978-94-024-1213-0_3

3.1 Water Situation in Canterbury

3.1.1 Significance of Water in Canterbury

Water is a critical ingredient for the Canterbury region, which allocates 58% of New Zealand's water, has 70% of the country's irrigated land, generates 24% of the nation's power through hydroelectricity, has 65% of the country's hydro storage and provides a high quality water supply to its major city without the need for treatment. Water also creates and sustains Canterbury's world-famous braided rivers, high country and coastal lakes, as well as lowland streams and wetlands (Jenkins 2007a).

Not only is Canterbury the region with the greatest allocation of water, it is also the region with the greatest Potential Evapotranspiration Deficit[1] and therefore with the greatest dependence on water (Table 3.1). It is part of the relatively dry east coast of New Zealand which is dependent on irrigation for increasing agricultural productivity. It has also developed an aquatic ecology sensitive to flow variability and low flows.

Water is crucial to both the economy and the ecology of Canterbury.

Table 3.1 Comparison of regional potential evapotranspiration deficit (Mullan, personal communication)

Council region	Averaged annual PED (mm)
Canterbury	322
Marlborough	288
Hawkes Bay	259
Otago	235
Auckland	231
Gisborne	209
Wellington	205
Northland	200
Manawatu	157
Bay of Plenty	128
Waikato	114
Tasman	111
Taranaki	86
Southland	72
West Coast	10

Note: Annual PED (mm) averaged over land below 500 m for the period 1972/3 to 2002/3

[1] Potential Evapotranspiration Deficit is the amount of water that would need to be added to keep pasture growing at its potential seasonal rate (NIWA 2016).

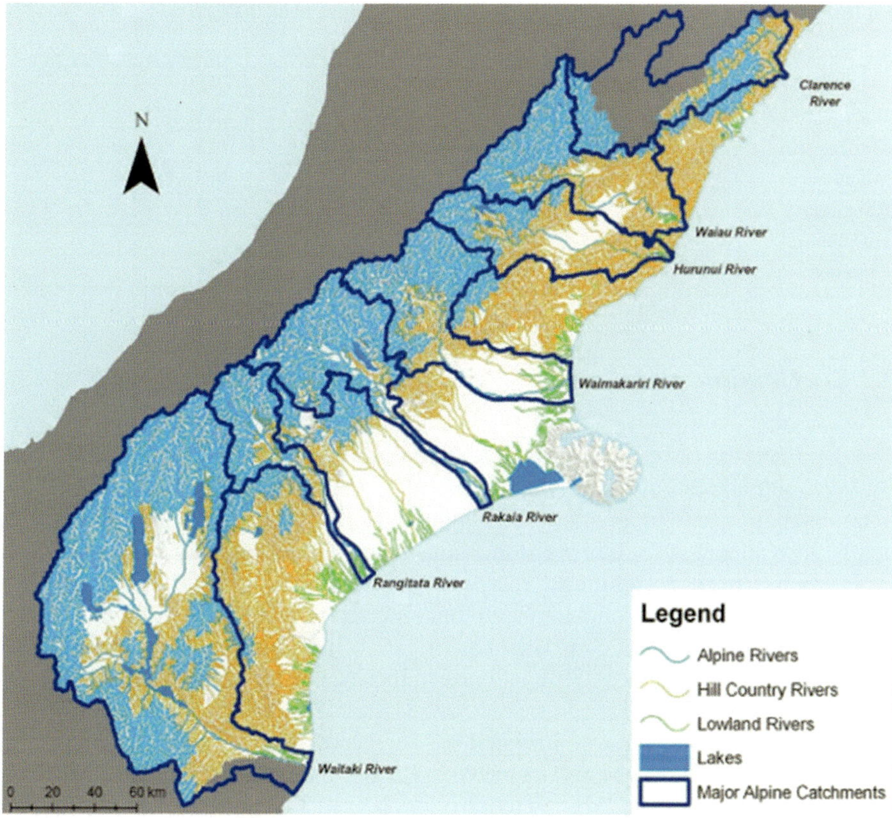

Fig. 3.1 River types in Canterbury (Environment Canterbury)

3.1.2 Types of River Systems

In Canterbury, there are three main types of river systems. Firstly, there are the alpine rivers with their upper reaches in the Southern Alps so that they are snow-fed with summer peak flows. Secondly, there are the foothill rivers with rain-fed catchments and winter peak flows. Thirdly, there are lowland streams that are fed from groundwater. Refer Fig. 3.1.

Table 3.2 sets out the mean flows in the major Canterbury rivers. The seven alpine rivers contribute 88% of the annual average flow and are an order of magnitude greater in volume compared to the foothill rivers. Lowland streams are even smaller in flow.

Table 3.2 Major Canterbury rivers mean flows (m^3/s) (Morgan et al. 2002)

Alpine rivers		Other major rivers	
Waitaki	373	Ashburton	15
Rakaia	221	Ashley	13
Waimakariri	120	Orari	11
Waiau	116	Opuha	10
Rangitata	100	Opihi	5
Hurunui	72	Waihao	4
Clarence	72	Pareora	4
		Waipara	3

3.1.3 Demand for Water

The dominant use of water is for hydroelectric power generation: more than 80% of surface water allocation is for hydro, nearly all in the Waitaki River catchment. This is an instream use. The greatest consumptive use is for irrigated agriculture with nearly 90% allocated for this purpose from surface and groundwater.

There has been a significant increase in irrigation in Canterbury in the last 30 years. Based on Environment Canterbury's consent records, in 1985 there were about 150,000 ha of land consented for irrigation in the region. In 2006, this has increased to 560,000 ha – a 270% increase in those 21 years (or 6.5% per annum on a compound growth basis).

Census data on actual area irrigated has increased from 297,108 ha in 2002 to 385,300 ha in 2007 and 444,800 ha in 2012. The dominant land use change has been conversions from dryland farming to dairying. Figure 3.2 shows the change in land use from 1996 to 2012 for Canterbury. While the region is predominantly pastoral around 3,000,000 ha, there has been an increase in dairying from 60,000 ha in 1996 to 280,000 ha in 2012.

3.1.4 Pressure on River Systems

The main pressure is coming on the lowland streams and foothill rivers in relation to water demand. Table 3.3 sets out the volume of water allocated as a proportion of the mean annual low flow (MALF). This shows the greater pressure on foothill rivers compared to alpine rivers.

Run-of-river takes are near sustainability limits. The degree of restriction is greater on lowland streams. Table 3.4 sets out the number of rivers on restriction during the dry period January 2006 for the different river types.

For Canterbury rivers to maintain their environmental values there is a need to protect the following types of flows (Biggs et al. 2008):

- Low flows – there is a need for restrictions on out-of-stream withdrawals so that there is not an increased frequency of low flows below the minimum flow to support instream values;

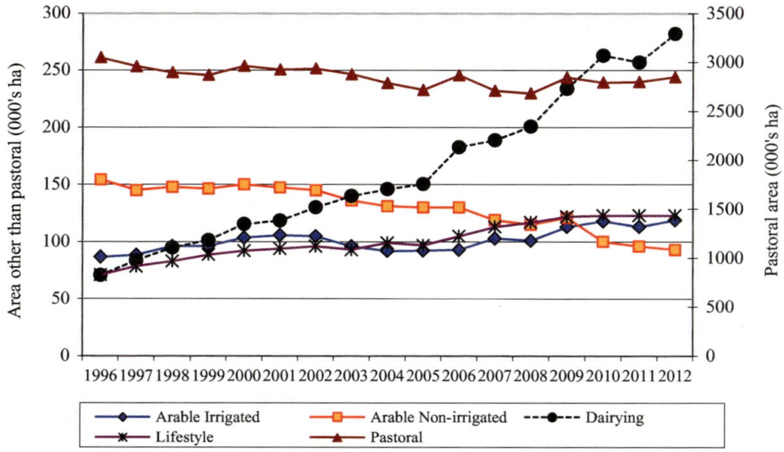

Fig. 3.2 Land use changes in Canterbury (Environment Canterbury)
Note the two scales: pastoral land use relates to the right-hand scale at 3,000,000 ha, while the other agricultural uses relate to the left-hand scale, e.g. dairying in 2012 is 280,000 ha

Table 3.3 Relative pressure on river systems: allocation as a proportion of MALF (Morgan et al. 2002)

Waipara (foothill)	12.8	Orari (foothill)	0.7
Maerewhenua (foothill)	2.0	Waihao (foothill)	0.63
Ashburton (foothill)	1.7	Hurunui (alpine)	0.47
Opihi (foothill)	1.7	Waiau (alpine)	0.45
Hakataramea (foothill)	1.0	Waimakariri (alpine)	0.36
Pareora (foothill)	1.0	Rakaia (alpine)	0.35
Selwyn (foothill)	0.9	Ashley (foothill)	0.34
Rangitata (alpine)	0.8	Waitaki (alpine)	0.26

Table 3.4 Rivers on flow restrictions during January 2006 (Environment Canterbury)

Lowland streams	28 of 57 on partial or full restriction	49%
Foothill rivers	13 of 36 on partial restriction	36%
Alpine rivers	2 of 7 on partial restriction	29%

- Flushing flows – there needs to be sufficient "flushing flows" (typically about three times the mean flow) to dislodge algae and prevent build-up of algae;
- Flood flows – there needs to be sufficient flood flows (greater than the one-in-one-year maximum flow) to ensure turnover of gravel in the river bed in order to maintain the braided character of major rivers.

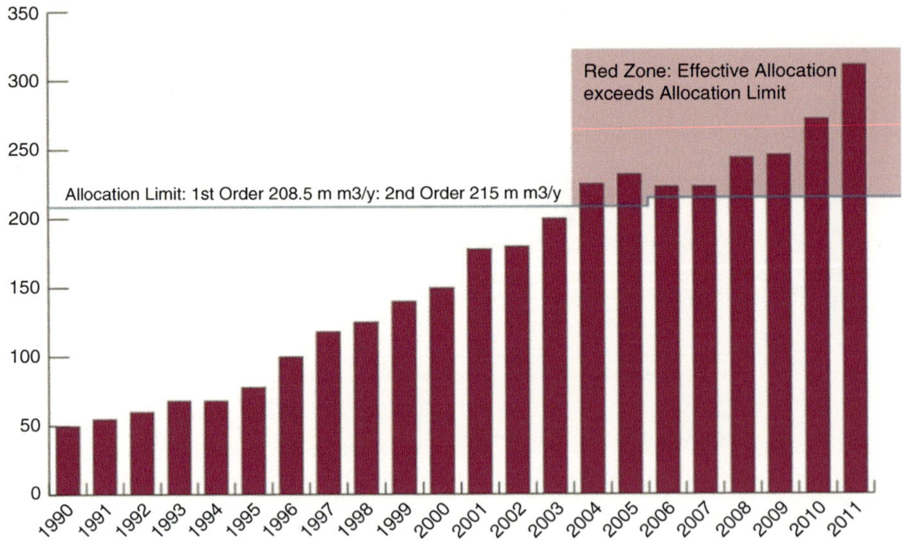

Fig. 3.3 Development in the Rakaia-Selwyn groundwater allocation zone since 1990 ($m^3 \times 10^6$/year) (Environment Canterbury)

3.1.5 Pressure on Aquifer Systems

Groundwater allocation limits and effective groundwater allocations were estimated for groundwater zones in the Canterbury region. When the effective allocation exceeded the allocation limit, the zone was considered to be fully allocated and defined as a "red zone". Unless better information was made available to indicate additional availability of water, it was recommended that no further allocations be granted in that zone.

Figure 3.3 shows the growth in groundwater allocated in the Rakaia-Selwyn zone in mid Canterbury. This zone was declared a red zone in 2004 when the first order allocation limit had been reached. There are now ten red zones in Canterbury and three "yellow zones", where effective allocation exceeds 80% of the allocation limit (refer Fig. 3.4).

3.1.6 Cumulative Effects

Consent conditions are incorporated into project approvals in order to address project specific effects. However, as sustainability limits are approached, there is the potential for cumulative effects. Management of cumulative effects requires a catchment-wide approach. However, with the RMA designed for managing the adverse effects of individual applications there are shortcomings in the legislative framework for the management of cumulative effects, such as the impact of groundwater withdrawals from Canterbury plains aquifers on the flows in spring-fed lowland streams.

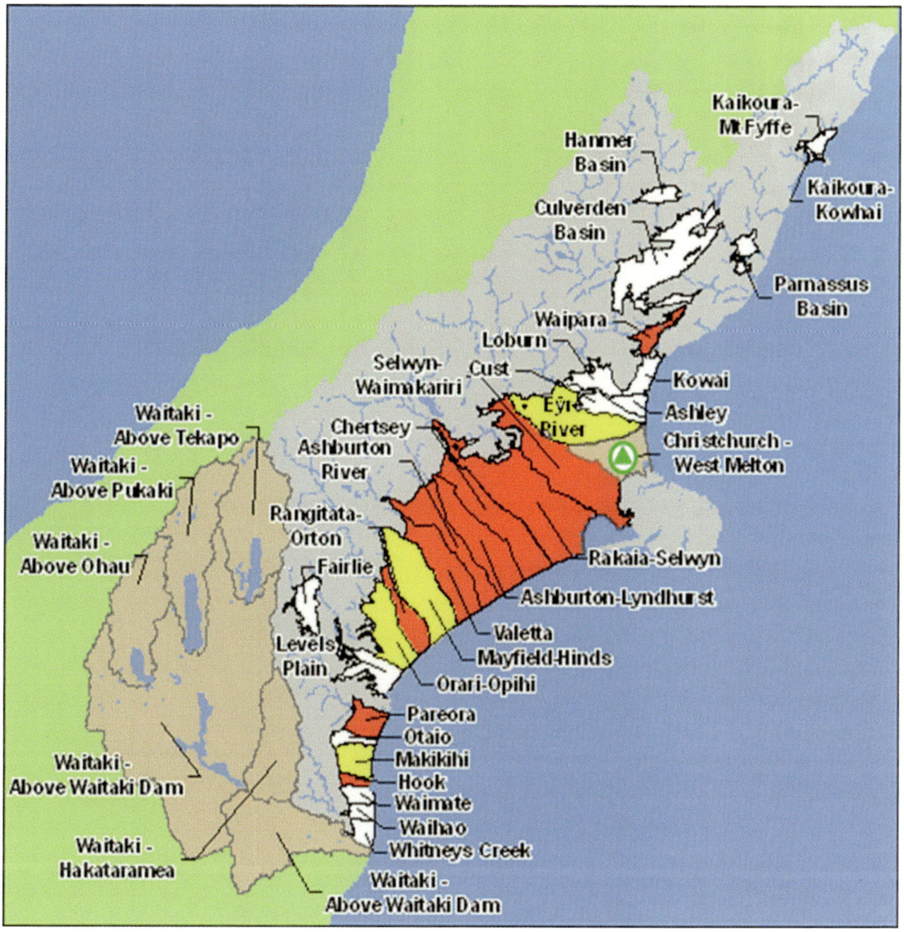

Fig. 3.4 Groundwater zones for Canterbury (Environment Canterbury)

Results of monitoring the aquatic ecosystem health in Canterbury lowland streams, indicates declining health (Fig. 3.5). This is due to a combination of a series of dry winters for the period 2000 to 2005, with low recharge of the aquifers which feed the lowland streams, and increasing levels of abstraction from groundwater.

One of the prime concerns with land use intensification is the potential for water quality impairment, in particular, nitrate contamination of groundwater. Environment Canterbury analysed the trends in nitrate levels in groundwater throughout the region to assess the long-term trends. The results of this analysis for the 212 monitoring wells in the region are shown in Fig. 3.6. Most wells (72%) show no long-term trend. However, 40 wells (19%) do show an increasing trend.

The possible causes of nitrate increases were investigated. The most significant increase was the land-based effluent disposal of the three Ashburton meat process-

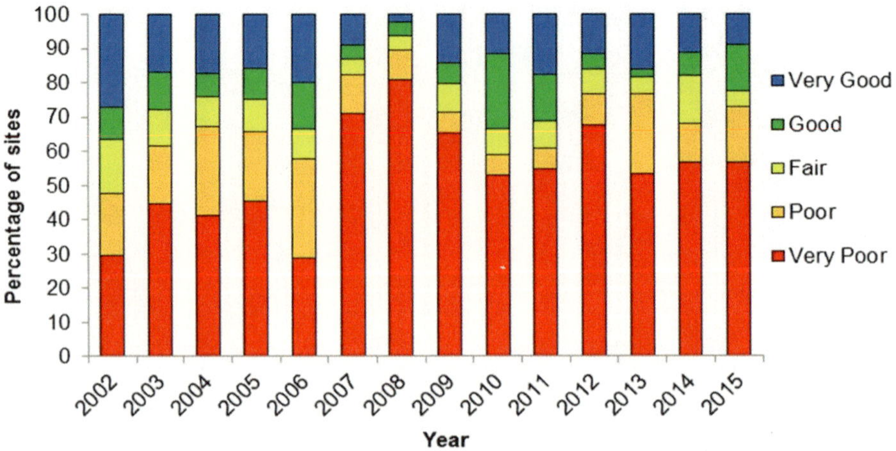

Fig. 3.5 Aquatic ecology health of lowland streams (Environment Canterbury)
Note: Percentage of sites graded in terms of ecological health assessed by abundance and diversity of macroinvertebrates

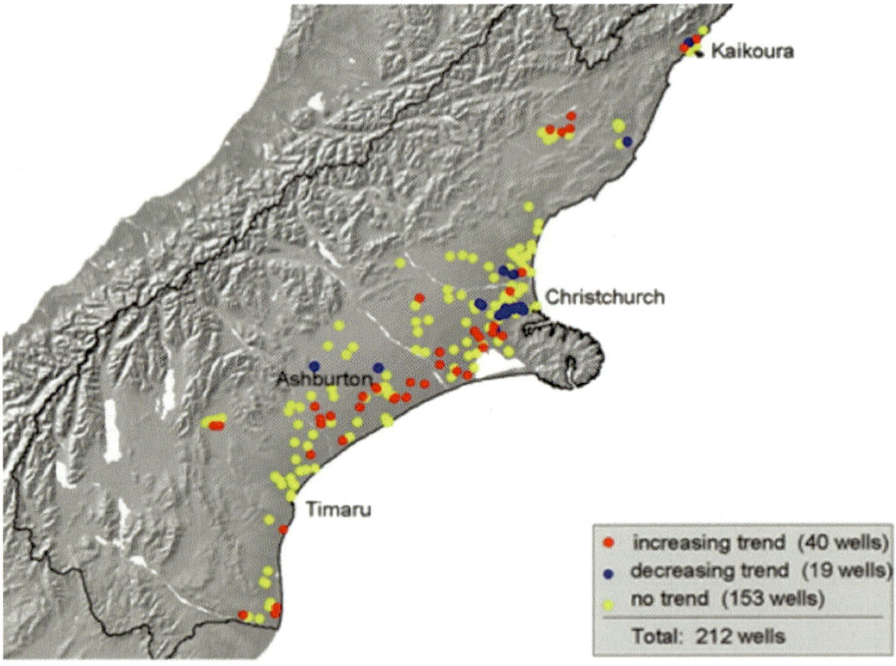

Fig. 3.6 Trends in nitrate nitrogen concentrations in Canterbury groundwater 1995–2004 (Environment Canterbury)

Table 3.5 Nitrogen leaching rates from agricultural land uses (Di and Cameron 2004)

Land use	Leaching rate (kgN/ha/year)
Forest	5–15
Cut grassland	6–49
Grazed pastures: Sheep	6–41
Grazed pastures: Dairy	47–110
Arable cropping	4–107
Ploughing of pastures	14–102
Horticulture	70–180

ing plants (which is being addressed through consent reviews). Other sources were leachate from the Burwood landfill, localised sources, such as offal pits and septic tanks. The distribution of nitrate increases did not appear to be highly correlated with the distribution of land use types. Generally, between rivers there were higher values from agricultural land use, whereas close to rivers there appears to be lower values due to higher groundwater flows for areas hydraulically connected to the rivers. There were also seasonal variations with winter maxima when soil moisture is available for nitrate leaching.

Table 3.5 sets out available data on measured nitrogen leaching losses from Canterbury and New Zealand research (Di and Cameron 2004). What is noticeable is the high degree of variation within land use types and the potential for arable and horticulture uses as well as dairying having high loadings.

3.2 Development of the Canterbury Water Management Strategy

3.2.1 Paradigm Shift Needed in Water Management

For Canterbury, it is recognised that there is a need for a paradigm shift in water management incorporating:

- Water allocation and availability which addresses sustainability limits and climate variability
- Management of cumulative effects of water takes and land use intensification
- A shift from effects-based management of individual consents to integrated management based on water management zones.

This recognition has led to the undertaking of the Canterbury Strategic Water Study (CSWS) and the development of a Canterbury Water Management Strategy (CWMS). This work has been undertaken in four stages:

- Stage 1: an initial study of water availability issues in Canterbury
- Stage 2: an investigation of potential storage sites

- Stage 3: a multi-stakeholder review of storage options
- Stage 4: the development of an integrated water management strategy.

The initial study was a technical study by government agencies while the subsequent stages were overseen by the Canterbury Mayoral Forum. The Forum comprises all of the mayors of the District and City Councils, the Chair of the Regional Council as well as their chief executives. A Steering Group of representatives of local and central government, tangata whenua as well as farming, environmental, industry and recreational interests provided advice to the Mayoral Forum.

In parallel with the strategic initiatives, there have been on-going water investigations and planning to address the water management issues facing Canterbury using existing RMA mechanisms and non-statutory processes.

3.2.2 Stage 1: Availability of Water

With increasing demand for water in Canterbury there has been increasing conflict over the allocation of water for abstraction and for the maintenance or improvement of instream values. In addition, there has been concern that ad hoc decisions by one group might foreclose on protection or development options that provide greater benefits to the wider community.

This led to the undertaking of the initial study by Environment Canterbury, Ministry of Agriculture & Forestry and Ministry for the Environment (Morgan et al. 2002) to provide fundamental information on:

- The potential long-term requirement for water;
- The capacity of the region to meet these requirements;
- The water resources that would come under the most stress;
- The reliability, over the long term, of water supplied from natural systems for abstractive uses.

The scope of the initial study was limited to water quantity issues.

Among the key findings of CSWS Stage 1 (Morgan et al. 2002) were the following:

- Irrigation is the dominant consumptive use now and for the future, with irrigation representing 89% of future potential peak demand, stock water 5%, municipal supplies 3%, industrial use 2% and plantation forestry 1%.
- There was approximately 1 million ha of land in Canterbury that could be productively irrigated (compared to current consented area of 560,000 ha).
- Surface water abstraction is placing pressure on the smaller foothill rivers (such as the Waipara, Maerewhenua, Ashburton and Opihi), while the larger alpine rivers (Waitaki, Rakaia and Waimakariri) are generally less pressured.
- Effective management of surface water resources requires abstraction limits and while minimum flows have been specified, the protection of other environmental flows (floods and flushing flows) is also required.

- Managing groundwater resources requires allocation limits and this is likely to be driven by requirements for flows in spring-fed lowland streams.
- Under low flow conditions current peak demand cannot be met by current abstraction methods. However, on an annual basis water is available to meet future demand but would require storage.
- Without the development of water storage, the irrigated areas in Canterbury can be expected to plateau well short of potential irrigated areas.
- There are few suitable storage sites and district councils need to work alongside Environment Canterbury to identify possible sites and ensure these sites are not foreclosed for future development by ad-hoc planning.
- The future development of Canterbury's water resources will require strategic integrated water resource management. However, there is no agency with the mandate to plan the long-term development of the region's water resources.

It is also important to note another significant finding that some catchments are water short – the foothill catchments have more irrigable land than water that could be provided from the catchment for irrigation. Whereas alpine river catchments are water rich and have volumes that could be made available to irrigate more land than the irrigable land within the catchment. Thus, full irrigation potential would require redistribution across the catchments. The issue of water availability therefore needs to be considered at a regional level rather than the catchment level.

Figure 3.7 shows the balance between supply and demand for water for the catchments and groundwater zones in Canterbury.

3.2.3 Stage 2: Potential Major Storage Sites

A second stage study was undertaken to determine whether it is practical to meet environmental needs and potential water demands through the use of storage as a core component of integrated management of surface water and groundwater in the Canterbury region. Stage 2 was also aimed at specifying a suite of practical options for meeting long term water demands in each part of Canterbury, including practical methods for sharing water equitably between neighbouring areas (Aqualinc Research Limited 2008).

Stage 2 not only drew upon the work of Stage 1 but also drew upon the work for the Natural Resources Regional Plan (Environment Canterbury 2004), particularly in relation to the rules for environmental flow requirements. The environmental flow requirements specify the allocation to instream uses which then enables the estimation of water available for out-of-stream uses.

The key output of Stage 2 was a suite of water supply options for each part of Canterbury. Each system option comprised the physical components such as water sources, storage and water conveyance facilities and management components such as river allocation rules. Hydrological performance was evaluated by computer simulations of the day-to-day operation of each system option over long periods using

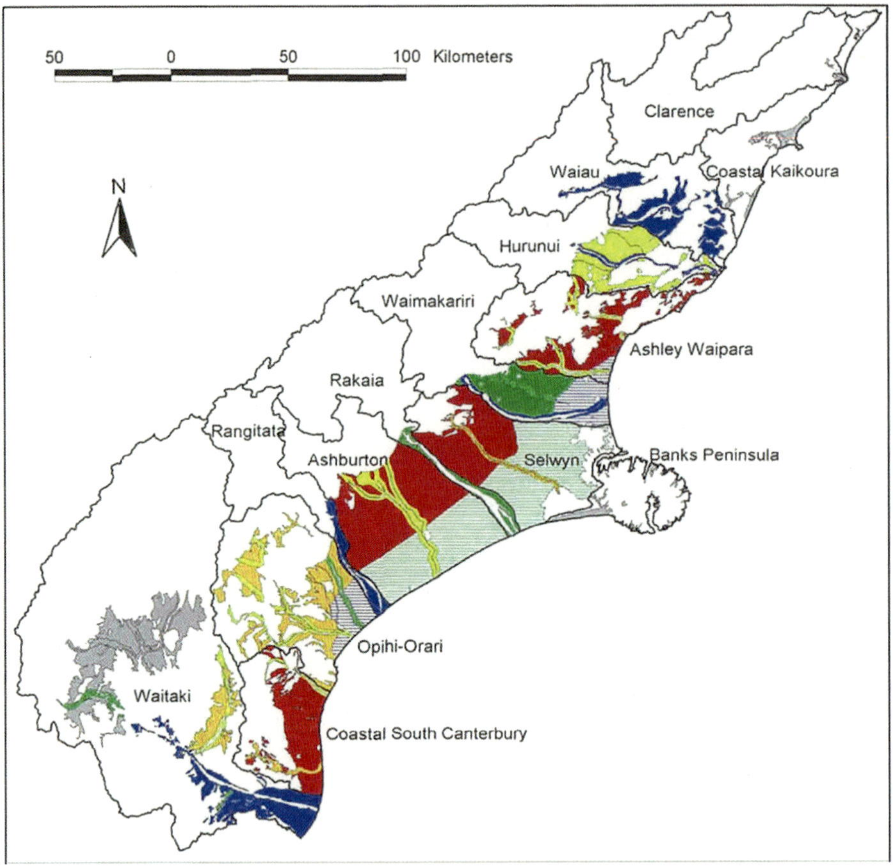

Fig. 3.7 Summary map of supply and demand situation in Canterbury (Morgan et al. 2002)
Legend for Fig. 3.7:

Striped blue: Demand can be reliably met from groundwater

Striped green: Demand can be reliably met from groundwater with the proviso that there is some upper plains irrigation which enhances recharge

Blue: Demand can be reliably met from run of river supply

Green: Unreliable run of river. Supply/demand ratio in worst irrigation season >1. Minimal storage needed

Yellow: Supply/demand ratio in worse case year >1. Moderate storage needed. Require river flows outside irrigation season to fully replenish storage

Orange: Average annual supply/demand ratio > 1. Storage possible but less likely. Large storage required which would not fully replenish every year

Red: Average annual supply/demand ratio < 1. No amount of storage replenished from within the zone can provide for the demand

Gray: There is insufficient supply data to compare with demand

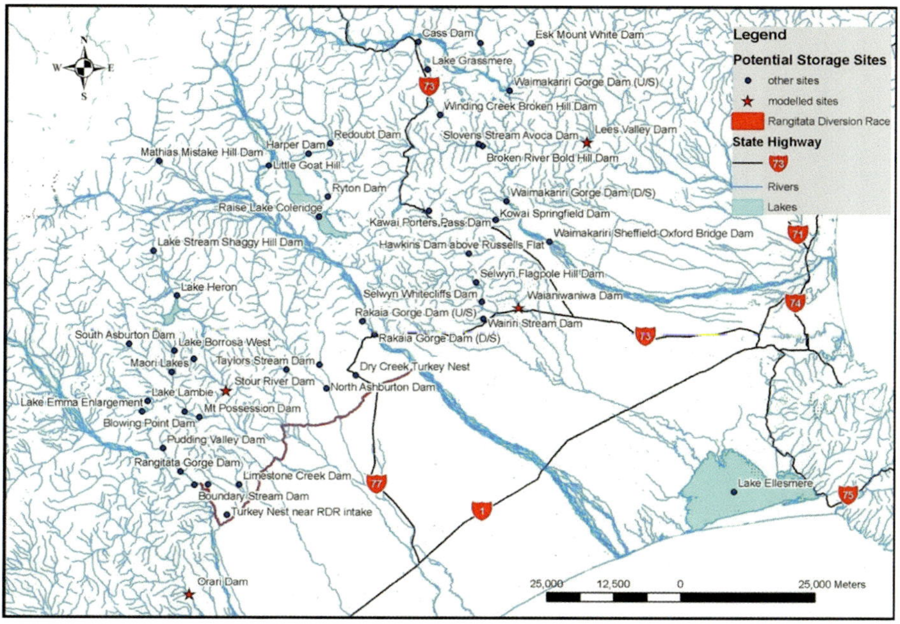

Fig. 3.8 Potential storage sites in Central Canterbury (Aqualinc Research Limited 2008)

historical data to test practicality and determine effects on river flow regimes. Figure 3.8 shows the range of potential sites considered in Central Canterbury.

Based on hydrologic performance, a small number of possible major storages were identified for the Canterbury region. These sites are shown in Fig. 3.9.

3.2.4 Stage 3: Multi-stakeholder Review of Storage Options

The CSWS Stage 3 work was undertaken in three phases: (i) formation and work of the regional reference group, (ii) core reference group members joined consecutively by north, south and mid Canterbury representatives in staged meetings of locality-based reference groups to evaluate storage options in their sub-region, and (iii) consultation with north, south and mid Canterbury interest groups based on the findings of the reference group evaluations (Whitehouse et al. 2008).

In the first phase, the multi-stakeholder core reference group was formed and their agreement sought about whether they were willing to work together as well as clarify roles, develop capability and build trust. This phase also involved the development of, and agreement to a sustainability framework for evaluating storage options. With the diverse views represented and the adversarial nature of water management issues in Canterbury this phase was essential if the Community Panel was going to be successful in evaluating options and identifying issues for further investigation.

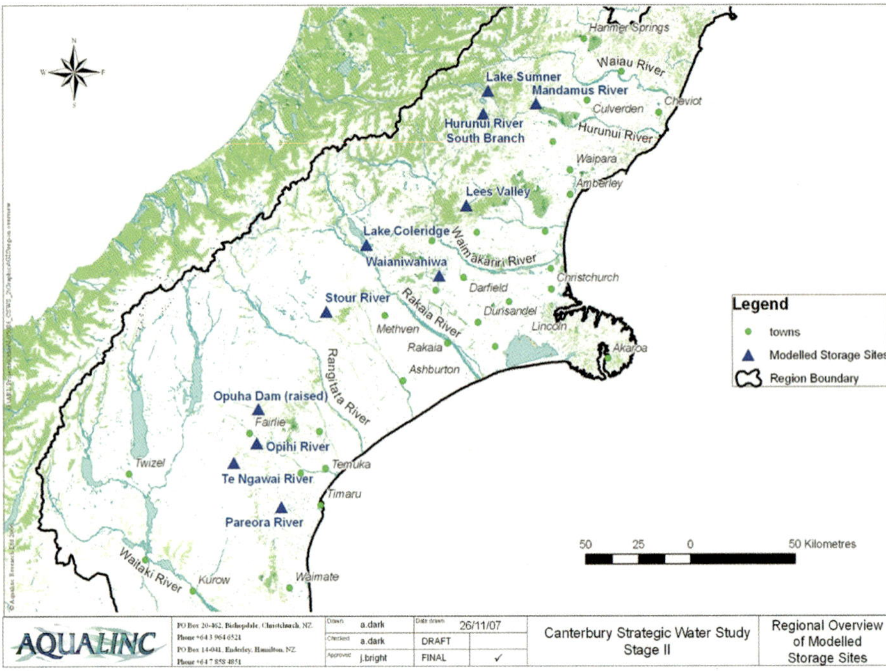

Fig. 3.9 Possible major storage sites for the Canterbury region (Aqualinc Research Limited 2008)

Experience during this first phase indicated that people entered the process with a belief that they were well versed in the issues of water management in Canterbury. What they found was that there were other issues that they were not familiar with. A critical point in the first phase was a two-day workshop covering key issues which provided all participants with a broader perspective on the range of issues to be addressed.

Over 20 options based on 12 storage reservoirs were evaluated in the second phase. The storages evaluated range from Hurunui River in the north to Opihi River in the south. No options were evaluated north of Hurunui River as the CSWS Stage 1 study had shown that potential demand in the Waiau area could be met from Waiau River without storage. South of Opihi River, the proposed Hunter Downs scheme (using water from Waitaki River) was not evaluated as the proposal was in the resource consenting process.

The evaluation of water storage options for Canterbury was done by multi-stakeholder groups of about 15–30 people in sub-regions – options for Hurunui River in four one-day workshops in September to December 2006, for South Canterbury in four workshops in February to April 2007, and for mid-central Canterbury in five workshops from May to July 2007. In early September 2007, a one-day workshop of participants from all three groups and from the CSWS Mayoral Forum Steering Group explored an integrated option for mid-central Canterbury.

The multi-stakeholder groups included people with a wide range of interests in water in Canterbury – irrigation, angling, farming, environmental concerns, community development, Ngāi Tahu, recreation, and conservation.

The multi-stakeholder groups provided a very robust evaluation of the water storage options, identifying the key features of each option and the beneficial and adverse impacts. There was generally a range of views with differences understood and respected by others in the group.

One of the features of the CSWS Stage 3 process has been the solution-seeking approach taken by the evaluation groups. Participants have suggested ways of overcoming or mitigating concerns. In some cases, these suggestions led to new options that were then modelled as variations from the options considered in CSWS Stage 2.

The main findings of the Stage 3 evaluation (Whitehouse et al. 2008) were:

- There was greatest acceptance of the Lees Valley and Lake Coleridge options.
- There was a need to address the water quality risk from land use intensification before any major storage option can be progressed.
- There was a need to adequately protect the flow variability of rivers.
- There was a desire for an integrated solution which minimises storage, puts water into lowland streams, protects flow variability and low flows in rivers and accounts for environmental and social impacts.
- There is a need for wider public consultation and engagement with interest groups that takes a solution-seeking rather than an adversarial approach.

3.2.5 Stage 4: Development of an Integrated Water Management Strategy

The outcome of the Stage 3 study was that a much broader scope than water availability was warranted. This led to the fourth stage which involved the development of an integrated water management strategy for the Canterbury region. The main elements of this stage were:

- Stakeholder and community engagement on option development and fundamental principles for a strategy
- Definition of strategic options by the Steering Group to the Mayoral Forum
- Community consultation on option preferences
- Strategic investigations of likely outcomes
- Sustainability appraisal of options at the regional level leading to sub-regional water management approaches
- Strategic approach to water management, environmental restoration, infrastructure requirements and governance arrangements.

A Strategic Framework document was released in 2009 (Canterbury Water 2009). Key findings from Stage 4 work are discussed below.

3.2.5.1 Fundamental Principles

One of the significant outcomes of the stakeholder engagement activities was the definition of a set of fundamental principles for the management of water in Canterbury. These have been refined into a set of primary and supporting principles.

The primary principles address:

- Sustainable management – where water is considered a common property managed under the sustainability principles of the Resource Management and Local Government Acts.
- Regional approach – involving a priority approach for allocation to the environment and drinking and stock water; and, a science-based approach that manages cumulative effects, incorporates efficiency of use, and integrated with land, biodiversity and water quality management.
- Tangata whenua – applying the Māori concepts of kaitiakitanga (stewardship of the resources).

The supporting principles comprise:

- Natural character – requiring the flow regimes, braided river processes, biodiversity and interdependence with coastal ecosystems are preserved and enhanced
- Indigenous biodiversity – flora, fauna and their habitats are protected and enhanced
- Access – maintaining public access to rivers, lakes and waterways
- Quality drinking water – protecting the quality of drinking water sources
- Recreational opportunities – ensuring contact recreation can be enjoyed and flows for recreational users are adequate
- Community and commercial use – water is available for efficient and effective use and the effects of the use do not compromise environmental quality.

3.2.5.2 Strategic Options

Using the "Strategic Choice" approach (Friend and Hickling 2005), the Steering Group developed four strategic options for the future of water management in Canterbury. These options are:

- Option A: Business As Usual – this option represents the current processes under the Resource Management Act based on applicant-driven proposals subject to effects-based assessment and resolution of conflict by adversarial hearings and court processes.
- Option B: Advance environmental protection and then proceed with infrastructure development – this option would address degraded waterways through formalising environmental limits, initiating restoration and improving water efficiency before further infrastructure development and land use intensification.

- Option C: Reconfigure consents and infrastructure to improve efficiency and reliability and enhance environmental quality – this option would take the opportunity to reconsider existing consents and operation of infrastructure in order to increase efficiency of currently allocated water while proceeding with complementary infrastructure and reducing environmental pressures.
- Option D: Advance infrastructure development with environmental repair and restoration – this option would create an infrastructure platform involving storage while incorporating environmental mitigation.

Unlike Option A, the other Options (B, C and D) would involve a significant degree of integration and evolution of governance arrangements. The differences between the other Options are largely around the priority given to primary focus of change. The options can be characterised as Environment-led (Option B), Efficiency-led (Option C) and Storage-led (Option D).

3.2.5.3 Community Consultation

Stage 4 included a programme of stakeholder engagement and structured public meetings to discuss water management options (Jenkins and Henley 2013). There was also a formal consultation and hearing process with public submissions. This included the request for feedback on the preferred option. Figure 3.10 displays the results for the preferences expressed for each of the strategic options.

From the responses, it is clear there is little support for Option A (Business as Usual). Option D (Storage-led) and Option B (Environment-led) were the most favoured. However, Option C (Efficiency-led) received considerable first preference support and was the dominant second preference.

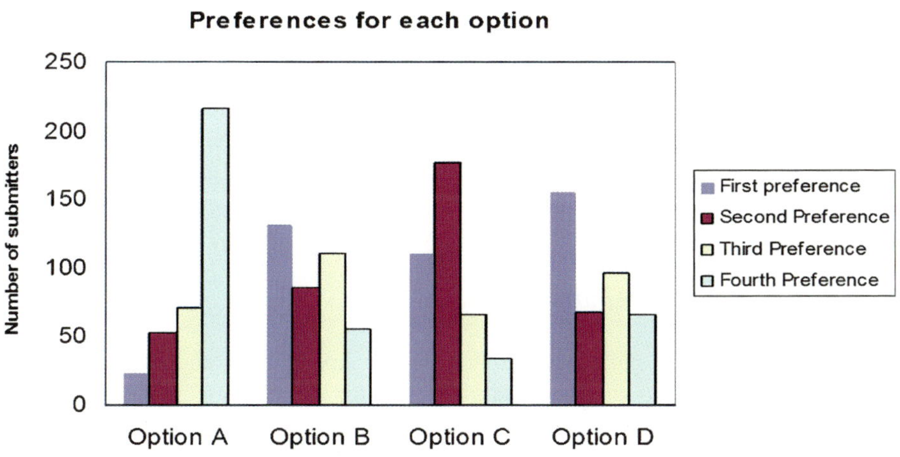

Fig. 3.10 Preferences for strategic options from public consultation (Environment Canterbury)

3.2.5.4 Strategic Investigations

There was a suite of strategic investigations being undertaken to inform the development and implications of the possible options for Canterbury water management. These included:

- Storage options that are most likely to be sustainable
- Efficiency and ecological enhancements through integrated water management
- Impact of land use intensification on water quality
- Identification of priority restoration programmes
- Integration of water for energy security and irrigation availability
- Economic modelling of production and ecosystem services
- Governance structures for sustainable management.

One of the major findings of this work has been the identification of significant gains that can be made from increased efficiency from existing allocations through water redistribution and infrastructure investment. For Mid Canterbury, the volume of storage needed to be able to supply all of the potentially irrigable area is half of a storage-led option if: piped distribution replaces canal distribution (which reduces losses and also enables irrigation to be applied when it is needed rather than when the system can supply it); water allocations are redistributed so that surface water is used in the upper part of the catchment and groundwater in the lower part of the catchment (which enhances aquifer recharge); high frequency, low application rate irrigation systems are installed (which reduces leakage to groundwater). This would also enable reduction in takes from the fully allocated Ashburton River.

Another significant finding has been the change in nitrate leaching if all the potentially irrigable areas in Canterbury are irrigated. Regional scale modelling of nitrate leaching from existing land use has been undertaken and correlates well with the field monitoring of nitrate levels in aquifers (Bidwell et al. 2009). The modelling indicates that there are exceedances of the drinking water standard for nitrate of 11.3 mg/L (refer to Fig. 3.11). The modelling of nitrate leaching if all potentially irrigable land was irrigated shows substantial areas exceeding the drinking water standard.

This means if further land use intensification of the Canterbury plains is to occur then there needs to be improvement in land use practices to manage nitrate leaching. These improvements are not only needed for new development but also for existing development.

3.2.5.5 Sustainability Appraisal

The four options were subject to a Sustainability Appraisal by the Steering Group and an Officials Group (technical advisors) using the Framework developed by Sadler and Ward (Sadler et al. 2008) to reflect New Zealand institutional

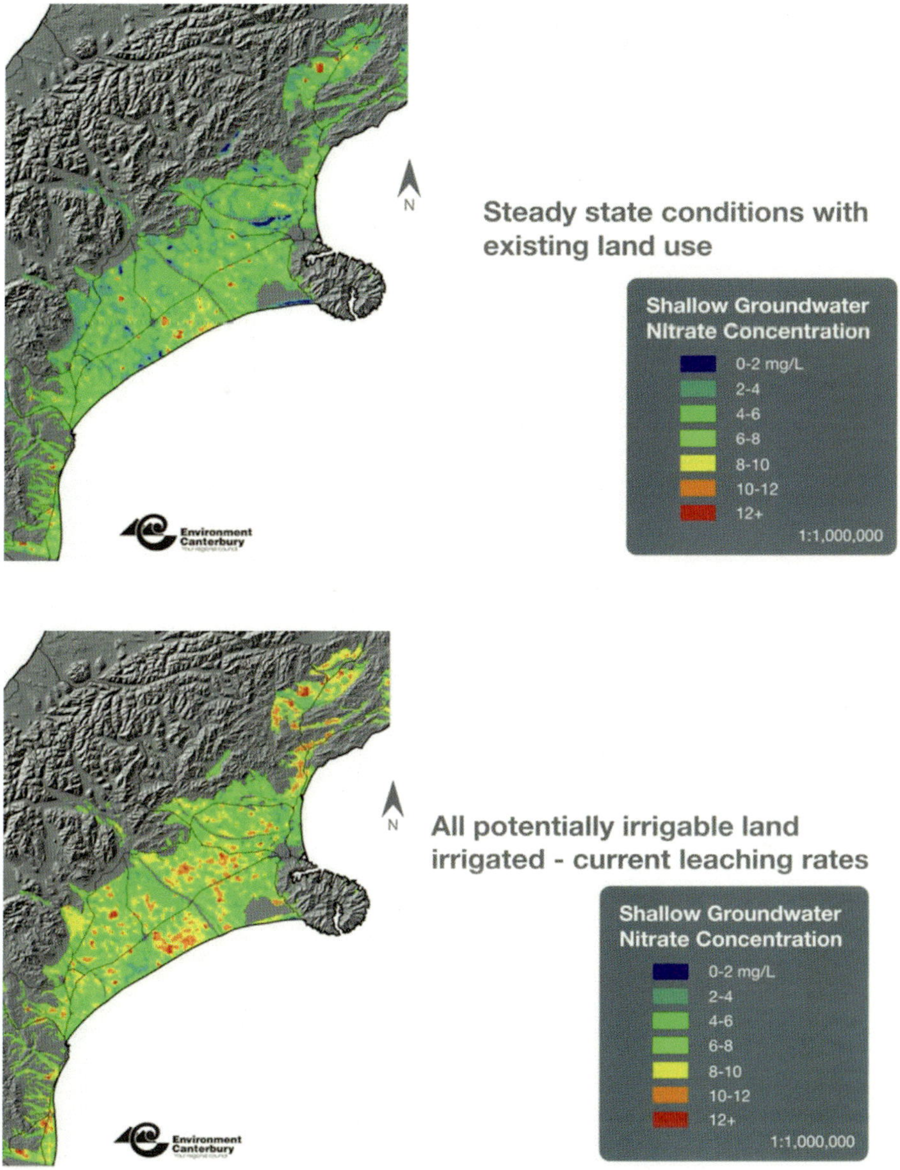

Fig. 3.11 Nitrate modelling in shallow groundwater (Bidwell et al. 2009)

arrangements. The Framework is founded on four pillars of sustainability (social, economic, environmental and cultural) which correspond to the four well beings of the Local Government Act.

The appraisal was conducted as a two-day workshop. Participants reviewed evaluation criteria and scale descriptions on a 5-point scale (from −2 strong negative impact to +2 strong positive impact with the neutral position 0 representing the status quo). Once the evaluation criteria had been amended, each group was asked to identify points on the five-point scale that represented an acceptable minimum position (quadruple bottom line) and a desirable objective position (quadruple top line).

The four options were then scored against the amended evaluation criteria. Figure 3.12 shows the comparative results. Some of the key findings of this appraisal were as follows:

- The bottom line is higher than Option A – Business as Usual
- Option B (environment-led) scores well on environmental criteria but is below the bottom line on economic criteria
- Option D (storage-led) scores well on economic criteria but is below the bottom line on environmental criteria
- Option C (efficiency-led) scores above the bottom line on nearly all criteria.

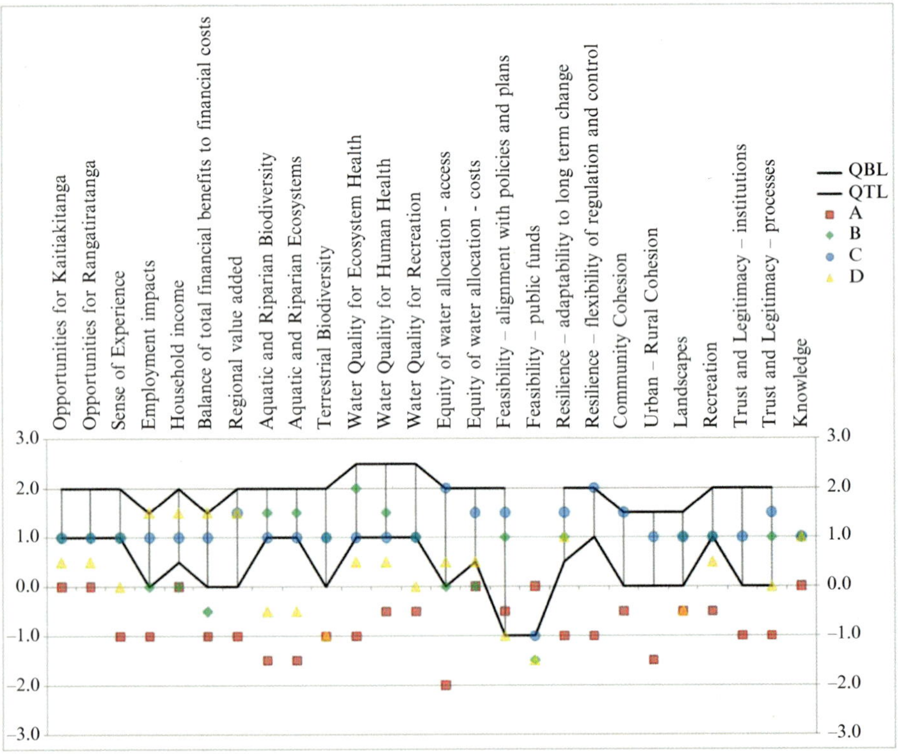

Fig. 3.12 Comparison of sustainability profiles for all options (Russell and Ward 2010)

The lower, thicker black line is the quadruple bottom line (QBL) representing an acceptable minimum position

The upper, thinner black line is the quadruple top line (QTL) representing a desirable objective position

The scoring of Option A against the evaluation criteria is shown by red squares

The scoring of Option B against the evaluation criteria is shown by green diamonds

The scoring of Option C against the evaluation criteria is shown by blue circles

The scoring of Option D against the evaluation criteria is shown by yellow triangles

When considered at the sub-regional level, the workshop participants considered that combinations of Options B, C and D were most likely to achieve sustainability at the sub-regional level.

3.2.5.6 The Overall Strategic Approach

A key point to emerge from consultation was the strong consensus in favour of a coordinated, collaborative approach that would combine the best features of the options which differed from Business-as-Usual. There was a need for a strategic approach that provided:

- Explicit recognition of environmental limits
- Programmes to restore ecological health and functioning to sustainable levels
- Development of infrastructure, technologies and practices to progressively deliver improved outcomes for Canterbury
- Evolution of governance structures to enable local government to better meet the challenges identified.

The desired outcomes were defined as targets across ten areas with goals set for the short term (2010 and 2015), medium term (2020) and long term (2040) (Canterbury Water 2009). The ten areas were:

- Drinking water
- Irrigated land area
- Energy security and efficiency
- Ecosystem health/biodiversity
- Water-use efficiency
- Kaitiakitanga
- Contribution to regional and national economies
- Natural character of braided rivers
- Recreational and amenity opportunities
- Environmental limits.

3.3 Implementation of the Canterbury Water Management Strategy

3.3.1 Implementation Framework

The implementation of the Canterbury Water Management Strategy was based on the parallel development of proactive implementation programmes to achieve the multiple targets defined in the strategy framework document.[2] A nested system of a regional implementation programme (RIP) and ten zone implementation

[2]This is in contrast to the approach envisaged under the RMA of relying on applicants' proposals for water resource development.

programmes (ZIPs) was created. The implementation programmes were not designed to be hierarchical. Rather the RIP was to address regional issues such as regional storage and distribution while the ZIPs were to address catchment issues such as land use practice improvements.

Zone Committees were joint committees of the regional council and the district and city councils in the zone. They comprised 6–7 community members who were locally based or had a special relationship with the zone, members of the rūnanga within the zone, and council representatives. The Regional Committee was a committee of the regional council with representatives of local government, central government, Ngāi Tahu, community, a member from each zone committee, and an independent chair.

Like the Canterbury Water Management Strategy, the recommended programmes of the committees were non-statutory. Statutory backing of the Strategy was provided by a new Regional Policy Statement (Environment Canterbury 2013a) and statutory backing for the implementation programmes is being provided by a new regional plan – The Canterbury Land and Water Regional Plan (Environment Canterbury 2015b) – which is a nested document to match the regional and zone implementation programmes with a regional component for region-wide requirements and specific sections for each zone.

The strategy implementation also led to organisation design changes within the regional council. A Strategy and Programmes Group was established to develop strategies for each target and co-ordinate implementation. A Water Executive was incorporated in the Strategy and Programmes Group with facilitators to service the regional and zone committees. The Planning Group was revised to undertake the preparation of the new Regional Policy Statement to provide statutory backing to the Canterbury Water Management Strategy and the new regional plan to provide statutory backing to the region and zone implementation programmes. A Resource Management Group was formed to integrate the resource care and regulatory functions around geographical zones.

Programmes were put in place to address key issues arising from the Strategy. One was the Land Use and Water Quality Project to develop an approach to manage cumulative nutrient loads and improve the science related to nutrient effects associated with land use change. Another was the Immediate Steps Biodiversity Programme to implement the Regional Biodiversity Strategy (Environment Canterbury 2008) and to work with the region and zone committees on the identification of priority biodiversity enhancement projects. A three-year programme was agreed with Ngāi Tahu to address freshwater management, governance arrangements, resource management, relationship building, institutional arrangements as well as the continuation of existing projects.

3.3.2 Implementation Progress

There has been progressive establishment of Zone Committees throughout the region commencing with the Hurunui Waiau Zone in July 2010 and concluding with the Christchurch West Melton Zone in November 2011. The Regional Committee was established in November 2010. Zone Implementation Programmes have been progressively generated by Zone Committees within 12 to 18 months of being established (Table 3.6). The Regional Committee produced a Regional Implementation Programme.

A programme of statutory instruments has also been implemented to provide legal backing to the Canterbury Water Management Strategy and the Implementation Programmes. The Canterbury Land and Water Regional Plan was notified as a draft in August 2012. The recommendations of the Hearing Panel were accepted in December 2013 but parts were subject to High Court challenge. The plan was made partly operative from September 2015 excluding the provisions under challenge.

Plans, that were already in process, also progressed. Following a long hearing process, the Natural Resources Regional Plan that had been notified in in July 2004, became operative in June 2011. A variation to the operative Waimakariri River Regional Plan became operative in June 2011. Plans for environmental flows and allocation limits from collaborative processes established prior to the Zone Committees were completed for the Pareora, Waipara and Orari catchments.

More recently a number of Zone Committees have prepared Addenda to their ZIPs. These documents have been focussed on "solution packages" for some of the more difficult issues (primarily water quality issues for lakes) in the respective zones that had not been addressed in detail in the initial ZIPs. These Addenda are generating Plan Changes to the Canterbury Land and Water Regional Plan to incorporate the statutory components of the agreements reached in the zones as set out in the Addenda.

There have also been Plan Changes to refine regional rules and to give statutory backing to the regional approach to nutrient management based on Farm Environment Plans using good management practices and predictions of nitrate leaching based on the model Overseer.

Table 3.6 Zone committee formation and preparation of zone implementation programmes Environment Canterbury 2016b

Zone	Formed	ZIP preparation	ZIP addendum
Ashburton	Sept 2010	Nov 2011	Mar 2014
Banks Peninsula	Sept 2011	Mar 2013	Nov 2014
Christchurch West Melton	Nov 2011	Mar 2013	
Hurunui Waiau	July 2010	July 2011	
Kaikoura	July 2011	Nov 2012	
Lower Waitaki	Oct 2010	Mar 2012	Sept 2014 July 2015
Orari-Opihi-Pareora	May 2011	Apr 2012	
Selwyn Waihora	Sept 2010	Dec 2011	Oct 2013
Upper Waitaki	Feb 2011	Apr 2012	July 2015
Waimakariri	Aug 2010	Dec 2011	

Note that source reference provides dates and documents

There have been significant changes in approaches to water resource management resulting from the Zone Committee programmes to implement the Canterbury Water Management Strategy. The Strategy has also encouraged changes in the private sector. Discussed below are noteworthy changes in the following areas:

- Provision of storage
- Water use efficiency
- Environmental flow adjustments
- Reduction of nutrients from land use intensification
- Farm Environment Plans and Audited Self-Management
- Water quality management for vulnerable freshwater bodies
- Biodiversity enhancements
- Kaitiakitanga.

Some trends and emerging issues are also identified.

3.3.3 Provision of Storage

While storage was seen as a key component of addressing water availability issues in Canterbury, there have been concerns with the sustainability some of the more cost effective forms of storage such as dams on the mainstems of braided rivers (Jenkins 2007b). Storage proposals like the dam on the south branch of the Hurunui River (part of the original consent application for the Hurunui Water Project) are in areas of high naturalness, modify downstream flow and sediment transport, have downstream effects on braided character and increased algal blooms, as well as affect recreational uses. Proposals on foothill rivers were also contentious, such as the Orari River (by Rangitata South) which would flood the Orari gorge, and, the Waianiwaniwa River (part of the consent application for the Central Plains scheme) (Fig. 3.13).

Collaborative decision processes led to different approaches to storage in the case of the Rangitata South and Hurunui Water Projects (Jenkins 2013). Both processes led to alternatives which were superior in terms of sustainable management, lower impacts and greater community acceptance. Instead of a dam on the Orari River, the alternative of an off-river storage involving the capturing of high flows from the Rangitata River evolved. This is 16.6 Mm^3 storage capacity (at a cost of \$82 m) to irrigate 14,000 ha with withdrawals when river flows exceed 110 m^3/s. In the case of the Hurunui Water Project the alternative of a series of storages on the Waitohi River (a tributary of the Hurunui River) with diversions from the Hurunui River (Fig. 3.14) was selected by a collaborative process. This has a possible capacity of 210 Mm^3 to irrigate 60,000 ha.[3]

[3] Recently a review of the Hurunui Water Project indicated the current demand for irrigation has reduced. While the Waitohi Storage will not be foregone, the HWP Board has approved an on-plains storage pond of 23 Mm^3 for an irrigated area of 21,000 ha fed by an intake canal from the Hurunui River (Pile and Robb 2017).

Fig. 3.13 Off River Storage adjacent the Rangitata River under Construction (John Bisset)

There have also been private investments in storage, usually at a smaller scale. There have been many on-farm storages, e.g., on a 779 ha dairy farm milking 1600 cows, a 2 ha storage pond capable of holding 40,000 m^3 of water has been constructed as insurance against weather and water restrictions. The water is enough to irrigate pasture with a 585 m centre pivot for 10 days. Irrigation schemes are also putting in storage to offset run-of-river restrictions. Mayfield Hinds Irrigation is constructing a 6.1 Mm^3 capacity pond at Carew to offset a 20% river restriction for 21 days. Waimakariri Irrigation Limited is seeking approval for an 8.2 Mm^3 storage at Wrights Road. This will hold enough water for 9 days of full irrigation flow to 18,000 ha of farmland. The design is to store water when river flows are high and irrigation demand is low, and used when abstraction is on restriction at times of low river flow. The additional storage would have made the scheme fully reliable for 27 of the past 42 years. Without storage, the scheme would have been fully reliable one year in 42 years. In the dry conditions of the 2013 summer an estimated $30 m of production was lost because of restrictions to irrigation supply. The proposal is under challenge in the Environment Court where the primary concern is the number of people in the likely flow path of the water released by catastrophic breach of the pond embankments (Newhook 2016).

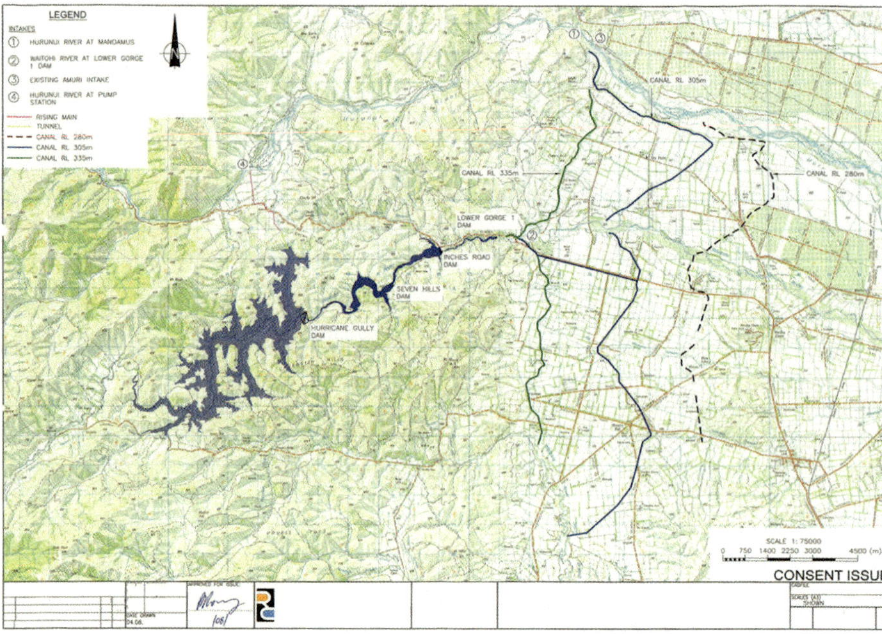

Fig. 3.14 Waitohi tributary storage proposal (Chris Hansen Consultants 2012)

3.3.4 Water Use Efficiency

The CWMS identified opportunities for water use efficiency improvements at the property scale, the scheme or delivery scale, and, at the catchment scale which would reduce storage requirements while increasing water availability (Canterbury Water 2009). Inefficiencies have been identified (Jenkins 2012) in:

- Irrigation methods, for example, the use of centre pivots need between half (for soils with PAW[4] 120 mm) and a quarter (for soils with PAW 60 mm) compared to border dyke irrigation;
- Application rates and macropore flow – where use of high application rates for irrigation cannot be retained in the soil profile and a substantial proportion passes through the soil to groundwater;
- Reliability of supply – where uncertainty of water availability encourages farmers to irrigate 'just in case' when water is available but not necessarily needed for crop requirements, rather than 'just in time' to meet crop requirements;
- Irrigation water distribution – where piped distribution can reduce water losses from open channel distribution;

[4]PAW is "Profile Available Water" which is the amount of water potentially available to plant growth that can be stored in the soil to 100 cm depth (Landcare Research 2016).

- Spatial application of surface and groundwater by irrigating only with surface water in the upper part of a groundwater catchment in order to enhance aquifer recharge and irrigating with groundwater in the lower part of a groundwater catchment.

Water use efficiency has not been considered in detail in the Zone Implementation Programmes and is not one of the "priority issues" in the Regional Implementation Programme despite its significance in the CWMS. In the private sector, there is a shift occurring to more efficient irrigation systems, for example, the Ashburton Lyndhurst Scheme was originally designed for border dyke irrigation now has 66% spray irrigation with a current conversion rate of 7% a year. However, little attention appears to have been given to the issue of application rates.

NIWA's work has highlighted the significance of reliability of supply (Duncan et al. 2010). One analysis involved two farms: one with on-farm storage and one without. The analysis compared "ideal" and actual irrigation for the two farms[5]. When the scheme was unable to deliver water because the Waimakariri River was on restriction, the farm without storage was unable to irrigate whereas the farm with storage could irrigate when required. The farm with reliable supply was better able to match the "ideal" pattern of irrigation and makes more effective use of irrigation water. For the farm without storage, soil moisture was below 50% field capacity for 10 out of 35 weeks of the irrigation season; whereas the farm with storage was only below 50% of field capacity for 4 of the 35 weeks.

The regional council commissioned an economic analysis of improved reliability (MRB 2011). The analysis was for a mix of pasture for dairy, dairy support, arable and mixed farming with a shift from 80% reliability in water availability to 95% reliability. For the dairy farm considered this achieved an increase from 10,430 kgDM/ha/year to 12,960 kgDM/ha/year. The increased production would generate a 12.1% return with a storage cost of $3750/ha or 7.9% return with a storage cost of $6250/ha.

In mid Canterbury, there have been projects to upgrade the original open channel system to a piped network to reduce conveyance losses. The Ashburton Lyndhurst Scheme has completed the first stage of a piped delivery system and is proceeding with a second stage. The initial stage (at a cost of $8 m) replaced 31 km of open channels with pipe servicing 3500 ha of irrigated land and enabling a further 550 ha to be irrigated with improved efficiency. A second stage (estimated to cost $95 m) involves more than 200 km of pipe to supply the remaining 21,000 ha of the scheme with the ability to supply a further 4000 ha and with 100 ha of land currently in channels returned to productive farmland. With the use of a pressurised pipe system there is a reduction in energy requirements for pumping irrigation water. A similar "pipe-replacement-of-open-channel" project has been completed for the Valetta Scheme (13,000 ha of irrigated land) (Valetta Irrigation Ltd 2016); while a proposal

[5] Ideal irrigation was assumed to be irrigation when soil moisture fell to 50% of PAW and the soil was either filled to 80% or 31.8 mm/week (whichever was the lower amount) and taking account of rainfall and PET at the sites based on NIWA's virtual climate network.

for the Mayfield Hinds Scheme (32,000 ha) was narrowly turned down with costs and uncertainty around development constraints to substantiate investment cited as the reasons for voting against the proposal (Tait 2015).

Changes in the spatial allocation of surface water and groundwater have not been incorporated in the regional implementation programme. However, there are outcomes being influenced by the relationship between irrigation, surface water and groundwater. One outcome is the amount of irrigation recharge beneath and down gradient of irrigation schemes. The regional council studied the effects of changes in groundwater level downstream of the Valetta Scheme due to shifts from border dyke irrigation to more efficient irrigation methods (Davey 2006). The study showed in periods of low winter rainfall many bores dropping in level over winter and recovering in summer due to infiltration losses from inefficient border dyke irrigation and rainfall on saturated irrigated paddocks. However, the study also showed that recharge from the Valetta Scheme was declining. With increasing groundwater use and declining recharge, the expectation is for further decline in groundwater levels. The extent of recharge is important because groundwater recharge from irrigation that has infiltrated past the root zone has been incorporated into the available allocation.

The recent Addendum to the Selwyn Waihora Zone Implementation Programme (Selwyn Waihora Zone Committee 2013) recommends the need for the Land and Water Plan to recognise the strong connection between groundwater and surface water in the Canterbury Plains by managing takes of groundwater and surface water as a combined resource. In particular, it is noted in the Addendum that Central Plains Water intends to use "alpine" water from the Rakaia and Waimakariri Rivers to irrigate 30,000 ha of dryland (i.e. new irrigation) and replace groundwater takes on 30,000 ha of currently irrigated land. This will improve flows in lowland streams and lower reaches of foothill rivers, and provide the opportunity to revise groundwater allocations downwards to address earlier overallocation decisions of independent commissioners. The Addendum also recommends use of managed aquifer recharge to maintain groundwater levels and flows in spring-fed lowland streams.

3.3.5 Environmental Flow Adjustments

Restrictions on the volume that can be taken from the rivers at different flows is the most significant mechanisms for maintaining instream environmental values, such as, aquatic ecology, recreation, natural character and cultural values. Ecologically important components include: minimum flows (the flow at which takes are restricted), flushing flows (flows needed for algae removal from the river bed), flood flows (flows needed for sediment transport and maintaining braided character), and allocation limits (the limit on the volume that can be taken at a particular flow). These components are given statutory backing in regional plans and consents.

Over the past ten years the regional council has been undertaking a review of environmental flow requirements of about 162 environmental flow monitoring points in the region. Many of the minimum flows had been set by catchment boards and have been found to be too low based on more detailed scientific assessments. Attempting to raise minimum flows and lower allocation limits through regulatory processes under the RMA had been contentious. However collaborative processes have achieved some success in addressing environmental flow requirements.

One example is the Pareora catchment. The minimum flow had been set by the catchment board at 300 L/s (which is 45% of the 7 day mean annual low flow) and the total allocation is 940 L/s (which is 142% of the 7 day mean annual low flow). Desirable environmental flow requirements are for a minimum flow of 600 L/s (90% 7DMALF), an A Block allocation limit of 198 L/s (30% 7DMALF), and a flushing flow of 4900 L/s (three times the median flow).

The outcome of the collaborative process (Environment Canterbury 2010) that commenced in 2005 was to:

- Set the A Block allocation to 30% 7DMALF (198 L/s compared to the existing 940 L/s) but apply the limit in 5 years' time;
- Create an alternative allocation at a higher flow (1600 L/s – median flow) for existing users as a source of water for storage;
- Establish a limited B Block allocation (2500 L/s) with a minimum flow above the flushing flow (5000 L/s);
- Increase the minimum flow from 300 L/s for total cessation of takes to 370 L/s and increase partial restriction from 400 L/s to 470 L/s;
- Encourage the major user, Timaru District Council, to secure its community supply from an alternative source during the critical period for fish (Oct–Nov) so that the minimum flow can be raised to 440 L/s and partial restriction to 540 L/s; and
- Foreshadow future reviews to further increase minimum flows.

These recommendations were incorporated in the draft Pareora Regional Plan, were effectively endorsed by the hearing commissioners, and, are now in the operative plan (Environment Canterbury 2012).

The collaborative approach has some interesting outcomes:

- There are improvements in environmental flow specifications compared to a previous overallocation. They are not to the full extent of desirable environmental flow requirements but come at a cost to existing users.
- Alternative water allocations at higher flows were incorporated to address the loss of access to water at lower flows for existing users but would require storage.
- The imposition of more restrictive requirements was delayed allowing existing users time to adjust their water infrastructure and management
- Provision has been made for future users but at flows above flushing flows and with low reliability of supply.

3.3.6 *Reduction of Nutrients from Land Use Intensification*

One of the key findings of the CWMS was that if there are to be substantial increases in land uses associated with nitrate leaching then there must be a corresponding decrease in nitrate leaching from existing land uses. Regional scale modelling indicated that it would only be possible to increase agricultural output while maintaining groundwater quality within acceptable limits if land management technologies that reduce nutrients and contaminants are applied across the region. To achieve this outcome would require existing users of water as well as new users to adopt the improved land management practices and technologies.

Two of the Zone Implementation Programmes (ZIPs) have addressed the issue of nutrients: the Hurunui-Waiau (Hurunui Waiau Zone Committee 2011) and the Selwyn-Waihora (Selwyn Waihora Zone Committee 2013). The more detailed analysis of nutrients associated with land use intensification has confirmed the earlier regional analysis.

The Draft Hurunui-Waiau ZIP considered the results of the more detailed analysis of the Land Use and Water Quality Project led by the regional council. It considered that the current water quality of the two sites on the mainstem of the Hurunui (confluence with the Mandamus River and SH1) was acceptable and should be maintained, i.e. average annual load limits should be set at current levels. The ZC also believed that nutrient guidelines should be established for the main tributaries based on dissolved inorganic nitrogen (DIN) concentrations being maintained below the nitrate toxicity level (1.7 mg/L) and dissolved reactive phosphorus (DRP) levels that existed in 1990–5. Annual load limits based on these criteria were compared with current estimated mean annual loads (Table 3.7).

Many of the proposed load limits are below current estimates. As reported by the Zone Committee, there was considerable anxiety amongst intensive land users in the Zone about the impact on their financial viability.

The subsequent Draft Hurunui-Waiau Regional Plan (Environment Canterbury 2011) included annual load limits for the mainstem sites with the allowance that nitrate levels could temporarily increase up to 20% prior to 2017. This was to provide some headroom for 100,000 ha of irrigation to occur. The tributaries are covered by narrative statements and a policy to progressively set nutrient limits.

At the plan hearings, the regional council submitted that land use change after 2017 that did not exceed 125% of the proposed nitrogen annual load limit or 110% of the proposed phosphorus limit should be a discretionary activity (i.e. require a consent). The dairy industry wanted a nitrogen load limit increased by 25% for 2012–2022 and by 50% after 2012. Other submitters argued that allowing any increase was inadvisable (Environment Canterbury 2013b).

With the Hurunui River considered to be phosphorus limited in terms of periphyton growth, the hearing commissioners recommended no increase in the phosphorus limit but allowed a 25% increase in the nitrogen limit. The increase in nitrogen limit of 25% would enable 18,600 ha to be converted from dryland sheep and beef farming to dairy. This would allow Ngāi Tahu's proposed conversion of Balmoral forest

Table 3.7 Nutrient load limits and current estimates for Hurunui catchment in draft ZIP (Environment Canterbury)

	Nitrate (DIN)			Phosphate (DRP)		
Location	Target concentration (mg/L)	Annual load limit (t/year)	Current estimate (t/year)	Target concentration (mg/L)	Annual load limit (t/year)	Current estimate (t/year)
Hurunui @ Mandamus	Current	40	40	Current	3.6	3.6
Hurunui @ SH1	Current	693	693	Current	10.2	10.2
Pahau @ Dazells	1.7	182 ± 23	196	0.0136	1.46 ± 0.19	2.2
Waitohi	1.7	86 ± 43	67	0.0056	0.28 ± 0.14	0.35
St Leonards	1.7	68 ± 3	133	0.012	0.48 ± 0.02	0.6
Dry Stream	1.7	53 ± 16	14	0.012	0.4 ± 0.12	0.5

to irrigated dairy (7000 ha) and Stage 1 of the Hurunui Water Project (15,000 ha) to proceed if all remaining border dyke irrigation was converted to spray irrigation.

The Selwyn-Waihora ZIP is at an earlier stage in the process. There has been an Addendum (21 July 2013) prior to regional plan formulation. The Addendum acknowledges the time delay between land use change and nitrate contamination reaching Te Waihora/Lake Ellesmere and indicates a 35% increase in the current load of nitrogen in the next 10–20 years because of the effects of recent land use intensification. Lake modelling predicts that a 50% decrease in the current load of both phosphorus and nitrogen is needed to achieve the objective of a trophic lake index of 6.0 or less. Further improvements are required for returning the lake from its phytoplankton (algae) dominated state to a self-sustaining macrophyte (aquatic plant) dominated state. These include a reduction in nitrogen loading to around 800tN/year, as well as addressing phosphorus retained in lake-bed sediments, improved management of lake levels and openings, and, reestablishment of macrophyte beds. The current and forecast nitrogen loads are set out in Table 3.8.

The current nitrogen load on the lake is estimated at 2650 t/year. With the delayed effect of recent intensification current land use is estimated to generate a nitrogen load of 4100 t/year. With additional 30,000 ha of irrigated land with Central Plains Water (CPW) and other intensification the load would rise to 5600 t/year. The load to meet Te Waihora targets for a macrophyte dominated lake is 800 t/year. The proposed solution package in the ZIP Addendum targets 4800 t/year. This represents a 12.5% improvement on "good management practice". This was considered to be "the contribution farming needs to make" and that the consequences of further reduction in terms of "the land use change required and the consequent social disruption unacceptable" (Canterbury Water 2013).

According to Dairy NZ, financial modelling indicates a 5% or less impact on farm productivity, reduction in milk production by 6–7% and a reduction in regional GDP of $30 m. This can be compared with the CPW and other intensification which is estimated to contribute about $310 m to regional GDP (Canterbury Water 2013).

Table 3.8 Agricultural load scenarios for Te Waihora catchment

Scenario	N Load (t/year)	Comments
Current	2650	Excludes lag effects
2011 baseline	4100	Includes lag effects
With CPW and other intensification	5600	30,000 ha CPW irrigation plus, other intensification
Proposed ZIP solution package	4800	12.5% less than good management practice
Te Waihora targets	800	Macrophyte dominated lake

Financial modelling was also undertaken of 18 farms representative of land use in the catchment (AgriBusiness Group 2012). This was based on an Overseer analysis for carrying out mitigation strategies and running the results through a financial model for effects on cash position and total equity. The Overseer modelling highlighted the importance of soil type, with light soils showing much higher leaching rates than heavy soils: 65–80 kgN/ha/year for irrigated dairy farms on light soils compared to 15–31 kgN/ha/year for heavy soils. It also showed high leaching rates for irrigated dairy support farms on light soils: 40–52 kgN/ha/year.

In terms of the range of mitigation strategies considered, active water management and reducing stocking rates showed the greatest reductions (57% less nitrate for 15% less cows on light soils, and, 38% less nitrate for soil moisture demand irrigation on light soils). In terms of cost effectiveness, DCD[6] use achieved 14% less nitrate with improved cash position and total equity. Active water management was achieved at low cost. Reduced stocking rates were achieved on improved cash position with reduced expenditure but reduced total equity.

Overseer 6.0 was used. This version incorporates soil drainage which was shown to be important for estimating leaching rates between light and heavy soils. However, use of monthly steps and average climate conditions as well as the inability to accommodate water use efficiency restricts the ability to model active water management strategies. The accuracy of the modelling of the farms was also highly dependent on data availability from the farmers involved.

3.3.7 Water Quality Management for Vulnerable Freshwater Bodies

Addenda to the Zone Implementation Programmes have focussed on "solution packages" to address some of the significant water quality issues in the zones.

[6] DCD is dicyandiamide, a nitrification inhibitor applied to farm soils to decrease nitrate leaching (Di et al. 2007).

3.3.7.1 Selwyn-Waihora Catchment

In the Selwyn-Waihora Zone, a package was recommended to improve cultural and environmental outcomes in the Selwyn Waihora catchment while maintaining farm viability and economic growth (Canterbury Water 2013). The key values underlying the package were the ecological and cultural health of Te Waihora/Lake Ellesmere, and, the contribution of agriculture to the economy. Key actions included (1) water allocation limits to provide for ecological and cultural flows; (2) restricting the nitrogen load from the catchment and managing nitrogen loss rates at the property level; (3) reducing phosphorus and microbial contaminants from the catchment; and, (4) lake interventions for improved lake level and margin management, addressing legacy phosphorus, restoring macrophyte beds and constructing lake margin and floating wetlands.

However, lake modelling predicts that to achieve a TLI of 6 for the lake would require a 50% decrease in the current load of both nitrogen and phosphorus. Achieving such large reductions in nutrient levels and rehabilitating Te Waihora/Lake Ellesmere is incompatible with current land use and further consented land use intensification in the catchment. With the time lag effect between land use intensification and groundwater migration of nitrate to the lake, the nitrogen load is expected to increase. Even with the reductions achieved by the solutions package a 50% increase in nitrogen load is predicted.

3.3.7.2 Hinds Catchment

There are four main parts to the solutions package in the Ashburton ZIP Addendum for the Hinds Catchment (Canterbury Water 2014a) which has elevated nitrate levels: (1) catchment scale actions, (2) local scale actions, (3) investigations, monitoring and review, and, (4) community engagement. Managing nutrients in the Hinds Plains area is the focus of the solutions package. It involves the interplay between the area of further land use intensification, the level of mitigation used to control nitrogen leaching from farms, and, the volume of clean water added through managed aquifer recharge.

The package of recommendations seeks to reduce the catchment nitrogen load by on-farm mitigation, resulting in a nitrate concentration of 9.2 mg/L in lowland water bodies. Dilution using up to 5 m^3/s of clean water through managed aquifer recharge is needed to reach the target concentration of 6.9 mg/L, a level consistent with the national bottom-line for nitrate. A trial of managed aquifer recharge is currently in progress (Environment Canterbury 2016c).

3.3.7.3 South Coastal Canterbury Streams

The Lower Waitaki ZIP Addendum for the South Coastal Canterbury Streams focussed on the health of the Wainono Lagoon ecosystem, a vibrant economy and improved water quality in the coastal streams (Canterbury Water 2014b). An integrated package of measures was recommended based on achieving a TLI of 6 for Wainono Lagoon and 90% protection level for nitrate toxicity levels in streams. This was through a catchment cap on nitrogen and an equitable allocation of nitrogen to farms through farm environmental plans. However, this is not sufficient to achieve a TLI of 6 for Wainono Lagoon with the addition of further land use intensification associated with the Waihao Downs and Hunter Downs irrigation schemes. The package also involves the augmentation of Wainono Lagoon with high quality water from the Waitaki River.

3.3.7.4 Upper Waitaki

The Upper Waitaki ZIP Addendum (Canterbury Water 2015) considered water quality improvements and economic outcomes. The principal water quality concern was maintaining the oligotrophic status of Lake Benmore. Water quality in Lake Benmore is also the predominant influence on water quality in the Waitaki River and downstream lakes, Lakes Aviemore and Waitaki. The main elements of the solutions package were adopting Good Management Practice for land and resource use and the need for Farm Environment Plans for agricultural properties, and, setting catchment load limits for nitrogen for agriculture, aquaculture and urban development.

The Haldon and Ahuriri Arms of Lake Benmore have distinct characteristics and values, and, require their own limits. For the Haldon Arm, a nitrogen-load limit of 1972 tN/a was recommended which provides for some further development. The Ahuriri Arm is more sensitive due to lower catchment inflows. The development scenarios that were explored as part of the limit setting process all predicted that the Ahuriri Arm would move out of the oligotrophic band. For the Ahuriri Arm the recommended load limit of 516 tN/a is based on the current consented load. Therefore, any new development needs to come from within this load and how the load is allocated.

3.3.7.5 Banks Peninsula

The Wairewa ZIP Addendum prepared by the Banks Peninsula Zone Committee (Canterbury Water 2014c) developed a specific implementation programme to address the poor health of Te Roto o Wairewa/Lake Forsyth. The lake has high cultural significance to Ngāi Tahu. The catchment is a nutrient "red zone", and, the Ōkana and Ōkuti valleys including the township of Little River have experienced significant flood events. The Addendum focuses on recommendations to improve the health of the lake and its catchment, and reduce the flood hazard in the catchment.

Deforestation in the catchment over the last 160 years has led to erosion with increased sedimentation of the lake with sediments naturally high in phosphorus. It has undergone eutrophication since the early 1900s. There are regular blooms of the toxic cyanobacteria, *Nodularia spumigena*, particularly in summer when deoxygenation associated with lake stratification leads to phosphorus release from the sediments adding to the phosphorus load in the water column. With limited vegetation cover and short, steep creeks the catchment is prone to flooding in periods of intense rainfall.

In relation to the management of cyanobacteria blooms in the lake, there were recommendations for actions in the catchment and the lake. A catchment limit for phosphorus of 1700 kg of phosphorus per annum was targeted to match the amount estimated to leave the lake through the constructed outlet to the sea. This represents an 85% reduction from the current catchment load which is estimated to be 11,300 kgP/a. At the property level, environmental plans to focus on preventing sediment and phosphorus into waterways were also recommended, such as sediment traps and stream bank stabilisation.

For the lake, the focus was on lake openings for sediment flushing, investigations of the cause of cyanobacteria blooms, investigations and trials of managing phosphorus-rich lake sediment (e.g. dredging, marginal wetlands, floating wetlands and macrophyte growth), and, a sediment retention basin at the head of the lake. Flood mitigation in the catchment was also recommended including drainage improvements, avoiding new development in flood-prone areas, and lake openings for allowing the passage of floodwaters.

3.3.8 Farm Environment Plans and Audited Self-Management

Plan Change 5 sets out the statutory basis for managing freshwater quality through requiring a Farm Environmental Plan as part of any application for resource consent to use land for a farming activity. Farm Environmental Plans are linked to the adoption of "Good Management Practices". Good Management Practices are articulated by farming industry sectors as a means of managing nitrogen and phosphorus losses from farms across the range of soils, climates and land uses (Foundation for Arable Research et al. 2015).

Farm Environment Plans describe the specific on-farm actions that will be undertaken by the farmer to implement Good Management Practices, the time frame within which these actions will be undertaken, and, how these actions will ensure progress towards management objectives and targets in Schedule 7 of Plan Change 5. Schedule 7 sets out seven management areas to be addressed in Farm Environmental Plans: (1) nutrient management, (2) irrigation management, (3) soil management, (4) collected animal effluent management, (5) water body management – riparian areas, drains, rivers, lakes and wetlands, (6) point sources – offal pits, farm rubbish pits, silage pits, and, (7) water use management – stock water and wash-down water. There are objectives and targets defined for each management area.

Additional requirements include a nutrient budget which shows the nitrogen baseline and nitrogen loss calculations for the farm, and a report from the "Farm

Portal"[7] showing the nitrogen baseline loss rate for the farm and the loss rate after adopting Good Management Practice.

Plan Change 5 also introduces a new approach to compliance. The Farm Environment Plans must be audited by a Certified Farm Environment Plan Auditor who is independent of the farm and has not been involved in the preparation of the Farm Environment Plan. As a minimum, the audit will assess the performance of the farming activity against the objectives, targets, Good Management Practices and timeframes in the Farm Environment Plan, the robustness of the nutrient budget, and the efficiency of water use for irrigated farms.

This approach to compliance is based on the concept of audited self-management (Jenkins 1996). The concept involves environmental performance requirements being set by the regulator (in this case through the Land and Water Regional Plan) but with industry being able to determine how to meet these requirements (through the Farm Environment Plan). Industry is required to have an environmental management system with certification by the regulator or independent certifier (in the case of the South Canterbury Coastal Streams through management plans of nutrient user groups). Industry is required to undertake measurements to demonstrate that environmental performance requirements have been met with the measurements audited by an independent auditor (in this case through the Certified Farm Environment Plan Auditor). This approach is discussed further in Sect. 8.3.

3.3.9 Biodiversity Enhancements

One of the issues in the CWMS was the decline in freshwater biodiversity. There has been on-going habitat loss and fragmentation of riparian habitat. Less than 10% of the region's previously extensive wetlands remain. Weeds have been replacing indigenous plants. The immediate steps biodiversity protection and enhancement project was launched in 2010 as an integral part of the CWMS with $2 m/year available for 5 years with two thirds from regional rates and one third from landowners and other stakeholders. Projects of regional significance have been selected by the Regional Committee and each Zone Committee recommends priority projects within their zone. The assessment criteria are based on the six goals of the Canterbury Biodiversity Strategy (Environment Canterbury 2008), e.g. to restore the natural character of degraded indigenous habitats and ecosystems, and, the ecological value of the project, e.g. ecological context: projects must provide a benefit to indigenous biodiversity and play an important role in the long term health of the wider ecosystem.

[7] Farm Portal is the nutrient management database accessed at www.farmportal.ecan.govt.nz and that is used to derive a Baseline loss rate and Good Management Practice loss rate for nitrogen losses from the farm.

At the regional level three projects are being supported: the Te Waihora/Lake Ellesmere enhancement project, enhancement of the upper catchments of the Rakaia and the Rangitata Rivers, and, the Wainono Lagoon project. At the zone level, smaller scale projects of fencing, riparian planting, willow control and stream crossings are in progress (Environment Canterbury 2016a).

3.3.10 Kaitiakitanga

In the CWMS, one of the first order priorities for water is customary use, one of the principles is tangata whenua, and, one of the outcome targets is kaitiakitanga. The practical goals in the CWMS include recognition of Ngāi Tahu Freshwater Policy on environmental flows, direct discharges, unnatural mixing of waters, and non-point source pollution control; involvement in restoration programmes for degraded wāhi taonga and mahinga kai waterways; having Iwi Management Plans in place; improving local government capability in kaitiakitanga; and, establishing co-governance arrangements for the management of Te Waihora and its catchment.

Some of the tangible progress in addressing kaitiakitanga includes:

- A restoration programme for Te Waihora/Lake Ellesmere – Whakaora Te Waihora (Ngāi Tahu and Environment Canterbury 2016) – with funding contributions from central government ($6 m) and regional government, the dairy industry and Ngāi Tahu ($5.6 m combined).
- Rūnunga representation on the Zone and Regional Committees for preparing the Zone Implementation Programmes.
- A relationship agreement between the regional council and Ngā Papatipu Rūnanga – known as Tuia – for on-going collaboration in water management was signed in February 2013.
- The Mahaanui Iwi Management Plan (covering the area from the Hurunui to the Hakatere/Ashburton) was released in March 2013: the document includes the objectives (ngā paetae), issues of significance (ngā take) and policies to guide freshwater management in a manner consistent with Ngāi Tahu cultural values and interests.
- The undertaking of operational "on the ground" biodiversity projects with each of the 10 Papatipu Rūnanga as part of the immediate steps biodiversity projects.

There are also techniques being developed to incorporate Māori water management concepts into western-style approaches to water management. This includes concepts like State of Takiwa reporting, Cultural Health Index, and, Cultural Opportunity Mapping, Assessment and Response. The minimum flows to protect cultural interests are determined to be those thresholds to protect values such as mauri, mahinga kai and wāhi taonga.[8] Some of the cultural flow recommendations

[8] *Mauri* means the life giving essence of a resource; *mahinga kai* means maintaining healthy populations of food species and their habitats; and, *wāhi taonga* means sites of significance.

are above the minimum flows considered sufficient to provide for instream ecological values. For example, Waikekewai and Taumutu Creek are of high cultural significance and it is considered inappropriate to be abstracting water for irrigation from the catchment because of wāhi tapu associations. The Selwyn Waihora ZIP Addendum recommends prohibiting, on expiry, of surface and groundwater takes that have a hydraulic connection to the creek, while enabling consent holders to move to deeper non-stream depleting groundwater sources (Canterbury Water 2013).

In addition to the regional council led restoration projects there are also collaboration with private interests. For example, the land owner of Minimoto Lagoon (near Amberley Beach), which has biodiversity and cultural significance, has recently withdrawn stock and fenced the wetland with support of the QEII Trust and Immediate Steps funding.

3.3.11 Trends and Emerging Issues

There has certainly been a significant change in the approach to water management in Canterbury with the introduction of collaborative processes for resolving water management issues. While it is still early days in terms of implementation, some different approaches have been identified.

Firstly, in relation to water storage, there have been some innovative ways to be able to store water to access alpine water but without storages on mainstems of alpine rivers. The off-river storage at Arundel and the tributary storage on the Waitohi represent changes in approach.

Secondly, in relation to water use efficiency, the replacement of distribution canals with pipe and the continuing shift from border dyke to spray irrigation are improving water use efficiency. However, there is insufficient attention to other aspects of water use efficiency, in particular, the use of soil moisture demand management and the spatial reallocation of surface and groundwater to enhance recharge.

Thirdly, with respect to environmental flows, collaborative processes have led to raising minimum flows and reducing allocations at low flows. These changes are not to the full extent of desirable environmental flows but they come at a cost to existing users. Collaborative outcomes have recognised the need for allocations at higher flows but involve storage for their effective use. There has also been the recognition of time needed for existing users to adjust.

The situation in relation to a fourth issue of nitrate levels is being shown to be problematic. Further irrigation will increase nitrate levels. Existing users will need to adopt better than good practice management and incur costs. However, the parallel targets of increased irrigated area and reductions in nitrate loads appear unlikely to be achieved. Even dilution approaches for the Hinds Catchment (with managed aquifer recharge) and for Wainono Lagoon (with diversion of alpine water) are at the margins of sustainability limits. Other emerging issues in relation to nitrate management include the high leaching rates of light soils and their suitability for intensification, and, the nitrate reduction potential of reduced stocking rates and improved irrigation management.

There is an opportunity cost associated with inefficient use of scarce resources with poor water use efficiency and with the constraints on new entrants from high nitrate loads from existing users. This is in addition to the $2.5b opportunity cost from "poor technology uptake" in the dairy sector identified in the briefing to incoming Minister of Agriculture (Ministry of Agriculture and Forestry 2011).

The target areas of biodiversity enhancement and kaitiakitanga are showing progress. However, target areas that have been taken off the regional priority list: drinking water, recreation and amenity opportunities, and, water use efficiency need attention.

It is noteworthy that the RMA processes to give statutory backing to projects and plans that implement the CWMS have been less contentious. In addition, hearing commissioners are making decisions which are marginal changes to the collaborative proposals.

An emerging issue is the need for improved integration of surface water and groundwater interaction. This includes the consideration of managed aquifer recharge as a form of storage, targeted recharge to maintain lowland stream flows, and spatial allocation of surface and ground water to enhance recharge.

Two other emerging issues from the CWMS implementation to date are the importance of modelling to predict outcomes both scientific and financial from the decisions being made, and, the related issue of the data available to operate and verify the models. With the need to manage more efficiently and to tighter limits predictive models and field measurement are essential. With increasing reliance on farm management plans and audited self-management foreshadowed in the CWMS and now being incorporated into plans and consents the need for modelling and measurement will escalate.

A further emerging issue is equity in allocation. This is arising in relation to the allocation of nutrient capacity. One type of issue is the allocation between existing users. A second type is between existing and future users: for further land use intensification to occur, existing users have to reduce their cumulative nutrient contribution below the specified limit(s) to create capacity (often referred to as "headroom") for future intensification. Box 3.1 compares the approaches taken in the Hurunui-Waiau Zone and the South Coastal Canterbury sub zone and the problems that are arising.

Box 3.1: Equity in Allocation of Nutrient Limits

Two significant issues have arisen in relation to the setting of nutrient limits in catchments where estimates of current discharges have been assessed to have reached or exceeded desired nutrient limits. One issue relates to the creation of "headroom" for further land use intensification.[9] This has implications for those responsible for existing discharges as well as new applicants. The second issue is the calculation and allocation of allowances for land uses creating nutrient discharges. This is heavily reliant on data availability and modelling of the effects of land use on waterways. Two examples are considered below: one from the Hurunui and Waiau Rivers, and the other from Wainono Lagoon.

Hurunui and Waiau Rivers

There had been a history of algal blooms from nutrient enrichment in the lower Hurunui River (Ausseil 2010). As part of the Hurunui-Waiau Zone Committee process involving extensive scientific input, there was an initial agreement to limit nutrient loads from land use in the catchment to current levels in order to maintain current water quality levels (Hurunui Waiau Zone Committee 2011). However in the process for gaining statutory backing for introducing catchment load limits for nutrients, the dairy industry group (Livestock Improvement Corporation and Dairy NZ 2015) argued for increased nitrogen loads to create headroom for further irrigation development on the basis that the river was considered phosphorus limited (DairyNZ 2011). The outcome of the RMA hearings on the proposed Hurunui and Waiau Regional Plan was to set the nitrogen limit for the Hurunui catchment at 963 tN/year (25% above the current estimated load), while setting the phosphorus limit at 10.7 tP/year (the current estimated load) (Environment Canterbury 2013c). This created headroom for two irrigation proposals: the Ngāi Tahu Properties Balmoral Forest proposal and Stage 1 of the Hurunui Water Project (Salmon et al. 2013).

Setting nutrient load limits was designed to control land use change that would increase nutrient loads. A change in land use on a property basis was determined as an increase greater than 10% in the long term average release of nitrogen and phosphorus to land which may enter water (Environment Canterbury 2013c).

However, the "10% rule" places greater constraints on sheep and beef farmers with low nutrient loss rates (e.g. 5 kgN/ha/year for dryland sheep and beef farms on poorly drained soils which allows 0.5 kgN/ha/year change) compared to an irrigated dairy farmer with high nutrient loss rates (e.g. 61 kgN/ha/year for irrigated dairy farms on extremely light soils which allows 6.1 kgN/ha/year change). Furthermore the 10% rule allowed the Amuri Irrigation Company to expand its irrigated area in the Hurunui catchment by offsetting increased nitrogen loads from converting dryland sheep and beef

(continued)

[9] Headroom is available when the current load is lower than the load limit. Headroom is equal to the difference between the load limit and current load.

Box 3.1 (continued)
farms to irrigated dairy farms, against decreased nitrogen loads from irrigation efficiency improvements on high loss rate farms, and secondly in the Waiau catchment by limiting expansion to 6.5% increase in nitrogen loadings (Environment Canterbury 2015d). Such conversion options were not available to individual sheep and beef farmers under the 10% rule.

The sheep and beef farmers expressed concern that the 10% rule did not provide sufficient flexibility for "normal" variations in sheep and beef farming, such as increased planting of fodder crops which could trigger the land use change rule. This led the regional council to issue an advice note that normal dryland farming changes would not be considered a change in land use under the 10% rule and that the regional council does not intend to undertake enforcement action against dryland farmers for normal variations in farming practice (Environment Canterbury 2015a).

Wainono Lagoon

Water quality in the Wainono Lagoon has a Trophic Level Index (TLI) of 6.5 and the South Canterbury Coastal Streams ZIP Addendum has set a goal of achieving a TLI of 6 (Canterbury Water 2014b). To achieve this there was a need to reduce the nitrogen loading in the catchment of the lagoon. A draft nitrogen load limit and allocation framework had been developed which allowed for a load for existing users at the current load of 605 tN/year and allowances for two consented irrigation schemes operating at Good Management Practice: Hunter Downs (182 tN/year) and Waihao Downs (71 tN/year) (Norton 2013). To meet a TLI of 6 for the lagoon, the increased load would be offset by a proposed augmentation of the flow to the lake with 1 m^3/s of high quality water from the Waitaki River using distribution infrastructure of the Hunter Downs Irrigation Scheme. Under the proposed Land and Water Regional Plan in catchments of lakes like Wainono Lagoon existing users would be constrained to their nitrogen baseline (the mean discharge of nitrogen below the root zone for the years 2009–13 calculated by a model called Overseer) or for low emitters to 10 kg/ha/year. If the Hunter Downs scheme did not occur and augmentation did not proceed then a "sinking lid" option similar to the proposal in the Selwyn Waihora catchment was proposed. This would require a nitrogen load limit of 514 tN/year for existing users, involving a 15% reduction in nitrogen load to maintain current water quality and a 30% reduction to achieve a TLI of 6 for Wainono Lagoon (Norton 2013).

However, a group of farmers expressed dissatisfaction with the nitrogen allocation framework in relation to the equitability of the framework for low

(continued)

Box 3.1 (continued)

emitters compared to high emitters. The concern was not about the need to set catchment load limits to achieve environmental outcomes but the method of allocation (Norton et al. 2014).

The Nitrogen Allocation Reference Group (NARG) was formed comprising a variety of farming interests, rūnanga representatives and general community interests. There was a series of meetings. The early meetings focussed on generating a common understanding of the information available. Meetings then assessed a range of nitrogen allocation options reducing from seven to a short-list of three. The final meetings were focussed on achieving a consensus.

Grandparenting[10] of current discharges (the Land and Water Plan proposal) was rejected. A cornerstone of the agreed framework was the requirement for all land users to achieve a minimum of Good Management Practice with respect to nutrient discharges so that poor performers were not rewarded with high nitrogen allocations. The main area of negotiation was the need to create head-room from improved management by high emitters to enable flexibility for nitrogen load increases by low emitters. "Maximum caps" were to be placed on high emitters according to soil type (35 kg/ha/year for light soils, 25 kg/ha/year for medium soils and 20 kg/ha/year for poorly drained soils) and that they be given a time period to adjust. "Flexibility caps" were set for low emitters. Initially these would be set at 10 kg/ha/year (excluding steep hill country farmers who would be assigned 5 kg/ha/year). Compliance with the maximum caps by high emitters and flow augmentation of the lagoon with the implementation of the irrigation schemes would allow the flexibility cap to be lifted to 15 kg/ha/year, and possibly 17 kg/ha/year based on assessing the effectiveness of mitigation and augmentation measures. If flow augmentation did not occur then the flexibility cap would need to remain at 10 kg/ha/year (Norton et al. 2014).

The agreement by the NARG was accepted by the Zone Committee, the regional council and the two district councils (Waimate and Waitaki) related to the South Canterbury Zone. The agreement was incorporated in the proposed plan change to the Land and Water Regional Plan (Environment Canterbury 2015e) to reflect the Zone Implementation Programme (Canterbury Water 2012) and its Addendum (Canterbury Water 2014b).

However, since the preparation of the proposed plan change there has been a revision of Overseer (the model used to estimate nitrogen loss rates for farms), adjustments to the leaching rates from the Look-Up Tables (the basis for estimating nitrogen leaching rates from farms with different soil types), concerns about the assumptions about denitrification in poorly drained soils,

(continued)

[10] Grandparenting is allocation directly related to historical discharges.

Box 3.1 (continued)
and revisions to soil mapping in the Wainono Lagoon catchment. The changes are likely to affect the calculations of catchment loads and maximum caps and thereby the flexibility caps. Interested submitters on the plan change were asked to caucus on the implications of these changes (Whiting et al. 2015). While there is agreement that the changes need to be addressed, the discussions reignited the debate about the appropriate nitrogen allocation methodology and the fairness of the allocations (Environment Canterbury 2015c).

References

AgriBusiness Group. (2012). *Selwyn Te Waihora nutrient performance and financial analysis*. Christchurch: Environment Canterbury.

Aqualinc Research Limited. (2008). *Canterbury strategic water study* (Stage II). Prepared for Environment Canterbury, Christchurch.

Ausseil, O. (2010). *Hurunui River – Influence of the middle reaches on water quality of the lower Hurunui River (2005–2008)*. Christchurch: Environment Canterbury.

Bidwell, V., Lilburne, L., Thorley, M., & Scott, D. (2009). *Nitrate discharge to groundwater from agricultural land use: An initial assessment for the Canterbury Plains*. Prepared for Steering Group of the CWMS, Christchurch.

Biggs, B., Ibbitt, R., & Jowett, I. (2008). Determination of flow regimes for protection of in-river values in New Zealand: An overview. *Ecohydrology Hydrobiology, 8*(1), 17–29.

Canterbury Water. (2009). *Canterbury Water Management Strategy Strategic Framework*. Christchurch: Environment Canterbury.

Canterbury Water. (2012). *Lower Waitaki South Coastal Canterbury zone implementation programme*. Christchurch/Waimate/Oamaru: Environment Canterbury/Waimate District Council/Waitaki District Council.

Canterbury Water. (2013). *Selwyn Waihora ZIP addendum*. Christchurch: Environment Canterbury.

Canterbury Water. (2014a) *Ashburton ZIP addendum: Hinds Plains area*. Christchurch/Ashburton: Environment Canterbury/Ashburton District Council.

Canterbury Water. (2014b). *South Coastal Canterbury ZIP Addendum*. Christchurch: Environment Canterbury.

Canterbury Water. (2014c). *Wairewa ZIP Addendum*. Christchurch: Environment Canterbury, Christchurch City Council.

Canterbury Water. (2015). *Upper Waitaki ZIP Addendum*. Christchurch: Environment Canterbury.

Chris Hansen Consultants. (2012). *HWP Waitohi irrigation and hydro scheme: Resource consent application*. Christchurch: Hurunui Water Project.

DairyNZ. (2011). *DairyNZ submission on the proposed Hurunui and Waiau River Regional Plan*. Hamilton: DairyNZ.

Davey, G. (2006). *The effects of border dyke irrigation recharge on groundwater levels in and below the Valetta Scheme* (Environment Canterbury Technical Report U06/11).

Di, H., & Cameron, K. (2004). Nitrogen leaching losses from different farms and rates of farm effluent applied to a Templeton soil in Canterbury, New Zealand. *New Zealand Journal of Agricultural Research, 47*(4), 429–437.

Di, H., Cameron, K., & Sherlock, K. (2007). Comparison of the effectiveness of a nitrification inhibitor, dicyandiamide, in reducing nitrous oxide emissions in four different soils under different climate and management conditions. *Soil Use and Management, 23*(1), 1–9.

Duncan, M., Srinivasan, M. S., & Diettrick, J. (2010). *Improved irrigation management on water metering information: 2009–10 irrigation season* (NIWA Client Report: CHC 2010–091). Christchurch.

Environment Canterbury. (2004). *Variation 1 to the Proposed Canterbury Natural Resources Regional Plan*. Christchurch: Environment Canterbury.

Environment Canterbury. (2008). *A Biodiversity Strategy for the Canterbury region*. Christchurch: Environment Canterbury.
Environment Canterbury. (2010). *Proposed Pareora Catchment Environmental Flow and Water Allocation Plan*. Christchurch: Environment Canterbury.
Environment Canterbury. (2011). *Proposed Hurunui and Waiau Regional Plan*. Christchurch: Environment Canterbury.
Environment Canterbury. (2012). *Pareora Catchment Environmental Flow and Water Allocation Regional Plan*. Christchurch: Environment Canterbury.
Environment Canterbury. (2013a). *Canterbury Regional Policy Statement 2013*. Christchurch: Environment Canterbury.
Environment Canterbury. (2013b). *Hearing commissioner recommendation to the Proposed Hurunui and Waiau River Regional Plan and Proposed Plan Change 3 to the Canterbury Natural Resources Regional Plan*. Christchurch: Environment Canterbury.
Environment Canterbury. (2013c). *Hurunui and Waiau River Regional Plan operative December 2013*. Christchurch: Environment Canterbury.
Environment Canterbury. (2015a). *Advice note – July 2015: Dryland farming and triggering the land use change rule in the Hurunui and Waiau River Regional Plan (HWRRP)*. Christchurch: Environment Canterbury.
Environment Canterbury. (2015b). *Canterbury Land and Water Regional Plan December 2015*. Christchurch: Environment Canterbury.
Environment Canterbury. (2015c). *Caucasing statement in relation to nutrient management provisions: Plan Change 3 to the partially operative Land and Water Regional Plan*. Christchurch: Environment Canterbury.
Environment Canterbury. (2015d). *Notice of resource consent decisions, Record No: CRC153154*. Christchurch: Amuri Irrigation Company Limited, Environment Canterbury.
Environment Canterbury. (2015e). *Proposed Variation 3 to the Proposed Canterbury Land and Water Regional Plan – Section 15 – Waitaki and South Coastal Canterbury*. Christchurch: Environment Canterbury.
Environment Canterbury. (2016a). *Canterbury Water Immediate Steps Fund*. http://www.crc.govt.nz/advice/biodiversity/funding/Pages/immediate-steps.aspx. Accessed 31 Dec 2016.
Environment Canterbury. (2016b). *Committees*. http://ecan.govt.nz/get-involved/canterburywater/committees/Pages/Default.aspx. Accessed 16 Aug 2016.
Environment Canterbury. (2016c). *Hinds/Hekeao MAR Pilot Site*. Environment Canterbury. http://ecan.govt.nz/publications/Council/hinds-mar-pilot-project.PDF. Accessed 25 Nov 2016.
Foundation for Arable Research, New Zealand Pork, DairyNZ, Beef+Lamb New Zealand, Horticulture New Zealand, & Deer Industry New Zealand. (2015). *Industry-agreed good management practices relating to water quality*. Christchurch: Environment Canterbury.
Friend, J., & Hickling, A. (2005). *Planning under pressure: The strategic choice approach* (3rd ed.). Oxford: Elsevier.
Hurunui Waiau Zone Committee. (2011). *Hurunui Waiau draft Zone Implementation Programme*. Christchurch: Environment Canterbury.
Jenkins, B. R. (1996, December 5–6). *Best practice environmental regulation – The Western Australian approach*. Paper presented at the Environmental Management Beyond 2000, Griffith University, Brisbane.
Jenkins, B. R. (2007a, March 27–30). *Water Allocation in Canterbury*. Paper presented at the NZ Planning Institute conference 2007, Palmerston North.
Jenkins, B. R. (2007b, November 19–20). *What role for storage in Canterbury?* Paper presented at the NZSOLD/ANCOLD conference "Dams – Securing Water for Our Future", Queenstown.
Jenkins, B. R. (2012, November 20). *Water issues in Canterbury*. Paper presented at the Agronomy Society of New Zealand Annual conference 2012, Lincoln University.
Jenkins, B. R. (2013). *The development of sustainable alternatives to applicant's proposals using collaborative approaches*. Paper presented at the New Zealand Planning Institute 2013 Annual conference, Hamilton, 30 April to 3 May 2013.
Jenkins, B. R., & Henley, G. (2013). *Collaborative management: Community engagement process as the decision making process*. Paper presented at the Environment Institute of Australia and New Zealand, Melbourne, October 23–24

Landcare Research. (2016). *SMap Glossary*. https://smap.landcareresearch.co.nz/glossary. Accessed 31 Dec 2016.

Livestock Improvement Corporation, & Dairy, N. Z. (2015). *New Zealand dairy statistics 2014–15*. Hamilton: LIC and Dairy NZ.

Ministry of Agriculture and Forestry. (2011). *Briefing for incoming ministers*. Wellington: Ministry of Agriculture and Forestry.

Morgan, M., Bidwell, V., Bright, J., McIndoe, I., & Robb, C. (2002). *Canterbury strategic water study*. Lincoln: Lincoln Environmental.

MRB. (2011). *On farm impact of variation in reliability*. Environment Canterbury.

Newhook, L. (2016). Environment Court Interim Decision 07–09-2016 in the matter of the appeal under the RMA by Eyre Community Environmental Safety Society Inc. Christchurch.

Ngāi Tahu, & Environment Canterbury. (2016). *Whakaora Te Waihora*. http://tewaihora.org/the-plan-2/. Accessed 31 Dec 2016.

NIWA. (2016). *Backgrounder*. https://www.niwa.co.nz/climate/nzcu/climate-update-73-july-2005/backgrounder. Accessed 31 Dec 2016.

Norton, N. (2013). *Revised working draft nitrogen load limit and allocation framework*. Christchurch: Environment Canterbury.

Norton, N., Harris, S., Scott, M., Lilburne, L., Robson, M., Stapleton, J., MacDonald, M., Newman, N., & Whitehouse, I. (2014). *Process and outcomes of the nitrogen allocation reference group (NARG) for the South Canterbury Coastal Streams area. R14/110*. Christchurch: Environment Canterbury.

Pile, C., & Robb, C. (2017, May 16). *Hurunui Water Project – Scheme update*. Presentation to Hurunui Waiau Zone Committee.

Russell, S., & Ward, M. (2010). *Sustainability appraisal: Application to the Canterbury Water Management Strategy*. Lincoln: Landcare Research.

Sadler, B., Ward, M., & Frame, R. (2008). *A framework for sustainability appraisal in New Zealand*. Lincoln: Landcare Research.

Salmon, P., Kirikiri, R., & van Voorthuysen, R. (2013). *Hearing commissioner recommendation to the proposed Hurunui and Waiau River Regional Plan and Proposed Plan Change 3 to the Canterbury Natural Resources Regional Plan*. Christchurch: Environment Canterbury.

Selwyn Waihora Zone Committee. (2013). *Addendum to Selwyn Waihora Zone Implementation Programme (21 July version)*. Christchurch: Environment Canterbury.

Tait, H. (2015). *Brief of evidence in the matter of the hearing of submissions on proposed variation 2 (Hinds Plain) to the proposed Canterbury Land and Water Regional Plan*. Christchurch.

Valetta Irrigation Ltd. (2016). *Valetta irrigation scheme conversion*. http://valetta.co.nz/scheme-info. Accessed 31 Dec 2016.

Whitehouse, I., Pearce, A., & McFaddern, G. (2008). *Final report – Canterbury Strategic Water Study* Stage 3. Christchurch: Environment Canterbury.

Whiting, G., Ellison, E., & Fenemor, A. (2015). *Minute 2 (minute and directions of hearing commissioners on expert caucasing): Proposed Plan Change 3 (South Coastal Canterbury Streams)*. Christchurch: Canterbury Regional Council.

Chapter 4
The Collapse of the Maya and the Development of a Sustainability Framework

Abstract A central theme of this book is the definition of potential failure pathways that can lead to socio-ecological system collapse and then the derivation of sustainability strategies to address these potential failure pathways so that a socio-ecological system has the capacity to manage system-threatening disturbances and maintain its structure and function, i.e. that the socio-ecological system is sustainable. The collapse of the Maya is of value to this theme because of the variety of theories that scholars have put forward for the reasons for the collapse based on the interaction between socio-economic and biophysical systems at several spatial scales as well as the significance of water management in the collapse.

Categories of failure pathways are distilled from these theories of collapse and then placed in the framework of nested adaptive systems. A framework for analysing the resilience of societal systems to major disturbances is then developed. Approaches to the development of sustainability strategies are described within the same framework based on the principles of ecosystem stewardship.

Keywords Maya collapse • Failure pathways • Sustainability strategies • Socio-ecological systems • Adaptive cycles • Nested adaptive systems

4.1 The Classic Maya Collapse

The Classic Maya (AD 250–900) reached intellectual and artistic heights which no others in the New World and few in the Old could match at the time. Maya civilisation reached its full glory in the early eighth century but in the century and half that followed all its magnificent cities had fallen into decline and ultimately suffered abandonment. This was surely one of the most profound social and demographic catastrophes of all human history (Coe 2005).

There have been many theories about the cause of the collapse of the Classic Maya. In considering the literature and associated research on the possible causes there has been an increasing sophistication in the analysis as the body of archaeological evidence has grown. Theories commenced as generalised speculation in the absence of evidence. The next step appears to be making inferences from site specific research evidence.

B.R. Jenkins, *Water Management in New Zealand's Canterbury Region*, Global Issues in Water Policy 19, https://doi.org/10.1007/978-94-024-1213-0_4

Models of societal collapse were also applied to the Maya situation to identify potential causes. There was also a shift from hypotheses based on single causes of collapse to a recognition of multiple factors associated with explanations of collapse and then causal linkages between these factors.

Increasing analytical sophistication was then achieved through efforts to predict effects for comparison with effects found in the archaeological record and efforts to simulate the pattern of collapse. There was also an increase in comparative analysis. This was not only in terms of comparisons of different Maya sites but also comparisons of collapse between Classic Maya and other societies.

From the extensive literature addressing the Classic Maya collapse, the discussion below draws on (1) Tainter's analyses of complex societies as a basis for a general explanation of collapse (Tainter 1988); (2) Diamond's multi-factor approach for the reasons for collapse (Diamond 2005); (3) Webster's analysis of the collapse of Copan (one of the Mayan cities that collapsed in the ninth century) by considering the range of possible causes (i.e. failure pathways) and the dynamics of change for each cause with the available archaeological evidence (Webster 2002); and (4) Lucerno's comparative analysis of water availability in Mayan cities (Lucerno 2002).

4.1.1 Tainter's Themes for Explaining Collapse

Tainter analyses the collapse of complex societies with the objective of developing a general explanation of collapse (Tainter 1988). He identifies characteristics that demonstrate collapse has occurred[1] and provides historical examples of collapse including the collapse of the Classic Maya of the Southern Lowlands. Tainter then examines why complex societies are established and considers two schools of thought: one focussing on the management of conflict in society, the other on the facilitation of integration of functions to meet societal needs.

From his review of the literature on collapse of complex societies, Tainter identified eleven major themes in the explanation of collapse:

- Depletion or cessation of a vital resource or resources on which the society depends[2]
- The establishment of a new resource base (or technology), which changes the social relationships in society
- The occurrence of some insurmountable catastrophe, such as hurricanes, volcanic eruptions, earthquakes or major disease epidemics
- Insufficient response to circumstances due to fundamental limitations of social, political and economic systems

[1] For example, "a lower degree of stratification and social differentiation" Tainter (1988), p. 4.

[2] Tainter considers a number of resource depletion explanations including deterioration due to human mismanagement, loss due to environmental fluctuation or climate shift, and, loss of trade networks for external resources and imported goods. Tainter (1988), pp. 44–51.

- Other complex societies leading to competition between societies with the demise of the weaker society
- Intruders leading to invasions and collapse of the invaded society
- Class conflict, social contradictions, elite mismanagement or misbehaviour where there is antagonism and conflicting goals between social classes
- Social dysfunction due to internal processes leading to integrative deficiencies
- Mystical factors such as decadence, loss of vigour or senility in society
- Chance concatenation of events with concurrent outbreaks of clusters of problems and weaknesses leading to societal collapse.
- Economic factors associated with declining marginal productivity of increasing complexity.

Tainter evaluates the explanatory power of these themes and argues they are insufficient because they do not provide the causal chain between the explanatory factor and the collapse outcome. He then develops the causal chain for the economic factor on the basis of productivity associated with increasing complexity in society. He argues that while there can be societal benefits from increasing complexity, there comes a point where there is a decrease in marginal productivity of further increases in complexity. Unless there is technical innovation, new energy source, or, expansion gaining further resources, then the society will collapse to a less complex form.

He then considers his causal chain of declining marginal productivity with increasing complexity in relation to the decline of the Roman Empire, the Maya collapse and the Chacoan collapse. While there is data to indicate the decline in economic viability of the western Roman Empire, there is insufficient data to test his arguments in relation to the Maya collapse.

4.1.2 Diamond's Multi-factor Approach

Diamond was interested in the question of why only some societies proved fragile and what distinguished those that collapsed from those that didn't (Diamond 2005). He had categorised processes by which past societies had undermined themselves by damaging their environment and distinguished those that collapsed from those that didn't.

In his comparative analysis, Diamond did not find cases for which a society's collapse could be attributed solely to environmental damage: there are always other contributing factors. He proposed a five-point framework of possible contributing factors in understanding any putative environmental collapse:

- Environmental damage
- Climate change
- Hostile neighbours
- Friendly trade partners
- Society's responses to its environmental problems.

Diamond uses the Maya Collapse as one of his case studies in considering his five point framework and considers that the Maya illustrate four of his points:

> They did damage their environment, especially by deforestation and erosion. Climate changes (droughts) did contribute to the Maya collapse, probably repeatedly. Hostilities among the Maya themselves did play a large role. Finally, political/cultural factors, especially the competition among kings and nobles that led to a chronic emphasis on war and erecting monuments rather on solving underlying problems, also contributed. (Diamond 2005)

Diamond considers the city of Copan in more detail (relying on Webster's analysis – see below). Population increased from the fifth century to an estimated 27,000 at AD 750–900, based on numbers of house sites. Later monuments record the arrival of a noble in AD 426. Construction of royal monuments glorifying kings was especially massive between AD 650 and 750. Nobles also began erecting their own palaces. The last date (of an incomplete building) is AD 822.

Farming began in the flat land with fertile alluvial soils along a river valley. By AD 650 people started to occupy hill slopes but these were only cultivated for about a century. Excavations of building foundations in the valley floor showed evidence of sedimentation in the eighth century suggesting erosion from the hill slopes. Pollen samples from lake sediment cores show that the pine forests originally covering the upper elevations of the hill slopes were cleared.

Skeletal analysis shows that the health of Copan's inhabitants deteriorated from AD 650 to 850. It is hypothesised that Copan's king would be blamed for the agricultural failure. Monument construction stopped in AD 822 and the royal palace was burnt around AD 850. The estimated population in AD 950 was around 15,000 (54% of the 27,000 peak). Population continued to decline; there were no signs of anyone in the valley by around AD 1250. There was also reappearance of pollen from forest trees in lake sediment cores.

Diamond also notes some complexities in the Maya collapse. First, there were a series of collapses: in AD 150 when El Mirador and other cities collapsed (the Pre-Classic collapse), in the late 6th and early seventh century collapse (the Maya hiatus), and the Post-Classic collapses such as Chichen Itza around AD 1250 and Mayapan around AD 1450. Second, the collapse was not complete because Maya population continued in reduced numbers in different locations. Third, the collapse in population was slower than the decline in cultural system. Fourth, there is also evidence of city decline due to military defeat by other cities: Copan grew in power until AD 738 when its king was captured and killed by its rival city Quirigua. Fifth, different parts of the Maya area rose and fell on different trajectories, for example the Puuc region (Northwest Yucatan Peninsula) after reaching a low population in AD 700, increased after AD 750 while southern cities were collapsing, peaked in population between AD 900 and 925 and then collapsed in turn between AD 950 and 1000.

Diamond highlights the role of warfare and drought. The archaeological record shows that wars became more intense and more frequent towards the time of the Classic collapse. This includes massive fortifications surrounding Maya sites, depictions of warfare on monuments and artefacts, and Maya writing of conquests.

This included wars between kingdoms, attempts at succession within kingdoms, and civil wars between rivals for being king.

Diamond notes the alignment of drought conditions with occurrence of collapse. There was a drought from AD 125 to 250 associated with the collapse of El Mirador and other sites. The build-up of Classic Maya sites was temporarily interrupted by a drought around AD 600 corresponding to a decline at Tikal and some other sites. There was an extreme drought starting around AD 760 with peaks in AD 810–20, around AD 860 and around AD 910 which coincides with three clusters of collapse dates for Maya centres around AD 810, 860 and 910.

In summary Diamond identifies five strands. One strand consisted of population growth outstripping available resources. A second strand was the effects of deforestation and hillside erosion reducing useable farmland. A third strand was the increased fighting. A fourth strand was the incidence of drought and the fifth strand is the failure of kings and nobles to recognise and resolve the problems undermining their society.

4.1.3 Webster's Analysis of Copan and Other Mayan Cities

Some of the impressive evidence of the reasons for collapse have come from combining the results of detailed field work with predictive simulations. In relation to the collapse of Copan, Webster (2002) brings together the work of Freter (1992) using obsidian hydration dating of 239 sites to develop a population history of Copan (Fig. 4.1) with the work of Wingard (1996) who simulated the agricultural productivity of the Copan valley using a soil analysis programme of the Department of Agriculture.

The population history from Freter's settlement data indicate a population of about 5000 in AD 600-650, increasing to around 28,000 about AD 750 (a growth rate of about 1% per annum for 150 years). The population level stays in the range 26,000–28,000 for about a century and then declines. There is still evidence of human presence as late as AD 1250.

Wingard modelled the population that could be supported by agricultural production. Starting from a population of 1000 he modelled what happened as numbers increased and land use strategies were applied to the landscape. He used the concept of carrying capacity as a variable and dynamic concept that could be changed by farming practice. The modelling indicated that the carrying capacity of the best alluvial soils (on the basis of long-cycle swidden farming) was reached by the mid-to-late sixth century with a population of 5000. Farmers on the valley floor then shifted to more intensive forms of cultivation and the excess population eventually spread to the less productive and more unstable uplands. By AD 800 at least 120 km^2 of the most usable lands in the valley had been colonised and deforested and Wingard predicted a major erosional event in the eighth century. Using the most intensive forms of cultivation that their technology allowed the population peaked

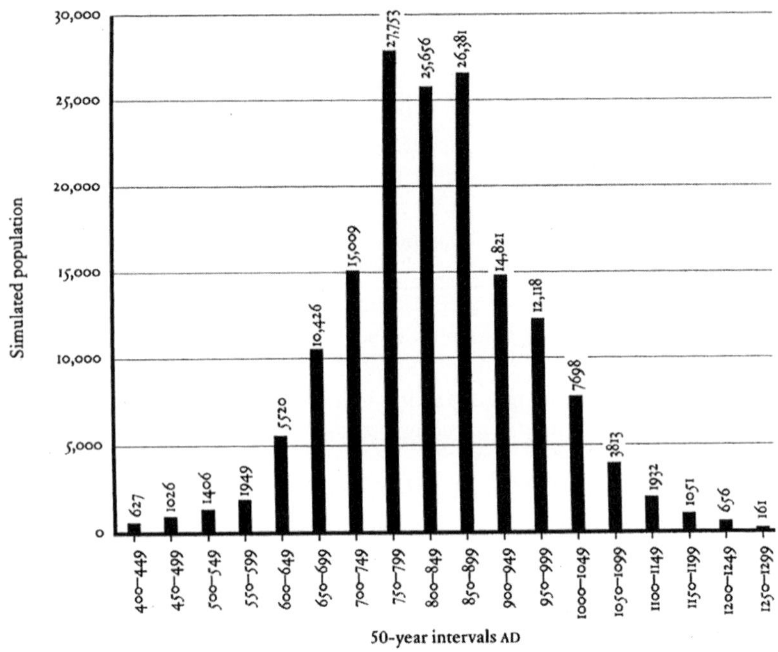

Fig. 4.1 Population history of Copan (Webster 2002). Drawing by Philip Winton. After an original graph by Timothy Murtha from information provided by David Webster. From "The Fall of the Ancient Maya" by David Webster, published by Thames & Hudson, London and New York

between AD 850 and 900 at about 22,000 then began to decline. The simulation was carried through to AD 1050 when the population level was about 14,000.

As Webster indicates the simulations prove nothing by themselves but the similar outcome from two independent approaches indicate the plausibility of the causal explanations.

4.1.3.1 Dynamics of Change

Webster provides a detailed review of the evidence associated with each of the potential causes of collapse to derive his conclusions on the cause of collapse. In his review Webster not only seeks to identify the range of possible causes, he also explores the dynamics of the changes associated with each potential causal explanation and tests the archaeological evidence for support (or otherwise) of each potential cause.

Webster, in summarising the range of theories put forward to explain the collapse of the Classic Maya in the southern lowlands of Yucatan Peninsula, identifies a longer list in two categories – collapse of the elite, and, collapse of the system.

Under collapse of the elite, Webster lists the following causes:

- Peasant revolts

- Internal warfare
- Foreign invasion
- Disputed trade networks

Under collapse of the system, he lists two subcategories – non-ecological and ecological causes. Under non-ecological causes, he includes:

- Collapse of trade networks, and
- Ideological pathology.

While under ecological causes, his list comprises:

- Earthquakes, hurricanes, volcanoes
- Climate change
- Epidemic diseases, and
- Agricultural degradation.

Based on his review of the archaeological evidence, Webster concludes in relation to the Maya collapse "that the collapse was fundamentally triggered by three interrelated and dynamic factors, in the following order of importance: one, a worsening relationship of the Maya population to their agricultural and other resources; two, the destabilising effects of warfare and competition; and three, the rejection of ideology and the institution of kingship. These in turn created or exacerbated a series of secondary stresses, including increased vulnerability to drought, peasant unrest, and disease" (Webster 2002).

In his analysis of Maya collapse, Webster identifies the significance of the failure of leadership to provide solutions to the crises facing Maya in the southern lowlands as a major factor in the collapse.

4.1.3.2 Comparative Analysis

Webster had tested the comparative circumstances facing a number of Maya centres. He reached the view that with the information currently available, the most convincing collapse explanation we have for the Tikal kingdom is overpopulation and agrarian failure with all of the attendant political consequences (Webster 2002). He reaches a similar view for Copan; citing one overarching cause that seems clear in the Copan collapse – too many people on a landscape deteriorating through overuse by humans.

However not all centres in the Southern Lowlands suffered total decline and abandonment. Webster provides the example of Lamanai which saw much construction during the Post Classic period. It was never sacked by its enemies, its sustaining river and lagoon never dried up, its agricultural fields still produced staple crops and no epidemic disease could have ravaged nearby kingdoms while leaving it untouched. It would appear that Lamanai was able to continue as it was not affected by major disturbance.

For other centres reviewed by Webster, the cause of collapse is less clear. For centres in the Petexbatum region there is evidence of conflict. There are also waxing and waning fortunes that are coincident with success and failure in local wars, and, with respect to allegiances to the major powers of Tikal and Calakmul and their respective periods of dominance as well.

4.1.4 *Lucerno's Analysis of Water Vulnerability*

There have also been some interesting comparative analyses in relation to the relative vulnerability of Maya centres to potential causes of collapse. Lucerno considered the scale of water control associated with the scale of Maya centres at three scales – regional, secondary and minor (Lucerno 2002). Regional centres such as Tikal, Calakmul and Caracol were located in upland areas with large pockets of dispersed fertile land difficult to monopolise and without permanent water sources. The need for an adequate water supply in the area of the regional centres is related to annual rainfall; it was typically less than at secondary and regional river centres. Tikal, Calakmul and Caracol each had artificial reservoirs constructed next to palaces and temples. Maya rulers became responsible for providing enough potable water to last through the dry season by organising the continual maintenance required to keep the reservoirs clean. Not all regional centres were located without lakes or rivers; Palenque and Copan are prime examples. More typical of most other ancient civilisations, these sites are found along rivers with concentrated alluvium that supported regional polities.

Secondary centres, such as Lamanai, Yalbec, Seibal, Piedras Engrams, Dos Pilas and Xunantunich, are typically found along rivers largely in upland areas with dispersed pockets of agricultural land that supported local polities. Because of dispersed agricultural soils, Maya farmers used scattered small-scale water systems including aguadas, dams, canals and drainage ditches. Their inconsistent distribution suggests that water systems had less of a political role in these areas.

Minor centres, such as Barton Ramie and Saturday Creek were located in lower elevations and had higher annual rainfall than the majority of regional centres. These centres were made up of relatively low densities of dispersed farmsteads.

4.2 Resilience Framework Based on Nested Adaptive Systems

4.2.1 *Sustainability as an Adaptive Cycle*

The Maya Collapse provides a fascinating example of the complexity that has to be considered in dealing with societal crises. With increasing sophistication of analysis and increasing levels of data focussed on critical issues, more refined findings have been possible. A key purpose of this chapter is to provide an improved framework for analysing the resilience of societal systems. To this end the Maya Collapse is considered in the framework of "nested adaptive cycles" of Gunderson and Holling. This placement can provide a systematic framework for the dynamics of system failure and approaches to system management to address the potential causes of system failure.

The framework builds on the work of Gunderson and Holling on understanding transformations in human and natural systems (Gunderson and Holling 2002). Their framework considers how human and ecological processes interact. The framework considers interactions over different time scales and interactions over different spatial scales.

A key component of the framework is the "adaptive cycle" which describes how an ecological or human system can be sustained both in obtaining resources for its ongoing survival and its ability to accommodate disturbance to the system and restructure. Sustainability can then be defined in relation to the maintenance of the relationships in adaptive cycles across different time scales and geographical scales.

The four phases of an adaptive cycle are considered to be:

- Exploitation – the use or harvesting of resources
- Accumulation[3] – the storage of material or energy in the system
- Release – disturbance of the system
- Reorganisation – restructuring of system after disturbance

The adaptive cycle can be sustained if the resources needed to maintain the cycle continue to be available and the system can be restructured after disturbance. If not, then systems may fail and adverse consequences can result. Gunderson and Holling depict the adaptive cycle as a Lissajous figure (Fig. 4.2). The cycle contains the four phases of exploitation, accumulation, release and reorganisation. There is the criti-

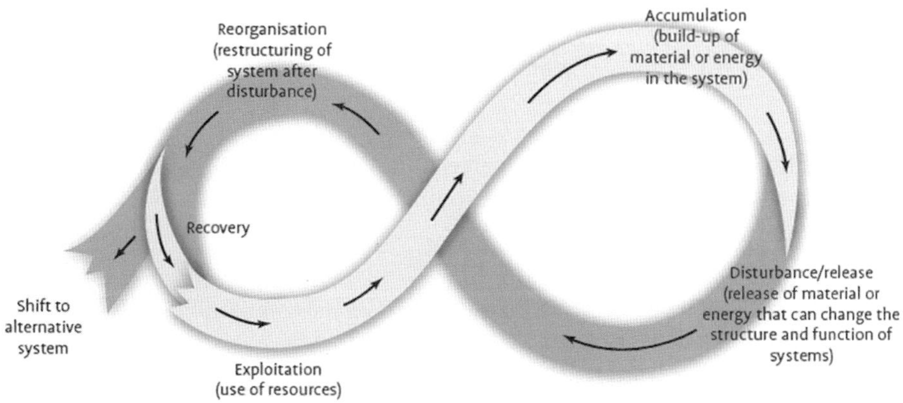

Fig. 4.2 Graphic representation of an adaptive cycle (Adapted from Gunderson and Holling 2002). From *Panarchy* edited by Lance H. Gunderson and C.S. Holling. Copyright © 2002 Island Press. Reproduced by permission of Island Press, Washington, DC

[3] Gunderson and Holling refer to "conservation" rather than "accumulation". Accumulation is considered to have wider application for dealing with both storage in systems for productive purposes and build-up of contaminants which can have adverse effects.

cal point after the restructuring of the system whether the adaptive cycle continues (recovery) or whether the system fails and shifts to an alternative system.

Taken as a whole the adaptive cycle has two opposing modes: a development or front loop of the exploitation and accumulation phases, and a back loop of the release and reorganisation phases. The front loop has relatively predictable dynamics while the back loop is characterised by uncertainty: it is the time for either destructive or creative change in the system (Walker and Salt 2006).

Some of the key properties for sustainability are:

- The potential for resources – there are limits on the resource available through use or harvesting
- Connectedness – the nature and degree of links between processes
- Resilience – the capacity of a system to absorb disturbance and still retain its basic function and structure

These properties change throughout the adaptive cycle. In the exploitation phase, potential and connectedness are initially low but growing while with resources available this growth phase has high resilience. In the accumulation phase, there is high potential and connectedness. However, the high resource use and the need to maintain connections make the system vulnerable to disturbance, i.e. it has low resilience. In the release phase after a disturbance there is a loss of potential and declining connectedness. In this destabilised phase resilience is low. In the reorganisation phase there is high potential but low connectedness. As the system recovers resilience begins to grow. This variation in properties throughout the adaptive can be plotted in three dimensions (Fig. 4.3) creating the pattern of a three dimensional Lissajous figure.

4.2.2 *Four Types of Sustainability Issues*

While the concept of sustainable management of adaptive cycles was originally developed for natural resource management, the concept is being used more widely for management of transformations in both human and natural systems. Paton and Johnston have applied the concept to the management of natural disasters. They recognised three types of sustainability issues:

- The capacity of a natural system to adapt to demands made upon it that are independent of human activity
- The impact of human activity on natural systems and whether the degree of disturbance alters the maintenance of the system
- The capacity of human systems to adapt to the effects of natural hazards.

Paton and Johnston develop the concept of "community resilience" as a basis for disaster preparedness of communities with respect to natural hazards (Paton and Johnston 2006). For example, in the immediate aftermath of an earthquake when urban infrastructure has been disrupted and civil defence are attempting to restore

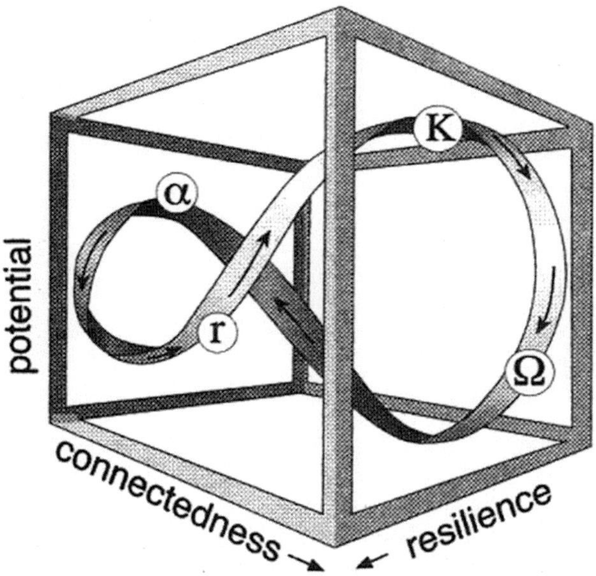

Fig. 4.3 Changes in properties throughout the adaptive cycle (Gunderson and Holling 2002). From *Panarchy* edited by Lance H. Gunderson and C.S. Holling. Copyright © 2002 Island Press. Reproduced by permission of Island Press, Washington, DC
Note: "r" is the exploitation phase; "K" is the accumulation phase; "Ω" is the release phase; and "α" is the reorganisation phase

services such as power and water supply, there is a need for householders to have sufficient water and power for survival for at least three days.

The broader linkages (i.e. connectedness) of biophysical systems in sustaining socio-economic systems have been identified in the Millennium Ecosystem Assessment (MEA) (Millennium Ecosystem Assessment 2005). The MEA identifies ecosystem services (provisioning, regulating and cultural) and their linkages to the constituents of human well-being (security, basic material for good life, health and good social relations). MEA also identifies "supporting" ecosystem services: these are equivalent to the capacity of a natural system to adapt to demands made upon it that are independent of human activity.

Conceptually there is a fourth type of sustainability issue:

- The capacity of a human system to adapt to demands made upon it.

The four types of sustainability issues can be depicted as a pair of Lissajous figures as shown in Fig. 4.4: one Lissajous figure representing biophysical systems and the second representing socio-economic systems. The four types of sustainability issues are shown numerically on Fig. 4.4 as:

1. capacity of the biophysical system to be maintained
2. capacity of linkages of the socio-economic system to the biophysical system
3. capacity of linkages of the biophysical system to the socio-economic system
4. capacity of the socio-economic system to be maintained

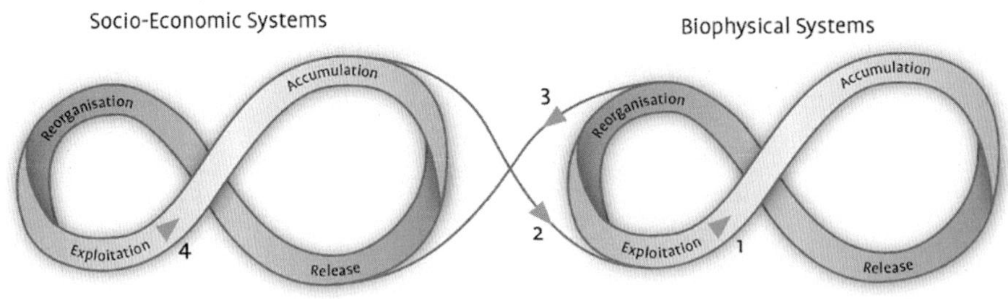

Fig. 4.4 Four types of sustainability issues (Jenkins 2016)

4.2.3 *Nested Adaptive Systems*

Ecosystem research has demonstrated that it is appropriate to consider biophysical processes at different time scales and geographical scales. For example, biophysical processes can be considered at a broad geographical scale – the bioregion, intermediate scale – the watershed, and, a small scale – local ecosystem. Each level can be considered as an adaptive cycle of exploitation, accumulation, release and reorganisation.

However, there are also connections between levels. For example, land clearance at a small scale can influence a watershed through erosion, sediment transport and deposition (Fig. 4.5). The influence can also be from the larger scale to the smaller scale with the bioregion providing the seed source for potential regrowth of eroded areas, depositional areas or cleared areas in the watershed. They can be considered a nested system of adaptive cycles with connections between different levels.

Socio-economic systems can be considered as adaptive cycles operating at multiple time/space scales. For example, with the society of the Classic Maya as the intermediate level socio-economic system, it can be considered in the context of a broader regional Mesoamerican system. At a lower level, it is possible to consider the individual or the extended family household within the Maya society as a socio-economic system. There are connections between these three levels, such as trade between the Maya and neighbouring societies, and the individual acceptance of the Classic Maya cultural practices. Freter (2004) puts forward a "multiscalar model of rural households and communities" in order to "appreciate fully the diversity of Maya social organization during the Late Classic/Terminal Classic transition" (Freter 2004).

4.2.4 *Societal Collapse in Adaptive Cycle Framework*

To establish a framework encompassing the range of collapse pathways a three level system is considered:

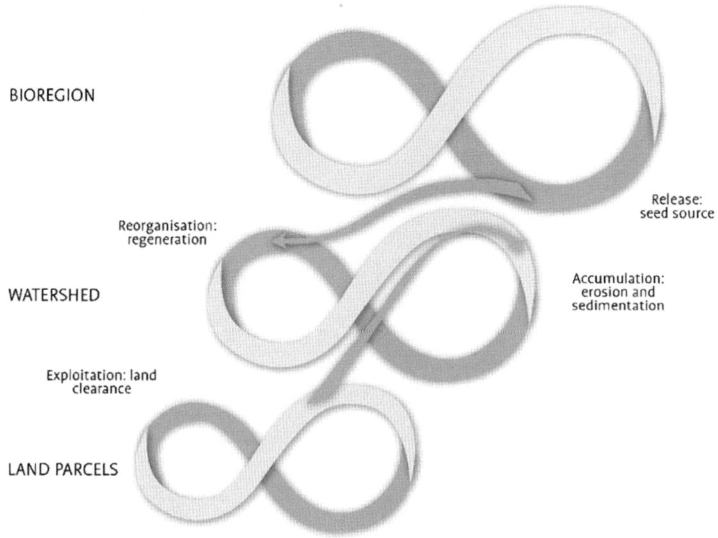

Fig. 4.5 Nested adaptive systems of bioregion/watershed/land parcels (Gunderson and Holling 2002). From *Panarchy* edited by Lance H. Gunderson and C.S. Holling. Copyright © 2002 Island Press. Reproduced by permission of Island Press, Washington, DC
Exploitation by land clearance leads to accumulated erosion and sedimentation at the watershed scale; release of seed source at the bioregion scale provides the potential for reorganisation through regeneration at the watershed scale

- one at the level of the society under consideration;
- a lower level at the scale of the individual within society; and
- a higher level of the broader region in which the society operates.

This is for both the socio-economic system and the biophysical system, thereby creating a pattern of six adaptive cycles with potential linkages to identify collapse pathways (Fig. 4.6).

At the geographical scale of the society under consideration the potential system collapse pathways identified by Diamond, Webster and Tainter can be related to an adaptive cycle framework as follows:

- the *environmental damage* (of Diamond), *agricultural degradation* (of Webster), and *resource deterioration due to human mismanagement* (of Tainter) pathways are represented by pathway 2 – socio-economic impact on the biophysical environment.[4]
- the *natural disaster* pathway (of Webster, i.e. hurricanes, earthquakes, volcanoes) and *insurmountable catastrophe* (of Tainter) are represented by pathway 3 – biophysical impact on the socio-economic environment.
- the *peasant revolt* and *internal warfare* pathways (of Webster) and *class conflicts* (of Tainter) are represented by pathway 4 – the maintenance of the socio-economic system with the peasant revolt pathway relating to the exploitation

[4]The pathway numbers chosen to match the numbering of the four types of issues in Figure 4.6.

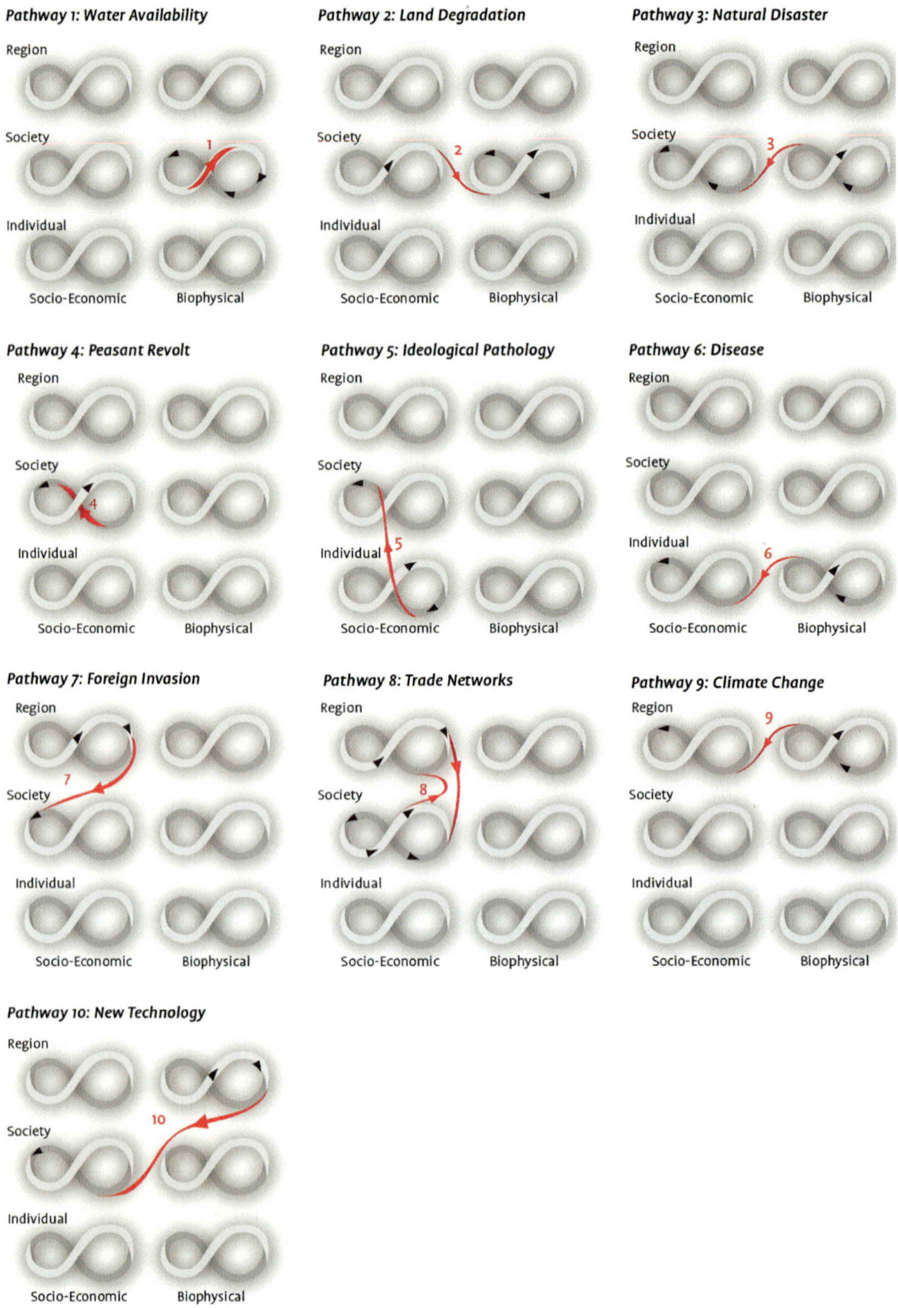

Fig. 4.6 Collapse pathways in a nested adaptive cycle framework

phase of the adaptive cycle (with concerns of the lower classes with the demands placed upon them by the societal leadership), while the internal warfare pathway relates to the release phase (with disputes over the distribution of the accumulated wealth of society).

- The *diminishing marginal productivity with increasing complexity* pathway of Tainter is also a failure to maintain the socio-economic system (i.e. pathway 4).[5] However this is a failure of integration of function rather than a failure to manage conflict in society.

Some of the collapse pathways relate to the individual scale in the nested system:

- the *ideological pathology* pathway (of Webster) and the *mystical factor* (of Tainter) are represented by pathway 5 relating to the release of the individual commitment to the society leading to societal reorganisation.
- the disease pathway (of Webster) is shown as pathway 6 which links the individual biophysical environment with the individual's socio-economic environment.

Other collapse pathways relate to the broader region in which the society operates:

- the *hostile neighbours* (of Diamond), *foreign invasion* (of Webster) and *intruders* (of Tainter) are represented by pathway 7 linking the release phase at the regional scale in the form of hostilities by external forces with the disturbance phase at the societal scale resulting in reorganisation of the local society[6]
- the *disruption/collapse of trade networks* (of Webster), *friendly trade partners* (of Diamond), and *loss of trade networks for external resources and imported goods* (of Tainter) are represented by pathway 8 – a two directional link of exchange between accumulated resources of the society being released to the region being traded for accumulated resources of the region being released to the society.
- the climate change pathways (of Diamond, Webster and Tainter) are represented by pathway 9 where climate disturbance (release phase) at the regional level leading to reorganisation at the regional level for both the socio-economic and biophysical environments.
- the *introduction of new technology* that changes relationships in society (of Tainter) is represented by pathway 10 where a regional resource or technology causes a disturbance to the socio-economic system at the society level leading to restructuring.

---

[5] As stated by Tainter: "After a certain point increased investments in complexity fail to yield proportionately increasing returns. Marginal returns decline and marginal costs rise. Complexity as a strategy becomes increasingly costly and yields decreasing marginal benefits" (Tainter 1988).

[6] Tainter's *other complex societies leading to competition between societies* is a 'cold war' version of this pathway.

One additional pathway that can be derived from Lucerno's analysis of water availability at different centres:

- the water availability pathway, represented by pathway 1 which relates to the limits of the biophysical system at the level of the society under consideration with respect to water supply. This is also a special case of Tainter's *resource depletion due to environmental fluctuation.*

In addition to these pathways there is an additional contributing factor (in Diamond's analysis) or explanatory theme (from Tainter's list) that is significant in relation to the maintenance of adaptive systems. In Tainter's list of explanatory themes it is "insufficient response to circumstances". In Diamond's contributing factors it is "society's responses to its environmental problems". In Webster's conclusions relating to the Maya Collapse it is "failure of leadership to provide solutions to the crises" facing the Maya.

In the graphic representation of the adaptive cycle in Fig. 4.2, it is the critical point after restructuring of the system whether societal responses enable the adaptive cycle to continue or whether the system fails (the "x" branch of the diagram in Fig. 4.2). This is considered further in the section below on system resilience and management intervention to prevent failure.

The different pathways can be categorised in terms of the four types of sustainability issues identified above (i.e. maintenance of the biophysical system, impact of the socio-economic system on the biophysical system, impact of the biophysical system on the socio-economic system, and, maintenance of the socio-economic system) and three geographical scales (regional, societal and individual). This is set out in Table 4.1.

Note that for the maintenance of the socio-economic system at the regional and societal level two types of pathways are identified; one relating to the failure to manage conflict, and, the other relating to the failure to maintain the benefits of integration.[7] Thus at the regional level there is a 'conflict' failure pathway from *external intrusion*, and an 'integration' failure pathway of *loss of trade networks*. At the society level, there is the 'conflict' failure pathway of *internal warfare* and the 'integration' failure pathway of *diminishing marginal productivity*.

At the individual level for maintenance of the socio-economic system the failure pathway has been labelled *individual commitment to society.* This concept captures the ideological pathology pathway of Webster. While Tainter is dismissive of what he calls "mystical factors" such as decadence, he recognises the "role of legitimising activities in maintaining a governing elite" and the "development of apathy to the well-being of the polity" as a factor in the collapse of complex societies (Tainter 1988).

The table has two pathways for climate change: one is *climate variability* which is concerned with fluctuations in climate due to natural processes (a type 1 sustainability issue relating the maintenance of the biophysical system); and, the other is *climate change* which is associated with human-induced climate effects (a type 2

[7]This is consistent with Tainter's two reasons for establishing complex societies: management of conflict, and, facilitation of integration.

Table 4.1 Categorisation of Failure Pathways by Type of Sustainability Issue and Geographic Scale (Jenkins 2016)

	Type of sustainability issue			
	Type 1	Type 2	Type 3	Type 4
Geographic scale	Biophysical maintenance	Socio-economic impact on biophysical	Biophysical impact on socio-economic	Socio-economic maintenance
Region	*Climate variability*	*Climate change*	*New technology changing social relationships*	Conflict: *External intrusion*
				Integration: *Loss of trade networks*
Society	*Cumulative depletion of natural resources*	*Cumulative environmental degradation*	*Natural disasters*	Conflict: *Internal warfare*
				Integration: *Diminishing marginal productivity*
Individual	*Local natural resource depletion*	*Local environmental degradation*	*Disease*	*Individual commitment to society*

sustainability issue of socio-economic impact on the biophysical system e.g. the enhanced greenhouse effect from increased levels of carbon dioxide and other greenhouse gases from anthropogenic activity). This separation has relevance to the possible responses by societies: for climate variability there is a need for an adaptation response while for climate change there is also the possibility of reducing the anthropogenic sources of human-induced climate effects.

It should also be noted that the *environmental degradation* pathway at the geographical scale of the society is a result of the cumulative effects of activities at the geographical scale of the individual. Thus the table shows the pathway at both the societal and individual scales with cumulative effects at the societal level and local effects at the individual level. Similarly, the *natural resource depletion* pathway at the geographical scale of the society (e.g. Lucerno's water availability pathway) has implications for activities at the geographical scale of the individual.

For type 3 sustainability issues (biophysical impacts on the socio-economic system) there are *new technology changing social relationships* at the regional level, *natural disasters* at the societal level, and, *disease* at the individual level.

4.2.5 Pathways and Connectedness

The main collapse pathway identified by Webster (Sect. 4.2.3) can be summarised in the following sequence of processes:

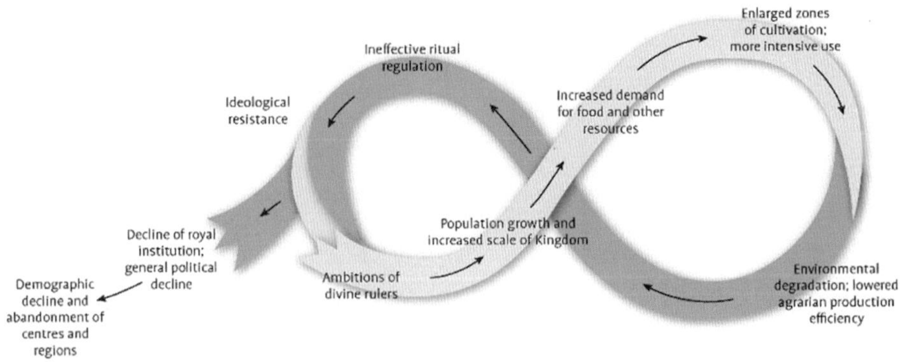

Fig. 4.7 Webster failure pathway (Adapted from Webster 2002)

1. the ambitions of divine rulers, leading to
2. population growth and increased scale of kingdoms, leading to
3. increased demand for food and other products, leading to
4. enlarged zones of cultivation and more intensive land use, leading to
5. environmental degradation and lowered agrarian production efficiency, leading to
6. ineffective ritual regulation, leading to
7. ideological resistance of the peasant community, leading to
8. decline of the royal institution and general political decline, leading to
9. demographic decline and abandonment of centres and regions.

Webster displays the links between the processes. In the language of adaptive cycles, he sets out the "connectedness" (i.e. the nature and degree of links between processes) within the Maya society, between the society and its members and between the society and the broader region. An example of how a failure pathway identified by Webster can be placed in the adaptive cycle framework is shown in Fig. 4.7.

The exploitation phase is generated by the "ambitions of divine rulers" needing resources to maintain their households and for construction of monuments. This leads to seeking an increased scale of kingdom and a desire for a greater population (i.e. labour force) to achieve these ambitions. This leads to an increased demand for food and other resources and as a consequence enlarged zones of cultivation and more intensive land use – the accumulation phase of the adaptive cycle.

However, when this demand exceeds the carrying capacity of the land to produce, it results in environmental degradation and a lowered agrarian production efficiency – the release phase of the adaptive cycle. In the case of the Maya in the southern lowlands the response to this crisis was greater ritual ceremonies to appease the gods – the reorganisation phase of the adaptive cycle.

The failure to provide an effective response leads to ideological resistance in the community and a decline in the belief of the institution of kingship. Ultimately this leads to demographic decline and abandonment of centres and regions and results in collapse of the Classic Maya society.

It is important to note that Webster identified multiple pathways leading to collapse. This includes increased warfare, increased disease through famine and vulnerability to drought. This is consistent with a framework based on a nested system of adaptive cycles.

4.2.6 Resource Potential

One of the critical issues in the Maya Collapse was the limitation of the land resource to produce sufficient food for the growing population. Furthermore, there was reduction in the productive capacity of the land resource through land degradation. In addition, water availability was an issue for centres with low rainfall and without access to river or groundwater even with storage of surface runoff.

Evidence of a cycle of destructive changes from increased deforestation to create farmland to feed an increasing population was found by Binford and his colleagues at Tikal (Binford et al. 1987). At Copan, Wingard undertook a simulation of agricultural productivity after analysing hundreds of soil samples from the area. As noted by Webster, central to Wingard's thinking was carrying capacity, however, it was treated as a variable and dynamic concept that could be changed (up to a point) by farmers (Wingard 1996).

These explanations are consistent with the characteristic of nested adaptive systems of resource potential and the presence of limits on the resource available through use or harvesting.

4.2.7 System Resilience and Management Intervention to Prevent Failure

Societal collapse can occur when there is a disturbance to the socio-economic or biophysical system which is beyond the adaptive capacity of the society or the environment upon which it depends. This is reflected in the critical property of an adaptive cycle of resilience – the ability of the system to adapt to disturbance or the vulnerability of the system to change.

The biophysical resilience of the Classic Maya was demonstrated to have been exceeded in relation to agricultural production and food requirements of the population. The vulnerability of key centres to water availability was also identified and there was evidence of the occurrence of drought conditions during the period of collapse. The vulnerability of the socio-economic system to warfare was also identified.

In the socio-economic system there is the ability to take actions to address potential collapse pathways. This is the society's response to its environmental problems as identified by Diamond, or Webster's leadership to provide solutions to crises. A critical factor in the sustainability of socio-economic and biophysical systems is the strategic response by societies and in particular its leadership to effectively address potential failure pathways.

As noted by Webster, the Late and Terminal Classic periods were increasingly perturbed by disorder resulting from many causes, and we have no clear evidence that kings, or for that matter their associated officials and nobility, ever took effective and practical managerial action to contend with developing crises (Webster 2002).

It is interesting to note that the areas of greatest land degradation have been found in the Petén region (Lakes Yaxha, Sacnab, Quexil, Macanche and Salpeten) (Binford et al. 1987). As Webster indicates for Copan, we have only sparse and highly localised evidence for agricultural terraces and none at all for drained fields. Both kinds of intensive agro-engineering are found in other parts of the Maya lowlands (Webster 2002). Evidence of land management has been found in the Petexbatun area (in the form of stone walls to form terraces) to the south west of the Petén. Agricultural terracing has been found in the Rio Bec region to the north and the Vaca Plateau to the east.

Dunning and Beach find it curious that in many areas where agricultural terracing would seem to have been a useful slope management tool, little evidence for its use has been found (Dunning and Beach 1994). Very little evidence of terracing was found during archaeological surveys of the central Petén despite apparently high regional population pressure during the Late Classic period AD 550–830 (Rice 1978).

Dunning and Beach note that the construction of Petexbatun dry slope terraces is consistent with the gradual unplanned manner in which terrace systems have been observed to evolve. There is no firm evidence that terrace construction was ever accomplished rapidly as part of a centrally directed agricultural system (Dunning and Beach 1994).

This can be contrasted to the spectacular monument construction of the Classic Maya. In addition, in the case of Tikal, where the quarries from monument construction, the causeways and nearby depressions (bajos) were used to create water storage there is clearly evidence of central coordination (Scarborough and Gallopin 1991).

Sharer endorses the analysis of Willey and Shimkin (Willey and Shimkin, 1973) that the Maya elite made no technological or social adaptive innovations which might have mitigated these agricultural degradation difficulties. In fact, the Maya elite persisted in traditional directions up to the point of collapse. The critical variables for addressing the threat were not identified. The most conspicuous response was the construction and renewal of the large ritual structures. Sharer concludes that reinforced by their belief system, the Maya elite probably saw increasing investment of labour and time in ritual and construction as essential to solving the crisis and, ultimately, to insuring the survival of the Maya way of life. The fact that it was logically counter-productive and potentially self-destructive was apparently not recognised (Sharer 1977).

A key aspect of maintaining nested adaptive systems is the concept of adaptive management. Where there is evidence of unexpected change which has the potential to disturb the viability of the system being managed, there is a need to respond to the unexpected change by adaptation or mitigation measures (Holling 1978). The ineffective response of the Maya to their changing circumstances appears to have led to a decline in system resilience and the collapse of Maya society. The need for management intervention to prevent failure is an important component of maintaining sustainability in a resilience framework based on nested adaptive systems.

4.2.8 Sustainability Approaches

Chapin and his colleagues have identified four approaches to foster sustainability:

1. Reduce vulnerability
2. Enhance adaptive capacity
3. Increase resilience, and
4. Enhance transformability (Chapin et al. 2009).

Table 4.2 Sustainability approaches

General approach	Stewardship strategies
1. *Reduce vulnerability*	
Address nature of stresses that cause change	Reduce exposure to stress.
Reduce sensitivity of system to stress	Sustain slow ecological variables that determine natural capital.
	Maintain components of well-being.
	Pay attention to vulnerable components.
Enhance adaptive capacity and resilience	
2. *Enhance adaptive capacity*	Foster diversity that provides building blocks for adjusting to change.
	Foster social learning of how the system works and why it is changing.
	Experiment to test understanding.
	Develop capacity to govern and implement.
3. *Increase resilience*	Adapt governance to changing conditions.
	Sustain legacies that provide pathways for rebuilding.
	Develop capacity to plan for uncertainty and change.
	Foster stabilising feedbacks.
	Foster innovation that creates opportunity for change.
4. *Enhance transformability*	Create new system.

Source: Chapin et al. (2009)

Table 4.2 sets out stewardship strategies associated with these approaches. The approaches are seen to be overlapping.

The strategies to reducing vulnerability are seen to be: (a) reducing the exposure to stresses that are causing change, (b) reducing the sensitivity of the system to stress by either sustaining the slow variables at a larger scale that determine natural capital, maintaining the components of well-being for the system, or, paying attention to the vulnerable components, and (c) enhancing the adaptive capacity and resilience to cope with stress (i.e. the second and third approaches to fostering sustainability).

Enhancing adaptive capacity (i.e. the capacity to respond to change in the system) can be achieved by (a) fostering diversity that provides the building blocks for adjusting to change, (b) fostering social learning of how the system works and why it is changing, (c) experimenting to test the understandings of the system, and (d) developing the capacity to govern effectively to select and implement appropriate solutions.

Strategies for increasing resilience (i.e. the ability of the system to adapt to disturbance or the vulnerability of the system to change), have been identified as (a) enhancing adaptive capacity (i.e. the second approach to fostering sustainability), (b) the capacity to adjust governance structures to address (i.e. similar to strategy d for enhancing adaptive capacity), (c) sustaining legacies that provide pathways for rebuilding, (d) fostering stabilising feedbacks that buffer the system against disturbance, and (e) fostering innovation that creates opportunities for change.

Enhancing transformability is improving the capacity to create fundamentally new systems with different characteristics. This overlaps with strategy for increasing resilience (i.e. fostering innovation).

In relation to natural resource management from a sustainability perspective there are three types of outcomes. One is the maintenance of the current biophysical system. A second is that sustainability thresholds are exceeded leading to an unintended transformation of the biophysical system to a degraded state.[8] The third is the active navigation to transform to a new biophysical system.

Chapin et al. (2009) provide examples of strategies to implement resilience-based stewardship and the process for purposeful navigation of transformations.

4.2.9 Sustainability Approaches in a Nested Adaptive System

The classification of Chapin et al. of sustainability approaches can be refined and extended by considering sustainability approaches in the context of nested adaptive systems. For a society and its natural resource base it is appropriate to consider the socio-economic system linked to the biophysical system in the manner described above in Sect. 4.2.2 and as displayed in Fig. 4.4 with four sustainability issues to be managed:

[8] This is the outcome of the collapse of the Classic Maya.

1. Capacity of the biophysical system to be maintained
2. Impact of the socio-economic system on the biophysical system
3. Impact of the biophysical system on the socio-economic system
4. Capacity of the socio-economic system

As noted above, in relation to the stewardship of the natural resource base, there are three general types of outcomes:

- Maintenance of the natural resource base
- Unintended transformation to a new, usually degraded, state
- Actively navigated transformation to a new state which is more resilient to the drivers of change.

This relationship is shown in Fig. 4.8.

The sustainability strategies identified by Chapin et al. can be related to these linked adaptive systems.

Reducing vulnerability by "reducing exposure to stresses" is equivalent to the reduction of the impact of the socio-economic system on the biophysical system. The second strategy for reducing vulnerability is by "reducing the sensitivity of the system to stress through sustaining the slow variables at a larger scale that determine natural capital vulnerability". This requires consideration of the cross-scale linkages to the larger biophysical system in the nested system as depicted in Fig. 4.5. The third strategy of "paying attention to the vulnerable components" requires consideration of the lower biophysical level in the nested system as depicted in Fig. 4.5.

In terms of enhancing adaptive capacity, the strategy of "fostering learning of how the system works and why it is changing" relates to increasing the accumulation of knowledge in the socio-economic system at the same level as the biophysical system being managed. However, the strategy of "developing the capacity to govern and implement" relates to the higher level of the socio-economic system. In con-

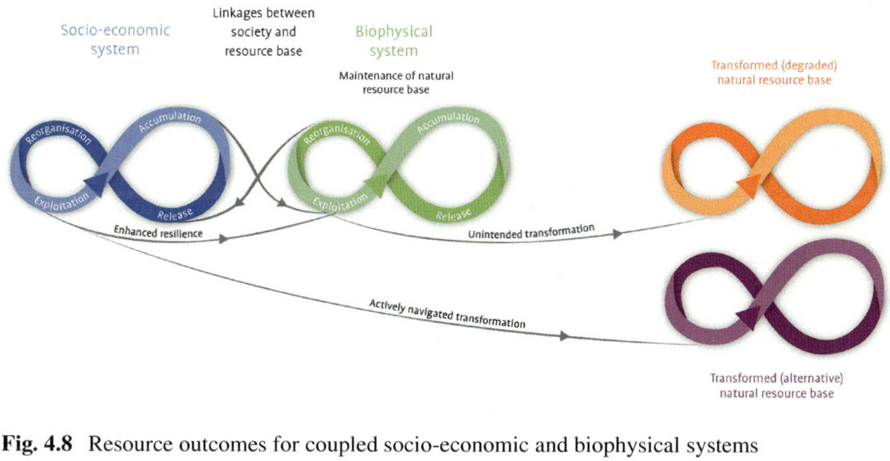

Fig. 4.8 Resource outcomes for coupled socio-economic and biophysical systems

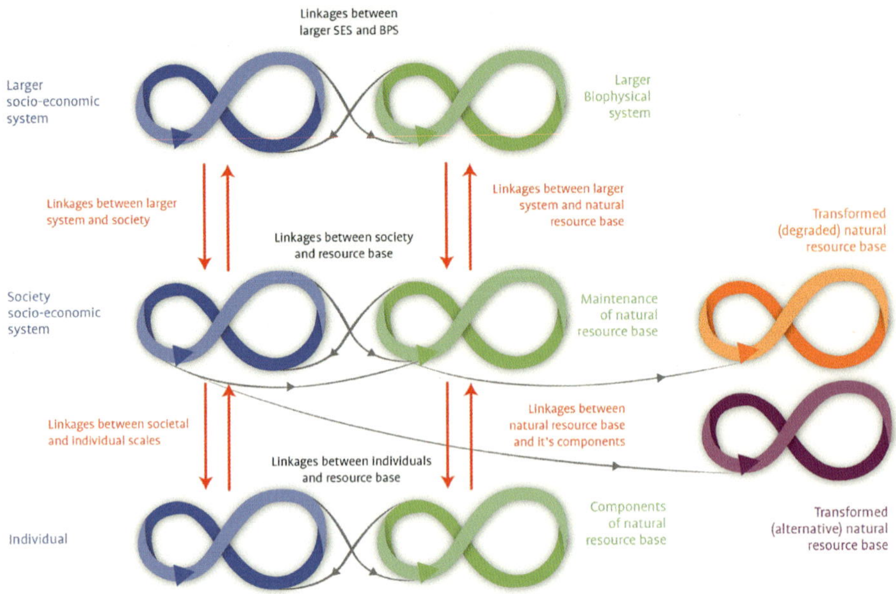

Fig. 4.9 Framework for classifying approaches to sustainability

trast, the strategy of "experimenting to test understandings" relates to the lower level linkages between individuals and the resource base.

For increasing resilience, the strategy of "fostering stabilising feedbacks" relates to the cross scale linkage between the larger socio-economic system and society for socio-economic system feedbacks, and between the larger biophysical system and the natural resource base of that society for biophysical system feedbacks.[9] The strategy of "sustaining legacies that provide seeds for renewal" and "developing the capacity to plan for uncertainty and change" are enhancing social capital in society for achieving resilience.

Figure 4.8 can be expanded to include the nested system with cross scale linkages. The expanded diagram is set out in Fig. 4.9. This is similar to the diagrams showing collapse pathways (Fig. 4.6). However, in this case it provides the framework for classifying approaches to achieving sustainability rather than threats to sustainability. Table 4.3 sets out the classification of the sustainability strategies identified by Chapin et al. that have been discussed above. In the vertical columns, the classification separates out strategies based on the socio-economic system, strategies based on the biophysical system, strategies based on the linkages between socio-economic and biophysical systems, and strategies for transformation. In the horizontal rows, the classification separates out strategies at different spatial scales, i.e. the scale of society and its natural resource base, a larger scale and a smaller scale, and strategies at the cross-scale linkages from the society and its resource base to a larger scale and to a smaller scale.

[9]The release of seed source at the bioregion scale to facilitate revegetation at the watershed and land parcel scale as depicted in Fig. 4.4 is an example of a stabilising feedback loop.

Table 4.3 Classification of sustainability strategies for a nested socio-economic and biophysical system

Geographic scale	Socio-economic system (SES)	Linkages between SES and BPS	Biophysical system (BPS)	Transformation
Larger scale	Adjust governance. Learn from multiple cultural perspectives.	Sustain cultural connections to land and sea. Exercise caution when perturbing larger system.	Sustain slow variables that determine natural capital. Explore system dynamics.	Enhance diversity that provides building blocks for change.
Cross scale linkage to larger scale	Stabilise feedback from socio-economic systems.		Stabilise feedback from natural system.	
Society and natural resource base	Enhance learning capacity. Sustain legacies that provide seeds for renewal.	Explore consequences of options. Reduce exposure	Renew functional diversity of degraded systems.	Create new systems
Cross scale linkage to smaller scale	Engage stakeholders			
Smaller scale		Experiment to test understandings	Pay attention to vulnerable components	Foster innovation to create opportunities for change

For cells shaded gray no entries are possible. For cells that are white with no entries indicate no strategies of this type were identified by Chapin et al.

The framework proposed here has similar characteristics to the "framework for vulnerability analysis in sustainability science" described by Turner and his colleagues. These include:

- The coupled human-environment system, whatever its spatial dimensions, constitutes the place of analysis
- Linkages to the broader human and biophysical conditions and processes operating on the coupled system in question
- Perturbations and stressors/stress that emerge from these conditions and processes, and
- The coupled human-environment system of concern in which vulnerability resides including exposure and responses (Turner et al. 2003).

In the language of Table 4.3, the words of Turner et al. can be rephrased as:

- Linked socio-economic and biophysical systems at multiple spatial scales form the analytical framework

- The socio-economic/biophysical systems that are being analysed are linked to socio-economic and biophysical systems at a larger scale
- Disturbances emerge to the socio-economic/biophysical systems that can become failure pathways

The resilience of linked socio-economic biophysical systems can be considered in relation to its vulnerability to disturbance and possible responses to foster sustainability.

References

Binford, M. W., Brenner, M., Whitmore, T. J., Higuera-Gundy, A., & Deevey, E. S. (1987). Ecosystems, paleoecology and human disturbance in Subtropical and Tropical America. *Quaternary Science Review, 6*, 115–128.

Chapin, F., Kofinas, G., & Folke, C. (2009). *Principles of ecosystem stewardship: Resilience-based natural resource management in a changing world*. New York: Springer.

Coe, M. D. (2005). *The Maya* (7th ed.). London: Thames and Hudson.

Diamond, J. (2005). *Collapse: How societies choose to fail or survive*. London: Penguin.

Dunning, N. P., & Beach, T. (1994). Soil erosion, slope management and ancient terracing in the Maya lowlands. *Latin America Antiquity, 5*(1), 51–69.

Freter, A. C. (1992). Chronological research at Copan: Methods and implications. *Ancient Mesoamerica, 3*, 117–133.

Freter, A. C. (2004). Multiscalar model of rural households and communities in late Classic Copan Maya society. *Ancient Mesoamerica, 15*, 93–106.

Gunderson, L. H., & Holling, C. S. (2002). *Panarchy: Understanding transformations in human and natural systems*. Washington, DC: Island Press.

Holling, C. S. (1978). *Adaptive environmental assessment and management*. Chicester: Wiley.

Jenkins, B. R. (2016). Sustainability analysis of the management approach for six New Zealand lakes. *Lake and Reservoir Management, 32*(2), 101–115. https://doi.org/10.1080/10402381.2015.1135491.

Lucerno, L. I. (2002). The collapse of the Classic Maya: What happened to Classic Maya civilisation? *American Anthropologist, 104*(3), 814–826.

Millennium Ecosystem Assessment. (2005). *Ecosystems and human well being: Synthesis*. Washington, DC: Island Press.

Paton, D., & Johnston, D. (2006). *Disaster resilience: An integrated approach*. Springfield: Charles C Thomas.

Rice, D. S. (1978). Population growth and subsistence alternatives in a tropical lucastrine environment. In P. D. Harrison & B. L. Turner (Eds.), *Pre-Hispanic Maya Agriculture* (pp. 301–344). Albuquerque: University of New Mexico Press.

Scarborough, V. L., & Gallopin, G. G. (1991). A water storage adaptation in the Maya lowlands. *Science, 251*, 658–662.

Sharer, R. J. (1977). The Maya collapse revisited: Internal and external perspectives. In N. Hammond (Ed.), *Social process in Maya prehistory*. London/New York: Academic.

Tainter, J. A. (1988). *The collapse of complex societies*. Cambridge: Cambridge University Press.

Turner, B., Kasperson, R., Matson, P., McCarthy, J., Corell, R., Christensen, L., Eckley, N., Kasperson, J., Luers, A., Martello, M., Polsky, C., Pulsipher, A., & Schiller, A. (2003). A framework for vulnerability analysis in Sustainability Science. *Proceedings of National Academy of Sciences, 100*(14), 8074–8079.

Walker, B., & Salt, D. (2006). *Resilience thinking: Sustainaing ecosystems and people in a changing world*. Washington, DC: Island Press.

Webster, D. L. (2002). *The fall of the ancient Maya: Solving the mystery of the Maya collapse*. London: Thames & Hudson.

Willey, G. R., & Shimkin, D. B. (1973). The Maya collapse: A summary view. In T. P. Culbert (Ed.), *The Classic Maya collapse* (pp. 457–502). Albuquerque: University of New Mexico Press.

Wingard, J. (1996). Interactions between demographic processes and soil resources in the Copan Valley, Honduras. In S. Fedick (Ed.), *The managed Mosaic*. Salt Lake City: University of Utah Press.

Chapter 5
Application of Sustainability Framework

Abstract This chapter has two main purposes. One purpose is to provide examples of how the sustainability framework can be applied. Because it is relatively novel approach, illustrations provide the reader with an understanding of what the application entails and valuable insights that can be gained by this approach. The second purpose is to describe the water management topics to which the framework is applied to in the remaining chapters of the book.

The first two sections address the identification of failure pathways: one covering illustrative examples; the second the chapters addressing water management failure pathways for Canterbury. The process of identifying water management failure pathways is illustrated for the Waimakariri catchment. This involves the interpretation of sustainability issues as adaptive cycles; the identification of critical variables and their thresholds for management; and how adaptive cycles for different spatial or time scales are nested and how they are connected. In addition, the example of maladaptive cycles is illustrated by the analysis of dryland salinity in the Western Australian agricultural region. In the second section, the failure pathways for water management are derived from the ten failure pathways identified in Chap. 4 and how they are addressed in the remainder of the book is explained.

The other two sections address the formulation of sustainability strategies. One illustration of a sustainability strategy compares the water availability elements of the Canterbury Water Management Strategy as an example of integrated water management, with the initial concept that was focused only on storage as the strategy for addressing water availability. A second illustration is the resilience analysis for Te Waihora/Lake Ellesmere which facilitated a rehabilitation strategy being implemented where multi-criteria effects analysis had failed to lead to any action. The last section describes how subsequent chapters are structured to discuss the steps in applying the sustainability framework to develop a sustainability strategy: sustainability assessments, sustainability decision making and sustainability appraisal. It also describes how the concluding chapter brings together the insights gained from failure pathway analysis to identify implications for water management in Canterbury.

B.R. Jenkins, *Water Management in New Zealand's Canterbury Region*, Global Issues in Water Policy 19, https://doi.org/10.1007/978-94-024-1213-0_5

Keywords Sustainability framework • Failure pathways • Sustainability strategies • Adaptive cycles • Nested adaptive systems

5.1 Failure Pathway Analysis

In order to undertake failure pathway analysis there is a need to interpret the water management issues as adaptive cycles with the four phases of exploitation, accumulation, disturbance and reorganization for each of the relevant spatial or time scales. This is needed to identify the critical variables for potential failure pathways and the thresholds for these variables that can alter the structure and function of the system. It is also necessary to determine how the adaptive cycles are nested and how the cycles at the different spatial or time scales are connected.

In this section three sustainable management issues for the Waimakariri catchment are presented (Jenkins 2015). The first issue is the water extraction from the river for irrigation. A second issue is the groundwater extraction for Christchurch's public water supply. The third issue is the extraction of gravel from the bed of the lower reaches of the river. Also, the issue of maladaptive cycles is discussed and the problem of dryland salinity management in Western Australia is provided as an example.

5.1.1 Management of Irrigation

The Waimakariri River is an alpine river in Canterbury (refer Fig. 3.1). Its headwaters are in the Southern Alps and its lower reaches cross the Canterbury Plains. It has a catchment area of 3564 km^2 and a mean flow of 120 m^3/s. Other flow characteristics are set out in Table 5.1.

The current major extraction from the Waimakariri River is an irrigation take for Waimakariri Irrigation Scheme with a consented allocation of 10.5 m^3/s subject to environmental flow restrictions. These restrictions are based on an allocation regime

Table 5.1 Flow statistics for Waimakariri River (Environment Canterbury 2011)

Flow statistic	Value	Comments
Median	90 m^3/s	Flow that is equaled or exceeded half the time
Mean annual flood	1495 m^3/s	Average of the highest instantaneous flow measurement from each year
7DMALF	40.2 m^3/s	7-day annual flow is the lowest flow at a given site in a year; annual values are averaged to give 7DMALF
FRE 3	15 per year	Number of occurrences of 3 times median flow (270 m^3/s)

Table 5.2 Waimakariri River Allocation Regime (Environment Canterbury 2011)

Allocation block	Allocation limit (m^3/s)	Minimum flow at Otarama (m^3/s)	Notes
AA permits	5	–	Stock and drinking water
A permits	17	46	Need to cease takes during a fresh that occurs after a period of 21 days of flow below A permit minimum flow; Includes 10.5 m^3/s for Waimakariri irrigation.
B permits	27	68	1:1 flow sharing; Need to cease takes during a fresh that occurs after a period of 21 days of flow below A permit minimum flow; Includes 25 m^3/s for Central Plains irrigation

Based on data record from 1967 to 2009

that has been defined for the river in a regional plan[1] (Environment Canterbury 2011) and the key elements are shown in Table 5.2.

AA permits are for stock and drinking water allocations and are limited to 5 m^3/s. They can be taken without restriction from the flow in the river. A permits are limited to 17 m^3/s (referred to as the "A Block") with no water able to be taken if the flow in the river is below 46 m^3/s as measured at Otarama (referred to as the "A Block minimum flow"). Above a river flow of 68 m^3/s the full consented allocation can be taken by an A permit holder. Between 46 and 68 m^3/s a proportion of the full allocation can be taken. A "B Block" of 27 m^3/s has been defined for future irrigation schemes with a minimum flow of 68 m^3/s and takes limited to half the available flow above 68 m^3/s (referred to as "1:1 sharing").

The Waimakariri Irrigation Scheme provides irrigation water to 212 shareholders to irrigate 18,000 ha over a command area of 44,000 ha. As an A permit holder, it is subject to the A Block minimum flow of 46 m^3/s and cannot take its full allocation until the river flow is above 68 m^3/s.

The flow in the Waimakariri River is highly variable and is typically low in February to April of the irrigation season. At the present time, the Waimakariri Irrigation Scheme is fully reliable in only 1 year in 42 based on simulations using past flow records.

For the irrigating farmer, irrigation can improve production when the potential evapotranspiration rate exceeds rainfall. As noted in Sect. 3.1.1, Canterbury is the region with the highest potential evaporation deficit in New Zealand. Water availability for irrigation can be constrained by low flows in the Waimakariri River, particularly in the latter part of the irrigation season. This can also be a time of lower rainfall. This means that for farmers reliant on scheme distribution of water there can be insufficient water to irrigate to maintain soil moisture levels.

The vulnerability for the irrigating farmer is from lost production from suboptimal soil moisture conditions. Waimakariri Irrigation Scheme farmers lost an estimated $30 million of production in the dry summer of 2013 because of the

[1] Refer to Sect. 2.2 for regional plan provisions under the Resource Management Act.

inability to irrigate. There is also a vulnerability to the environment from excessive irrigation saturating the soil resulting in leakage to groundwater and contamination of groundwater quality especially by nitrates.

There are three geographical scales to be considered for the sustainable management of the Waimakariri Irrigation Scheme:

- the Waimakariri catchment as the source of water for the irrigation scheme
- the irrigation scheme as the source of water for the farmer
- the individual farm where the irrigation water is applied.

For each of these levels the adaptive cycle of exploitation/accumulation/disturbance-release/reorganisation can be described and the critical variables identified.

The key components of the adaptive cycle for the Waimakariri River catchment in relation the management of the Waimakariri Irrigation Scheme are:

- exploitation: the rainfall that falls on the catchment that generates runoff
- accumulation: the accumulation of runoff that generates river flow
- disturbance-release: the volume of water that is extracted from the river for irrigation
- reorganisation: the adequacy of the remaining flow to sustain the river ecology and uses downstream of the irrigation take.

The sustainability threat for the river is the adequacy of the flow downstream of the irrigation take. If this flow is inadequate to maintain river ecology and instream uses then the river will have been degraded through the disturbance of water extracted for irrigation.

The adaptive cycle can be drawn as a Lissajous figure as shown in Fig. 5.1.

The critical variables for river management are the flow requirements for the uses downstream of the irrigation take and the restrictions placed on the amount that can be extracted at particular river flows for irrigation. These include the low flows when aquatic systems are likely to be under the greatest stress, smaller floods and

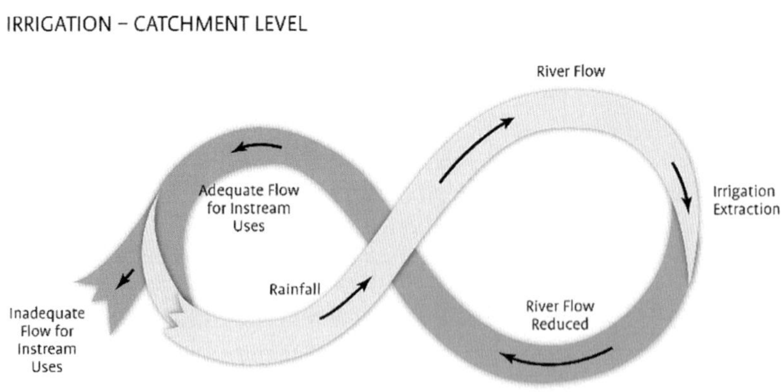

Fig. 5.1 Adaptive cycle for irrigation at the catchment level (Jenkins 2015). Courtesy of WIT Press from 8th International Conference on Sustainable Management, A Coruna Spain 15-17 June 2015

freshes to mobilise sediment and remove periphyton, and large floods to maintain the river's braided character (Jowett and Biggs 2006).

The allocation regime can be compared with the default values for the proposed National Environmental Standard for ecological flows (Ministry for the Environment 2008). The default value for minimum flow for rivers with mean flows greater than 5 m^3/s is 80% of 7DMALF or 32.2 m^3/s for the Waimakariri River compared to the minimum flow in the allocation regime of 46 m^3/s. The default value for the allocation limit is 50% of 7DMALF or 20.1 m^3/s for the Waimakariri River compared to the combined AA and A permits of 22 m^3/s in the allocation regime.

For the B block allocation, the critical issues are the need to have flows in the range 55–96 m^3/s during September to December for riverbed nesting bird breeding and to have flow in the range 60–100 m^3/s during December to April for salmon angling. If flows were sufficient for these activities then there would be sufficient flow for salmon passage, kayaks and jet boats (Duncan 2008). Abstractions are likely to reduce the magnitude and frequency of small freshes capable of flushing sediment and periphyton. In order to mitigate this effect consent conditions are required to maintain the frequency of flows greater than 80 m^3/s and preferably greater than 130 m^3/s after a period of low flows of sufficient duration to potentially allow the growth of periphyton to nuisance levels (Duncan 2008). These issues are addressed through 1:1 flow sharing[2] and the need to cease takes during a fresh after a period of 21 days below the A Block minimum flow.

At the next geographical scale is the irrigation scheme. The adaptive cycle for the irrigation scheme can be defined as follows:

- exploitation: irrigation take subject to river flow restriction
- accumulation: water for distribution from irrigation canals or from storage
- disturbance-release: release of irrigation water to farmers
- reorganisation: further irrigation supply from river or storage

The vulnerability of the irrigation scheme is the reliability of the irrigation supply to have water available when needed to supplement rainfall.

The adaptive cycle can be depicted as a Lissajous figure as shown in Fig. 5.2.

The critical variable for the irrigation scheme is the reliability of supply which at full reliability for 1 year in 42 is very low. A proposal is being implemented to provide a scheme storage of 8.2 million m^3 which provides for 9 days of storage for supplying the entire scheme. If surplus consented water is used to refill the storage throughout the irrigation season the scheme would be fully reliable for 23 of the past 42 years.

The third geographical scale in this nested system is the farm which uses irrigation water from the scheme. The adaptive cycle for on-farm irrigation can be described as:

- exploitation: irrigation to supplement rainfall
- accumulation: soil moisture levels on irrigated farmland
- disturbance-release: evapotranspiration from the soil

---

[2] 1:1 Flow sharing: as river flow increases 50% of the additional flow can be extracted while 50% is specified to remain in the river

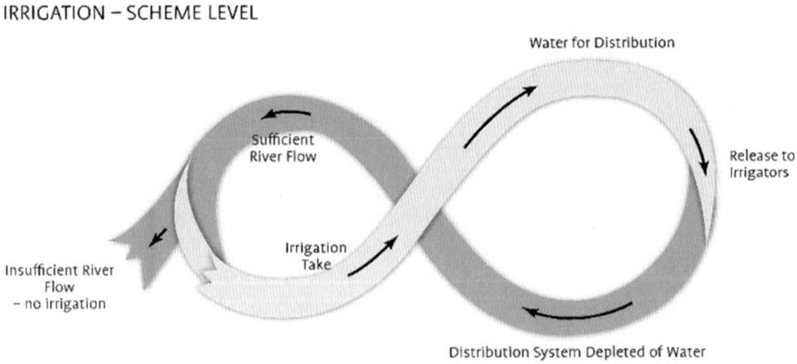

Fig. 5.2 Adaptive cycle for irrigation at the irrigation scheme level (Jenkins 2015). Courtesy of WIT Press from 8th International Conference on Sustainable Management, A Coruna Spain 15-17 June 2015

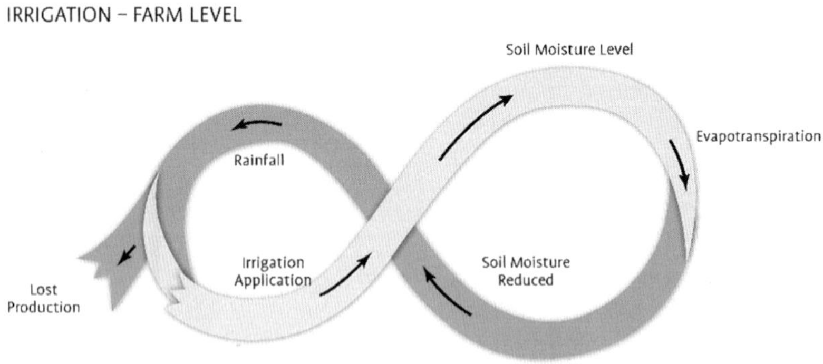

Fig. 5.3 Adaptive cycle for irrigation at the farm level (Jenkins 2015). Courtesy of WIT Press from 8th International Conference on Sustainable Management, A Coruna Spain 15-17 June 2015

- reorganisation: maintain soil moisture in the 50–80% range for optimum production.

The vulnerability for the irrigating farmer is also reliability of supply as effective irrigation involves irrigation application rates to compensate for the exceedance of potential evapotranspiration rates over rainfall.

Figure 5.3 shows the adaptive cycle as a Lissajous figure.

The critical variable for irrigating farmers is the potential for lost production from inadequate soil moisture. This involves managing irrigation application rates to meet soil moisture requirements. This involves both matching the soil moisture deficit with the volume of irrigation (soil moisture demand management) but also having water available for irrigation (reliability of supply). On-farm storage provided farmers with capacity to continue irrigation when the scheme was unable to provide water due to flow restrictions. On-farm storage gave farmers greater flexibility to match irrigation application with soil moisture deficit.

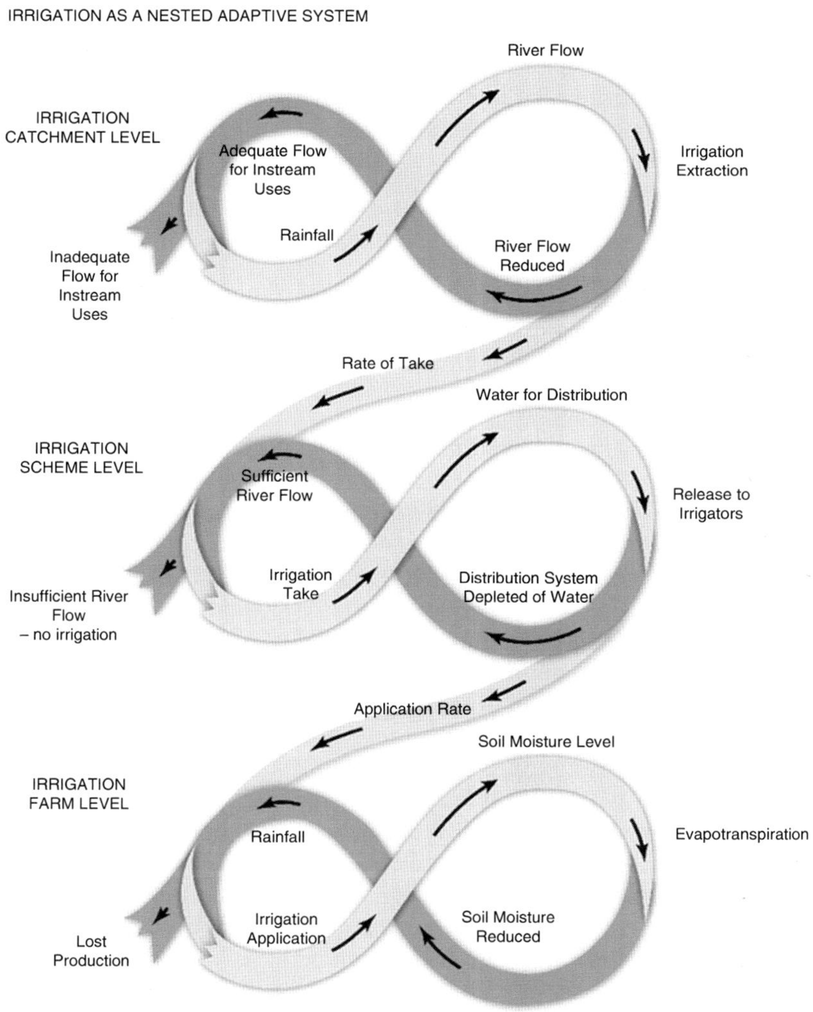

Fig. 5.4 Irrigation as a nested adaptive system (Jenkins 2015). Courtesy of WIT Press from 8th International Conference on Sustainable Management, A Coruna Spain 15-17 June 2015

It is possible to combine the adaptive cycles from the three different geographical cycles into a nested system. Figure 5.4 shows the three adaptive cycles as a nested system. The link between the catchment scale and the irrigation scheme is the irrigation extraction from the disturbance-release phase of the catchment cycle as an input to the reorganisation phase of the irrigation scheme cycle. The link from the irrigation scheme cycle is the release to the farm for the irrigation scheme distribution system into the reorganisation phase of the on-farm irrigation system.

5.1.2 Public Water Supply for Christchurch City

There are three geographical scales that are relevant to the use of groundwater for the public water supply for Christchurch urban area:

- the Waimakariri catchment as the main source of recharge to groundwater
- the groundwater system and the groundwater – surface water interactions
- the groundwater extraction for public water supply and its implications.

Each geographical scale is described as an adaptive cycle in relation to sustainable management of groundwater with respect to public water supply for Christchurch. The three levels are then integrated as a nested adaptive system.

5.1.2.1 Waimakariri Catchment as a Source of Recharge to Groundwater

The Waimakariri catchment, as with other alpine river catchments in the region, has a strong precipitation gradient from the headwaters to the mouth. At the headwaters, the precipitation ranges from 2000 to 5000 mm, in the foothills the range is from 1000 to 2000 mm, and on the plains it is less than 1000 mm.

Thorpe describes the geological setting of the Canterbury Plains which defines the groundwater system in the Waimakariri catchment (Thorpe 1992). The Canterbury Plains were built up from coalescing alluvial fans of gravel originating in the Southern Alps. In the interglacial periods when sea level rose, marine silts and clays were deposited over gravels on the coastal margins. The result near the coast is a sequence of gravel aquifers separated by fine-grained marine deposits that form confining layers so that Christchurch sits on at least four aquifers.

The Waimakariri River is incised across the upper plains and does not lose flow to groundwater. However, in the lower plains surface water seeps from the gravel river bed recharging the unconfined aquifers of the Canterbury Plains. In areas of the plains away from the river the recharge to groundwater is from excess rain infiltrating through the soil.

The groundwater recharge from the Waimakariri River can be described as an adaptive cycle:

- exploitation: the rainfall that falls in the catchment that generates runoff
- accumulation: the accumulation of runoff that generates river flow
- disturbance-release; the seepage of surface water to groundwater
- reorganisation: the reduced river flow downstream of the reach subject to flow loss.

The vulnerability for groundwater recharge is the maintenance of flow in the Waimakariri River. The critical variable is the seepage rate from the river to groundwater.

The adaptive cycle can be shown as a Lissajous figure (Fig. 5.5).

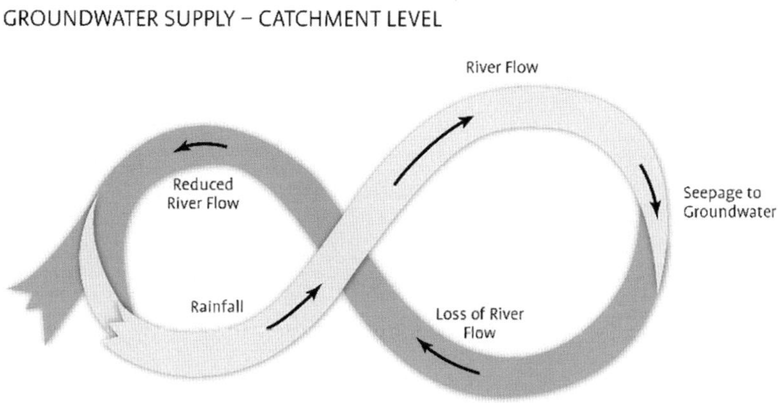

Fig. 5.5 Adaptive cycle for groundwater supply at the catchment level (Jenkins 2015). Courtesy of WIT Press from 8th International Conference on Sustainable Management, A Coruna Spain 15-17 June 2015

5.1.2.2 The Groundwater – Surface Water Interactions

White and his colleagues have undertaken an assessment of the interaction between groundwater and surface water in the Waimakariri River (White et al. 2012). Daily groundwater outflow from the Waimakariri River bed to the Springston Formation aquifer was estimated using water budgets. For the overall reach from the Waimakariri Gorge to the Old Highway Bridge the groundwater outflow to the Springston Formation gravels beside the river was estimated to be 11.7 m^3/s for average flows (120 m^3/s). There is also a net movement from groundwater (i.e. groundwater from the gravel river bed and the Springston Formation) of 1.5 m^3/s to the river from Crossbank to the Old Highway Bridge (Fig. 5.6).

Analysis of river flow and outflow to groundwater shows a relationship with high flows. For example, daily outflow to groundwater is greater than 15 m^3/s when the river flow is greater than 200 m^3/s. However, below river flows of 120 m^3/s groundwater outflow is relatively uniform. Analysis of taking full irrigation takes indicated a reduction in groundwater outflow of 0.2 m^3/s (White et al. 2012).

In addition to the seepage from the Waimakariri River, there is also inflow to the aquifers from rainfall recharge over the area of the confined aquifer. About 30% of rainfall infiltrates to groundwater.

The recharge to the unconfined aquifer can be described as an adaptive cycle:

- exploitation: seepage from the Waimakariri River and rainfall infiltration entering groundwater
- accumulation: increased aquifer inflows from the river seepage and rainfall infiltration
- disturbance-release: aquifer recharge which increases groundwater levels
- reorganisation: increased aquifer flow in the unconfined aquifer.

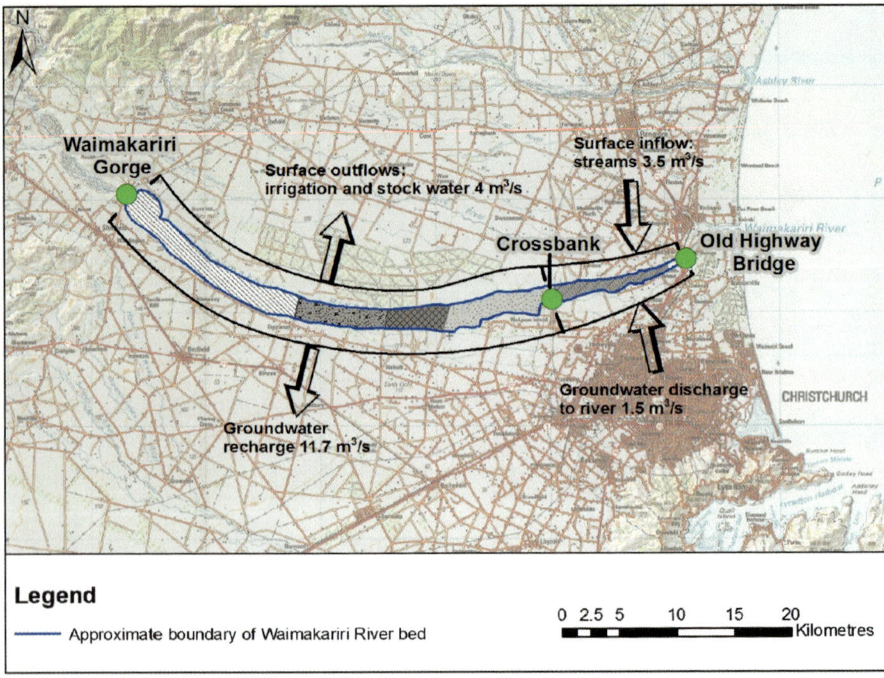

Fig. 5.6 Main inflows and outflows for the lower reaches of the Waimakariri River (White et al. 2011)

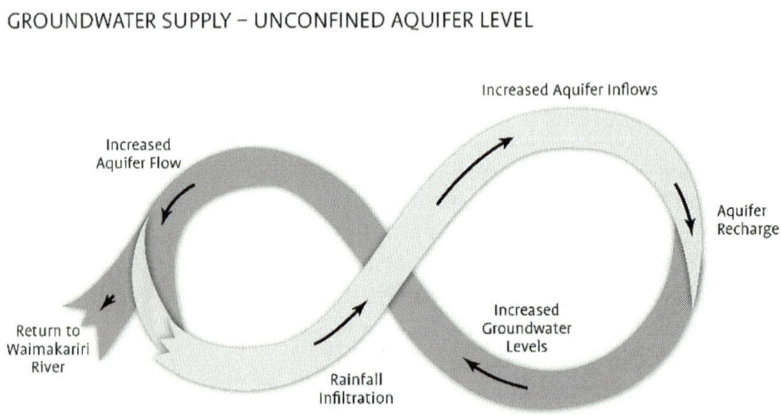

Fig. 5.7 Adaptive cycle for groundwater supply at the unconfined aquifer level (Jenkins 2015). Courtesy of WIT Press from 8th International Conference on Sustainable Management, A Coruna Spain 15-17 June 2015

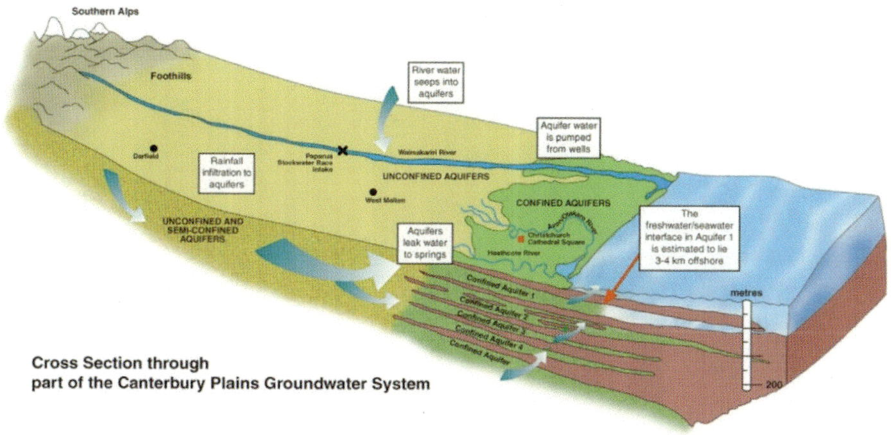

Fig. 5.8 Schematic of groundwater system (Environment Canterbury 2000)

The vulnerability of the recharge to the unconfined aquifer is from the amount of seepage from the river and the amount of rainfall over the unconfined aquifer. These are the critical variables. There is also a feedback loop for groundwater – surface water interactions. Further downstream heightened groundwater levels can lead to return flow from groundwater to the Waimakariri River.

The adaptive cycle can be depicted as a Lissajous figure (Fig. 5.7).

Figure 5.8 shows a cross section through the Canterbury Plains groundwater system. It shows the unconfined aquifer of the Canterbury Plains and the interbedding of confined aquifers and confining layers on the coast. It highlights the inputs of river seepage and rainfall recharge to groundwater for Christchurch and the main outputs of leakage to spring-fed lowland streams and pumping from wells penetrating the aquifers (Environment Canterbury 2000).

5.1.2.3 Groundwater Extraction for Public Water Supply

Groundwater is the source of drinking water for the city of Christchurch. Current extraction is about 47 million m^3 per year (approximately 1.7 m^3/s). It is a high quality supply as most of the groundwater flow comes from the Waimakariri River. It does not require treatment to meet drinking water quality standards. It is under artesian pressure which provides protection from contaminant leakage from land use activities.

In the Christchurch – West Melton groundwater zone there is also extraction for agriculture, horticulture, industry as well as commercial and other activities. Groundwater provides the baseflow for spring-fed lowland streams such as the Avon-Ōtākaro and Heathcote/Ōpāwaho Rivers. The estimated annual inflows and outflows for the Christchurch – West Melton groundwater system from 1965 to 1999 are shown in Figs. 5.9 and 5.10.

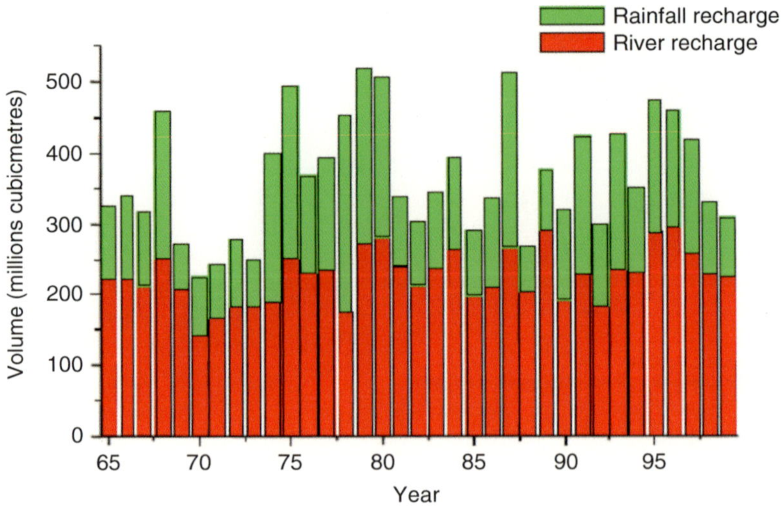

Fig. 5.9 Inflows to the Christchurch – West Melton aquifer (Environment Canterbury 2000)

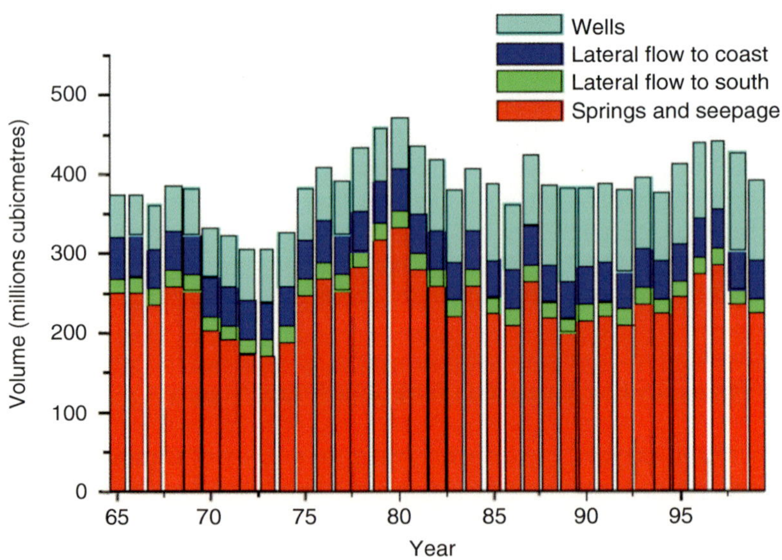

Fig. 5.10 Outflows from the Christchurch – West Melton aquifer (Environment Canterbury 2000)

Groundwater extraction lowers groundwater levels and the upward artesian pressure in the aquifers. This can lead to reduced baseflow in lowland streams from lowered groundwater levels, increased risk of contamination from contaminant leakage from land use activities, and, if the groundwater levels fall below sea level, increased risk of saltwater intrusion.

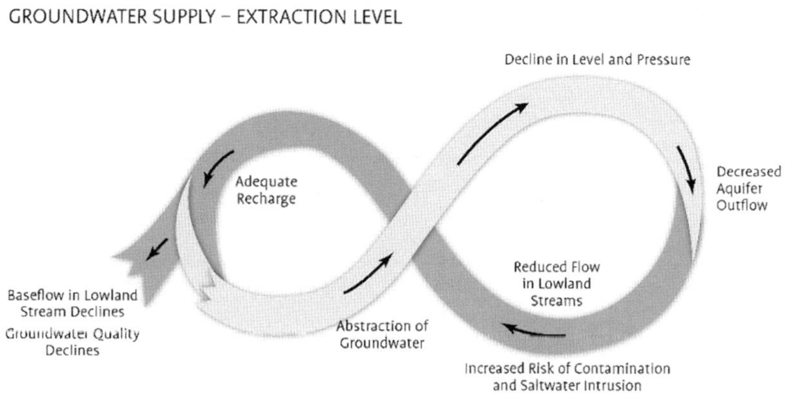

Fig. 5.11 Adaptive cycle for groundwater supply at the extraction level (Jenkins 2015). Courtesy of WIT Press from 8th International Conference on Sustainable Management, A Coruna Spain 15-17 June 2015

This can be represented as an adaptive cycle as follows:

- exploitation: abstraction of groundwater
- accumulation: decline in groundwater level and artesian pressure
- disturbance-release: reduced flows in lowland streams and increased risk of contamination from land use and saltwater intrusion
- reorganisation: maintenance of adequate recharge.

The sustainability risks in relation to groundwater extraction for public water supply for Christchurch are the maintenance of baseflows in lowland streams and the maintenance of quality as a drinking water supply. There is also the issue of further allocations to other uses affecting the allocation limit to public water supply. Critical variables include the demand for drinking water, groundwater quality with respect to drinking water standards, groundwater levels and pressures, baseflow in lowland streams, and allocation to other uses.

Figure 5.11 shows the adaptive cycle for groundwater extraction for public water supply.

5.1.2.4 Nested Adaptive Cycles for Christchurch Water Supply

These different geographical scales can be linked as nested adaptive cycles (Fig. 5.12). Seepage to groundwater is the link between the catchment scale and the unconfined aquifer. The link is from the disturbance-release phase of the catchment adaptive cycle to the exploitation phase of adaptive cycle for the unconfined aquifer.

The aquifer inflow is the link between the unconfined aquifer adaptive cycle and the adaptive cycle for groundwater extraction for public water supply. The link is from the reorganisation phase of the unconfined aquifer adaptive cycle to the reorganisation phase of the adaptive cycle for public water supply.

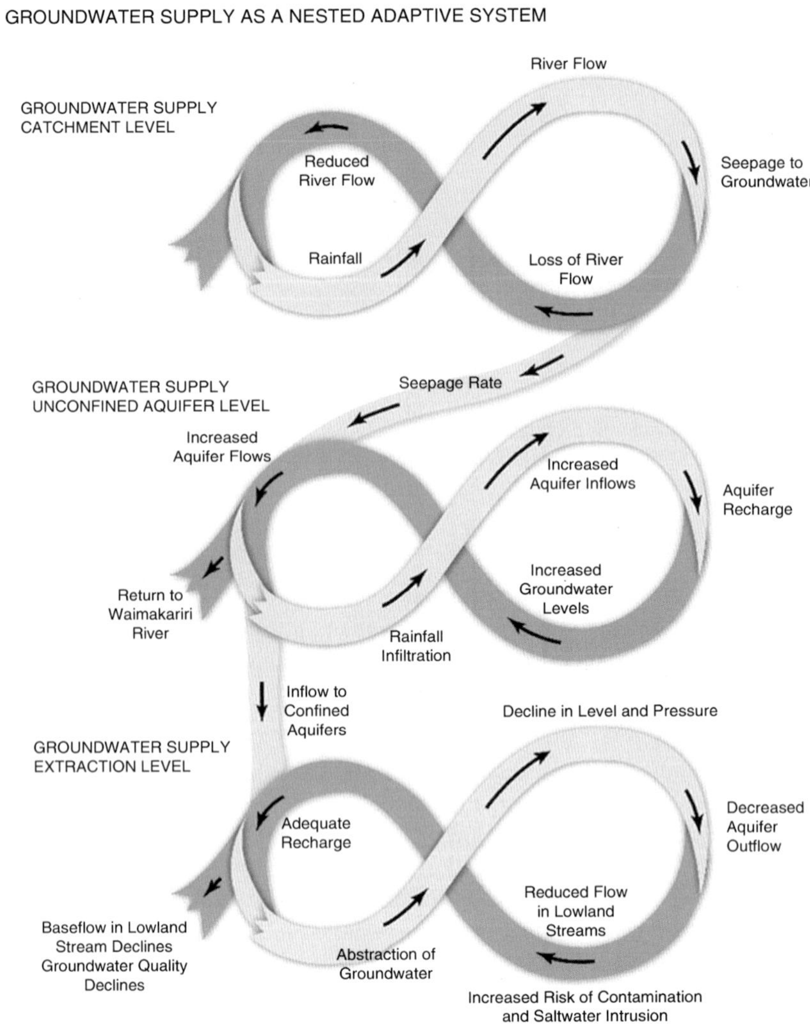

Fig. 5.12 Groundwater supply as a nested adaptive system (Jenkins 2015). Courtesy of WIT Press from 8th International Conference on Sustainable Management, A Coruna Spain 15-17 June 2015

5.1.3 Extraction of Gravel from the Bed of the Waimakariri River

Three geographical areas are relevant to considering the gravel extraction for the bed of the Waimakariri River and its sustainable management:

- the Waimakariri catchment as the source of gravel
- the river reach from which gravel is extracted
- the coastal zone around the mouth of the Waimakariri River.

An adaptive cycle is defined for each of these geographical areas. The nesting of the three areas is also described.

5.1.3.1 Adaptive Cycle for the Waimakariri Catchment for Gravel Extraction

Sediment yield for a catchment largely depends on rainfall and geology. The Waimakariri catchment with its high rainfall in the upper catchment in the Southern Alps and greywacke[3] rocks has a relatively high suspended sediment yield of 989 t/km^2/a (Hicks 2004).

Sediment is transported by a river when the entrainment velocity threshold is exceeded for the grain size of the sediment with finer material held in suspension (suspended sediment) and coarser sand and gravels moving along the river bed (bedload transport) (Hicks 2004). The Waimakariri River has an estimated suspended sediment load of 3.1 Mt/a and a bedload of approximately 260,000 m^3/a (Hudson 2005). The gravel bed in the river finishes 2.5 km upstream of the river mouth.

The high sediment load creates the braided character of the Waimakariri River. Aggradation of sediment in the river bed reduces the channel capacity which increases flooding risk and can eventually lead to overflows creating new river channels.

This sequence of processes can be described as an adaptive cycle:

- exploitation: the erosion in the catchment and riverbed that generates sediment in the river
- accumulation: the accumulation of sediment in the river bed
- disturbance-release: the movement of sediment when the river reaches threshold velocity for suspended sediment and bedload transport
- reorganisation: sediment removal from the bed to maintain channel capacity for the river flow.

The vulnerability of the existing river channel is the aggradation of the river bed which eventually leads to overflows forming a new channel.

The adaptive cycle at the catchment scale can be depicted diagrammatically as shown in Fig. 5.13.

The critical variables for the processes are the sediment yield for the catchment, the suspended sediment and bedload transported by the river, and the rate of aggradation in the riverbed.

[3] Greywacke is a slightly metamorphosed sandstone and mudstone with structural characteristics (joint and fractures) that make it susceptible to being broken up by water action (Environment Canterbury 2006).

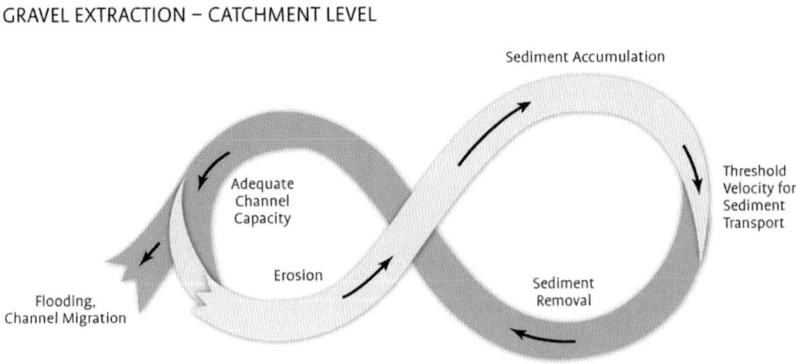

Fig. 5.13 Adaptive cycle for gravel extraction at the catchment level (Jenkins 2015). Courtesy of WIT Press from 8th International Conference on Sustainable Management, A Coruna Spain 15-17 June 2015

5.1.3.2 Adaptive Cycle for River Reach where Gravel is Extracted

Gravel extraction occurs from the lower reaches of the Waimakariri River. The historical record of extraction over the last 15 years has been between 300,000 and 700,000 m^3/a (Measures 2012). This extraction rate exceeds the bedload transport rate which has led to a 1 m bed level reduction between 1985 and 2008. Modelling of sediment transport predicts that if no extraction had taken place that bed levels would be over 2.5 m higher in some reaches (Measures 2012).

Gravel extraction can therefore help maintain the floodway capacity of rivers like the Waimakariri River. However excessive gravel extraction can pose a threat in relation to the undermining of flood protection and erosion control works as well as bridge abutments. The sustainable supply of gravel from the Waimakariri River has been assessed to be 250,000 m^3/a (Environment Canterbury 2006). Recent bed level investigations indicate that some of the historical gravel extraction consents have minimum bed levels that are below the currently recommended bed levels (Environment Canterbury 2009).

In addition to the direct effect on bed levels, the lowered bed levels from gravel extraction can affect the gravel transport rates for different reaches of the river. Modelling indicates that extraction has resulted in an increase in gravel transport rate (up to 60,000 m^3/a) in the excavated area and a decrease downstream of the excavated area. Furthermore the location of increased deposition corresponds to the reach where the greatest gravel extraction has taken place (Measures 2012).

In relation to reorganisation after gravel abstraction, the river adjusts to the new bed levels. After a year there is a decline in bed levels within about 1 km of the excavated area. After 20 years a 0.5 m excavation has reduced to 0.1–0.2 m in depth and there are reduced bed levels within 4 km upstream and downstream of the excavated area.

The adaptive cycle for these processes can be set out as follows:

- exploitation: gravel extraction from the river bed
- accumulation: reduction in river bed levels

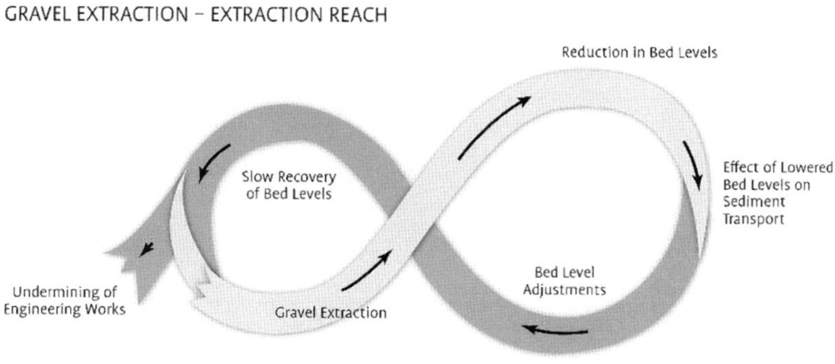

Fig. 5.14 Adaptive cycle for gravel extraction at the extraction level (Jenkins 2015). Courtesy of WIT Press from 8th International Conference on Sustainable Management, A Coruna Spain 15-17 June 2015

- disturbance-release: the effect of lowered bed levels on sediment transport rates
- reorganisation: bed level adjustment with reductions in level upstream and downstream, and infilling of the excavated area.

The vulnerability for gravel extraction is the lowering of bed levels to the point where engineering works are undermined. Critical variables in the cycle are the bedload transport rate, the gravel extraction rate and bed levels.

The gravel extraction process can be depicted as a Lissajous figure as shown in Fig. 5.14.

5.1.3.3 Adaptive Cycle for the Coastal Zone Near the Waimakariri River Mouth

River sediment is one of the major sources of sediment with respect to the sediment budgets for coastal systems. Excavation of gravel from a river bed forms pits within the channel profile which can trap much of the incoming bedload sediment preventing it or slowing it from reaching the coastline.

The Waimakariri River is the largest contributor of sediment to Pegasus Bay and represents about 77% of the sediment supplied to Pegasus Bay. Estimates of suspended sediment load vary from 2.78 Mt/a to 5.36 Mt/a. If the beach sand fraction is between 20 and 40%, then this gives a beach sand load of between 0.6 and 2.1 Mt/a (Hicks 1998).

With respect to gravel extraction, the river is essentially a closed system with gravel deposition ceasing about 2.5 km from the coast.[4] This means there should be little or no effect of river gravel extraction on the coast. The sandy beaches nourished by the Waimakariri River are accretionary which indicates the sand fraction reaching the coast is sufficient to maintain the stability of the southern Pegasus Bay

[4] Technically, gravel refers to a specific size range of rock fragments: 2–64 mm)

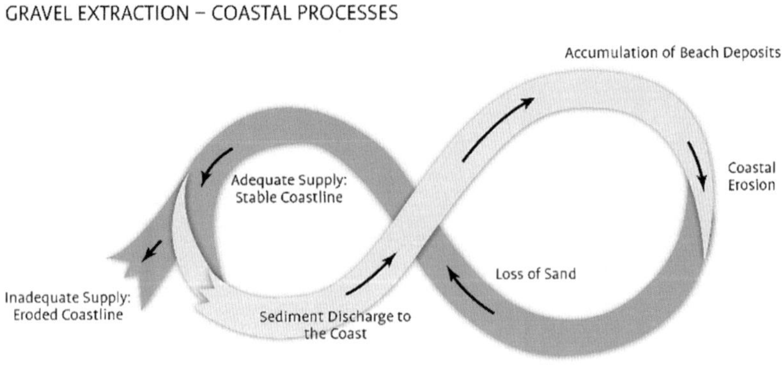

Fig. 5.15 Adaptive cycle for gravel extraction at the coastal level (Jenkins 2015). Courtesy of WIT Press from 8th International Conference on Sustainable Management, A Coruna Spain 15-17 June 2015

shoreline (Environment Canterbury 2006). Modelling by Hicks predicts an advancement of the shoreline by 40 m over the next 50 years (Hicks 1993).[5]

Any changes to the sediment contribution from the Waimakariri River would have significant implications for the coastal sediment budget. Modelling by Hicks of a 50% reduction in the supply of river sand over 50 years predicts shoreline retreat of 80 m inland and extending over 8 km around the river mouth (Hicks 1993).

This interaction between gravel extraction and coastal processes can be expressed as an adaptive cycle:

- exploitation: sediment discharge to the coast
- accumulation: accumulation of sediment as beach deposits
- disturbance-release: coastal erosion leads to loss of sediment
- reorganisation: sediment supply maintained so that coastal sediment budget is maintained.

The vulnerability of the coastal system is in relation to the ongoing sediment supply for the river. Critical variables are the sediment supply to the coast and coastal erosion rates.

The adaptive cycle can be depicted diagrammatically as shown in Fig. 5.15.

5.1.3.4 Nested Adaptive Cycles for Gravel Extraction

Figure 5.16 shows the three adaptive cycles as a nested system. The linkage between the catchment level and the extraction reach is the river transport rate, in particular, the bedload transport of gravel. The linkage between the extraction reach and the coastal processes is the sediment supply to the coast and in this instance the sand supply.

[5]This modelling assumed the current rate of sea level rise of 1.8 mm/year measured at the Port of Lyttelton. The modelled effect of an additional 0.5 m sea level rise over 50 years would be a reduction in the rate of shoreline advance.

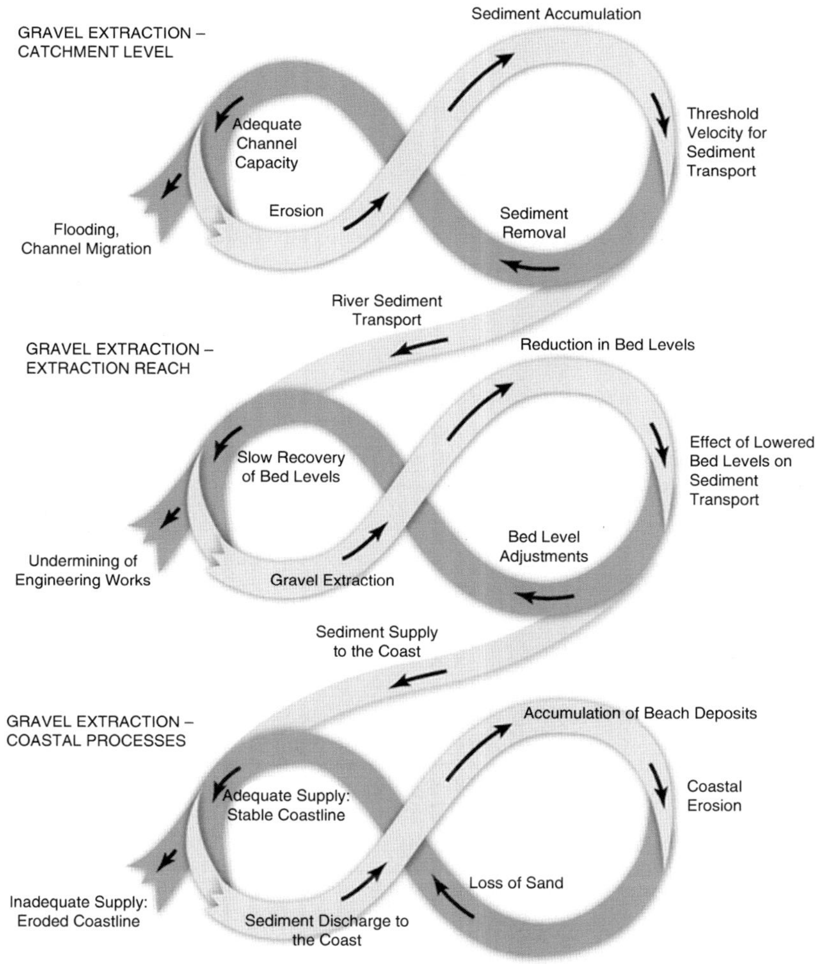

Fig. 5.16 Gravel extraction as a nested adaptive system (Jenkins 2015). Courtesy of WIT Press from 8th International Conference on Sustainable Management, A Coruna Spain 15-17 June 2015

5.1.4 Key Points from Examples

The nested adaptive systems approach provides a framework for the definition of sustainability limits. The approach considers adaptive cycles (i.e. exploitation/ accumulation/disturbance-release/reorganisation) at multiple spatial scales. The approach identifies critical variables that can cause the adaptive cycle to fail and identifies the linkages between different spatial scales that need to be maintained. The critical variables and the linkages that determine the limits will depend on the socio-ecological system under consideration. A systems approach with multiple spatial scales can identify additional management interventions compared to problem-focussed approaches which typically are the basis of current management actions.

In drought-prone Canterbury, irrigation is valuable in maintaining agricultural production. Following an analysis of future water demand and potential supply (Morgan et al. 2002), the way forward was seen as further extraction or storage on rivers like the Waimakariri with their headwaters in the Southern Alps. These alpine rivers account for 88% of annual flow in the region. As shown in the nested adaptive system analysis for irrigation, setting environmental flows in the river (catchment scale), scheme storage for reliability of supply (irrigation scheme scale), and on-farm storage and soil moisture demand irrigation (on-farm scale) were critical. There has been a major debate in relation to the allowable irrigation extraction from the Waimakariri River between out-of-river users (irrigators) and in-river users (fishermen, kayakers) on how much flow needs to remain in the river. Taking a nested adaptive systems approach has broadened the debate to consider the provisions of irrigation scheme storage and on-farm storage (Canterbury Water 2009).

In relation to Christchurch's groundwater supply the initial controls were based on drawdown interference effects and low groundwater level effects. To address concerns about drinking water availability for public water supply, restrictions were placed on abstraction from private bores based on groundwater level. Progressive restrictions were defined in a series of steps as groundwater levels declined (Pattle Delamore Partners and Canterbury Regional Council 1997). Evidence of the drying of the headwaters of groundwater-fed streams (Daly 2006) indicated that a management system based on groundwater interference was insufficient. The nested adaptive systems analysis for public water supply from an aquifer primarily fed from river leakage, indicated the critical variables were maintenance of river flow (catchment scale), seepage from the river (groundwater – surface water interaction scale), and, maintenance of groundwater-fed streams and maintenance of aquifer pressure to prevent contaminant leakage (groundwater extraction scale).

For gravel extraction, the initial management strategy was to encourage extraction as a key method of increasing and maintaining flood capacity because of the high rate of sediment accumulation in Canterbury rivers. However with increasing demand, increased gravel abstraction near convenient access points in rivers, led to concerns about river bank stability and the need for a more comprehensive approach for sustainable gravel extraction in order to manage the cumulative effects of multiple extractors (Environment Canterbury 2006). A review was undertaken to quantify the scale of the regional gravel resource and relate sustainable supply to regional demand. The review led to a change in the way gravel should be managed. The review showed increasing demand while the sustainable supply of river gravel was relatively small, highlighting the need to plan for land-based extraction (Environment Canterbury 2012). In the nested adaptive system analysis for gravel extraction from the Waimakariri River, river avulsion from aggradation (catchment scale), maintaining bed levels to avoid infrastructure damage (gravel extraction reach scale), and maintaining sand supply, rather than gravel supply, to the coast for shoreline stability (coastal zone scale) were the critical variables.

While the nested adaptive system analysis indicates that gravel extraction from the Waimakariri River is not a sustainability issue in relation to coastal processes, there are other segments of the Canterbury coastline where this is a concern.

Amberley Beach on the North Canterbury coast is between two rivers, the Waipara and the Kowai, which do supply gravel to the coast. Following over a century of accretion, Amberley Beach has suffered retreat since the 1980s (Geotech Consulting 2000). Nourishment works added 10,000 m^3 of gravel in 2004 and a further 6000 m^3 in 2007. Influence on coastal processes is now a sustainability management criterion for gravel extraction (Environment Canterbury 2012).

The analysis identified that there are some sustainability issues, such as the reliability of supply for the Waimakariri irrigation scheme and the bed levels in the river due to gravel abstraction. It also identified issues that are not a current problem, such as the seepage from the Waimakariri River to the aquifer, and the sediment supply to the coast. Furthermore, it highlighted some potential future problems, such as the impact of further extraction for public water supply. Some of the critical variables were beyond reasonable management influence, such as bedload transport in the river, and coastal erosion rates along the coast. However other critical variables could be influenced by management intervention to enhance sustainability, such as provision of scheme and on-farm storage to improve reliability of supply, and placing bed level restrictions on gravel extraction from the river.

The nested adaptive system approach identifies multiple failure pathways for which additional management interventions might be needed. It can predict in advance the issues that need to be addressed and for which limits need to be set such as the need to develop a sustainability strategy for Christchurch water supply from groundwater in relation to maintaining flows in lowland streams and aquifer pressure for protection from land use contaminants; this is particularly the case as the preferred means of providing additional capacity is a storage on the Waimakariri. As noted in the dredging analysis, such a storage would reduce sand supply to the coast leading to coastal erosion.

5.1.5 Collapsing Panarchies and Maladaptive Cycles

There may not always be recovery in an adaptive cycle and there can be maladaptive cycles that can be sustained. Gunderson and Holling identify two possibilities – the poverty trap and the rigidity trap (Gunderson and Holling 2002). The poverty trap is where the potential and diversity have been eradicated by misuse or an external force leading to an impoverished state of low connectedness, low potential and low resilience. Biophysical examples are the major extinction events leading to loss of species that never recovered.

The rigidity trap is where there is high potential, high connectedness and high resilience. This enables the system to resist external disturbances and persist beyond the point where it is adaptive and creative. Large bureaucracies are socio-economic examples of rigidity traps.

Allison and Hobbs in their analysis of the Western Australian agricultural region identified a third – the "lock-in trap" which is characterized by low potential for change, high degree of connectedness, and, a high resilience to change (Allison and

Hobbs 2004). The region is predominantly a wheat growing area where there has been clearance of the primary native vegetation. This has led to reduced evapotranspiration and rising saline groundwater tables – the problem of dryland salinity. Sixteen per cent of land had developed soil salinity by 2000 with predictions that 33% would be salinized by 2050.

Allison and Hobbs found that increasing wheat yields were only sufficient to offset declining terms of trade and the increasing area of unproductive salinized land. With no profit, there was little capacity for change. They found a high connectedness between producers and the agro-industry they support. The driving forces in the WA agricultural region were macroeconomic. The adaptive cycle for the industry was synchronous with Kondratiev long-wave economic cycles.[6]

However, the relatively fast-moving economic variables are largely decoupled from the slow-moving ecological variables. The ecological system is becoming severely impoverished. However, the socio-economic system is becoming more tightly connected thereby increasing the resilience of the socio-economic system. This is because the system has reached such a depauperate state that it is extremely stable.

Allison and Hobbs have labelled this state as "lock-in" based on the description from economics where an industry has so much "sunk costs" that it may continue to degrade its resource base. From an economic perspective, the adaptive cycle can be maintained, but from a natural resource management perspective it is a maladaptive cycle.

5.2 Chapters on Failure Pathways

Chapters 6, 7, 8, 9 and 10 focus on failure pathways as they relate to water management in the Canterbury region:

- Chapter 6 addresses failure pathways associated with cumulative effects at the catchment scale
- Chapter 7 considers biophysical failure pathways related to climate change and the natural hazards of drought and floods
- Chapter 8 addresses socio-economic failure pathways of governance, institutional arrangements and individual commitment
- Chapter 9 examines water-related health failure pathways, and
- Chapter 10 considers regional level failure pathways related to economics, the social effects of changing technology, and foreign intrusion.

[6] Kondratiev cycles are cycles in the world economy with a cycle period of forty to sixty years characterised by four phases of prosperity, recession, depression and recovery which can be related respectively to the exploitation-accumulation, accumulation-disturbance, disturbance-reorganisation, and reorganisation-exploitation sequences of the adaptive cycle.

Sustainability strategies are also considered where they introduce other failure pathways. One of the important aspects of the nested adaptive systems analysis is not only addressing the failure pathway of the current issue under consideration but also the potential failure pathways of possible strategies to address the current issue. Failure pathway issues for Canterbury are also considered in the context of international failure pathway issues.

5.2.1 *Cumulative Effects at the Catchment Scale*

The examples from the Waimakariri catchment in Sect. 5.1 above are examples of sustainability issues at the catchment scale. Chapter 6 focuses on cumulative effects at the catchment scale for the Canterbury region – the cumulative depletion of natural resources (failure pathway 1 in Sect. 4.2.4 and Fig. 4.6) and cumulative degradation of the environment (failure pathway 2 in Sec. 4.2.4 and Fig. 4.6).

Water availability (failure pathway 1) in Canterbury is a significant issue as demand, principally for increased irrigation, outstrips supply for the current methods of abstraction. Water scarcity is an international issue. Irrigation for agriculture accounts for about two thirds of global water use. Increased food production from irrigation has been an important method for feeding the world's growing population. However, world irrigated area per capita peaked in 1978 and is now declining (Postel 1992).

In addition, land use intensification associated with water use from increased irrigation has led to water quantity impacts of reduced river flows and lowered groundwater tables, and, water quality impacts mainly from increased nutrients, sediment and bacterial contamination (failure pathway 2). Abstraction effects and the impacts of pollution are also global issues (Postel 1992).

Covich summarises the range of global issues for impacts of freshwater use on ecosystem and biodiversity (Covich 1993). He considers physical variables of flow and temperature and chemical variables of dissolved oxygen, salinity, nutrients and acidity. For Canterbury changes to flow and nutrients are the dominant issues. Covich also identifies six examples of persistent biological problems: acidification, sustainable fisheries yields, non-native species transfers, regulated rivers and reservoirs, deforestation and catchment land use, and, global climate change. Catchment land use and river regulation are the significant issues for Canterbury that are considered in Chap. 6. The implications of climate change (failure pathway 9) are considered in Chap. 7.

5.2.2 *Biophysical System Failure Pathways at the Regional Scale*

Chapter 7 addresses the biophysical system failure pathways of climate change (failure pathway 9) and natural disasters (failure pathway 3). Climate change is a global issue but with regional and local water management implications for temperature, precipitation, soil moisture, snowfall and snowmelt, storm frequency and intensity, and, runoff (Gleick 1993). The chapter includes projections of climate change for New Zealand and their implications for the Canterbury region. In addition, catchment-specific implications are presented.

The most significant water-related natural disasters for Canterbury are droughts and floods. The high potential evaporation deficit for the region makes dry conditions an issue even in average years. Even with increased irrigation in the region, dryland farming predominates (refer Fig. 3.2). Adaptation to drought conditions is a significant sustainability issue for Canterbury's agricultural economy.

The alpine rivers of the region have variable flows and aggrading river beds (refer Sect. 5.1). Christchurch is the highest value economic asset at risk of flooding in New Zealand (Logan, 2008). A resilience approach has been taken to address this issue from a sustainability perspective.

5.2.3 *Socio-economic Failure Pathways*

Chapter 8 describes the socio-economic failure pathways at the societal level (failure pathway 4) and individual level (failure pathway 5). Failure pathway 4 relates to the institutional arrangements that have been put in place to manage conflict and facilitate integration between different interests while failure pathway 5 relates to the achievement of individual commitment to societal outcomes.

The institutional arrangements in New Zealand were put in place with the major reforms of natural resource management in the late 1980s (refer Sect. 2.1). These arrangements emphasized a regulatory role for government in water resource management. The approach of achieving integration was through defining environmental limits while allowing users to pursue their own interests within those limits. While effective in dealing with point source discharges, these arrangements have not been effective in managing diffuse sources and cumulative effects of multiple sources. Conflict resolution was through legal processes and the Environment Court. This has led to a legalistic and adversarial approach for resolving differences between conflicting interests.

However, since then there have been significant developments in institutional arrangements in forms of democracy (Dryzek 2010; Keane 2009), for managing common pool resources (Ostrom 1990), models of planning (Innes and Booher

2003), government's role in environmental management (Bleischwitz 2007; Jenkins 2009) and concepts of the post-regulatory state (Scott 2005). The collaborative governance approach of the Canterbury Water Management Strategy which is aligned with these developments represented a new paradigm for institutional arrangements for water management in New Zealand.

The regulatory model has also not been fully effective in achieving individual commitment to consent conditions. Alternative approaches of voluntary compliance and audited self-management have been considered in Canterbury.

Consideration has also been given to two behavioral models in relation to achieving individual commitment. One is the evaluation process model (Dornbusch and Scott 1975) which applies to the exercise of authority, such as a regulatory approach to compliance. The second is the expectancy theory of motivation (Lawler and Porter 1967) which can be applied to voluntary approaches.

5.2.4 Waterborne Disease Pathways

Disease pathways (pathway 6) associated with contamination of water are considered in Chap. 9. Ingestion of contaminated drinking water is the main waterborne disease pathway. Other pathways are water contact recreation leading to ingestion of contaminated water, contact with toxic algae blooms, and shellfish gathering in contaminated water. Because of the significance of waterborne disease there are public health management regimes in place for each of the pathways. Thus it is not only the contamination of water that needs to be considered but also the efficacy of the management system.

Public health is managed at the national level with New Zealand adapting World Health Organisation approaches to local circumstances. Water Safety Plans are required for drinking water supplies for communities greater than 500 under Health Act amendments in 2007 and provisions for Catchment Protection Zones under RMA amendments. Implementation of these requirements are in progress. However large areas of rural Canterbury have water supplies graded as marginally unsatisfactory to unacceptable in terms of health risk. Higher incidence of waterborne disease is associated with small supplies. Examples of improved water supply from increasing scale of technical management are Akaroa and Franklin District. Affordability of water supply and treatment infrastructure is a concern for small communities. The international example of Cochabamba where affordability concerns led to civil unrest is discussed.

Management regimes for water contact recreation, toxic algae blooms, and public harvesting of shellfish are limited to public warnings of risk and do not require the cause of contamination to be addressed. In Canterbury only 67% of freshwater bathing sites meet water contact recreation criteria and the incidence of toxic algal blooms appears to be increasing. Commercial marine farms have comprehensive management regimes based on the "hazard analysis and critical control points" approach. However, aquaculture is a relatively small industry in Canterbury.

5.2.5 Regional Socio-economic Failure Pathways

Chapter 10 describes the socio-economic failure pathways at the higher spatial scale of the broader region in which the society operates. These comprise external intrusion (pathway 7), the loss of trade networks (pathway 8) and new technology changing social relationships (pathway 10).

For export-oriented economies like the Canterbury region and New Zealand as a whole, regional and national economic analysis of agricultural intensification based on irrigation is one key component of the trade network analysis. Also, the externalities of agricultural intensification in relation to natural resources and the environment are also assessed. However, in considering water in the context of trade networks, it is not the water per se that is traded rather it is the water required to produce exported or imported goods that is relevant – what is referred to as "virtual water". The hidden nature of virtual water can obscure the significance of this potential failure pathway. The example of groundwater depletion in Saudi Arabia associated with the export of wheat which required irrigation is provided as a specific instance of this failure pathway.

For an island-based country like New Zealand and catchment-based regions like Canterbury, the vulnerability to an external intrusion failure pathway is quite low. Border issues between regions can occur and the Canterbury-Otago dispute over the Waitaki catchment is discussed. Also, foreign purchase of land in New Zealand and the approvals of overseas investment in land has been contentious. To demonstrate the complexities of the external intrusion failure pathway, water conflict in the Jordan River Basin is described in some detail. In addition, the international foreign investment in farmland is described to demonstrate the broader implications of this form of foreign intrusion failure pathway.

5.3 Sustainability Strategy Development

The development of sustainability strategies is the key to sustainable management. Chapter 3 described the progression of the Canterbury Water Management Strategy from the initial concept of storage on alpine rivers to address water availability concerns to the final integrated water management strategy. The first part of this section translates this progression into the classification of sustainability strategies as set out in Table 4.3.

The second part of this section summarises the development of the restoration programme for Te Waihora/Lake Ellesmere. The lake has suffered both human and natural disturbances. While significantly degraded it still retains multiple values. However little action had been taken to improve values of the lake because of the complexity of the system and because any action to improve one component had negative consequences for components. The development of a rehabilitation programme (*Whakaora Te Waihora)* became possible when there was a shift to management based on nested adaptive systems with a focus on vulnerable components and use of resilience analysis to identify priorities for restoration without affecting other values.

5.3.1 *Canterbury Water Management Strategy in the Sustainability Framework*

The Canterbury Water Management Strategy (CWMS) was developed using the concepts of nested adaptive socio-ecological systems (Jenkins 2007). This involved issues at four spatial scales:

- the regional level, where the key issues are water availability and land use intensification;
- the catchment level, at which the sustainability levels of water use and its effects, cumulative impacts of water use, and the reliability of supply are the main issues;
- the subcatchment level, where environmental flow requirements in river reaches, and management of streams and their riparian margins are the most significant issues; and
- the property level, where the land use practices that influence water quantity and quality are defined.

The CWMS evolved from the initial concerns of earlier strategic studies about water availability and the perceived need for major storages on alpine rivers. Table 5.3 sets out the components of the storage "solution" in the framework for classifying approaches to sustainability as described in Sect. 4.2.9 and Table 4.3. The regional transformation based on *Storage on alpine rivers* is added to existing supply strategy of *Run-of-river supply* within *Allocation limits* at the catchment

Table 5.3 Components of a storage approach to water availability in Canterbury

Geographic scale	Socio-economic system (SES)	Linkages SES and BPS	Biophysical system (BPS)	Transformation
Canterbury Region				Storage on alpine rivers
Linkage to region				
Catchment		Allocation limits	Run-of-river supply	
Linkage to subcatchment				
Subcatchment			Environmental flow for river reach	
Linkage to individual				
Individual	Consent			

For cells shaded gray no actions are possible. For cells that are white with no entry indicates no strategy component related to this cell.

Table 5.4 Key components of a sustainability strategy for water availability in Canterbury

Geographic scale	Socio-economic system (SES)	Linkages SES and BPS	Biophysical system (BPS)	Transformation
Canterbury Region	Regional Committee	Kaitiakitanga	Regional infrastructure	Water use efficiency; Storage associated with alpine rivers
Linkage to region	Regional implementation		Inter-basin transfers	
Catchment	Zone Committee; Moratorium on consents	Allocation limits; Reallocation of consents.	Run-of-river supply	Groundwater storage
Linkage to subcatchment	Zone implementation		Piped distribution	
Subcatchment	Water User Group	Real time monitoring of total take	Environmental flow for river reach; nutrient caps	Irrigation scheme storage
Linkage to individual	Brokerage		Real time monitoring of take	
Individual	Consent	Soil moisture monitoring	On farm irrigation efficiency	On farm storage; improved land and water management

level, *Environmental flow for river reach* at the subcatchment level, and, C*onsents* at the property level.

The range of strategy components for water availability in CWMS is more comprehensive. Table 5.4 sets out some of the key components in relation to water availability for the CWMS in the framework for classifying approaches to sustainability. The CWMS includes a wider range of components, for example, the addition of *water use efficiency* measures (e.g. *Piped distribution*, *Soil moisture monitoring*), and *Real-time monitoring of takes* for improved management. Storage on mainstems of alpine rivers is excluded because of concerns about the impacts on braided rivers, but a wider range of storage options is included, for example, *On-farm storage*, *Irrigation scheme storage*, *Groundwater storage*, and *Storage associated with alpine rivers* (but not on alpine rivers) such as off-river storages and diversions to tributary storage. Significantly, the CWMS includes socio-economic components, for example, *Regional Committee* and *Zone Committees* to develop implementation strategies (Jenkins 2011)[7]. It also includes additional linkages such as *Kaitiakitanga* (Māori

[7]These socio-economic components were based on Ostrom's approach of collaborative governance through 'self-managed communities'(Ostrom 1990).

concepts of stewardship) at the regional scale, *Reallocation of consents* at the catchment scale, and, *Real time monitoring of the total take* at the subcatchment scale.

5.3.2 Resilience Analysis of Te Waihora/Lake Ellesmere

Te Waihora/Lake Ellesmere is New Zealand's fifth largest lake in area (around 20,000 ha). It is a brackish "bar lagoon" type lake at the coastal margins of a largely agricultural catchment of 256,000 ha (Fig. 5.17). The northerly longshore drift of gravel in the Canterbury Bight has created a barrier bar (Kaitorete Spit) leading to the formation of the lake with limited connection to the sea. It is shallow (maximum depth of 2.5 m) and its salinity ranges between 5 to 10 ppt (15 to 30% of sea water). It is hypertrophic (TLI in the range 6 to 8) and turbid (Seechi depths less than 0.2 m).

It is of recent formation and about 500 years in its current geological form. Its water quality has degraded dramatically from oligotrophic (TLI 2 to 3) since clearance of the catchment for agriculture about 150 years ago. There has been drainage of the surrounding swamp and lowering of the lake level to enable farming. The lake level is controlled by artificial lake openings to the sea. It has been subject to multiple disturbances both human-induced (e.g. nutrient loading) and natural (e.g.

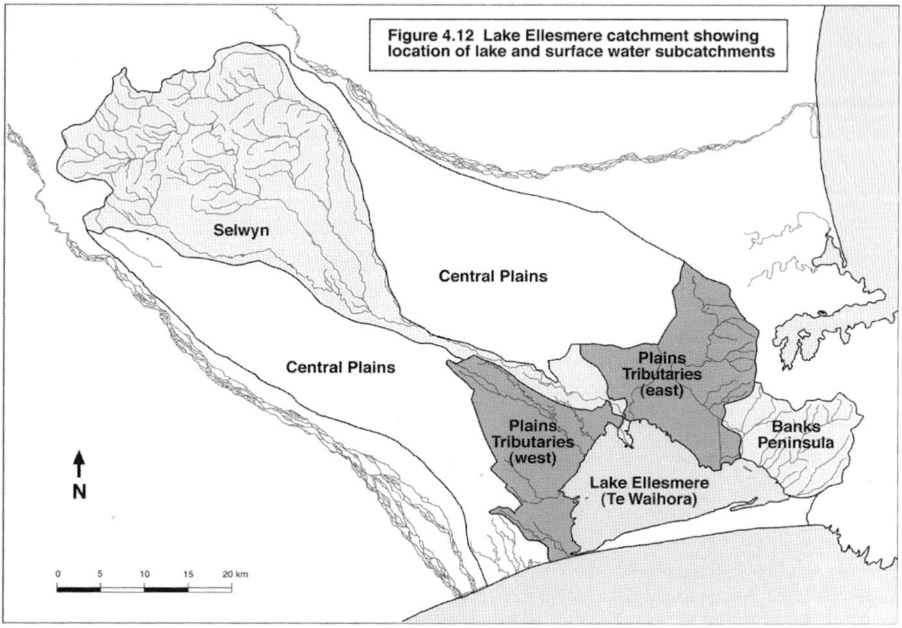

Fig. 5.17 Lake Ellesmere catchment showing location of lake and surface water catchments (Taylor 1996)

the Wahine storm[8] in 1968 uprooting macrophyte beds that have not re-established).

Even in its degraded state, the lake has multiple environmental values. These values include its:

- avifauna: especially waterfowl, migratory and indigenous waders, swamp birds, as well as oceanic, wetland and terrestrial species;
- fisheries: in particular, longfin and shortfin eels, flatfish (yellowbelly, black and sand flounder), yelloweye mullet, brown trout and whitebait;
- vegetation: with high value areas of freshwater (such as the raupo stands of the west bank of the LII) and brackish wetland communities (such as the saline lagoon margin on sandy soil at Greenpark Sands).

There are also high socio-economic values. The lake is a taonga[9] to Ngāi Tahu and provides a major source of mahinga kai and mana; the lake is also known as Te Kete Ika a Rākaihautū – the fish basket of Rākaihautū, who is credited in Māori mythology with creating the South Island lakes. As part of the Ngāi Tahu Claims Settlement Act 1998, the Crown vested the bed of the lake with Ngāi Tahu.

Farming production around the lake has an estimated worth of $34 million per annum. The lake maintains important customary and commercial fisheries. The lake is rated as a nationally significant water body for recreation including waterfowl hunting, fishing, cycling, water sports (water skiing, kayaking) and bird watching.

In 1947, agreement was reached in relation to the lake levels when openings to the sea would be made for flood management of lakeside farms. In 1986, a Water Conservation Order (WCO)[10] was sought to protect the outstanding wildlife habitat values of the lake, in particular its birdlife. As there had been adaptation to the agreed lake levels in 1947, the lake level and opening regime was retained when the Order was gazetted in 1990.

A major report on the lake was prepared in 1996 by the regional council (Taylor 1996). The report indicated the multiple values of the lake and the continuing decline in lake water quality and ecology. The report also indicated the complexity of the different systems and the incompatibility of achieving the optimum outcome for all values associated with the lake. The report highlighted the interdependencies between different components of the lake system. It is therefore difficult to establish a management framework based on the optimum outcome for each component.

Insights from resilience analysis have enabled the development of a rehabilitation programme for some of the most vulnerable components of the lake system. Key elements of nested adaptive systems as a management framework were:

---

[8] The Wahine storm is the biggest storm to strike New Zealand in the last 50 years. It was associated with Cyclone Giselle but is commonly known as the Wahine storm as the inter-island ferry *Wahine* sank during the storm with the tragic loss of 54 lives (NIWA 2017).

[9] *Taonga* means a treasure, considered to be of value including socially and culturally valuable resources.

[10] Refer to Sect. 2.2 for Resource Management Act legislative provisions.

- the environmental, economic, social and cultural values to be considered;
- the definition of the adaptive cycle related to each value;
- the points of intervention in the adaptive cycle for possible management actions;
- the tolerance range of sustainability parameters to ensure resilience of the adaptive cycle.

As a nested system, the lake has three main spatial levels:

- catchment scale (100s of kilometres) with issues such as land and water use, runoff to streams, recharge to groundwater, groundwater levels and lowland stream flow;
- lake scale (10s of kilometres) with issues related to lake level and volume, and lake water quality;
- lake margins (kilometres) with issues such as fringing vegetation, mudflats and shoreline erosion.

There are also multiple time scales:

- geomorphological (1000s of years) for issues such as filling of the lake with sediment;
- climate change (100s of years) for issues like sea level rise and reduced freshwater inflow;
- climate variation (10s of years) for issues such as rainfall variability;
- annual changes (years) for issues such as seasonal changes in rainfall and evaporation, and bird migrations.

Two of the key adaptive cycles for the lake are the cycles for water quantity and water quality. These are set out in Tables 5.5 and 5.6. For both adaptive cycles land and water use in the catchment and lake openings are significant drivers of water quantity and water quality.

The lake has been subject to many disturbances that have led to reorganisation of the lake. Some of the significant disturbances and reorganisations include:

- the Waimakariri River avulsion: the Waimakariri River used to discharge to the ocean through the lake; the lake has changed from an estuary to a coastal lake;
- the clearance of forest from the catchment for agriculture: this increased sediment and other contaminants shifting the trophic status from oligotrophic to hypertrophic;
- the Wahine storm in 1968 removed the ruppia beds which have not regrown: with ruppia removal and increased nutrients there has been a shift from a macrophyte-dominated lake to a phytoplankton-dominated lake;
- the increased groundwater abstraction in the catchment: this has lowered groundwater levels and reduced flows in groundwater-fed lowland streams into the lake

Some components of the lake system have remained resilient to these changes. For example, freshwater wetland vegetation fringing the lake has increased in area (from 452 ha in 1983 to 555 ha in 2007) despite a reduction in freshwater inflow. Other components have declined. In relation to fish, common bullies and shortfin

Table 5.5 Water quantity adaptive cycle

Adaptive cycle component	Biophysical processes
Exploitation –catchment water resources and use	Natural and Artificial Inflows less Abstraction –surface water inflow (rainfall and irrigation runoff) –groundwater inflow (recharge and irrigation leakage) –abstraction from surface and groundwater
Accumulation –water in storage in the lake	Lake Water Balance –inflows from surface and ground water –rainfall on lake surface –evaporation from lake surface –seepage to ocean through Kaitorete Spit
Release –lake opening to the sea	Channel Flow –freshwater exiting the lake –seawater entering the lake
Reorganisation –closure of opening to the sea	Return to Lake Condition –lowered lake level after lake opening
Resilience/Vulnerability –sustainability measures	Lake level Inundation frequency and magnitude on lake margins Timing of openings for fish passage

Table 5.6 Water quality adaptive cycle

Adaptive cycle component	Biophysical processes
Exploitation –effects of human use and natural processes	Water Quality Impacts of Land Use and Sea –added sediments, nutrients and bacteria –overtopping barrier with seawater
Accumulation –as sink for the catchment	Retention of Contaminants in Lake and Lake Ecosystems –build-up of sediment, nutrients and bacterial levels –nutrient uptake by plants
Release –lake opening to the sea	Channel Flow –contaminants removal during lake discharge –seawater incursion during lake opening
Reorganisation –closure of opening to the sea	Return to Lake Condition –reductions in sediment and nutrients –increase in salinity levels
Resilience/Vulnerability –sustainability measures	Lake trophic status Aquatic ecological health Water quality ranges

Table 5.7 Bird guilds and lake levels

Guild	Beneficial change	Harmful change
Open water divers	Permanent high lake	Permanent low lake
Deep water waders	Seasonally adjusted levels	Permanent high lake
Shallow water waders	Seasonally adjusted levels	Permanent high lake
Dabbling waterfowl	Permanent high lake	Permanent low lake
Aerial hunting gulls	Permanent high lake	Permanent low lake
Swamp specialists	Permanent high lake	Permanent low lake
Riparian wetland species	Permanent high lake	Permanent low lake

eels were maintaining numbers whereas trout and longfin eels were declining. Black swans that fed on the ruppia beds have reduced from around 70,000 prior to the Wahine storm to around 4000 after the storm that decimated the ruppia beds. There are also invasive components such as willow displacing native vegetation and Canada geese displacing native bird species.

There are also some changes that reflect the complex interactions within the lake system. Salinity levels in the lake have declined despite freshwater inflows decreasing. This is because there have been fewer lake openings and therefore less seawater entering the lake. Dissolved inorganic nitrogen in the lake has declined even though the nitrogen loadings in the catchment have increased. This is because inflows to the lake have decreased. Nitrogen concentrations in the inflows to the lake are higher but because the flows are lower the total nitrogen load entering the lake has reduced.

The lake reflects the challenges facing those wishing to undertake management interventions in complex systems. One is the changing values over time. Initially the lake was managed for land drainage. Then management of lake levels for birdlife was added to the values to be managed. More recently ecosystem health and cultural values have been added.

A second challenge is that the lake has been subject to multiple disturbances. Some of the significant disturbances were noted above.

The third challenge is that the reasons for the changes in the lake have multiple causes. There is not a simple intervention that can deliver an improved outcome.

The fourth challenge is that proposals for interventions may not resolve an issue because they don't address the dominant cause, may have unanticipated consequences for other values, or, may have conflicting requirements for different values. Table 5.7 shows the implications of different lake level interventions for different bird guilds that use the lake and its surrounds as habitat. The differences highlight the problem of defining interventions that achieve improvements for all components of the lake system.

Using a nested adaptive system framework, the following steps were taken to identify interventions that could improve vulnerable components of the lake system without compromising other components:

- a forum was established of agencies that have a statutory role in lake management, Ngāi Tahu representatives and representatives of the Waihora Ellesmere

Trust (a community organisation associated with the lake) to reflect the range of values to be considered, to coordinate activities, and, to resolve conflicts;
- resilience analysis was undertaken to identify priorities for restoration without adversely affecting other values;
- models were developed to enable system predictions of the effects of possible interventions;
- a series of public symposia were held that were open to all parties to be informed of findings and to be involved in decisions about possible interventions;
- a restoration programme to address the most vulnerable components of the lake system was agreed, funding sought and implementation initiated;
- statutory instruments including the Water Conservation Order (WCO) and the resource consent for lake opening were changed to reflect the proposed interventions.

One of the vulnerable species in the lake is the longfin eel (*Anguilla dieffenbachia*). It is one of the largest eels in the world and is only found in New Zealand. The Department of Conservation lists the species as "At risk/Declining" (Alliborne et al. 2010). Longfin eels breed only once at the end of their lifetime. They spend most of their life in freshwater. When ready to breed they head out to sea on high river flows in autumn to undertake a 5000 km journey to the tropical Pacific Ocean. Larvae return to New Zealand on tidal currents in spring to freshwater environments (Jellyman 2012). Lake openings coincident with eel migrations (i.e. adults leaving for breeding in autumn and larvae returning in spring) are important for maintaining the eel life cycle.

Changing the lake opening regime to specifically allow for openings during eel migration was modelled. Of specific concern was the potential for reduced lake levels in summer. Low lake levels in summer can decimate the lake-edge wetlands which is a key habitat for international migratory waders.

The original WCO allowed for openings when the lake level exceeded 1.05 metres above sea level (masl) between 1 August and 31 March, or 1.13 masl between 1 April and 31 July. It also allowed for openings any time between 15 September and 15 October. Lake level modelling was undertaken for a 38 year historical period comparing the WCO opening regime with two alternative scenarios: one with a spring recruitment opening between 15 September and 15 October so long as the lake level exceeded 0.9 masl; the second with a spring recruitment opening and an autumn migration opening (1 April to 15 June if lake level exceeded 0.9 masl) (Horrell 2011).

Table 5.8 compares the lake openings for the original WCO opening regime and the two alternative scenarios – spring recruitment openings; and spring recruitment and autumn migration openings. The spring recruitment scenario doubles the number of openings between 15 September and 15 October but decreases the openings between 1 April and 15 June compared to the original WCO regime. The spring recruitment/autumn migration scenario increases the frequency of both migration openings by about 70%.

Table 5.8 Modelling results for lake openings (Horrell 2011)

Opening regime	Autumn migration Openings between 1 April and 15 June	Spring recruitment Openings between 15 Sept and 15 Oct
Original WCO	18	13
Spring recruitment	16	26
Spring recruitment and Autumn migration	30	22

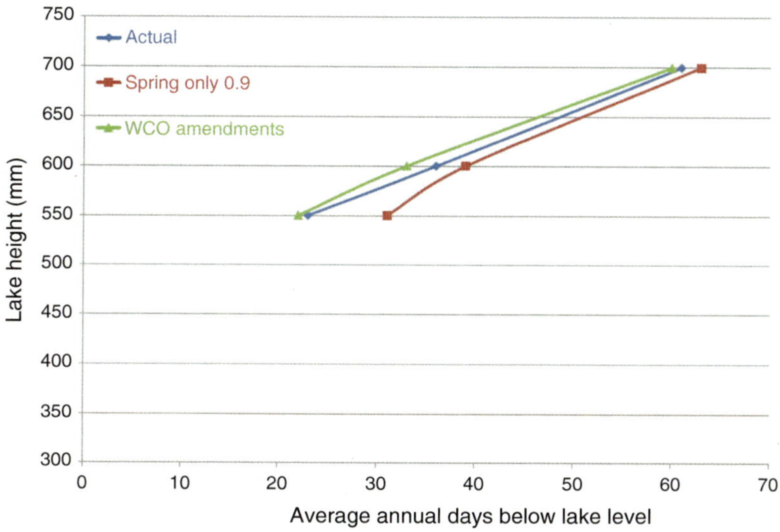

Fig. 5.18 Days lake level is low between December and April (Horrell 2011)

The scenarios were also compared in relation to summer lake levels (Fig. 5.18). The spring recruitment scenario lowers summer lake levels compared to the original WCO opening regime while the spring recruitment/autumn migration scenario increases summer levels. This analysis indicated that allowing openings for eel migration in spring and autumn would not adversely affect wader habitat.

The decision was made to change the WCO to allow openings based on the timing of eel migration. Also, the values for lake management were expanded from wildlife habitat to include fish, indigenous vegetation as well as Ngāi Tahu history, mahinga kai and customary fisheries. Decisions about actual openings are based on a protocol of the consent holders (the regional council and Ngāi Tahu) consulting eight stakeholder groups – the Taumutu Rūnanga, the farmers on the lake margins, the lake fishermen, the Waihora Ellesmere Trust, the two local authorities (Selwyn District Council and Christchurch City Council), and two statutory authorities (Department of Conservation and Fish & Game) – to ensure all significant interests are represented in opening decisions.

Where findings from the resilience analysis indicated areas of improvement for the lake and its ecosystems they were incorporated into a lake restoration programme, e.g. the re-establishment of ruppia beds, and, removal of invasive willow and replanting native species. The restoration programme *Whakaora Te Waihora* incorporated four major components:

- co-governance arrangements for lake and catchment management
- land purchase and protection around the lake margins
- enhancing ecosystem connections and buffering the lake from agriculture
- introducing innovative solutions to deliver sustainable benefits

5.4 Chapters on Sustainability Strategies

Chapters 11 and 12 focus on the development and decision making around formulating sustainability strategies, while Chaps. 13 , 14 and 15 brings together the findings of all chapters in terms of implementing a sustainability strategy for water management in the Canterbury region.

5.4.1 Sustainability Assessments

The development of the restoration programme for Te Waihora/Lake Ellesmere in Sect. 5.3.2 is an example of resilience analysis that identifies the vulnerable components of the lake system for which management interventions are needed in order to devise a sustainability strategy for the lake. Chapter 11 is a comparative sustainability assessment of six New Zealand lakes, including Te Waihora/Lake Ellesmere, with respect to water quality management.

The sustainability assessment comprises two parts: (1) a vulnerability assessment (or resilience analysis) to identify adaptive cycles and failure pathways with the critical variables and thresholds associated with those failure pathways, and, (2) sustainability strategy development to identify management interventions that adequately address the vulnerable components without compromising other components of the lake systems.

The catchments of the six lakes are all subject to agricultural land use intensification. However, the vulnerability assessments indicate different failure pathways related to water quality degradation for each lake. This means different management interventions are needed to formulate sustainability strategies for the different lakes. The current management interventions are then analysed in relation to their adequacy for achieving the water quality outcomes for each lake.

While the levels of intervention are less than that required to achieve the desired water quality outcomes, the types of interventions and the institutional changes adopted indicate the patterns of sustainability strategies that are likely to be effective. These patterns are summarized as potential management intervention pathways to achieve sustainability.

5.4.2 Decision Making for Sustainability

The management of Te Waihora/Lake Ellesmere also demonstrates the challenges in making decisions where there are multiple values, multiple stakeholders and information uncertainty at the catchment scale. Chapter 12 discusses how these issues were addressed in the formulation of the Canterbury Water Management Strategy (CWMS) at the regional scale. It explores the difference between multi-stakeholder decision processes and the two dominant existing paradigms of planner-led technical decision making and process-led legal decision making.

It describes the community engagement process for the CWMS as the basis for decision making. It also describes the decision-making technique of *Strategic Choice* and the evaluation process of *Sustainability Appraisal* that facilitated the strategy development and selection for the CWMS.

5.4.3 Implications for Water Management in Canterbury

The implementation of the CWMS is work-in-progress. There have been significant shifts in approach to water management in Canterbury and the approach is still evolving. Chapter 13 examines the biophysical limits in relation to sustainable management of water resources in Canterbury. It discusses the evolution of strategic thinking from a focus on water availability to enable further irrigation development through storage on alpine rivers, to consideration of the sustainable management of multiple water uses of importance to the community. Furthermore, water use efficiency was identified as a more effective approach to enhance water availability. Different forms of storage are now being implemented such as off-river storage, managed aquifer recharge and on-farm storage. While some aspects of improved efficiency and increased storage are being progressed, there are still implementation issues around climate change (both emission reduction and adaptation), infrastructure investment and coordination, and measurement and management of water use efficiency. The analysis highlights the spatial variation in water management issues across the region, the innovations in management practices, as well as the increasing need for modelling and monitoring of cumulative effects. The analysis demonstrates the management to specified limits is not enough to achieve sustainability. Rather using nested adaptive cycles provides a more robust basis for management.

Chapter 14 discusses the implications associated with socio-economic issues. One is the evolution of collaborative governance in Canterbury from an experimental stage, to the regional strategy development, to the zone implementation programme preparation and now the operational stage. Another is the inadequate institutional arrangements for infrastructure development and coordination. A third issue is the absence of funding arrangements for the solutions packages for improving water quality. A related issue is the absence of a mechanism for the internalisation of the externalities associated with water resource development. The question of the affordability of environmental improvements is limiting their implementation. A further issue is that while collaborative governance is leading to water management improvements there is a need to ensure all water management target areas, in particular environmental and recreational outcomes, are adequately addressed to maintain engagement of all interests, particularly environmental and recreational interests. The RMA legislative framework no longer provides an adequate basis for water management at sustainability limits nor the changing and more proactive role of government in sustainable development. Water framework legislation is needed to provide regional sustainability strategies to guide water resource management.

Chapter 15 summarises the main implications of the sustainability analysis for water management in Canterbury.

References

Alliborne, R., David, B., Hitchmough, R., Jellyman, D., Ling, N., Ravenscroft, P., & Waters, J. (2010). Conservation status of New Zealand freshwater fish 2009. *New Zealand Journal of Marine and Freshwater Research, 44*(4), 271–287.

Allison, H., & Hobbs, R. (2004). Resilience, adaptive capacity, and the "Lock-in Trap" of the Western Australian agricultural region. *Ecology and Society, 9*(1).

Bleischwitz, R. (2007). *Corporate governance of sustainability: A co-evolutionary view on resource management*. Cheltenham: Edward Elgar.

Canterbury Water. (2009). *Canterbury Water Management Strategy strategic framework*. Christchurch: Environment Canterbury.

Covich, A. P. (1993). Water and ecosystems. In P. H. Gleik (Ed.), *Water in crisis: A guide to the world's fresh water resources*. New York: Oxford University Press.

Daly, A. (2006). *Christchurch Rivers Survey September 2006: Report of a survey on waterways appearance and flows*. Christchurch: Environment Canterbury.

Dornbusch, S. M., & Scott, W. R. (1975). *Evaluation and the exercise of authority*. San Francisco: Jossey-Bass.

Dryzek, J. (2010). *Foundations and Frontiers of Deliberative Governance*. Oxford: Oxford University Press.

Duncan, M. (2008). *Waimakariri River: B/C Block allocation review*. Prepared for Environment Canterbury, Christchurch.

Environment Canterbury. (2000). *Our water in the balance: Christchurch – West Melton rivers and groundwater*. Christchurch: Environment Canterbury.

Environment Canterbury. (2006). *Regional gravel management report*. Christchurch: Environment Canterbury.

Environment Canterbury. (2009). *Waimakariri River bed level investigation*. Christchurch: Environment Canterbury.

Environment Canterbury. (2011). *Waimakariri Regional Plan*. Christchurch: Environment Canterbury.

Environment Canterbury. (2012). *Canterbury Regional River Gravel Management Strategy*. Christchurch: Environment Canterbury.

Geotech Consulting. (2000). *Hurunui district engineering lifelines project: Natural hazards assessment*. Christchurch: Environment Canterbury.

Gleick, P. H. (1993). Water in the 21st century. In P. H. Gleik (Ed.), *Water in crisis: A guide to the world's fresh water resources*. New York: Oxford University Press.

Gunderson, L. H., & Holling, C. S. (2002). *Panarchy: Understanding transformations in human and natural systems*. Washington, DC: Island Press.

Hicks, M. (1993). *Modelling long-term change of the Pegasus Bay Shoreline*. Christchurch: NIWA.

Hicks, M. (1998). *Sediment budgets for the Canterbury coast – A review, with particular reference to the importance of river sediment*. Christchurch: NIWA.

Hicks, M. (2004). Stream sediment load and organic matter. In J. Harding (Ed.), *Freshwaters of New Zealand*. Wellington: NZ Hydrological Society and NZ Limnological Society.

Horrell, G. (2011). *Te Waihora/Lake Ellesmere water balance modelling*. NIWA Client Report CHC 2011-104, Christchurch.

Hudson, H. (2005). *Waimakariri River: Status of gravel resources and management implications*. Christchurch: Environment Canterbury.

Innes, J., & Booher, E. (2003). Collaborative policymaking: governance through dialogue. In M. Hajer & H. Wagenaar (Eds.), *Deliberative policy analysis: Understanding governance in the network society*. Cambridge: Cambridge University Press.

Jellyman, D. (2012). *The staus of longfin eels in New Zealand – An overview of stocks and harvest*. Christchurch: NIWA.

Jenkins, B. R. (2007, March 27–30). *Water Allocation in Canterbury*. Paper presented at the NZ Planning Institute conference 2007, Palmerston North.

Jenkins, B. R. (2009, February 26). *Are regional councils past their use-by date?* Paper presented at the Environmental Defence Society conference – Reform in paradise: Threat or opportunity? Auckland.

Jenkins, B. R. (2011, February 3). *Regulatory innovations in New Zealand water management*. Paper presented at the Australian National University Regulatory Institutions Seminar, ANU Canberra.

Jenkins, B. R. (2015, June 15–17). *Sustainable water management analysis using nested adaptive systems*. Paper presented at the 8th international conference on Sustainable Management, A Coruna, Spain.

Jowett, I., & Biggs, B. (2006). Flow regime requirements and biological effectiveness of habitat-based minimum flow assessments for six rivers. *International Journal of River Basin Management, 4*(3), 179–189.

Keane, J. (2009). *The life and death of democracy*. London: Simon & Schuster.

Lawler, E., & Porter, L. (1967). Antecedent attitudes of effective managerial performance. *Organizational Behavior and Human Performance, 2*, 122–142.

Logan, R. (2008). *Waimakariri: An illustrated history*. Christchurch: Phillips and King.

Measures, R. (2012). *Modelling gravel transport and bed level change in the Waimakariri River*. Christchurch: NIWA.

Ministry for the Environment. (2008). *Proposed National Environmental Standard on ecological flows and water levels*. Wellington: Ministry for the Environment.

Morgan, M., Bidwell, V., Bright, J., McIndoe, I., & Robb, C. (2002). *Canterbury Strategic Water Study*. Lincoln Environmental: Lincoln.

NIWA. (2017). Looking back at the *Wahine* Storm. https://www.niwa.co.nz/our-science/coasts/research-projects/all/physical-hazards-affecting-coastal-margins-and-the-continental-shelf/news/winds. Accessed 3 July 2017.

Ostrom, E. (1990). *Governing the commons: The evolution of institutions for collective action*. New York: Cambridge University Press.

Pattle Delamore Partners and Canterbury Regional Council. (1997). *Christchurch-West Melton groundwater: Existing management practices*. Christchurch: Canterbury Regional Council.

Postel, S. (1992). *Last Oasis: Facing water scarcity*. New York: W.W. Norton & Company.

Scott, C. (2005). Regulation in the age of governance: the rise of the post-regulatory state. In J. Jordana & D. Levi-Faur (Eds.), *The politics of regulation: Institutions and regulatory reforms for the age of governance*. Cheltenham: Edward Elgar Publishing.

Taylor, K. J. W. (1996). *The natural resources of Lake Ellesmere (Te Waihora) and its catchment*. Christchurch: Canterbury Regional Council.

Thorpe, H. (1992). The Canterbury plains. In P. Mosley (Ed.), *Waters of New Zealand*. Wellington: NZ Hydrological Society.

White, P., Kovacova, E., Jebbour, N., & Tschritter, C. (2011). *Waimakariri River bed and groundwater – Surface water interaction* (GNS Science report 2009/41). Avalon: Institute of Geological and Nuclear Sciences.

White, P., Kovacova, E., Zemansky, G., Jebbour, N., & Moreau-Fournier, M. (2012). Groundwater – Surface water interaction in the Waimakariri River, New Zealand, and groundwater outflow from the river bed. *Journal of Hydrology (NZ), 51*(1), 1–24.

Part II
Failure Pathway Analysis

Chapter 6
Cumulative Effects at the Catchment Scale

Abstract Two of the significant failure pathways at the catchment scale are the cumulative depletion of natural resources and cumulative degradation of the environment. These two crucial issues relate to the sustainability limits and cumulative effects of water extraction and use at the catchment scale. There is now water scarcity in Canterbury as the abstraction demands for human use exceed the biophysical capacity of the water resource system to regularly supply water. Supply/demand comparisons are made for run-of-river abstraction indicating lack of reliable supply. However annual comparisons indicate potential capacity but involve storage. Storage investigations indicate that there are limited opportunities for options that were environmentally and economically sustainable. Other investigations indicated the potential for greater water availability by more efficient use of currently allocated water. Consideration of failure pathways and a broader range of sustainability approaches led to an integrated water management strategy for Canterbury instead of a storage strategy.

The cumulative effects of water use comprise the impact on river flows (including flow variability) from abstraction; the management of groundwater zones; the effects on water quality from land use intensification (especially nutrients, bacterial contamination and sedimentation); and seawater intrusion.

One of the major challenges in water management where there are multiple potential failure pathways is the determination of the significance of different pathways when adverse effects occur. One example is the teasing out the respective contributions of climate variability (in particular historically low winter rainfall and hence low aquifer recharge) and increased groundwater abstraction to the decline in the aquatic health of groundwater-fed lowland streams.

Keywords Cumulative effects • Sustainability limits • Water availability • Water quality impacts • Natural variations

B.R. Jenkins, *Water Management in New Zealand's Canterbury Region*, Global Issues in Water Policy 19, https://doi.org/10.1007/978-94-024-1213-0_6

6.1 Water Availability

Canterbury's dependence on irrigation makes it vulnerable to drought. As noted in Sect. 3.2.2, water scarcity in the dry periods of the late 1990s prompted an analysis of future demand and availability of supply. As set out in the first subsection, this analysis considered water availability on a weekly basis for the run-of-river schemes of Canterbury indicating current demand cannot be reliably met. However, considering water availability on an annual basis indicates future demand can be met but would require major storage. This led to the identification of possible major storage options. The second subsection indicates some of the sustainability constraints on three of the short-listed storage sites while the third subsection discusses the adverse effects of storages on the mainstems of alpine rivers. These constraints and the investigations of the Canterbury Water Management Strategy led to the consideration of improving water use efficiency at the farm, irrigation scheme and regional scale as a central component of an integrated water management strategy as well as the potential for groundwater storage. These alternative approaches to improving water availability are discussed in the fourth subsection.

6.1.1 Supply/Demand Analysis

The dominant form of surface water abstraction for irrigation in Canterbury is direct withdrawal from rivers. Stage 1 of the Canterbury Strategic Water Study provided an analysis of available supply from allocable flow from surface water based on mean annual low flow conditions over 7 days (175 m^3/s) and allocable flow from groundwater (16 m^3/s) (Morgan et al. 2002). This allocable supply (191 m^3/s) was compared with current water demand (in 2001) for peak 7-day allocation (290 m^3/s). This demonstrates that under low flow conditions current peak demand cannot be met by current abstraction methods. With future peak weekly demand projected to increase to 569 m^3/s in 2021 the shortfall in water availability is further compromised (Table 6.1).

Table 6.1 Potential for resources on a weekly basis (Morgan et al. 2002)

Weekly supply	
Allocable flow from surface water (under mean annual low flow conditions)	175 m^3/s
Allocable flow from groundwater (assuming no irrigation discharge)	16 m^3/s
Total allocable flow	191 m^3/s
Weekly demand	
Current water demand 2001 (peak 7-day allocation)	290 m^3/s
Future water demand 2021 (peak 7-day demand)	569 m^3/s

Table 6.2 Potential for resources on an annual basis (Morgan et al. 2002)

Annual supply	
Average annual flow from surface water	594 m^3/s
Allocable flow from groundwater	16 m^3/s
Total allocable flow	610 m^3/s
Annual demand	
Current water demand 2001 (average annual demand)	81 m^3/s
Future water demand 2021 (average annual demand)	229 m^3/s

However, if potential supply is considered on an annual basis, the average annual allocable surface water is 594 m^3/s. When added to allocable flow from groundwater (16 m^3/s) provides a total allocable flow of 610 m^3/s. This can be compared to current average annual demand of 81 m^3/s and future water demand of 229 m^3/s. Thus, on an annual basis, water is potentially available to meet future demand but would require storage (Table 6.2).

6.1.2 The Search for Sustainable Storage Options

The initial response to the supply/demand analyses was to consider storage options (Canterbury Strategic Water Study Stage 2, Aqualinc (2008)). However, finding sustainable surface water storage options proved challenging. The review of hydrologic performance identified a short-list of potential sites (refer Fig. 3.9). However further analysis of these sites indicated sustainability concerns for many of them. The examples of the raising Opuha Dam, storage options for the Hurunui/Waipara catchments, and the Waianiwaniwa proposal are discussed below.

6.1.2.1 Raising Opuha Dam

Opuha Dam is a current storage in South Canterbury (refer Fig. 3.9) which supports about 16,000 ha of irrigation. However, it has a high degree of unreliability as a water source. It does not meet the strategic target of no more than 10 consecutive days of restriction during the simulation period (using the data record for 1972–2001). The simulation indicated 36 restriction periods of 10 days or more for takes by the lower Opihi irrigators. There were also 3 restriction periods of between 81 and 90 days. In some periods, the storage would not refill over winter. In the simulation in the drought of the late 1980s the storage was depleted in 1988 and did not return to full supply level until 1992 (Aqualinc Research Limited 2008).

The raising of the dam wall by 6 m was tested for hydrologic feasibility. This enabled a small increase in irrigable area (from 16,000 ha to 18,810 ha) but achieved only a marginal improvement in reliability with 22 restrictions of 10 days or more (for takes by the lower Opihi irrigators) and 2 restrictions in excess of 100 days. In hydrologic terms, it is a problem of dependence on a foothill river with insufficient rainfall in dry years. Climate variability is the failure pathway for supply reliability.

6.1.2.2 Hurunui/Waipara Storage Options

In North Canterbury a range of storage options were investigated in the Hurunui catchment including:

- Lake Sumner: managing the outflows of the natural lake with a control gate structure;
- Hurunui South Branch: a 75 m dam on the river for irrigation;
- Mandamus: a dam on a tributary upstream of its junction with the Hurunui River;
- Waitohi: a series of dams on a small tributary with pumped flows from the Hurunui; and
- Pahau: a 35 m dam on a tributary of the Hurunui.

An economic comparison of the alternatives is set out in Table 6.3. This shows Lake Sumner and the Hurunui South branch as the cheapest options. In combination these two storages could supply the potential irrigable area in the Hurunui and Waipara catchments. There is an estimated irrigable area of 74,671 ha of which 7336 ha is currently irrigated, leaving 67,335 ha of potentially irrigable land. About 67,900 ha could have been irrigated with water from the Lake Sumner and Hurunui South Branch storages with a reliability of 91% (Aqualinc Research Limited 2008).

However, there was considerable opposition to the storages. Lake Sumner on the Hurunui North Branch was in a conservation area. The Hurunui South Branch was in an area of high naturalness where storage was proposed to be prohibited in the draft regional plan. Both branches of the Hurunui were highly valued whitewater kayaking reaches. Societal conflict was the failure pathway for these storages.

As a result of a collaborative approach (see Sects. 3.3.3 and 12.4.1), the Waitohi option was chosen as the preferred alternative because of the environmental and recreational values that it preserved. However, because of its greater cost there was

Table 6.3 Economic comparison of storage options in the Hurunui catchment (Aqualinc Research Limited 2008)

Storage option	Volume (Mm^3)	Capitalised Cost ($m)	Unit Cost ($m/$Mm^3$)
Lake Sumner	37	3.0	0.08
Hurunui South Branch	96	32.8	0.34
Mandamus	35	16.2	0.46
Waipara North Branch	30	20.1	0.67
Waitohi (pumping from Hurunui)	130	94.5	0.73

a concern that the project may not be affordable (i.e. its marginal productivity may have been its failure pathway). An affordability analysis was undertaken indicating a cost at the farm gate in the order of $7000 per hectare. This was considered to be at the upper end of affordability but still acceptable.

6.1.2.3 Waianiwaniwa Storage

The Waianiwaniwa storage proposal was part of the original Central Plains Water Enhancement Scheme. It was designed with a storage capacity of 280 Mm^3 to receive water from the Waimakariri and Rakaia Rivers to supply 60,000 ha of irrigation in the Central Plains between the two rivers. The Commissioners hearing the consent application for the scheme proposal recommended against the Waianiwaniwa storage component because of its adverse effects (Milne et al. 2009).

The major concerns of the Commissioners were in relation to the adverse social impacts on the Coalgate township which was half a kilometre downstream of the proposed storage, and, the adverse social and ecological impacts in the Waianiwaniwa Valley. They also noted the availability of alternatives with less impact.

The social impacts in relation to Coalgate concerned the construction effects of dust, noise and construction activity as well as the visual and landscape effects of a 55 m high and 2 km long dam in close proximity to the town. In addition, the Commissioners were concerned about the division created in the community, the community feeling that their town was being sacrificed for the benefit of others, the perceived risk of dam failure, and, the change in community perception of their town.

Inundation of 12 km^2 of the Waianiwaniwa valley would result from the proposed storage. This would affect the habitat of the Canterbury mudfish, a species endemic to the region with the valley the most important remaining habitat. About 25% of the known habitat would be lost. Furthermore, the water from the two rivers would introduce predators of the mudfish in what was currently a predator-free area.

There was also concern for the loss of 29 properties and displacement of 45 residents, many with multi-generational association with the area. Furthermore, the best farming soils would be lost.

As a failure pathway, the storage proposal represented new technology changing social relationships.

6.1.3 Effects of Storages on the Mainstems of Alpine Rivers

With 88% of the mean annual flow of Canterbury Rivers in alpine rivers, and with foothill and lowland streams under the greatest pressure from extraction, the initial focus was for storage on the mainstems of alpine rivers. However, there are significant sustainability issues with mainstem storages.

Both under the Resource Management Act and the Canterbury Regional Policy Statement, there is a policy to provide for water bodies which should be sustained as far as possible in their natural state. Water bodies under Water Conservation Orders or having high naturalness (in terms of ecosystem value, landscape, vegetation, fauna habitat or amenity) have been identified (Fig. 6.1). This includes much of the high country of the Southern Alps in Canterbury. The policy makes mainstem storage a non-complying or prohibited activity.

The character of Canterbury's braided rivers can be affected by dams. Maintenance of the geomorphology of the braids is dependent on flood flow. With the damming of the Waitaki, the number of braids has decreased with the reduction in flood flows. Figure 6.2 compares the pre-dam braided section at Ferry Road (10 km from the coast) with the same area in recent times. The width of the river and the number of braids has approximately halved at this point.

In addition to damping flood flows the hydro dams and control structures significantly reduce the supply of sediment and bed material from the upper catchment. Sediment entrapment in the hydro lakes has reduced the supply of bedload by approximately 75%. Some of this deficit has been recovered by the river scouring gravel from its own bed and floodplain but this will be limited by a concurrent reduction in the river's bedload potential due to flood damping. This means the supply to the coast should have been decreased (Hicks 2007). Erosion rates north of the Waitaki River mouth are about 0.5 m/year (Environment Canterbury 2008). Historical results show a high natural variability making it difficult to distinguish accelerated erosion due to sediment entrapment associated with the Waitaki hydro dams (Hicks 2007).

Algal growth in braided rivers is controlled by flushing flows (flows about three times the mean flow) which can dislodge algae building up on the gravels in the river bed. The environmental flow regime downstream of the dam was set as a minimum flow which had to be maintained. In the Opuha River there was a significant build-up of periphyton under stable flows (8–14 m^3/s). In the summer of 2005/6 there was 10–30% cover of thick *Phormidium* sp. mats (>3 cm thick) and long filamentous green algae (>2 cm long).

An experimental programme was conducted to introduce pulses of flushing flows (18–22 m^3/s) from the dam to reduce algal cover. Flushing flows were most effective at removing nuisance periphyton from the river bed near the dam. Percent removal of nuisance periphyton after each flushing flow decreased with distance from the dam. The programme indicated that higher flushing flows are needed to increase the effectiveness of periphyton reduction (Arscott et al. 2007).

In temperate regions like Canterbury, lakes can enter a stable stratified state in summer with a layer of warm water (epilimnion) overlaying cooler more dense water (hypolimnion). Stratification leads to the development of a density discontinuity (thermocline) that separates the epilimnion from the hypolimnion (Fig. 6.3). This acts as a barrier to effective mixing and inhibits diffusion of oxygen to the hypolimnion.

In the early years of operation thermal stratification and water quality impairment were evident in the lake formed by Opuha Dam. Lake Opuha has only a single

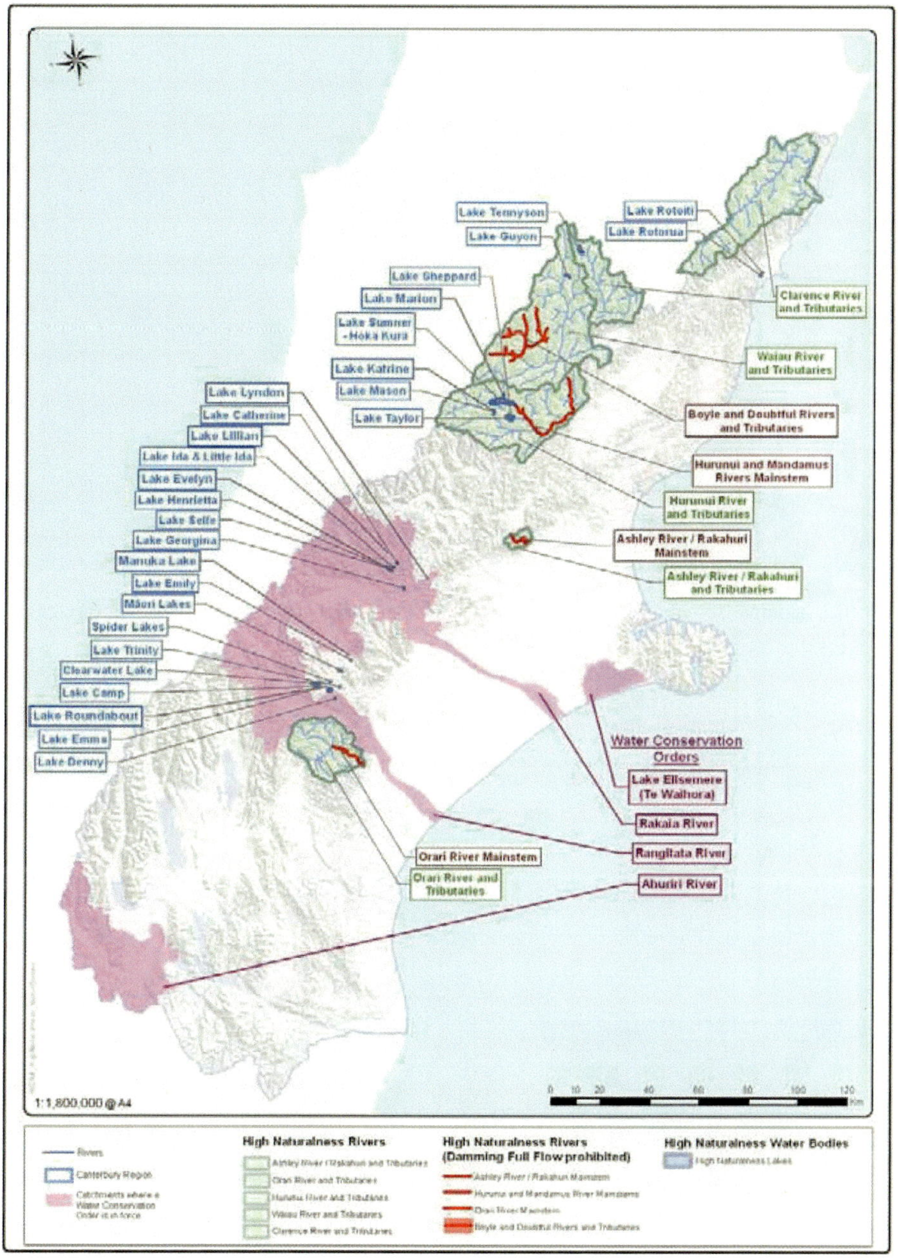

Fig. 6.1 Natural state and high naturalness areas (Environment Canterbury)

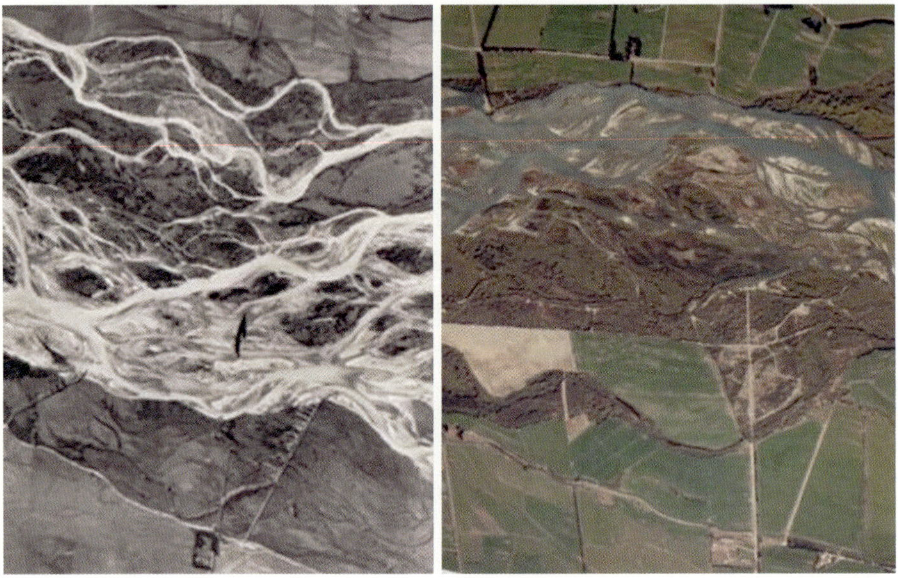

Fig. 6.2 Comparison of Waitaki River in 1943 (left) and 2001 (right) at Old Ferry Road (Jenkins 2007)

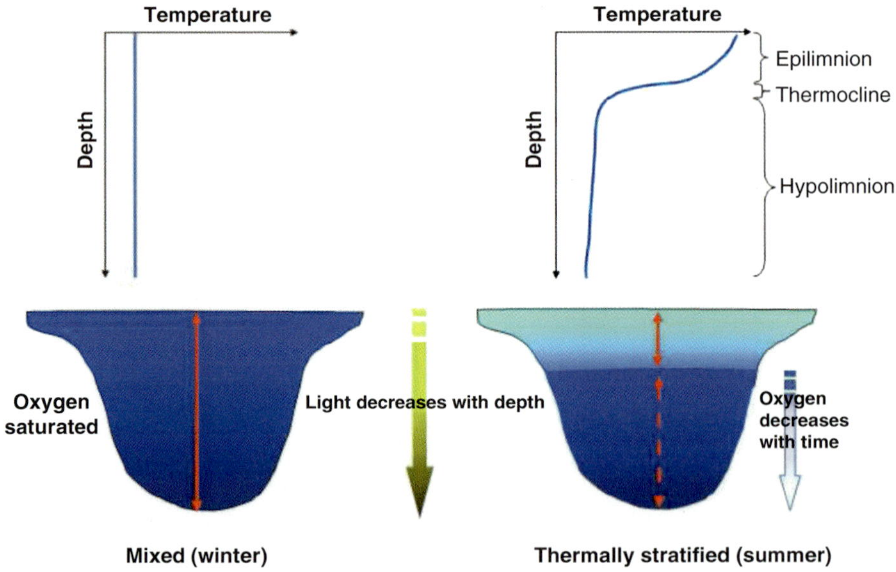

Fig. 6.3 Thermal stratification in lakes (Gibbs and Hickey 2012)

fixed draw off point near the bed of the lake (in the hypolimnion) resulting in downstream water quality impacts. Downstream measurements and observations indicated significant concentrations of contaminants still present and biological effects in the downstream river persisting including depauperate macroinvertebrate communities, pollution-tolerant periphyton communities, and humic iron precipitates and flocs in the river bed below the dam. Also notable were very high concentrations of predominantly dead or decaying zooplankton in drift samples in the river below the dam (Meredith 1999).

The resource consent was amended requiring the installation of lake aeration equipment. The aeration system was to be operated when the dissolved oxygen levels fell below 40% saturation. The results of the effect on dissolved oxygen at 30 m for the summer of 2003/4 are shown in Fig. 6.4. Lake aeration with 1000 cfm of compressed air restored the lake dissolved oxygen levels to above 70% within 5 days of continuous operation.

Land use intensification in catchments is also resulting in the occurrence of the aquatic weed *Lagarosiphon major* in Canterbury and Otago lakes. For example, *Lagarosiphon* was discovered in the Ahuriri Arm of Lake Benmore in March 2003 and subsequently treated with Diquat on an emergency resource consent. Annual treatment will be required to control further spread (Landward Management 2006).

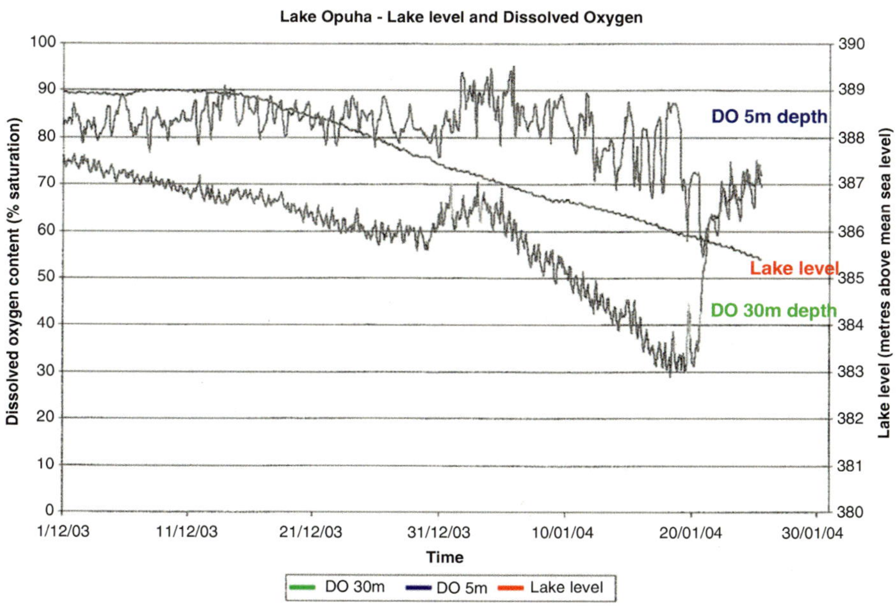

Fig. 6.4 Dissolved oxygen levels in Lake Opuha in Summer 2003/4 (Opuha Dam Limited 2004)

6.1.4 *Alternatives to Major Surface Water Storage*

Not only were there sustainability issues with many storage options there was also a high cost of water from storage. Furthermore, the strategy development work indicated that the cheapest water is water that is currently allocated but used inefficiently. Strategic investigations showed inefficiencies in:

- Irrigation methods
- Application rates and macropore flow
- Reliability of supply
- Irrigation water distribution
- Spatial application of surface water and groundwater.

These inefficiencies are discussed below.

In addition, the propensity on the Canterbury Plains for leakage to groundwater from irrigation indicated the potential for managed aquifer recharge and the use of groundwater storage. This is also discussed below as an alternative to surface water storage and the failure pathways to be addressed.

6.1.4.1 Irrigation Methods

Measurements of irrigation systems showed variable levels of water use efficiency. Table 6.4 sets out a summary of the results. With different irrigation technologies there are different delivered water requirements to meet crop water requirements. Table 6.5 shows the volume differences. The water use of centre pivots is between

Table 6.4 Water use efficiency of different irrigation systems (Environment Canterbury)

Irrigation type	Efficiency range (%)	Average (%)
Laser level (timber sill)	24–93	48
Laser level (grass sill)	37–92	62
Contour border	28–62	44
Travelling irrigator	62–96	85 roto-rainer
		67 gun

Table 6.5 Crop water requirements for different irrigation technologies (Environment Canterbury)

PAW class (mm)	Centre Pivot (mm)	Roto-rainer (mm)	Border dyke (mm)
60	729	1200	2880
90	684	1000	1800
120	663	950	1272

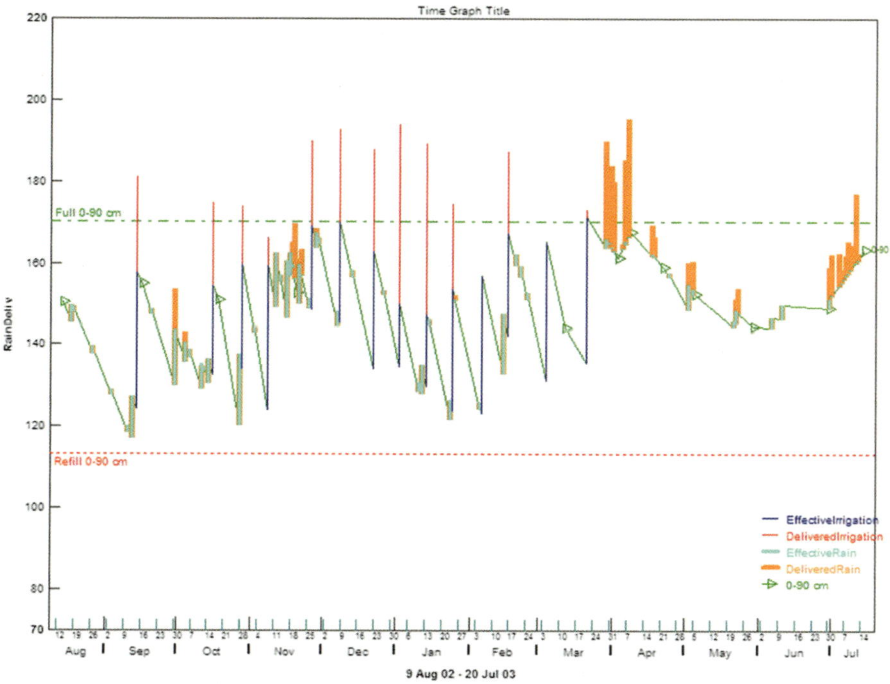

Fig. 6.5 Soil moisture profile for Camden Farm property (Davoren 2008)

half (for soils with PAW[1] 120 mm) and a quarter (for soil with PAW 60 mm) compared to border dyke irrigation. Shifts to more efficient irrigation technologies means agricultural productivity can be achieved with reduced allocations.

6.1.4.2 Application Rates and Macropore Flow

Lysimeter results for irrigation using roto-rainers on Canterbury's gravel soils indicate that with high application rates that not all of the applied water remains in the soil profile. Much of the water passes through the soil (macropore flow). Figure 6.5 shows the soil moisture changes (green line) for a Camden Farm property for the 2002/3 irrigation season. Only a proportion of the applied irrigation (vertical red/blue lines) is effective in increasing soil moisture (blue line). The remainder (red line) is lost by macropore flow through the soil profile. There is a similar pattern with rainfall (yellow/green columns) with the green column showing the rainfall retained in the soil profile and the yellow column showing the macropore flow.

The implications for irrigation efficiency are shown in Fig. 6.6. A summary of five irrigation seasons is shown for the total irrigation (purple columns), the effective

[1] PAW is "Profile Available Water" which is the amount of water potentially available to plant growth that can be stored in the soil to 100 cm depth (Landcare Research 2016).

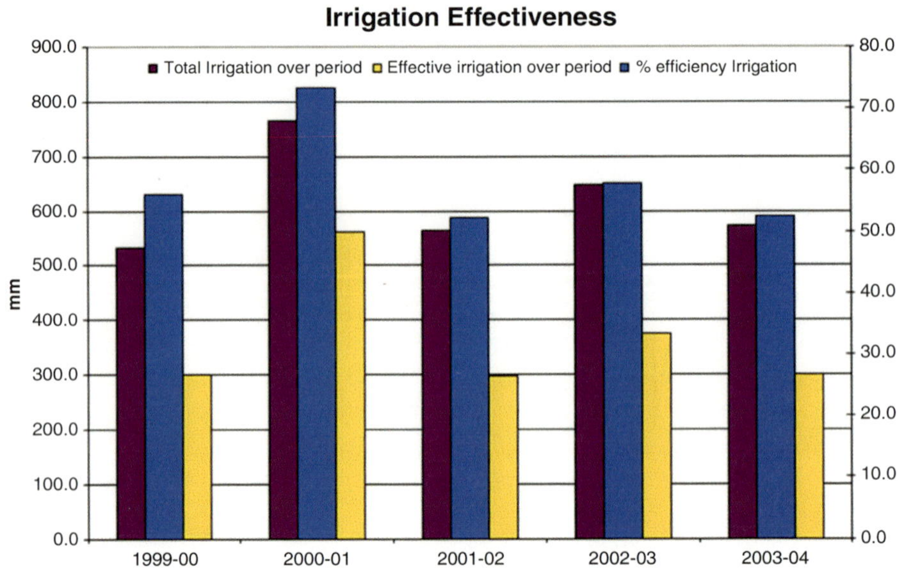

Fig. 6.6 Effect of macropore flow on irrigation effectiveness (Davoren 2008)

irrigation (yellow columns), and the percentage efficiency (blue columns). This shows the effect of macropore flow making application only 52–72% efficient. This can be overcome by using lower application rates.

6.1.4.3 Reliability of Supply and On-Farm Storage

NIWA undertook analyses of a number of farms in the Waimakariri Irrigation Scheme (Duncan et al. 2010). One analysis involved two farms: one with on-farm storage and one without (Fig. 6.7). The analysis compared "ideal" and actual irrigation for the two farms.[2] When the scheme was unable to deliver water because the Waimakariri River was on restriction, the farm without storage was unable to irrigate whereas the farm with storage was able to irrigate when required. The farm with reliable supply was better able to match the "ideal" pattern of irrigation and makes more effective use of irrigation water. For the farm without the pond, soil moisture was below 50% of field capacity for 10 out of 35 weeks of the irrigation season; whereas the farm with the pond was only below 50% of field capacity for 4 of the 35 weeks (MS Srinivasan, personal communication).

For efficient irrigation, irrigators need to have soil moisture measurement, to have access to water at times of soil moisture deficit, to match irrigation application

[2] Ideal irrigation was assumed to be irrigation when soil moisture fell to 50% of PAW and the soil was either filled to 80% of PAW or 31.8 mm/week was applied (whichever was the lower amount) and taking account of rainfall and PET at the sites based on NIWA's virtual climate network.

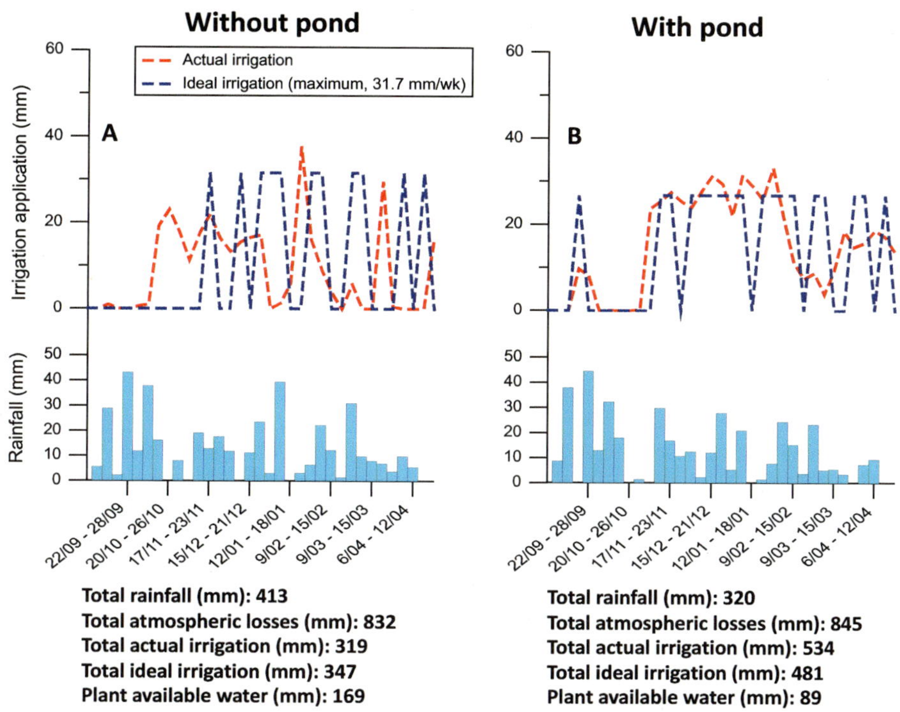

Fig. 6.7 Actual and ideal irrigation for farms with and without storage (Duncan et al. 2010)

to times and degree of soil moisture deficit while leaving capacity for rainfall. The irrigation scheme needs the ability to deliver water to meet soil moisture demand, to ensure reliability of supply, to have a method of distribution to meet farmer demand and, to have scheme or farm storage to cover times when there are run-of-river restrictions.

6.1.4.4 Irrigation Water Distribution

For the Central Plains Water scheme an analysis was made of water distribution comparing a piped network with a traditional open channel design (Ritso Society 2007). The cost comparison included capital costs, annual costs of operation and maintenance, costs of energy for on-farm pumping, value of water saving, and the difference in easement and land footprint needs. For designs providing the same level of service, the capital cost of the pipe network was nearly double the open channel design ($123 m compared to $64 m). However, the higher capital cost was more than offset by lowered operational costs, water and energy savings, as well as smaller land requirement and no land purchase required. The pipe network is cheaper in net present value terms ($102 m compared to $132 m at 8% discount rate) and an estimated water savings of 20%.

Table 6.6 Storage reductions with efficiency and reallocation gains (Dark 2010)

Scenario	Storage	Irrigated area (ha)	Storage (Mm^3)
Storage only	Coleridge	63,500	158
	Coleridge + Stour	146,300	312
Storage + Efficiency	Coleridge	129,000	158
	Coleridge + Stour	146,300	187
Storage + Reallocation	Coleridge	107,400	158
	Coleridge + Stour	146,300	206
Storage + Efficiency + Reallocation	Coleridge	138,000	158

6.1.4.5 Application of Surface Water and Groundwater

One of the strategic investigations looked at whether water availability can be enhanced by only irrigating with surface water in the upper part of a groundwater catchment (Dark 2010). Irrigated land results in greater leakage to groundwater. With current irrigation practice, there is the potential for about a 30% increase in water availability.[3] Reallocation of consents so that surface water is used in the upper part of the catchment and groundwater is used in the lower part of the catchment would mean that the amount of storage requirements to meet future demand can be reduced.

An analysis was made for the area between the Rakaia and Rangitata Rivers. The recharge from the upper zone was balanced with the net use of groundwater in the lower zone. To irrigate the entire potentially irrigable area between the Rakaia and Rangitata Rivers would require sufficient additional water to irrigate 143,600 ha and there would be 103,500 ha irrigated by groundwater.

Without incorporating reallocation or efficiency improvements the storage requirements would be 312 Mm^3 (refer Table 6.6). In CSWS Stage 2, the storage was achieved with increased storage from Lake Coleridge and a new storage on the Stour River. If increased efficiency is incorporated then the storage requirement is 187 Mm^3. If reallocation of consents is undertaken then the storage required is 206 Mm^3. If both efficiency improvements and reallocation are incorporated then the storage required is 158 Mm^3; i.e. about half the storage required without efficiency and reallocation.

If the re-allocation and efficiency gains are achieved then the following can occur on the Canterbury Plains:

- All household, municipal and industrial water requirements can be met from groundwater
- All reasonable irrigation and stockwater requirements can be met either from groundwater or from alpine rivers backed up by storage
- All existing irrigation takes from the Selwyn, Ashburton and Hinds Rivers can cease

[3] Note that with more efficient irrigation the potential increase in water availability would be reduced.

- The area of land consented for irrigation from groundwater on the upper part of the plains can be reduced by about 100,000 ha (with a corresponding reduction in pumping energy requirements).

6.1.4.6 Managed Aquifer Recharge

Managed Aquifer Recharge involves supplementing natural recharge to an aquifer system under controlled conditions by diversion of water into recharge wells or infiltration of water through the floor of infiltration basins, galleries or river beds. The resulting increase in groundwater storage in aquifer systems can then be utilised for consumptive purposes or to enhance environmental values associated with groundwater such as stream baseflow or groundwater dependent ecosystems. The most prospective opportunity in Canterbury would appear to be the use of river water at times of high flow to augment naturally occurring aquifer recharge (SKM 2010).

The capacity for artificial recharge has been demonstrated by the leakage that occurs from irrigated land resulting in groundwater recharge. However, as irrigation efficiency improves the recharge from irrigation declines (refer Sect. 3.3.2 in relation to the Valetta groundwater zone). Furthermore, recharge from irrigation brings with it contaminants from land use, such as nitrates and bacterial contamination. Managed Aquifer Recharge can achieve the replenishment of groundwater levels with water free of land use contaminants.

While there are international examples of successful Managed Aquifer Recharge, there have only been limited trials of aquifer recharge in Canterbury and New Zealand. One is the Eyre River aquifer recharge trial for the Waimakariri Irrigation Scheme. Recharge was through the Eyre River bed when the river was naturally dry. Approximately 2.7 m^3/s was discharged over 24 days. This resulted in a rise of groundwater levels of more than 2.5 m over an area of more than 4000 ha. Groundwater remained elevated for 2–3 months. This indicates that augmentation of surface flow can be an effective method for conveyance and storage of water. However, the main limitation highlighted by the trial was the limited residence time of the water in the aquifer system due to the highly permeable nature of the unconfined aquifer and high rate of groundwater through flow (SKM 2010).

Trials of infiltration basins near West Melton were undertaken in the early 1990s. One trial indicated the formation of a groundwater mound of at least 0.2 m over an area of about 4 km^2 after 3 weeks of recharge at 110 L/s. The second trial at a separate location did not result in measurable rise at depth: this was attributed to the presence of a lower permeability clay-bound gravel strata between the water table and the underlying aquifer. A subsequent test indicated the formation of a groundwater mound in the shallow perched aquifer. There was also evidence of silt accumulation during the trial reducing infiltration capacity.

These trials indicate the potential for managed aquifer recharge but also the need to investigate further the aquifer hydraulics of potential sites, the potential for clogging in surface soils, and the ability to recover the artificial recharge. These factors can influence the location and design of an appropriate managed aquifer recharge system for the Canterbury Plains.

A concept design using 200 m depth injection bores was undertaken as an alternative to the Waianiwaniwa storage for the proposed Central Plains scheme (SKM 2010). An economic assessment indicated a net present value cost of $470 m for managed aquifer recharge compared to surface water storage of $560 m. This is a result of the lower capacity headworks and head race for a Managed Aquifer Recharge scheme compared to a surface water scheme of the same delivery capacity and the reduced need for water distribution infrastructure with reliance on the groundwater system for distribution rather than pipework or canals needed for surface water distribution. Furthermore, it is easier to stage the development of managed aquifer recharge systems to meet demand whereas surface water storage requires a significant capital investment at the initial stage of the project.

6.2 Cumulative Environmental Effects of Use

Cumulative effects are the additive effects of proposals that individually are of minor or moderate significance but in total have an effect of major significance. There are cumulative effects from water abstraction and use in Canterbury with respect to (1) river flows, (2) groundwater levels, and (3) water quality.

6.2.1 River Flows

Water extraction from rivers reduces river flows and groundwater abstraction lowers groundwater tables which reduces the flows in groundwater-fed streams. As mentioned in Sect. 3.1.4, for Canterbury rivers there are three critical variables relating to the ecologically important components of a hydrological regime:

- The magnitude and duration of low flows. These are the main determinants of habitat quantity and influence the connectivity to other habitats.
- The magnitude, frequency and duration of high flows ("freshes"). These flows are sufficient to move fine particles and algal growth and maintain habitat quality and are typically three to six times the median flow.
- The magnitude, frequency and duration of flood flows. These are flows large enough to move the armour layer and erode banks. These maintain the character of braided rivers. They are typically ten times the mean flow (about 40% of the mean annual maximum).

In terms of the cumulative effects of multiple takes from rivers there are conflicting concerns of maintaining flows in rivers to sustain ecosystems and instream uses, and of maintaining reliability of supply to water users.[4] The approach to managing these effects in New Zealand has been to restrict takes at times of low flows and to allow the passage of freshes, and, to limit the total volume that can be abstracted at different flow rates in the river.

Water allocation regimes typically consist of minimum flow specifications and allocation block specifications. Minimum flow is the flow at which abstractions from a water body must cease. An allocation block is the amount of water either set as a flow rate or set as a volume that is set aside for abstraction where all users allocated a proportion of that water will be subject to the same management controls. Small rivers (like Waikuku Stream) may only have one minimum flow and one allocation block, whereas larger rivers (like the Rakaia River) may have a series of bands. The intent is to define combinations of minimum flows and allocation blocks to limit the effect on instream values and meet specified levels of reliability.

Allocations at the lowest flows (referred to as the "A Block") have the greatest reliability but also the greatest impact on the magnitude and duration of low flows in the river and therefore the quantity of freshwater habitat. Specification of minimum flows for the A Block take into consideration the effects on issues such as aquatic ecology, aquatic and riparian vegetation, landscape values and tangata whenua values. Figure 6.8 shows the "weighted usable area" (WUA)[5] curves for relevant aquatic species from the application of Instream Flow Incremental Methodology (IFIM) to the environmental flow review for the Pareora River (Golder

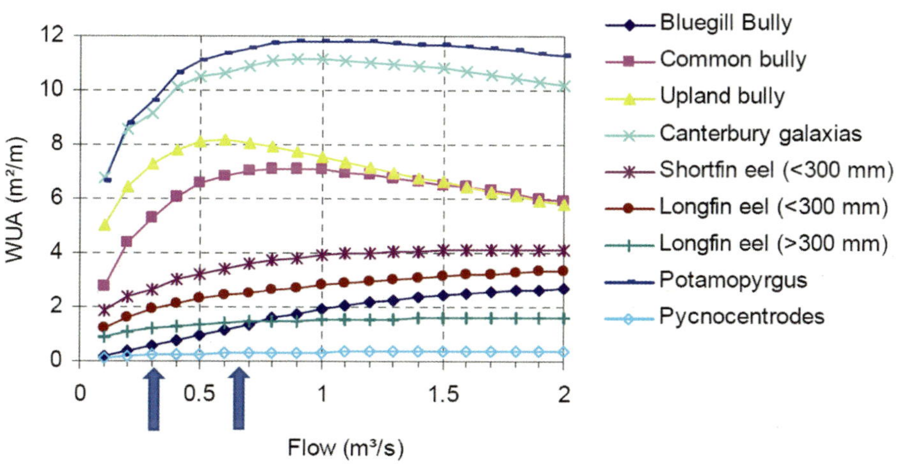

Fig. 6.8 WUA curves for native fish and invertebrates at the Pareora River Huts Site (Golder Associates 2008)

[4] Refer Sect. 5.1.1 for the management of irrigation in relation to flow restrictions for extraction from the Waimakariri River and the implications for reliability of supply for irrigators in the Waimakariri Irrigation Scheme.

[5] Weighted useable area is an index of habitat availability based on the wetted area of a stream weighted by its suitability for use by aquatic organisms.

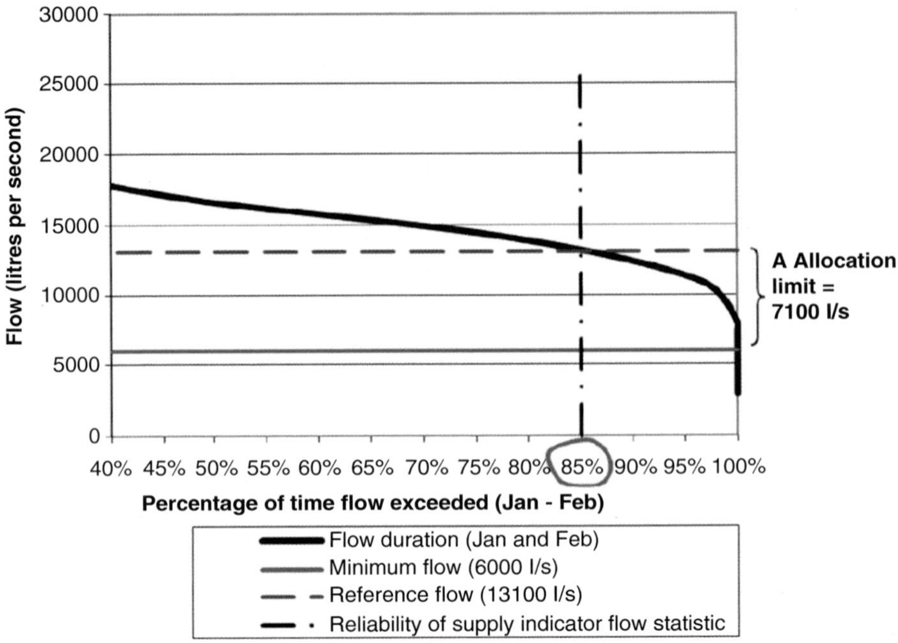

Fig. 6.9 Determination of A block allocation limits (Environment Canterbury)

Associates 2008). The objective is to set the minimum flow at the point where the WUA curve begins to flatten at 90% of the maximum WUA. The arrow on the left indicates the minimum flow of 0.3 m^3/s that had previously been set for the Pareora River, while the arrow on the right shows the recommended minimum flow from the review of 0.66 m^3/s based on aquatic ecology considerations.

The size of the A allocation block is based on the specified reliability of supply. A typical metric is 85% availability of supply in the months of January and February (the time of peak demand for irrigation in Canterbury). The reference flow for this availability is determined from the flow duration curve for January and February. In Fig. 6.9, this is 13,100 litres per second (L/s) – the flow exceeded 85% of the time in January and February. The allocation limit is the reference flow minus the minimum flow. In Fig. 6.9, with the minimum flow of 6000 L/s, the A Block allocation limit is 7100 L/s. To manage for cumulative effects relating to freshwater ecology and reliability of supply, the sum of the allocations is limited to 7100 L/s and extractions are restricted when the river flow drops to 6000 L/s or less (Environment Canterbury 2011).

Further allocations above the reference flow need to accommodate the maintenance of freshes (to maintain habitat quality) and other instream uses. In Sect. 5.1.1 the example of setting the B block for the Waimakariri River was described with allowance made for river bird nesting, salmon angling, fish passage, kayaks and jet boats as well as flushing flows for sediment and periphyton. B Blocks have lower reliability of supply and require off-river storage to be effective for irrigation.

With the principle of first-come, first-served in relation to water allocation and the application of the common law principle of non-derogation of the permissions of existing consent holders, banding of consents has been introduced to protect the reliability of supply of existing consent holders for the Rakaia River. Early applicants were placed in a band with minimum flow conditions that had been set for the river. Subsequent consents were grouped in a new band with higher minimum flow conditions. Seven bands have been established for the Rakaia River (Dysart et al. 2008).

Figure 6.10 illustrates the banding system and water allocation approach for the Rakaia River. The minimum flow was set at 124 m^3/s. Above that half the flow is available for extraction (referred to as “one to one flow sharing”) thereby making an allowance for retaining freshes. The allocation limit is 70 m^3/s which is available for flows above 264 m^3/s. At flows just above 124 m^3/s consent holders in Band 1 can abstract. As the river flow increases bands with higher minimum flows can abstract.

There is also another characteristic of many Canterbury rivers that is affected by the cumulative effects of water abstraction. Coastal lagoons occur at river mouths. One form that is rare internationally but common for Canterbury are referred to as “hapua” which are coast-parallel bodies of predominantly freshwater impounded by a long narrow spit of coarse sediments formed by longshore drift resulting in the river outlet being offset from the main channel alignment (Kirk and Lauder 2000). Figure 6.11 shows the hapua at the mouth of the Ashburton River.

Hapuas are a result of the interaction between fluvial and coastal processes. River flow delivers sand and gravels to the coast. Longshore drift creates a spit across the river mouth. At moderate flows the river outlet is offset forming a lagoon on the landward side of the spit. At low flows this can lead to closure of the river

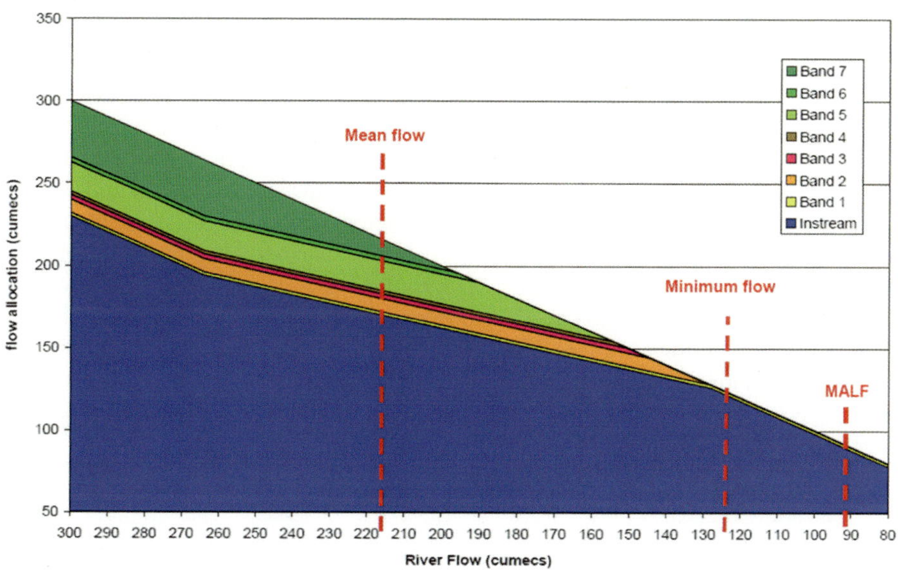

Fig. 6.10 Rakaia River water allocation and banding system (Dysart et al. 2008)

Fig. 6.11 Hapua at Mouth of Ashburton River (Hart 2011)

mouth. At high flows the spit is breached and the river flows directly out to sea. The river mouth cycles through stages of offset outlet, closure and breaching depending on river flow, coastal conditions and antecedent barrier conditions (Hart 2009).

Rivers with large flows like the Rakaia (mean flow 221 m^3/s) rarely close[6] while rivers with small flows like the Pareora (mean flow 3.7 m^3/s) are frequently closed. One of the cumulative effects of extraction of water from rivers with hapua-type river mouths is the reduction in flows to maintain the river outlet (increasing closure frequency) and reduced frequency of flows sufficient to breach the barrier spit to re-establish river flow to the sea (increasing closure duration) (Horrell 2011).

Increased frequency and duration of closure affects fish passage and water quality (through increased retention times). In rivers like the Ashburton, minimum flows sufficient to meet instream fish habitat requirements can be less than the flow to maintain river outlets to the sea. Recreational angling data was collated for the October 2007/March 2008 fishing season showing periods of low flow leading to river closure and the absence of salmon catch (Webb 2009). The minimum flow at that time (3.5 m^3/s) met instream fish habitat requirements but as indicated by studies (Todd 1992) that 6 m^3/s is needed to maintain an open mouth while mouth closures occur when flows are less than 4 m^3/s.

The cumulative effects of abstraction from rivers on river mouth closure needs greater consideration in setting and managing environmental flow regimes for Canterbury rivers.

[6] Two short term closures (several days) have been recently recorded for the Rakaia in March 2010 and May 2012 under conditions of relatively low river flow (less than 104 m^3/s), neap tidal cycle, and following storms with significant wave height greater than 4 m (Ball 2012).

6.2.2 *Groundwater*

With individual groundwater bores the main concern is with the interference effect of groundwater extraction on neighbouring bores. In Canterbury with groundwater-fed streams there can also be concerns with reduced flow in lowland streams. With permeable soils, groundwater abstraction can draw water from hydraulically connected surface streams. Near the coast there can also be a concern with sea water intrusion.

The management of interference effects for individual bores can be achieved by constraints on bore location and rate of abstraction. The objective in the regional plan is that there is no significant adverse effect in conjunction with other bores on neighbouring bores that adequately penetrate the aquifer (Schedule 12). The policy is that interference effects are no more than 20% of available drawdown. In managing natural variability in groundwater levels for supply reliability, the 80% available drawdown needs to be a groundwater level exceeded 80% of the time during the period of proposed use. The issue of cumulative effects brings in issues at a larger scale, in particular, what is the total volume that can be abstracted from a groundwater zone and still maintain adequate flows in lowland streams and not deplete water availability. There are also cumulative effects in the interaction between surface water and groundwater, and between sea water and groundwater.

This subsection considers (1) the use of groundwater zone allocation limits to address cumulative effects in groundwater zones, (2) accounting for the interactions between surface water and groundwater use, and (3) the management of sea water intrusion.

6.2.2.1 Management of Groundwater Zones

In order to manage the cumulative effects of groundwater abstraction groundwater zones were identified and zone allocation limits were set with the primary aim of maintaining flows in groundwater-fed streams. A key issue for setting groundwater allocation limits is the availability of information to assess the availability of groundwater and the effects of groundwater abstraction. Depending upon the data available three approaches were taken to assess groundwater allocation limits:

- a first order approach of 50% of rainfall recharge: this required estimating rainfall, the recharge from that rainfall, and then setting the allocation limit as half of that recharge;
- a second order approach of 50% of land surface recharge: this required identifying soil zones, developing a soil moisture model to estimate recharge and making half of that amount available for allocation;
- a third order approach based on the effects on groundwater-fed streamflow: this required a groundwater model to predict lowland streamflow response to abstraction (Veltman et al. 2004).

The policy was established to restrict takes from groundwater when the effective allocation of groundwater exceeded the groundwater zone allocation limit. First and second order assessments were considered to be "interim allocation limits" as they represented estimates of resource availability which were to be refined as further information was obtained.

The effective allocation of groundwater was based on the consented volumes of groundwater users. Box 6.1 sets out the example of the Chertsey Groundwater Zone. As noted in Sect. 3.1.5, most groundwater zones in Canterbury have been declared "red zones" where the effective allocation of groundwater exceeds the groundwater zone allocation limit (see Fig. 3.4).

With the introduction of cumulative effects management there has been a need to retrofit additional conditions to existing consents granted when the main concern was the drawdown at neighbouring wells with only limits on the rate of pumping. Additional conditions were for annual volume limits on all consents in fully allocated zones, metering of use, and restrictions on wells with hydraulic connection to lowland streams.

Box 6.1: Allocation Status for the Chertsey Groundwater Zone

The allocation status for the Chertsey Groundwater Zone is based on a second order approach (Scott 2004). This involves estimating the groundwater zone allocation limit based on land surface recharge and then comparing it with the effective allocation associated with the different types of consents. Table 6.7 sets out the estimated land surface recharge based on three components: (1) rainfall recharge, (2) groundwater irrigation, and (3) surface water irrigation.

Land surface recharge is calculated for dryland and irrigation conditions using rainfall and evapotranspiration data for a 30-year climate record and soil properties relating to water holding capacity. Irrigated areas are based on the consent data base. A simple soil moisture model was used to estimate drainage under dryland and irrigated conditions.

The effective allocation of groundwater was calculated based on consented volumes in the following way: (1) public water supply and commercial/industrial users: 100% of consented volume over 365 days; (2) irrigation users on a daily rate were assumed to use 60% of their consented rate over a 150-day irrigation season; (3) irrigation consents with an annual or seasonal limit were assumed to use 90% of their consent limit; and (4) irrigation consents based on the seasonal irrigation demand standard (Schedule WQN9 of the regional plan) were assumed to use 85% of their annual volume limit. Account was also taken of metering data and stream depletion effects.

Effective allocation is estimated to be 122 million m^3/year (as at 26 September 2011) (Environment Canterbury 2016). This is 109% of the groundwater allocation zone limit.

Table 6.7 Groundwater allocation limit for the Chertsey Zone

Zone Area	Mean annual rainfall	Recharge (dryland)	Recharge volume
68,570 ha	776mm	34.3%	200.7 m^3 x 10^6
Groundwater irrigation	Drainage		
13,871 ha	168mm		23.3 m^3 x 10^6
Surface water irrigation	Drainage		
1,097 ha	73mm		0.8 m^3 x 10^6
		TOTAL RECHARGE	224.8 m^3 x 10^6
		ALLOCATION LIMIT (50% land surface recharge)	112.4 m^3 x 10^6

In relation to applications for new consents in groundwater zones where the interim allocation limit is already fully allocated, the policy intent was that applications be declined unless there is information available indicating further taking of groundwater would not result in adverse environmental effects. Such was the demand for water that many further applications were received. With the permissive nature of the RMA, independent hearing commissioners accepted information from applicants against the advice of the regional council and granted more consents leading to over-allocation of the groundwater resource in many zones.

The approach to calculating groundwater zone allocation limits is based on historical data and average conditions. There is considerable inter-annual variability in rainfall recharge which is reflected in variability in groundwater levels and lowland stream flows. It is desirable in managing for cumulative effects in avoiding adverse effects in low rainfall/low groundwater level/low lowland streamflow years. Also, there is the potential to access water in high rainfall/high groundwater level/high lowland streamflow years.

In consent applications in the Rakaia Selwyn groundwater zone after it had reached 'red zone' status, adaptive management approaches were considered to integrate climate variability and abstraction. One approach was based on groundwater trigger levels (Davoren 2006) and another based on antecedent recharge (Williams et al. 2008).

The trigger level method was based on the groundwater level at the commencement of the irrigation season (September 1). A trigger level was defined on the basis that the combination of recession due to natural discharge (the "seasonal environmental discharge"), the recession due to abstraction drawdown of existing consents (the "seasonal abstractive discharge"), and the recession from the drawdown of the new consent (calculated as 150 day drawdown in a virtual bore 750 m from the applicant bore) would not result in the groundwater level going below the "environmental flow safeguard" level at the end of the irrigation season (April 30).

The environmental flow safeguard level was based on the groundwater level of April 30, 2006. The recession due to natural discharge and existing users was based on the change in groundwater level in the 2005/6 irrigation season (the "total groundwater recession"). The groundwater level on September 1, 2005 is defined as the "reserved water level". For water to be available for the new consent in the current irrigation season, the groundwater level on September 1 needs to exceed the reserved water level (i.e. the level at September 1, 2005). The excess is compared to the calculated drawdown associated with the new consent. If the excess exceeds the calculated drawdown then the full allocation on the new consent is available for use for that irrigation season. If there is no excess then no water can be used. If the excess is a proportion of the drawdown then that proportion of the consented volume can be used. Box 6.2 provides an example of the seasonal assessment for the 2015/6 irrigation for an individual consent.

The antecedent recharge proposal is based on the premise that in periods of below average recharge it is not possible in a fully allocated zone to both provide the consented volumes and maintain flows in groundwater-fed lowland streams. The concept is to restrict entitlements based on the recharge that has occurred in previous years. Groundwater consent holders would receive a "base entitlement" as a fixed percentage of their consented annual allocation, and an "adaptive entitlement" as a variable amount based on recent recharge history (Williams et al. 2008).

Figure 6.12 shows a simulation of how antecedent recharge allocation would be applied for the rainfall recharge record from 1960 to 2008. In this example the average annual recharge for the groundwater zone is 400 million m^3, with 50% available for allocation resulting in an annual allocation limit of 200 million m^3. An exponentially weighted moving average (EWMA)[7] of rainfall recharge is used as the critical variable. When the EWMA is above the long-term average recharge, the full adaptive component is available. When the EWMA is below half of the long-term aver-

Box 6.2: Application of Trigger Levels to New Consents

1. Reference parameters based on 2005/6 irrigation season:

The reserved water level (spring high water level on 1 Sep 2005) is 2.18 m BGL and the environmental safeguard level (autumn low water level on 30 Apr 2006) is 1.678 m BGL. The difference between these levels, 0.502 m, is the total groundwater recession (i.e. the sum of the seasonal abstractive discharge and the seasonal environmental discharge).

The drawdown of the new consent, calculated as the drawdown at a virtual bore 750 m from new consented bore resulting from abstracting the consented volume (66,780 m^3) 150 days, is 0.050 m.

2. Seasonal assessment for the 2015/6 irrigation season

The trigger level (i.e. the groundwater level at 1 Sep 2015) is 2.19 m BGL. This is 0.01 m in excess of the reserved water level. Thus, the allowable drawdown from the new consent is 0.01 m. This is 19% of the calculated drawdown for full consented volume. Thus, the water allocation for the 2015/6 irrigation season is $0.19 \times 66{,}780 = 12{,}689$ m^3.

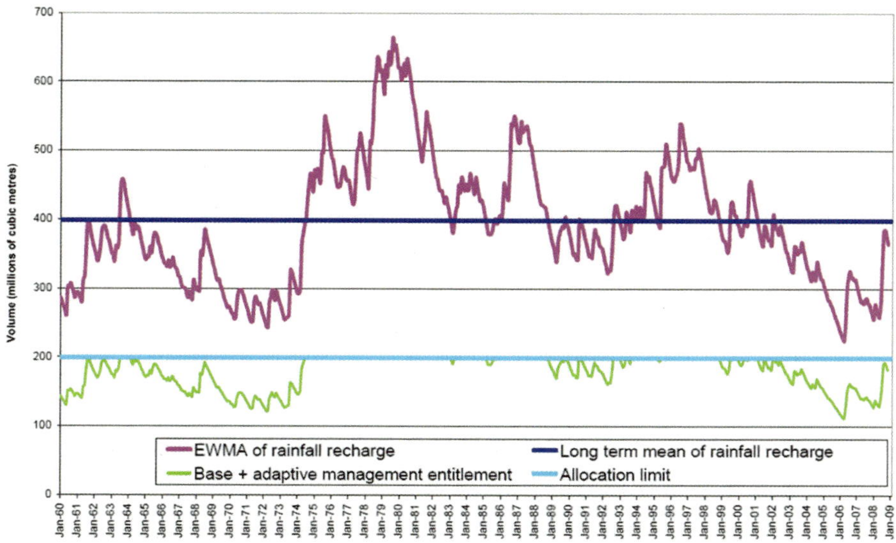

Fig. 6.12 Simulation of antecedent recharge allocation for Rakaia-Selwyn groundwater zone 1960–2008 (Williams et al. 2008)

age recharge, no adaptive component is available. When the EWMA is between the long-term average and half the long-term average, a proportion of the adaptive component is available.

While the trigger level method was adopted by hearing commissioners for those new consents in the Rakaia-Selwyn Zone (Environment Canterbury 2007), the antecedent recharge method has a number of advantages from a sustainable management perspective. The first is that it is based on data that is earlier in the impact pathway for potential adverse effects. The second is that it is independent of interference effects of other bores. The third is that it is well suited to addressing the implications of climate change (refer Sect. 7.1.6). Finally, it is a predictive approach rather than a reactive approach.

6.2.2.2 Groundwater: Surface Water Interaction

The two main forms of interaction between groundwater and surface water are: (1) streams gain water from groundwater through the stream bed when the groundwater table is higher than the stream bed, and, (2) streams lose water to groundwater by seepage when the groundwater table is lower than the stream bed. With the high permeability of the sediments of the Canterbury Plains streams can have gaining reaches (inflow with groundwater) and losing reaches (outflow to groundwater). Section 5.1.2 discusses the example of the Waimakariri River where, upstream of Crossbank, seepage is the main source of aquifer recharge to the Christchurch-West

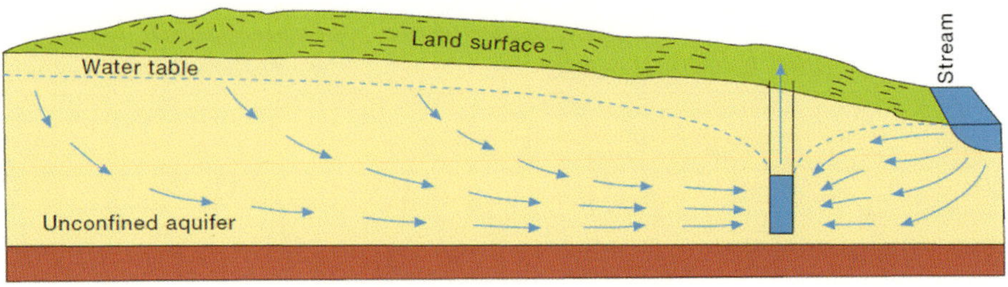

Fig. 6.13 Stream Depletion Effect (Pattle Delamore Partners 2000)

Melton Aquifer – the source of Christchurch's drinking water. There is also return flow from the aquifer to the river downstream of Crossbank (refer Fig. 5.6).

One of the most potentially significant forms of groundwater-surface water interaction is the effect of groundwater pumping on surface waterways. This is commonly referred to as the "stream depletion effect" (Fig. 6.13). The critical variables in relation to stream depletion are: (1) the pumping rate from the well, (2) the separation distance between the well and the stream (where separation distances are greater than 2000 m stream depletion effects are unlikely to be significant), (3) the length of time groundwater is pumped, (4) the transmissivity of the aquifer (a measure of aquifer permeability where values less than 10 m^2/day are unlikely to be sufficiently permeable to cause significant depletion effects),[8] (5) the storage coefficient of the aquifer (an indication of how much water is stored in the strata with values between 0.005 and 0.3 associated with stream depletion, and (6) the streambed conductance (which is related to the vertical hydraulic conductivity of strata in the streambed and the width of the streambed and inversely related to the thickness of the streambed, with values in the range 0.1–5000 m/day associated with stream depletion. From these variables the proportion of water drawn from the stream can be calculated (Pattle Delamore Partners 2000).

To address stream depletion effects the regional plan has provisions to incorporate groundwater withdrawals that are hydraulically connected to streams into the allocation limits for streams. Based on modelling of 7 days and 150 days continuous groundwater abstraction, the degree of stream depletion effect is defined as either (1) "direct" where 7 days pumping has a 90% or more stream depletion effect, (2) "high" where 150 days pumping has a 60% or more stream depletion effect, (3) "moderate" where 150 days pumping has a 40–60% stream depletion effect or exceeds 5 L/s, or, (4) "low" where 150 days has less than 40% and less than 5 L/s. The amount of the groundwater use included in the surface water allocation limit is on a sliding scale: 100% for direct, 75% for high, 50% for moderate, and none for low. The remainder is included in the groundwater allocation limit: 0% for direct, 25% for high, 50% for moderate and, 100% for low. Direct and high hydraulic connections are also included in surface water restrictions while moderate and low hydraulic connections are not.

[8] Canterbury Plains groundwater transmissivities are in excess of 1000 m^2/day (Scott and Thorpe 1986).

While this approach deals with individual wells it doesn't address some of the catchment scale cumulative effects on streams evident in Canterbury from groundwater abstraction. One example is the Selwyn catchment. The headwaters of the Selwyn River are in the foothills of the Southern Alps. The river crosses the alluvial Central Plains and discharges into Te Waihora/Lake Ellesmere. In the upper reaches the river loses surface water to groundwater, while in the lower reaches when the groundwater table intersects the elevation of the river bed, the river gains flow from groundwater. The middle reach is ephemeral (Larnad et al. 2007). A longitudinal section of the river is shown in Fig. 6.14.

In relation to water extraction for irrigation water demand has increased with the conversion of dryland farms to irrigation, mainly for maintaining pasture for dairying. The principal source of water is groundwater. Groundwater withdrawal lowers the groundwater table which increases the ephemeral reach of the Selwyn River and reduces the connectivity between the upper and lower reaches of the river. Figure 6.15 shows the change in flow permanence from Whitecliffs in the headwaters of the catchment to the outlet into Te Waihora/Lake Ellesmere 55 km downstream for no abstraction (0% use), current abstraction, and full use of consented abstractions (100% use).

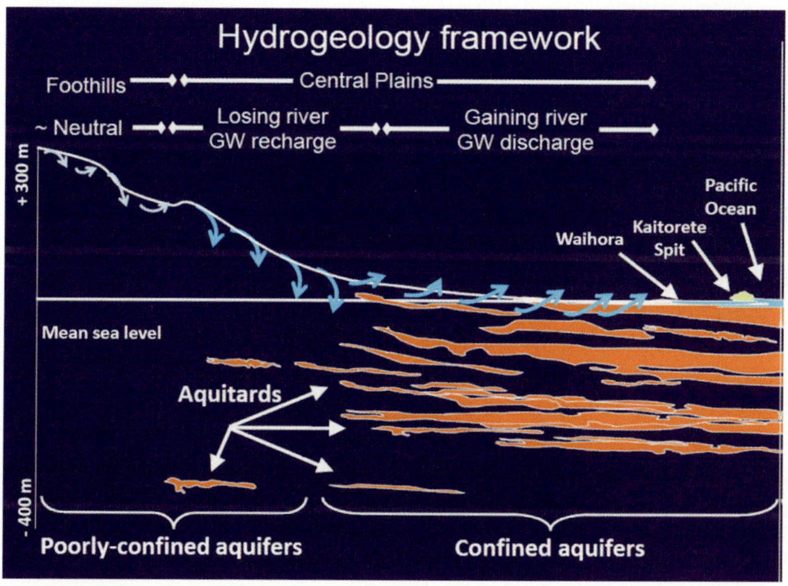

Fig. 6.14 Longitudinal section of gaining and losing reaches of the Selwyn River (Larnard 2013) Note: The river profile is from the foothills across the Central Plains to the coastal lake (Waihora) where there is a barrier bar (Kaitorete Spit) which is artificially opened to allow discharge to the Pacific Ocean. Across the Central Plains there is initially a losing reach with recharge to groundwater (shown by the downward arrows). At lower elevations the river changes to a gaining reach with groundwater discharging into the river (shown by upward arrows). The groundwater system is unconfined in the upper reaches of the Central Plains. Bands of confining layers are present closer to the coast.

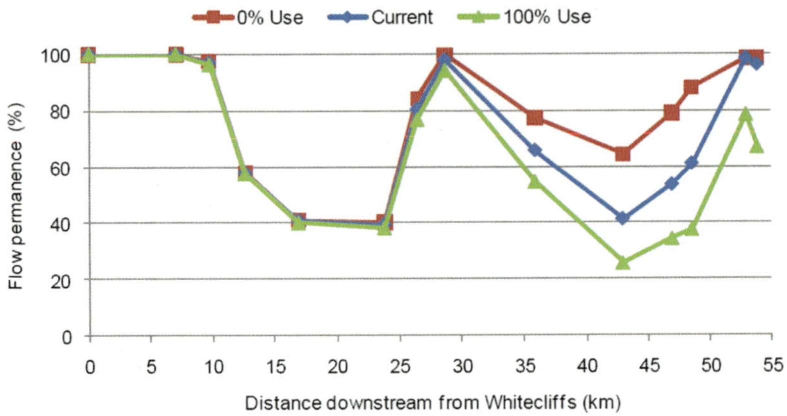

Fig. 6.15 Effect of groundwater abstraction on flow permanence in the Selwyn River (Burrell 2011)

The flow permanence begins to decline 10 km downstream of Whitecliffs as the river reaches the Canterbury Plains and river flow seeps into groundwater. It rises again as the tributary, Hororata River, joins the Selwyn River. For this upstream segment of the Selwyn River flow permanence is unaffected by groundwater abstraction. Downstream of the confluence of the Hororata River, the Selwyn River continues to lose flow until the groundwater table reaches the stream bed and the river gains flow from groundwater. With increasing groundwater abstractions and lowering of the groundwater table the flow permanence is estimated to drop from 62% to 22% about 43 km downstream of Whitecliffs.

This reduced connectivity has implications for fish ecology. Investigations indicate there appears to be a threshold when the flow permanence is below 60% that few fish species remain and fish are unable to recolonise (Larnard 2013). Reduced flows also affect invertebrate diversity (Burrell 2011).

There is a need to manage the cumulative effects of groundwater-surface water interactions at the catchment scale. A sustainability analysis of the Selwyn catchment (Jenkins 2017) highlighted flow permanence as a critical variable to be managed and identified possible management interventions, such as reducing groundwater abstraction, substituting surface water (external to the catchment) for groundwater for irrigation, or, managed aquifer recharge.[9] Interventions also need to consider implications for other issues, such as water quality.[10]

There are subtleties in the management of the groundwater-surface water interface. The Pareora case in Box 6.3 is an example where lowering the river bed level through gravel extraction appears to have caused the lowering of the groundwater

[9] Refer Sect. 1.4 for further discussion of the spatial application of surface water and groundwater, and, managed aquifer recharge.

[10] Sustainability decision making is discussed further in Chap. 12 and Selwyn catchment issues are considered in Box 12.1.

Box 6.3: Decline in the Pareora Groundwater Table
Long-term groundwater levels in several shallow wells within the Quaternary alluvium in the Pareora Valley have declined by more than 0.5 m over the period 1969–2014. There is no evidence of land surface recharge or river flow measurements that explain the decline. Nor is there evidence in groundwater discharge data (e.g. spring flows or well abstraction) that can provide an explanation. However, evidence from river cross section surveys indicate that between 1967 and 2013 the mean elevation of the river bed has been lowered by 0.56 m due to gravel extraction. In comparison, the average groundwater decline was 0.69 m. The potential for change in groundwater level due to change in river bed level was tested by numerical modelling with other parameters unchanged. There was a modelled effect on adjacent groundwater level consistent with the observed effect (Aitchison-Earl and Alkhair 2014).

table. This illustrates the need to treat water management as the management of a complex system.

Seawater Intrusion
There is another important groundwater interface which is between freshwater aquifers and saline groundwater due to the occurrence of seawater in areas close to the coast. Sea water can move inland and contaminate freshwater aquifers due to either a landward movement of sea water, or a lowering of the hydraulic head in freshwater aquifers due to the cumulative effect of groundwater abstraction (Pattle Delamore Partners 2011).

Two failure pathways for seawater contamination of freshwater aquifers have been identified for Canterbury: (1) lateral intrusion of sea water into an aquifer adjacent the sea; and (2) downwards seepage from a saline surface waterway that overlies an aquifer (Scott and Wilson 2012). Lateral intrusion affects a coastal bore in an unconfined aquifer in Makikihi, South Canterbury. Saltwater contamination has been observed over the past 30–40 years in a confined aquifer in the Woolston-Heathcote area of Christchurch.

Christchurch groundwater is generally considered relatively low risk for lateral intrusion. The upper aquifer is 2–4 m above sea level and the freshwater/saltwater interface is estimated to be at least 3 km offshore. However, in the Woolston/Heathcote area there were substantial industry and public water supply withdrawals for groundwater. In this area sea water, in the form of the Avon-Heathcote Ihutai Estuary, overlies a confined freshwater aquifer. Due to a localised pressure reversal caused by groundwater abstraction, the Woolston/Heathcote area has seen groundwater levels drop below mean sea level allowing the downward migration of saline water from the estuary into the freshwater aquifer (see Fig. 6.16) (Hertel 1998; Scott and Wilson 2012). Thus, the failure pathway is downwards seepage of sea water rather than lateral intrusion of sea water into the aquifer.

Chloride concentrations in the most contaminated well (M36/1159) increased from approximately 170 mg/L in 1979 to 1600 mg/L in 1994 exceeding the drinking

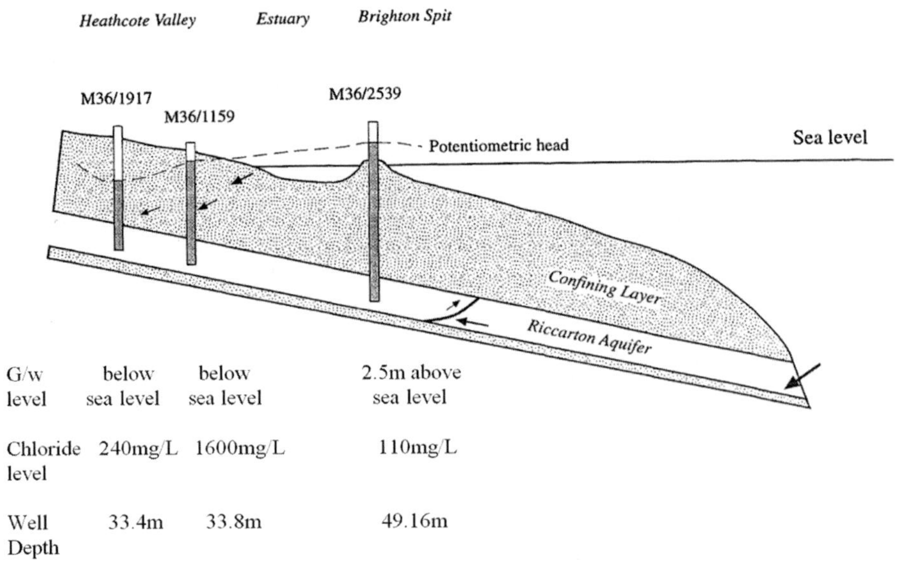

Fig. 6.16 Cross section of downward seepage of sea water into Woolston/Heathcote Groundwater (Hertel 1998)

water standard of 250 mg/L. The background chloride level elsewhere in the aquifer is around 5 mg/L (Hertel et al. 1998).

To address these effects a Users Group of all groundwater abstractors in the Woolston/Heathcote area was formed in 1998 to reduce water usage and increase water efficiency. For the critical variable of groundwater level drawdown by the cumulative effect of multiple abstractions a number of thresholds for different time scales were identified. One time scale is the tidal cycle and the thresholds for groundwater levels associated with low tide, mean tide and high tide. A second time scale is the groundwater usage pattern over a weekly cycle. Groundwater use was greatest during the working week with declining levels from Monday to Friday and some recovery on weekends (Hertel et al. 1998).

The most damaging groundwater drawdown threshold is for groundwater levels below the low tide mark (−0.5 mean sea level datum (MSL)). This is because for groundwater levels below the low tide mark the direction of saltwater flow is downward over a period longer than a tidal cycle. Allowing drawdown below low tide should be avoided. In 1998, the monitoring bore was below the low tide mark for 17% of the time. A second threshold is that groundwater is that groundwater drawdown should stay "on average" above mean estuary level (+0.25 MSL). In 1998 the monitoring bore was below mean estuary level for 56% of the time. The low point on weekly drawdown cycle should be above mean estuary level. Long term protection needs the groundwater drawdown level to be above high tide (+1.0 MSL) (Hertel et al. 1998).

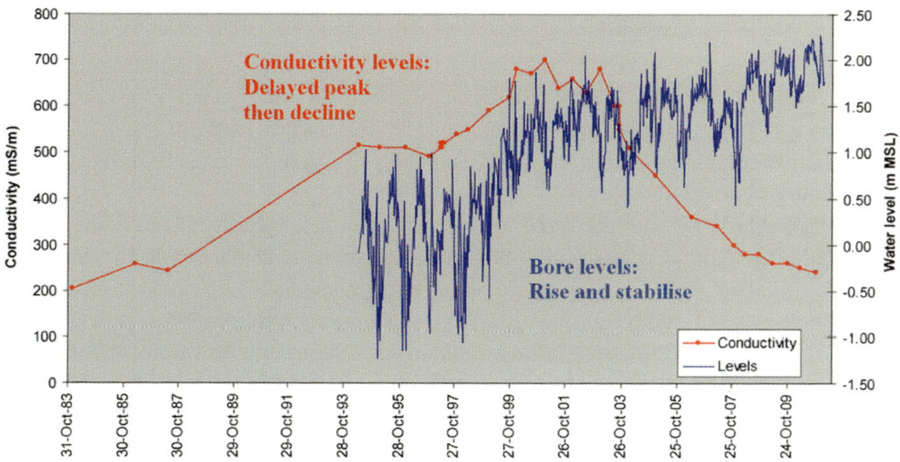

Fig. 6.17 Recovery of bore levels and conductivity in Woolston/Heathcote Groundwater (Scott and Wilson 2012)

In the groundwater users group, three trigger levels were agreed for the monitoring bore between the area of abstraction and the estuary for the key time scales:

- 1.0 m above datum (MSL) as the average over a 365-day period;
- 0.25 m above datum (MSL) as a 14-day moving mean (based on at least hourly readings); and
- 0.5 m below datum (MSL) over a 24-h period (based on at least hourly readings).

These trigger levels were incorporated in the regional plan (Environment Canterbury 2004) and reflected in resource consents of the groundwater users. If one trigger level was reached then the allowable consented take was reduced by a third. If two trigger levels were reached then the allowable take was reduced by two thirds. If all three trigger levels were reached then the take must cease. There is also a provision for a water sharing agreement among all users to provide flexibility among multiple users. The restrictions then apply to the combined takes of all users rather than to each user.

The outcomes of the management of the cumulative effects of groundwater abstraction are shown in Fig. 6.17. The groundwater levels have remained above sea level since about 1999 but it was not until 2003 that conductivity started to reduce (Scott and Wilson 2012).

6.2.3 Water Quality

The cumulative effects of land use intensification on water quality of freshwater systems is related to the nature and degree of links between land use and hydrologic processes (i.e. connectedness)[11] and the capacity of the freshwater system to absorb disturbances (i.e. resilience).[12]

It is useful to consider the three main river types in Canterbury (refer Sect. 3.1.2 and Fig. 3.1). Alpine rivers have their upper catchment in the Southern Alps and have relatively narrow catchments across the Canterbury Plains which have been intensively developed for agriculture. Alpine rivers are predominantly snow-fed and have high flows. Foothill rivers have their upper catchments in the foothills of the Southern Alps. They are predominantly fed by rainfall runoff and have moderate flow. Lowland streams are mainly groundwater-fed with small flows in the lower part of groundwater basins on the coastal margins of the Canterbury Plains.

It is also useful to distinguish between coastal lakes at the downstream end of some river catchments, and, high country lakes in the upper reaches of alpine rive catchments. Furthermore, among coastal lakes there are two distinctly different kinds. One kind is the hapua discussed in Sect. 6.2.1 and illustrated in Fig. 6.11. These are coast-parallel bodies of predominantly freshwater impounded by a long narrow spit formed of coarse sediments by longshore drift with the river mouth offset from the main river channel alignment. The second kind is the shallow lakes that develop landward of barrier beaches with intermittent connection to the sea, referred to as "Waituna-type lagoons" after the example of Waituna Lagoon in Southland (Kirk and Lauder 2000).

Groundwater is mainly unconfined in the sediments of the Canterbury Plains and directly exposed to leakage from surface land use. On the coastal margins, there are confining layers of marine sediments from sea incursions in the recent geological past which limit leakage from surface land use.

In addition, the consideration of the potential for water quality impairment (the potential for resources)[13] is related to the intensification of land use. Table 6.8 shows the major historical land use changes in Canterbury.

The change in land use over the last 20 years was shown in Fig. 3.2. It shows the significant increase in dairying during that time but also the continuing dominance of dryland pastoral farming in the Canterbury region.

---

[11] Connectedness is one of the key properties of adaptive cycles (refer Sect. 4.2.1).

[12] Resilience is a second key property of adaptive cycles (refer Sect. 4.2.1).

[13] The potential for resources is a third key property of adaptive cycles (refer Sect. 4.2.1). Note that "resources" in this instance are in the negative sense of potential sources of water quality contaminants.

Table 6.8 Major land use changes in Canterbury

Land use change	Time period (years ago)	Land use	Vegetation change
Māori arrival	500–800	Hunting, forest fires	Some deforestation from burning
Early pastoral	100–170	Sheep grazing, fires, clearing	Modified to grass/shrubland
Cropping and pastoral	20–100	Sheep/beef, grains, clearing	Soil conservation, shelter belts
Intensive irrigation	0–20	Dairy conversions	Irrigated pasture

Source: Wilmhurst (2007) and Dynes et al. (2010)

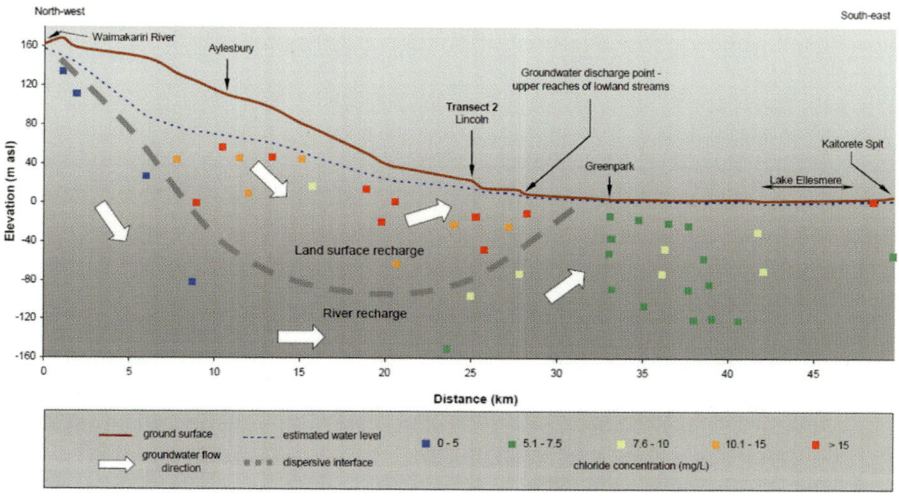

Fig. 6.18 Interpretative cross section of groundwater for the Central Canterbury Plains (Hanson and Abraham 2009)

6.2.3.1 Water Quality in the Groundwater System

A detailed investigation of the depth and spatial variation in groundwater chemistry of the Central Canterbury Plains has been undertaken (Hanson and Abraham 2009). This investigation concluded that groundwater came from two different recharge sources – seepage from alpine rivers and soil drainage from the land surface. Groundwater that is recharged from the alpine rivers flows into the deeper parts of the aquifer system then re-emerges near the coast. Groundwater recharged from the land surface remains in the shallower parts of the aquifer system and discharges in the upper reaches of the lowland streams. An interpretative cross section from the base of the Southern Alps to the coast based on chloride concentrations (Fig. 6.18) shows the two sources of recharge and their pattern of movement through the aquifer. The depth of the interface between the two recharge sources varies. Groundwater

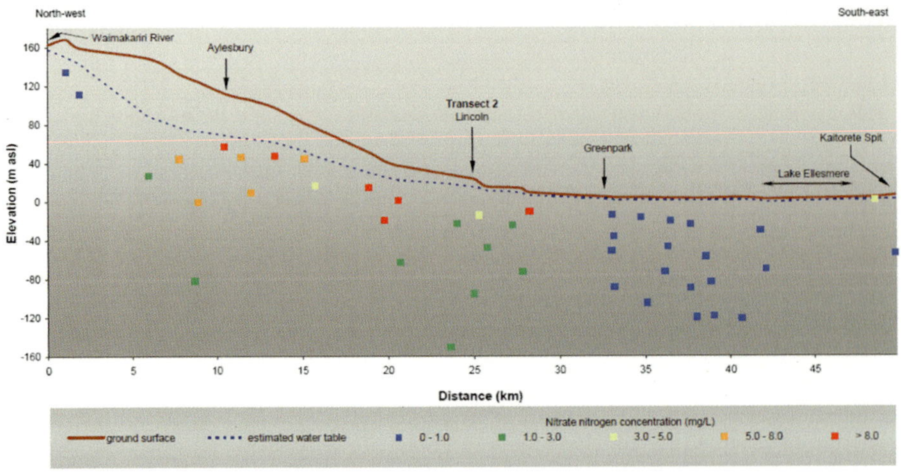

Fig. 6.19 Cross section of nitrate-nitrogen in groundwater in the Central Canterbury Plains (Hanson and Abraham 2009)

chemistry is affected by land surface recharge to depths of over 100 m below the water table in some parts of the Central Plains. Figure 6.19 shows nitrate-nitrogen concentrations greater than 8 mg/L in groundwater associated with land surface recharge.

This represents a cumulative impact of land use intensification as background levels are less than 3 mg/L. Nitrate is soluble and can readily be leached through the soil and transported into the underlying groundwater. In contrast, phosphorus, the other major nutrient tends to be occluded within iron oxides in the soil and is transported to surface water through soil erosion.

Figure 6.20 shows the distribution of nitrate-nitrogen concentrations in Canterbury groundwater in a 2012 survey. Samples from 11% of the wells had nitrogen concentrations that exceeded the maximum acceptable value (MAV) for drinking water of 11.3 mg/L. Figure 3.6 shows the results of a ten-year trend analysis from 2003 to 2012. Nitrate nitrogen concentrations have been increasing in about 30% of wells.

Shallow groundwater is also vulnerable to faecal contamination from land use intensification. Figure 6.21 shows sites where *E. coli* (as an indicator of faecal contamination) were detected. Of the detections, 24% of samples from wells less than 20 m had detections compared to 3% of samples from wells deeper than 20 m.

6.2.3.2 Water Quality in Rivers

An analysis of water quality in Canterbury rivers identified as a result of land use intensification issues with nutrient enrichment, faecal contamination, excessive sediment inputs and siltation of river beds, and potentially nitrate toxicity (Stevenson et al. 2010).

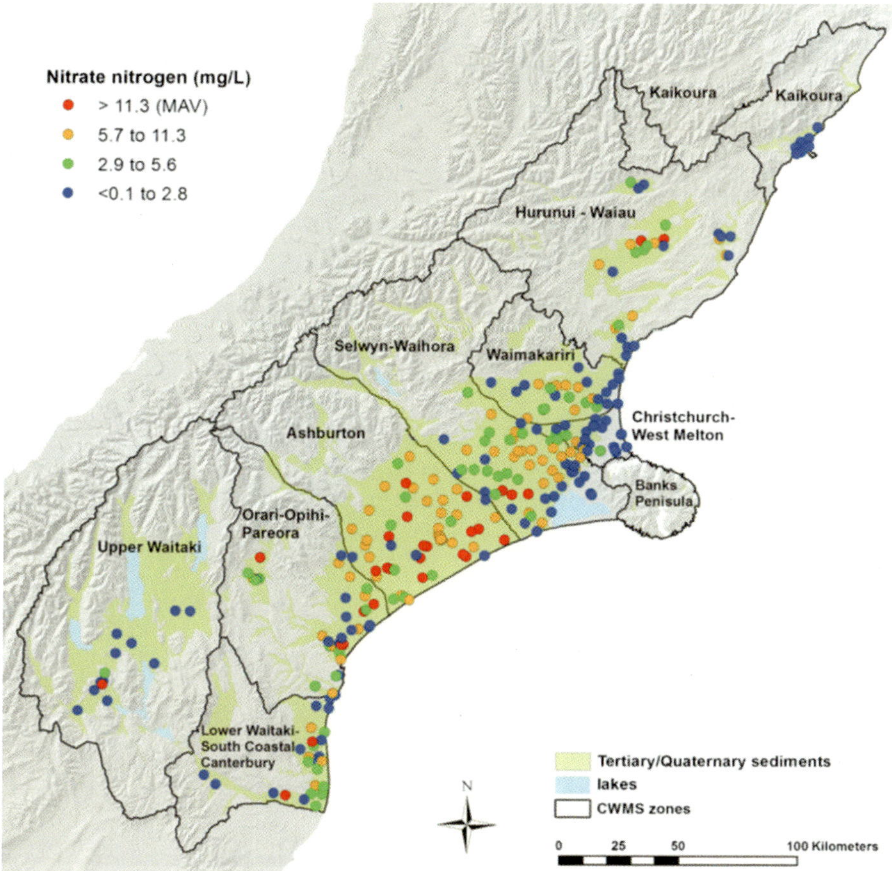

Fig. 6.20 Nitrate-nitrogen concentrations recorded in 2012 survey (Environment Canterbury 2012)

The analysis found that lowland streams, both urban and rural, are the most degraded. The lower reaches of foothill rivers and alpine rivers also show the effects of nutrient enrichment and faecal contamination. Rivers and streams in the high country have generally good water quality but increasing intensification of land use in some catchments is having noticeable deleterious effects.

In relation to dissolved inorganic nitrogen (DIN),[14] the main pathways for contamination of surface water are runoff, leaching to groundwater, fertiliser and effluent applications, wastewater discharges and livestock intrusion. Upland alpine and foothill rivers have median DIN values below the concentration indicative of enrichment in Canterbury (0.44 mg/L). The lower reaches of alpine and foothill rivers were consistently higher than upper reaches highlighting cumulative effects of

[14] Dissolved inorganic nitrogen is the sum of nitrite (NO_2), nitrate (NO_3) and ammonia. In Canterbury rivers, nitrate is the dominant component.

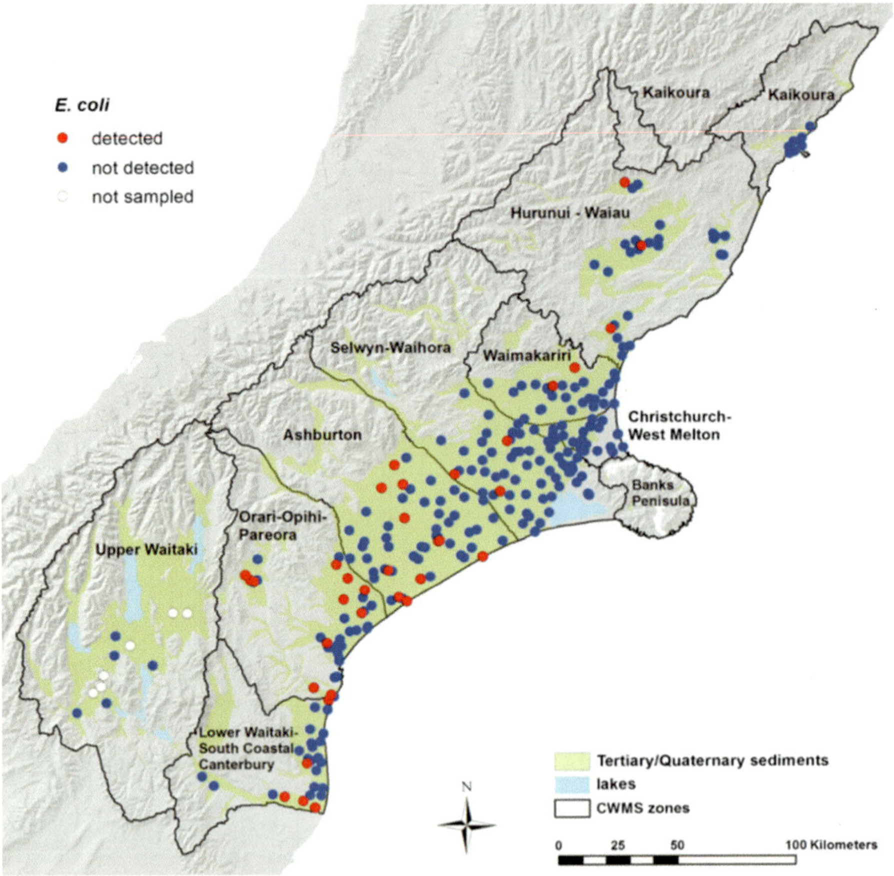

Fig. 6.21 Detection of Faecal contamination in Groundwater (Environment Canterbury 2012)

intensive land use on nitrogen concentrations. Nearly all lowland streams are in the enriched (0.44 to 2.0 mg/L) or excessive (>2 mg/L) categories for DIN concentrations (Stevenson et al. 2010) (refer Fig. 6.22).

In terms of nitrate toxicity, upland sections of foothill and alpine rivers lie below the 99% protection threshold of 1 mg/L (median).[15] However, in the lower reaches of alpine and foothill rivers there are exceedances of the 95% protection threshold of 2.4 mg/L (median).[16] However, for lowland streams, there are exceedances of the 80% protection threshold (6.9 mg/L median).[17]

[15] Indicative of pristine environments with high biodiversity and conservation values (Hickey 2013).

[16] Indicative of environments which are subject to a range of disturbances from human activities but with minor effects (Hickey 2013).

[17] Indicative of environments that are measurably degraded (Hickey 2013).

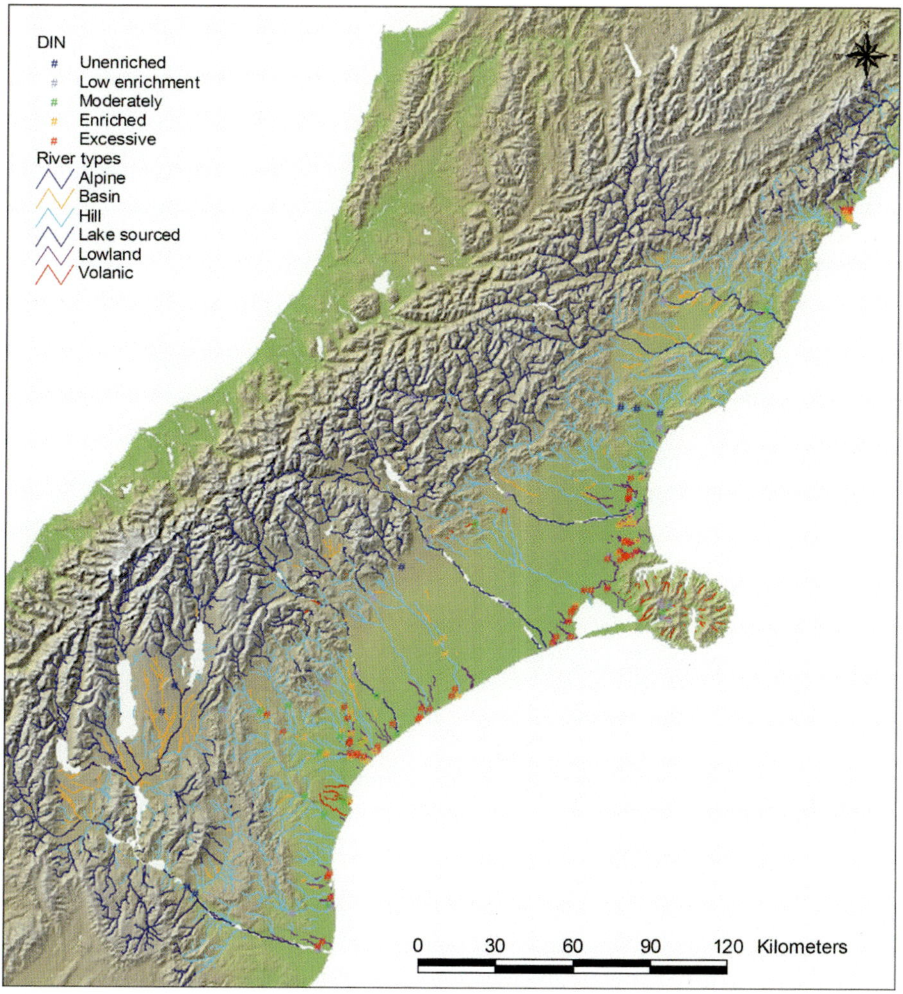

Fig. 6.22 Enrichment status for Canterbury Rivers for dissolved inorganic nitrogen (Environment Canterbury)

For the other major nutrient, phosphorus, phosphates bind to the soil and sediment particles, so overland flow is the dominant pathway for input to waterways from phosphate fertilisers, wastewater, animal manure and breakdown of phosphate rock. With phosphate-rich volcanic soils, Banks Peninsula streams have elevated dissolved reactive phosphorus (DRP),[18] with median values of 0.025 mg/L. Upland alpine and foothill rivers have median DRP concentrations less than the unenriched threshold for Canterbury (0.003 mg/L). The lower reaches of alpine and foothill

[18] Dissolved reactive phosphorus is a measure of the dissolved (soluble) phosphorus compounds that are readily available for use by plants and algae.

rivers showed higher concentrations, with median of 0.004 mg/L for alpine rivers and 0.005 mg/L for foothill rivers. Lowland streams showed results with 80% of sites considered enriched (0.009 to 0.030 mg/L) or excessive (>0.030 mg/L) levels. The median for urban streams (0.018 mg/L) was higher than for rural streams (0.015 mg/L) (Stevenson et al. 2010).

The ratio of nitrogen to phosphorus can indicate if one of the nutrients is limiting plant or algae growth. Ratios above 20 indicate phosphorus limitation while ratios below 4 indicate nitrogen limitation. Ratios between 4 and 20 indicate co-limitation. The nitrogen-enriched lowland streams and lower reaches of foothill rivers show the greatest skew towards phosphorus limitation. Upland reaches and Banks Peninsula streams indicate co-limitation (Stevenson et al. 2010).

Algal growth is related to nutrient availability and time between flushing flow events. Based on predictive models for chlorophyll *a* as an indicator of algal blooms, current nutrient concentrations are well in excess of the thresholds for both DIN and DRP with respect to acceptable chlorophyll *a* outcomes. Monitoring sites for periphyton in Canterbury are part of NIWA's national water quality network (Quinn and Raaphorst 2009). For all monitored Canterbury rivers,[19] the maximum values for periphyton mats and filamentous algae for the period 1990–2006 exceeded the Ministry for the Environment aesthetic and recreational guidelines, i.e. less than 30% of visible bed for filamentous algae and less than 60% for periphyton mats (Biggs 2000). The water quality network had a number of paired upstream and downstream sites. There was an overall pattern of downstream sites having more periphyton cover than upstream sites indicating periphyton is responding to land use pressures. However the paired sites for Canterbury (i.e. Hurunui, Waimakariri and Opihi Rivers) has similar periphyton cover (Quinn and Raaphorst 2009).

Microbial data collected from waterways are highly variable in space and time. Faecal contamination can occur from agricultural activities including direct stock access and pasture runoff while wildfowl can also contribute to faecal inputs. In urban areas stormwater, sewage overflows and wildfowl are the main sources. Lowland streams, both urban and rural, and Banks Peninsula streams are the most affected with 95 percentile concentrations exceeding the recreational water quality standard, i.e. 550 MPN/100 mL for *E. coli*, (2900 for urban, 2400 for rural and 2000 for Banks Peninsula streams). The lower reaches of alpine and foothill rivers also exceed the recreational standards (1600 and 820 MPN/100 mL respectively). Upper reaches of alpine and foothill rivers are close to the standard (460 and 570 MPN/100 mL respectively) (Stevenson et al. 2010).

The trigger value[20] for turbidity for aquatic ecosystems is 5.6 NTU (ANZECC 2000). The median values for all river types were below the 5.6 NTU trigger value.

---

[19]Monitoring was undertaken at the Hurunui, Waimakariri, Opihi, Opuha, Hakataremea and Waitaki Rivers.

[20]Trigger values are derived from ecosystem data for a reference site considered to have unmodified or slightly modified ecosystems. The trigger value is the 80th percentile value (i.e. 80% of results are below this value and there is a 20% chance that a single value will exceed the trigger value). Trigger values are used to assess risk of adverse effects in ecosystem types and represent an early warning mechanism to alert managers to potential problems (ANZECC 2000).

However, the upper quartile data for alpine rivers and lowland streams exceeded 5.6 NTU. For alpine rivers, widespread active erosion in the Southern Alps cause a high rate of sediment transport leading to sustained reductions in water clarity. Lowland streams have high clarity source water from groundwater but their small size, low gradient and proximity to high intensity land use has resulted in high sediment inputs from rural runoff or urban stormwater discharges (Stevenson et al. 2010).

6.2.3.3 Water Quality in Lakes

Coastal lakes have highly degraded water quality. Te Waihora/Lake Ellesmere, Wairewa/Lake Forsyth and Wainono Lagoon are hypertrophic: they are high in nutrients (nitrogen and phosphorus), high in suspended sediments (and low in water clarity), and prone to algal blooms. However, the degradation is not a recent phenomenon. There is an historical record of an algal bloom in Wairewa/Lake Forsyth in 1907. Te Waihora/Lake Ellesmere changed from a macrophyte-dominated lake to an algae-dominated lake in 1968 after the Wahine storm uprooted ruppia beds in the lake which have not recovered since then.

Analysis of sediment cores from the lakes indicate significant changes in trophic state from the early pastoral period of land use change involving forest clearance and the establishment of pasture grasses. Analysis of a sediment core from the centre of Te Waihora/Lake Ellesmere for pollen (indicating vegetation in the catchment) and diatoms (indicating the water quality environment of the lake) shows the onset of eutrophic conditions coincident with the decline of podocarp forest and increase in introduced grasses. Diatom changes also indicate increasing salinity (consistent with increased lake openings from human intervention) and increasing nitrogen levels (consistent increased fertilisation and land use changes) (Kitto 2010).

Analyses of a core from Wairewa/Lake Forsyth indicate a natural transition from a tidal embayment to a brackish coastal lake as the barrier bar, Kaitorete "Spit" closed the embayment entrance. Overlain on this natural coastal evolution are the human-induced effects from deforestation, land use intensification and artificial openings. Deforestation in the catchment from 1860 to 1890 led to increased sediment and nutrient input to the lake coincident with a decline in podocarp pollen and an increase in charcoal and the appearance of grasses in the sediment cores. A change in invertebrate composition indicated a shift from oligo-mesotrophic status to a highly productive (i.e. eutrophic) lake system. Invertebrate composition changes also indicated that salinity reduced from the shift from tidal embayment to coastal lake. However artificial openings have increased salinity tolerant invertebrates in recent decades (Woodward and Shulmeister 2005).

Wainono Lagoon was part of a vast swampland when Europeans arrived. Diatoms and macrofossil remains from sediment cores indicate that the lake was predominantly freshwater at that time. In the 1860s and 1870s there was fire-assisted conversion to pasture. There is evidence of the influx of fine then coarse soil particles and charcoal to the wetland area. However, the lake remained a freshwater system with little change to biological communities due to the buffering effect of fringing wetlands. In 1910 the

Waihao Box was installed and provided an intermittent lake outflow to the sea. This lowered water levels and introduced saline waters to the lake causing a shift to more estuarine communities. The opening to the sea allowed the drying, draining, burning and conversion of fringing wetlands into pasture. This increased the sediment and nutrient loading to the lake so that by the 1970s the lake was eutrophic and brackish with much reduced water depth. Loss of macrophytes in the 1980s–1990s led to increased turbidity and hypertrophic conditions (Schallenberg and Saulnier-Talbot 2014).

Figure 6.23 shows the monthly TLI for Te Waihora/Lake Ellesmere which is typically around 7 classed as hypertrophic. During 2013 the lake was open to the sea for two prolonged periods of time. This facilitated flushing of contaminants from the lake and the increasingly saline environment was unlikely to be favourable to the cyanobacteria characteristic of the lake. This led to a significant reduction in TLI. However since 2014, the lake has been dominated by potentially toxic picocyanobacteria. The bloom has prevailed year round raising the TLI above 7 (Lomax et al. 2015). While openings can improve the trophic status of the lake, the associated increases in salinity mean that macrophyte recovery is further compromised. Reductions in salinity and nutrients are needed to achieve a return to a macrophyte rather than algae-dominated lake.

Figure 6.24 shows the monthly TLI for Wairewa/Lake Forsyth from 1999 to 2015. The TLI is highly variable reaching peaks of over 9, regularly dropping to between 5 and 6, and going as low as 4. The index and variability have declined in recent years and may reflect the higher lake level over summer due to lake level management and from openings removing contaminants (Environment Canterbury 2015b). However year-round algal blooms persist.

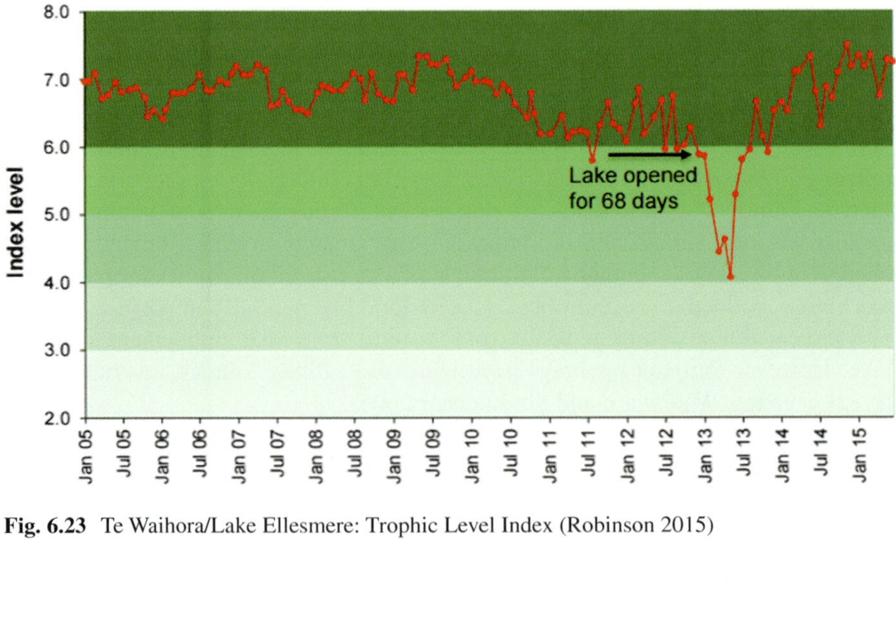

Fig. 6.23 Te Waihora/Lake Ellesmere: Trophic Level Index (Robinson 2015)

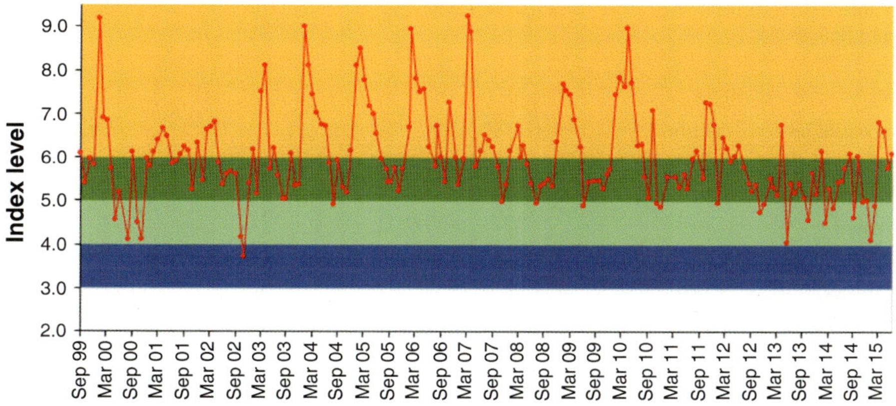

Fig. 6.24 Wairewa/Lake Forsyth: Trophic Level Index (Environment Canterbury 2015b)

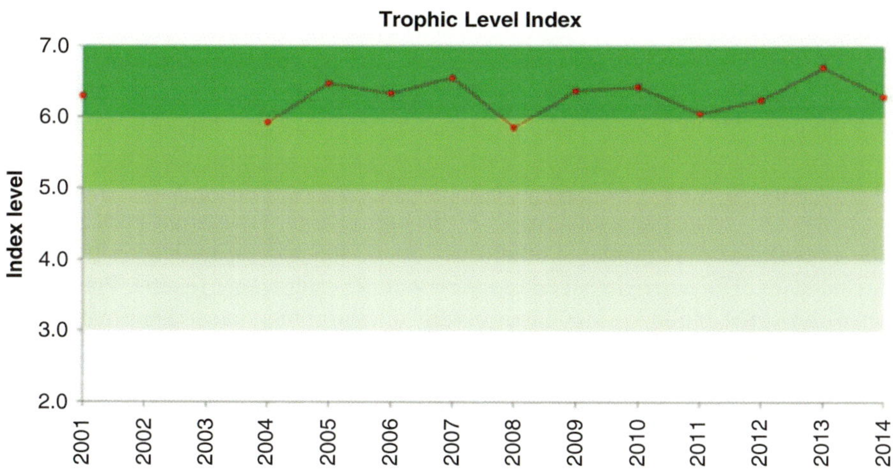

Fig. 6.25 Annual Trophic Level Index for Wainono Lagoon (Norton and Robson 2015)

Results for the annual TLI for Wainono Lagoon from 2001 to 2014 are shown in Fig. 6.25. The current TLI is 6.5 indicating hypertrophic status. The lagoon is turbid and has low clarity, predominantly determined by the amount of suspended sediment in the water column. The nutrient-enriched state of the lagoon increases the risk of cyanobacteria. Nuisance blooms have been reported.

Coastal lakes are the recipients of flows from foothill rivers, lowland streams and groundwater, all of which have been enriched by land use intensification (refer subsection on water quality in rivers above). Also with limited connection to the sea, nutrients and sediments can accumulate in the lake. With shallow depth the sediments can be resuspended by wind action and with artificial openings salinity levels have increased.

The water quality for hapuas has not received the same level of monitoring and analysis as the Waituna-type coastal lakes. A preliminary investigation (Samuel and

Table 6.9 Comparison of water quality for Hapua and Waituna-type Coastal Lagoons (Samuel and Jenkins 2012)

	Median water quality parameters			
Coastal lake	Salinity (mS/m)	Turbidity (NTU)	Total Phosphorus (mg/L)	Total Nitrogen (mg/L)
HAPUA				
Rangitata	6.2	3.3	0.012	0.31
Opihi	53	1.7	0.015	0.86
WAITUNA				
Te Waihora	990	82.5	0.21	1.60
Wainono	285	41.0	0.18	0.97

Jenkins 2012) compared the water quality of two hapuas (Rangitata and Opihi) and two Waituna-type coastal lakes (Te Waihora/Lake Ellesmere and Wainono Lagoon) from 2006 to 2011 (refer Table 6.9). Hapuas have significantly lower salinities, turbidity and total phosphorus, and slightly lower total nitrogen levels. Hapuas have no tidal prism[21]: they discharge to the sea or are closed (which severs the link between the river and the sea). They are less exposed to seawater intrusion compared to Waituna-type coastal lakes. There is also less opportunity for sediment deposition and wind resuspension of sediment compared to Waituna-type coastal lakes which is reflected in lower turbidity. With phosphorus bound to sediment the higher phosphorus levels would appear to be associated with the higher turbidity in Waituna-type coastal lakes.

Table 6.10 shows the results for TLI[22] for high country lakes from 2004 to 2015. These results are much lower in relation to trophic status compared to Waituna-type coastal lakes. The large glacial lakes in the Waitaki catchment (Tekapo, Pukaki and Ohau) have the highest water quality with microtrophic status. They have large volumes, and catchments with little land use intensification. They do experience increases in phosphorus levels associated with sediment after high flow events, shifting the TLI into the oligotrophic range above the outcomes set in the regional plan. High turbidity associated with rock flour from erosion in the Southern Alps does limit aquatic plant growth so that the desired "excellent" grading for LakeSPI[23] is not achieved at Lake Tekapo (Clarke 2015). The smaller lakes in the Ashburton catchment have the poorest water quality particularly in the shallower lakes. The most recent TLI results place the lakes in the mesotrophic category (3.0 to 4.0). However, Lake Emma has been in the hypertrophic category while Lake Clearwater has been in the eutrophic category.

[21] Tidal prism is the volume of seawater that enters an inlet or estuary during a tidal cycle.

[22] Note that TLI for high country lakes is based on TN, TP and Chlorophyll *a*. With the use of helicopters for sampling, seechi disc measurements were not feasible. Turbidity measurements were taken instead.

[23] LakeSPI refers to Lake Submerged Plant Indicators: LakeSPI Index is a synthesis of components from both the native condition and invasive condition of a lake and provides an overall measure of the lake's ecological condition. The Native Condition Index captures the native character of vegetation in a lake based on the diversity and quality of native plant communities. The Invasive Impact Index captures the invasive character of vegetation in a lake based on the degree of impact by invasive weed species.

Table 6.10 High country lakes: Trophic Level Index (Environment Canterbury)

Source	2004-06	2007	2008	2009	2010	2011	2012	2013	2014	2015
Hurunui catchment										
Loch Katrine	2.22	2.78	3.03	3.26	2.95	2.75	2.66	2.45	2.35	2.62
Lake Sumner	1.44	1.60	2.13	2.28	2.56	2.18	2.13	1.60	2.44	1.83
Lake Taylor	2.09	2.25	2.39	2.64	2.46	2.49	2.40	2.24	2.46	2.34
Waimakariri catchment										
Lake Sarah	2.62	2.84	2.82	3.24	3.42	4.04	2.88	2.81	3.19	2.93
Lake Grasmere	2.44	2.92	3.09	3.31	3.21	3.65	3.36	3.11	3.11	3.22
Lake Pearson	2.04	3.02	3.28	3.09	2.67	2.85	2.76	2.64	3.31	3.99
Lake Hawdon	2.53	2.83	2.98	3.18	2.65	2.94	3.19	3.26	2.70	3.40
Rakaia catchment										
Lake Lyndon	2.36	2.40	2.56	3.15	2.82	2.82	2.83	2.96	2.62	2.99
Lake Georgina	2.86	4.64	4.25	3.59	3.54	3.65	3.35	4.04	4.94	3.85
Lake Ida	2.18	2.37	2.43	2.59	2.78	2.76	2.45	2.54	3.19	3.11
Lake Selfe	2.19	2.28	2.78	2.90	2.94	2.82	2.01	2.00	2.00	3.04
Lake Coleridge	1.21	0.95	1.85	2.26	1.83	1.55	1.36	2.12	1.13	1.30
Lake Heron	2.31	2.66	2.76	3.03	2.89	3.11	2.92	2.78	2.92	2.64
Ashburton catchment										
Lake Emma	3.90	4.64	6.20	5.74	4.31	4.78	3.45	3.76	3.75	3.46
Lake Camp	2.67	3.18	3.09	3.36	2.86	3.19	3.18	3.38	3.01	3.23
Lake Clearwater	3.19	3.60	4.16	3.83	4.27	4.08	3.89	3.83	3.10	3.84
Waitaki catchment										
Lake Alexandrina	2.71	3.15	3.12	3.32	3.09	2.93	3.00	2.96	3.15	3.14
Lake Tekapo	1.40	1.50	1.15	1.79	2.17	1.94	1.54	1.43	1.25	1.22
Lake Pukaki	1.40	1.60	1.72	1.92	2.00	2.01	1.59	1.32	0.99	1.84
Lake Ohau	1.40	1.90	1.10	1.90	2.08	1.96	1.46	2.07	0.87	1.12
Lake Benmore - Haldon Arm	1.30	1.43	1.54	1.71	2.17	2.06	1.79	2.04	1.69	1.83

Ultramicrotrophic (0.0-1.0)
Microtrophic (1.0-2.0)
Oligotrophic (2.0-3.0)
Mesotrophic (3.0-4.0)
Eutrophic (4.0-5.0)
Supertrophic (5.0-6.0)
Hypertrophic (6.0-70)

The ecological integrity of the small to medium high country lakes has also been identified as being under threat with 13 of 24 lakes being graded as "moderate" (LakeSPI 25–50%): this is below the minimum grade of LakeSPI for freshwater outcomes for this type of high country lake of "high" (LakeSPI 50–75%) (Kelly et al. 2014). These smaller lakes are more vulnerable to land use intensification. The process of tenure review[24] has facilitated the shift in land use from low intensity dryland sheep and beef farming to fodder cropping and dairy support. In the Mackenzie Basin irrigation has been introduced to enable dairying.

There is a high level of inter-annual variability in water quality but the overall trend is for worsening water quality. There is limited data on catchment loads and their relationship to lake water quality. To investigate the cumulative effects of catchment land use, the annual loads of TN and TP to high country lakes were estimated using a nutrient transport model combined with the regionally-based hydrological regression model (CLUES version 10) (Kelly et al. 2014). The TN and TP loading derived for shallow and deep lakes, modelled separately, explained between 49% and 73%, respectively, of the variation of in-lake TN and TP concentrations.

---

[24] Tenure review is the process of reviewing leasehold tenure of some high country land. It involves individual lessees selling their leasehold interest to the Crown and negotiating to buy back freehold title to productive land while the Crown retained land of conservation and recreational value (Land Information New Zealand 2015).

The results also indicated mainly high in-lake N:P ratios suggesting that P or co-limitation was most prevalent in high-country lakes in relation to phytoplankton productivity. Kelly et al. (2014) emphasise that retention of macrophytes is a key element of the ecological health of small lakes. Macrophytes mediate the nutrient cycling process, provide habitat for other species and comprise a significant biodiversity component. Aquatic macrophytes are strongly linked to nutrient loads due to their sensitivity to either increased phytoplankton abundance (reducing light availability) or epiphyte growth on macrophytes. Macrophyte collapse is considered a significant resilience threshold. Lake Emma appears to have undergone a recent macrophyte collapse with TP concentrations around 30 mg/m^3. This is low by international comparisons.

While the modelling showed strong relationships between catchment loading and in-lake nutrient levels, it was recognised that other sources such as groundwater, and other processes such as in-lake nutrient cycling and wind re-suspension influence in-lake nutrient concentrations. Some lakes were outliers (Kelly et al. 2014).

A detailed investigation has been undertaken of Lake Clearwater which provides further insight into the sustainable management of high country lakes (Wadworth-Watts 2013). Nitrogen and phosphorus levels have increased threefold over the last decade. Average TN increased from 210 mg/m^3 in 2004 to 590 mg/m^3 in 2012, while average TP increased from 4 mg/m^3 in 2004 to 12 mg/m^3 in 2012.

Thresholds that can be defined for nutrients for Lake Clearwater and its catchment are set out in Table 6.11. These include: (1) the contaminant loss rate for farmland of 10 kg/ha/year for nitrogen, based on the Land and Water Regional Plan (Environment Canterbury 2015a) – there is no loss rate limit for phosphorus; (2) the ANZECC trigger values for instream contaminants for upland rivers of 295 mg/m^3 (median) for TN and 26 mg/m^3 (median) for TP (ANZECC 2000); and (3) in-lake concentrations from the threshold between mesotrophic (TLI 3) and eutrophic (TLI 4) of 337 mg/m^3 for TN and 20 mg/m^3 for TP (Burns et al. 2000). These thresholds are compared with monitoring results for Lake Clearwater (Wadworth-Watts 2013).

The comparison shows that Lake Clearwater exceeds the TN limit for lake concentration for TLI, and is close to the TP limit for lake concentration. However, the contaminant loss rate is only 20–30% of the limit in the regional plan and there is no phosphorus contaminant loss rate limit in the regional plan. To be effective in lake water quality management there is a need to reduce the allowable contaminant loss rate for TN and to introduce a limit for contaminant loss rate for TP. The absence of a phosphorus loss rate limit is of particular significance as Lake Clearwater is P limited in relation to phytoplankton production.

Table 6.11 Nutrient loadings for Lake Clearwater (Wadworth-Watts 2013)

	TN limit	TN actual	TP limit	TP actual
Contaminant loss farmland (kg/ha/year)	10	2–3	–	0.09–0.12
Contaminant in-stream median (mg/m^3)	295	220	26	13
Catchment load (kg/year)	–	1375	–	76
Lake concentration median (mg/m^3)	337	562	20	16

6.3 Natural Variations and Cumulative Effects

Water as a biophysical system is affected by multiple pathways. For effective management it is necessary to distinguish the contribution to effects by different pathways. This requires an understanding of cause-effect relationships. For making management interventions it also requires the ability to predict the impacts of changes on multiple pathways and to monitor and test the validity of predictions as well as the veracity of the predictive models. There is also a need for a management framework to reflect the biophysical systems and the impact pathways.

One of the critical variables for groundwater management is the ecological health of lowland streams. The decline in health is associated with reduced flows in these groundwater-fed streams. However, flow reductions due to lowered groundwater levels can result both from reduced rainfall recharge and from increased abstraction.

6.3.1 Groundwater Variation

An example is the Selwyn catchment and groundwater zones which have seen a significant increase in groundwater abstraction. (Refer Fig. 3.3 for the history of the development of groundwater in the Rakaia-Selwyn groundwater zone of the Selwyn catchment.) The catchment has also been subject to rainfall variability. Figure 6.26 shows the monitoring results for the Courtenay Road bore (L36/0092) from 1951 to 2016. The changes in groundwater levels prior to 1990 are predominantly related to climate variation as there was very little abstraction at that time. There are marked

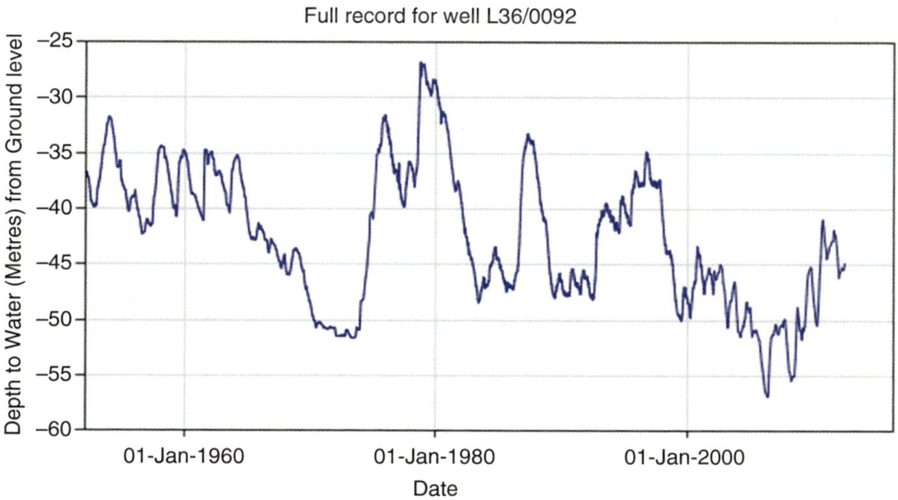

Fig. 6.26 Groundwater depth variations for the Courtenay Road monitoring bore (1951–2016) (Environment Canterbury)

low levels between 1970 and 1974 (around 51 m below ground) and marked high levels between 1978 and 1980 (up to 26 m below ground). There is also an increase in variations between winter (time of recharge) and summer (time of abstraction) levels from 2000 when rates of abstraction began to significantly increase. Drought effects in 1998–99 can be seen. The record low is in 2005 at the end of a 5-year period of particularly low winter rainfall and at a time of high abstraction.

6.3.2 Climate Variation

Canterbury rainfall is influenced by the El Nino Southern Oscillation (ENSO) with a negative Southern Oscillation Index[25] (El Nino) associated with drier conditions in the east of New Zealand and, wetter and cooler conditions in the west of New Zealand. The Southern Oscillation Index is shown in Fig. 6.27 showing El Nino conditions for 2001 to 2005. Figure 6.28 shows the rainfall deviation from the long term mean for rainfall monitoring sites across Canterbury for 2005 (an El Nino year) and for 2006 (a La Nina year).

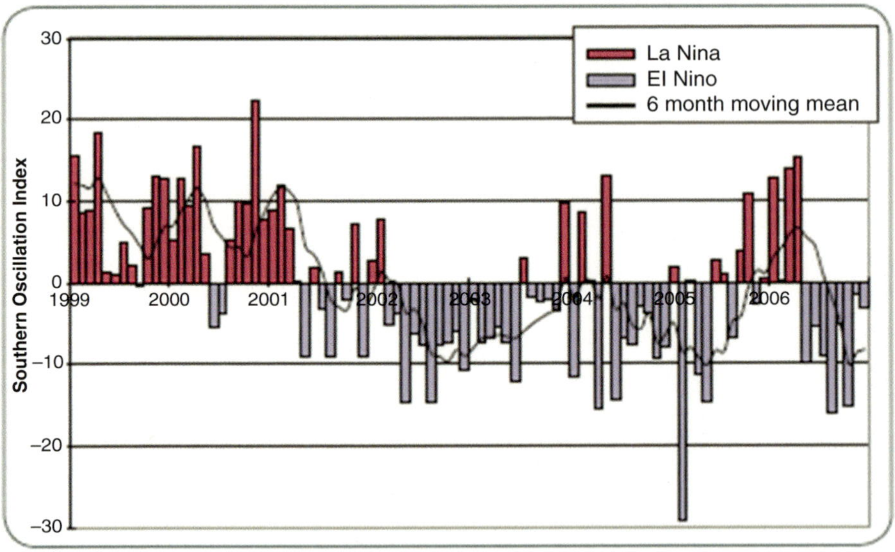

Fig. 6.27 Southern Oscillation Index from 1999 to 2007 (Environment Canterbury 2008)

[25] The Southern Oscillation Index is based on the pressure difference between Tahiti and Darwin. Negative values (El Nino) are associated with sustained warming of the central and eastern Pacific Ocean.

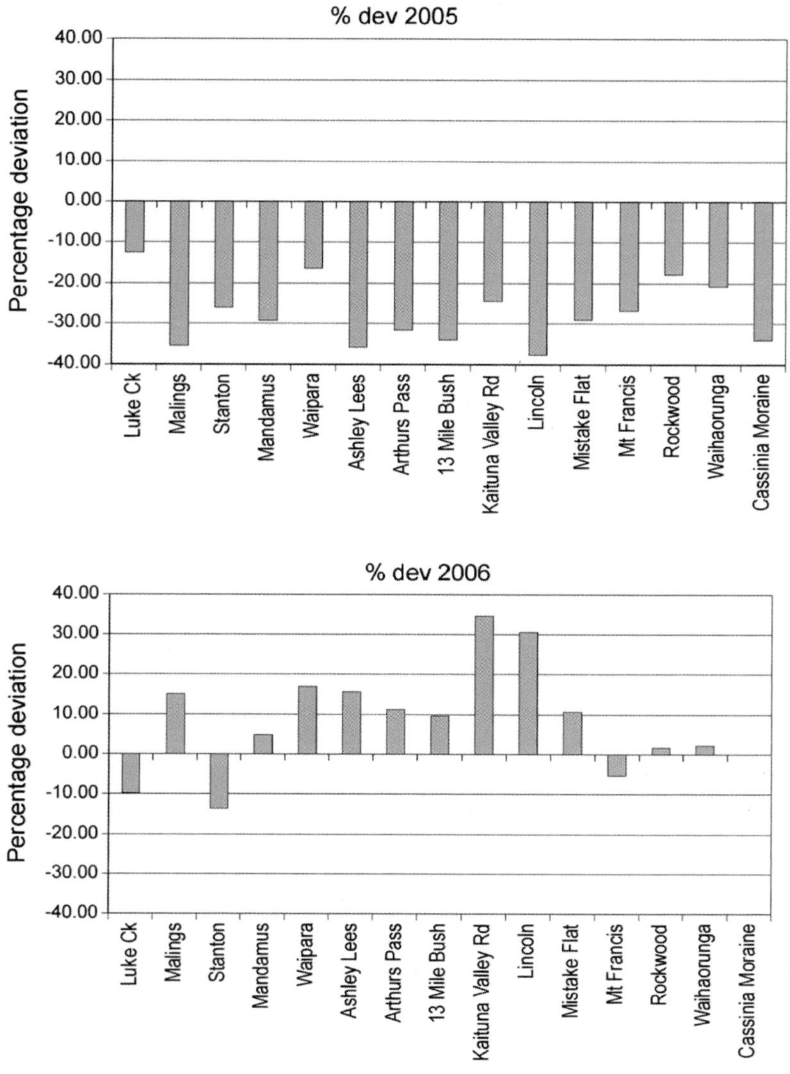

Fig. 6.28 Rainfall deviation (%) from long-term mean for 2005 (an El Nino year) and 2006 (a La Nina year) for Canterbury rainfall measurement sites (Environment Canterbury 2008)

6.3.3 Effects of Abstraction and Climate Variability

To distinguish between the effects of abstraction and climate variability in relation to the reduction in groundwater levels a time series finite-difference modelling tool (Eigenmodel) was developed (Bidwell 2003). The tool uses monthly measured or estimated land surface recharge values (1972–2006), estimates of groundwater use (for the period 1990–2006), and the monthly groundwater monitoring record. Values

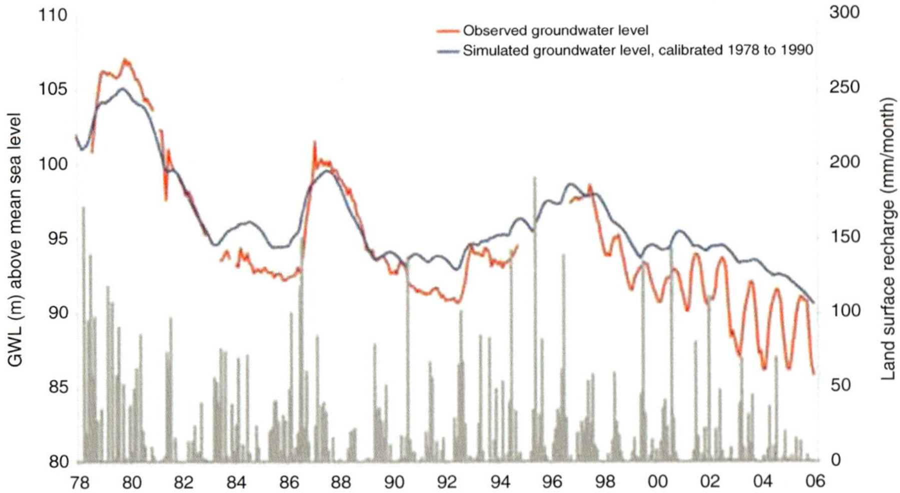

Fig. 6.29 Actual and modelled groundwater levels for monitoring bore L36/0023 (Environment Canterbury 2008)
Note: Grey bars represent monthly recharge in mm.

of land surface recharge and estimated use are calculated for each month for an entire groundwater allocation area and converted to millimetres/month. A typical plot of a monitoring record that has been modelled using the Eigenmodel method is presented in Fig. 6.29. The modelling process is an iterative one that runs on a spreadsheet, where the model is trained or calibrated, using the recharge and monitoring record over a limited period of time, such as 1972–1990. The model then predicts the likely groundwater level over the remainder of the record, 1990 onwards. The reason why the model is trained only over the early period is that during that time, little abstraction was occurring. The model is, therefore, measuring aquifer parameters associated with a purely climatic response. Figure 6.29 shows that for the period after 1990, there is a progressive difference between the actual and modelled groundwater levels; this difference is due to groundwater abstraction. The reduction in groundwater level is due to both climate variation and groundwater abstraction with a marked drawdown in summer due to abstraction.

References

Aitchison-Earl, P., & Alkhair, F. (2014, November 24–28). *Where's my water? What is causing a declining trend in shallow groundwater of the Pareora Valley, South Canterbury?* Paper presented at the New Zealand Frehwater Sciences Society Lim Soc 2014, Blenheim.

ANZECC. (2000). *Australian and New Zealand guidelines for fresh and marine water quality*. Australian and New Zealand Environment and Conservation Council, and Agriculture and Resource Management Council of Australia and New Zealand Canberra.

Aqualinc Research Limited. (2008). *Canterbury Strategic Water Study (stage II)*. Prepared for Environment Canterbury, Christchurch

Arscott, D., Kelly, D., & Tuck, E. (2007). *Flushing flows for the control of nuisance periphyton growth in the Opuha-Opihi River system (2004–2006)*. NIWA Client Report CHC 2007-034. NIWA: Christchurch.

Ball, S. (2012). *Rakaia Hapua: An investigation into observed temporary closures*. Christchurch: University of Canterbury.

Bidwell, V. (2003). *Groundwater management tools: Analytical procedure and case studies*. Wellington: Ministry for Agriculture and Forestry.

Biggs, B. (2000). *New Zealand periphyton guideline: Detecting, monitoring and managing enrichment of streams*. Wellington: Ministry for the Environment.

Burns, N., Bryers, G., & Bowman, E. (2000). *Protocol for monitoring trophic levels of New Zealand Lakes and reservoirs*. Wellington: Ministry for the Environment.

Burrell, G. (2011, November 15–16). *Ecological flows for rivers in the Te Waihora/Lake Ellesmere Catchment*. Paper presented at the Living Lake, Changing Catchment Symposium 2011, Lincoln University.

Clarke, G. (2015). *The current water quality state of lakes in the Waitaki catchment*. Christchurch: Environment Canterbury.

Dark, A. (2010). *Selwyn and Ashburton zone summaries*. Christchurch: Aqualinc.

Davoren, A. (2006). *Evidence presented at a hearing for 69 groundwater take applications in the Rakaia-Selwyn groundwater allocation zone*. Christchurch: Environment Canterbury.

Davoren, A. (2008, June 18). *The Donkers Site*. Paper presented at the WQN9 Workshop, Christchurch.

Duncan, M., Srinivasan, M. S., & Diettrick, J. (2010). *Improved irrigation management on water metering information: 2009–10 irrigation season*. NIWA Client Report: CHC 2010–091, Christchurch.

Dynes, R., Burggraaf, V., Goulter, C., & Dalley, D. (2010). Canterbury farming: production, processing and farming. In *Proceedings of the New Zealand Grassland Association* (Vol. 72, pp. I–VIII).

Dysart, M., Miller, M., & Pascoe, W. (2008). *Rakaia catchment water allocation issues: A review of water allocation administration, stream depletion management, and other issues within the Rakaia catchment and options for formalizing allocation arrangements*. Christchurch: Environment Canterbury.

Environment Canterbury. (2004). *Variation 1 to the Proposed Canterbury Natural Resources Regional Plan*. Christchurch: Environment Canterbury.

Environment Canterbury. (2007). *Decision of the Commissioners regarding 69 groundwater takes in the Rakaia-Selwyn groundwater allocation zone*. Christchurch: Environment Canterbury.

Environment Canterbury. (2008). *Canterbury Regional Environment Report 2008*. Christchurch: Environment Canterbury.

Environment Canterbury. (2011). *Natural Resources Regional Plan*. Christchurch: Environment Canterbury.

Environment Canterbury. (2012). *Annual groundwater quality survey 2012*. Christchurch: Environment Canterbury.

Environment Canterbury. (2015a). *Canterbury Land and Water Regional Plan December 2015*. Christchurch: Environment Canterbury.

Environment Canterbury. (2015b). *Section 32 report – Plan Change 6 to the partly operative Canterbury Land and Water Regional Plan*. Christchurch: Environment Canterbury.

Environment Canterbury. (2016). *The groundwater allocation zone summary report for Chertsey*. http://previous.ecan.govt.nz/services/online-services/monitoring/groundwater-allocation/zone-summary-reports/Pages/chertsey.aspx

Gibbs, M., & Hickey, C. (2012). *Guidelines for artificial lakes. NIWA: Before construction, maintenance of new lakes and rehabilitation of degraded lakes*. NIWA Client Report HAM 2011–045, Hamilton.

Golder Associates. (2008). *Pareora River aquatic ecology and minimum flow requirements.* Christchurch: Environment Canterbury.

Hanson, C., & Abraham, P. (2009). *Depth and spatial variation in groundwater chemistry – Central Canterbury Plains.* Christchurch: Environment Canterbury.

Hart, D. (2009). Morphodynamics of non-estuarine Rivermouth lagoons on high-energy coasts. *Journal of Coastal Research Special Issue, 56*(V2), 1355–1359.

Hart, D. (2011, August 3). *Rivermouth Geomorphology.* Paper presented at the River Mouths Seminar: Science-informed Management, University of Canterbury.

Hertel, I. (1998). *Groundwater contamination in the Heathcote/Woolston area, Christchurch, New Zealand.* Christchurch: University of Canterbury.

Hertel, I., Griffiths, E., Ettema, M., & Daly, A. (1998). *Background report for the Woolston-Heathcote groundwater users meeting of 7 September 1998.* Christchurch: Environment Canterbury.

Hickey, C. (2013). *Updating nitrate toxicity effects on freshwater aquatic species.* Hamilton: NIWA.

Hicks, M. (2007). *Brief of evidence in the matter of: An application for resource consent CRC071029 by South Canterbury Irrigation Trust and Meridian Energy Limited to take and use water from the Waitaki River.* http://files.ecan.govt.nz/public/south-canterbury-streams/research/hicks-hdi-evidence-coastal-processes.pdf

Horrell, G. (2011). *River mouth wellbeing – A key factor in integrated catchment management.* Christchurch: NIWA.

Jenkins, B. R. (2007, November 19–20). *What role for storage in Canterbury?* Paper presented at the NZSOLD/ANCOLD Conference "Dams – Securing Water for Our Future", Queenstown.

Jenkins, B. (2017). Sustainable water management analysis of the Selwyn catchment using nested adaptive systems. *International Journal of Sustainable Development and Planning, 12*(6), 1061–1072.

Kelly, D., Robertson, H., & Allen, C. (2014). *Nutrient loading to Canterbury high-country lakes for sustaining ecological values.* Nelson: Cawthron Institute.

Kirk, R. & Lauder, G. (2000). *Significant coastal lagoon systems in the South Island, New Zealand: Coastal processes and lagoon mouth closure.* Science for Conservation 146. Department of Conservation, Wellington.

Kitto, S. (2010). *The environmental history of Te Waihora – Lake Ellesmere.* MSc thesis, University of Canterbury, Christchurch.

Land Information New Zealand. (2015). *Tenure review; A detailed guide.* Christchurch: Land Information New Zealand.

Landward Management. (2006). *South Island lake aquatic weed control: Annual report 2006/7, Unpublished report prepared for Land Information New Zealand.* Landward Management Dunedin: New Zealand.

Landcare Research. (2016). *SMap glossary.* https://smap.landcareresearch.co.nz/glossary. Accessed 31 Dec 2016.

Larnad, S., Hicks, M., Schmidt, J., Davey, A., Dey, K., Scarsbrook, M., Arnscott, D., & Woods, R. (2007). The Selwyn River of New Zealand: a benchmark system for alluvial plain rivers. *River Research and Applications, 24*(1), 1–21. https://doi.org/10.1002/rra.1054.

Larnard, S. (2013, November 7–8). *The Selwyn River/Waikirikiri.* Paper presented at the Living Lakes Symposium, Lincoln University.

Lomax, A., Johnston, K., Hughey, K., & Taylor, K. (Eds.). (2015). *Te Waihora/Lake Ellesmere state of the Lake 2015.* Tai Tapu: Waihora Ellesmere Trust.

Meredith, A. (1999). *Water quality of Lake Opuha: Limnological issues in a newly formed lake.* Environment Canterbury, Christchurch: Unpublished report.

Milne, P., Fenemor, A., O'Callaghan, R., & Nixon, B. (2009). Minute 10 of the Commissioners: Preliminary recommendation and summary of conclusions and reasoning in relation to the proposed Waianiwaniwa reservoir and dam and upper Waimakariri intake. In the matter of an application by Central Plains Water Trust for resource consents. Environment Canterbury, Christchurch.

Morgan, M., Bidwell, V., Bright, J., McIndoe, I., & Robb, C. (2002). *Canterbury Strategic Water Study*. Lincoln: Lincoln Environmental.

Norton, N., & Robson, M. (2015). *South Canterbury coastal streams (SCCS) limit setting process predicting consequences of future scenarios: Overview report*. Christchurch: Environment Canterbury.

Opuha Dam Limited. (2004). *Report on the performance of lake aeration network*. Washdyke: South.

Pattle Delamore Partners. (2000). *Quantifying stream depletion effects*. Christchurch: Environment Canterbury.

Pattle Delamore Partners. (2011). *New Zealand guidelines for the monitoring and management of sea water intrusion risks in groundwater*. Christchurch: Pattle Delamore Partners.

Quinn, J., & Raaphorst, E. (2009). *Trends in nuisance periphyton cover at New Zealand National River Water Quality Network sites 1990–2006*. Hamilton: NIWA.

Ritso Society. (2007). *Comparison of piped and open channel distribution water supplies*. Wellington: Sustanable Farming Fund.

Robinson, K. (2015, November 18–19). *State of the Water: Te Waihora/Lake Ellesmere and Catchment*. Paper presented at the Living Lake Symposium 2015, Lincoln University.

Samuel, L., & Jenkins, B. (2012). *An initial assessment of river mouth water quality in Canterbury: Hapuas, lagoons and estuaries*. Christchurch: University of Canterbury (unpublished manuscript).

Schallenberg, M., & Saulnier-Talbot, E. (2014). *Recent environmental history of Wainono Lagoon (South Canterbury, New Zealand)*. Christchurch: Environment Canterbury.

Scott, D. (2004). *Groundwater allocation limits: Land-based recharge estimates. U04/97*. Christchurch: Environment Canterbury.

Scott, D., & Thorpe, H. (1986). *Ground water resources between the Rakaia and Ashburton Rivers*. Christchurch: Hydrology Centre.

Scott, M., & Wilson, N. (2012). *Seawater intrusion network review*. Christchurch: Environment Canterbury.

SKM (2010). *Managed aquifer recharge feasibility study*. Prepared for Environment Canterbury, Auckland.

Stevenson, M., Wilks, T., & Hayward, S. (2010). *An overview of the state and trends in water quality of Canterbury's rivers and streams*. Christchurch: Environment Canterbury.

Todd, D. (1992). River mouth and coastal processes of the Ashburton River Mouth. In T. Dons & D. Stringer (Eds.), *Natural resources of the Ashburton River and catchment*. Christchurch: Canterbury Regional Council Christchurch.

Veltman, A., Miller, M., Glennie, J., & Talbot, J. (2004). *Draft Natural Resources Regional Plan groundwater allocation policy implementation – Analysis and implications*. Christchurch: Environment Canterbury.

Wadworth-Watts, H. (2013). *A hydrological and nutrient load balance for Lake Clearwater catchment, Canterbury, New Zealand*. Christchurch: University of Canterbury.

Webb, M. (2009). *Hearing evidence for Fish and Game New Zealand*. Christchurch: Canterbury Natural Resources Regional Plan.

Williams, H., Scott, D., & Bidwell, V. (2008). *Adaptive management of groundwater in the Rakaia-Selwyn groundwater allocation zone: Technical and implementation issues*. Christchurch: Environment Canterbury.

Wilmhurst, J. (2007). *Human effects on the environment – Pre-European deforsestation*. Te Ara – The Encyclopedia of New Zealand. http://www.teara.govt.nz/en/human-effects-on-the-environment/page-2. Accessed 30 Dec 2016.

Woodward, C., & Shulmeister, J. (2005). A Holocene record of human induced and natural environmental change in Lake Forsyth (Te Wairewa) New Zealand. *Journal of Paleolimnology, 34*, 481–501.

Chapter 7
Biophysical System Failure Pathways at the Regional Scale

Abstract This chapter addresses water management failure pathways of climate change, and, the natural hazards of drought and floods. In terms of greenhouse gas emissions, the use of water for hydro-electric generation reduces New Zealand's greenhouse gas emissions from energy compared to other developed countries. However, agriculture is the largest source producing 47.2% of total emissions, primarily from methane and nitrous oxide contributions. In addition, further forest clearance mainly for dairying is reducing greenhouse sinks.

Projected changes in climate have significant implications for water availability in Canterbury. Increased temperature increases the potential evapotranspiration deficit and therefore irrigation demand. Reduced winter rainfall on the Canterbury Plains reduces groundwater recharge and hence reduces flows in lowland streams. Reduced winter rainfall in the foothills reduces flow in foothill rivers. Increased rain on the Southern Alps but reduced snow means that while annual flow increases in alpine rivers, the peak flows shift from spring and summer towards winter and the reliability of supply for the irrigation season declines.

New Zealand's response to climate change has been minimal with emissions continuing to increase. However, there are actions that could be taken through mitigation measures and offsets. Furthermore, better use could be made of economic instruments and environmental impact assessment procedures to manage emissions.

Drought can be defined in biophysical terms, i.e. meteorological and hydrological droughts but more relevant from a nested adaptive system perspective is the definition of agronomic drought because it focuses on damage from drought and can incorporate the socio-economic response. Drought adaptation responses are aligned with the sustainability approaches.

The case study of taking a resilience approach to management of the flood risk to Christchurch from the Waimakariri River is described. Rather than designing protection works for a flood of specific return period, the design incorporates the consequences of failure. A secondary stopbank system is provided to capture floodwaters if the primary system fails or is overtopped. The system also allows for return of floodwaters to the river. The international example of Hurricane Katrina and its flooding of New Orleans is also described. The inadequacies of the approach of designing just for a specific return period hazard has led the US Army Corps of Engineers to change to a comprehensive systems approach.

B.R. Jenkins, *Water Management in New Zealand's Canterbury Region*, Global Issues in Water Policy 19, https://doi.org/10.1007/978-94-024-1213-0_7

Keywords Climate change • Natural hazards • Resilience approaches • Sustainability strategies

7.1 Implications of Climate Change

7.1.1 New Zealand's Greenhouse Gas Emission Profile

Based on the official annual report of all human-caused emissions of greenhouse gases in New Zealand (Ministry for the Environment 2013) in 2011, New Zealand's total greenhouse gas emissions were 72.8 million tonnes of carbon dioxide equivalent (Mt. CO_2-e) and net removal associated with forestry was 16.8 Mt. CO_2-e. Unlike most developed countries, agriculture is the largest contributing sector representing 47.2% of total emissions.[1] This is primarily due to methane emissions from ruminant livestock and nitrous oxide emissions from the use of fertiliser. Agricultural emissions continue to increase (12.1% since 1990) particularly associated with the expansion of the dairy sector.

New Zealand's per capita rate of 16.4 t CO_2-e per person is fifth highest among the 40 Kyoto Protocol Annex 1 countries. However, its carbon dioxide only emissions (7.6 t CO_2-e per person) are relatively lower (22nd among Annex 1 countries). This reflects the high proportion of renewable generation in the electricity sector (77%) with hydro-generation producing 22,639 GWh (52.8%) of New Zealand's 42,900 GWh generated in 2012 (Ministry of Economic Development 2013).

Also New Zealand has a relatively high level of net removals from afforestation, reforestation and deforestation at 13.5 Mt. CO_2-e (18.5% of total emissions). However net removals from forestry have decreased due to increased harvesting of plantation forests as a larger proportion of the estate reaches harvest age, and forest being converted to pasture. Between 2003 and 2012, New Zealand's planted forest has declined from 1,827,333 to 1,719,501 ha (6% decline), while in Canterbury the planted forest has declined from 122,773 to 110,055 ha (10% decline) (Ministry of Agriculture and Forestry 2004) (Ministry for Primary Industries 2013). Deforestation intention surveys indicate 86% conversion from forestry to dairying (Manly 2013).

In relation to water management and New Zealand's emissions profile, there are the following significant linkages:

- The role of hydro-generation in reducing carbon dioxide emissions from the electricity sector: this is particularly significant for Canterbury with 65% of the country's hydro capacity;
- The role of agriculture, particularly dairying, in increasing methane and nitrous oxide emissions: this is also significant for Canterbury where the major growth in dairy conversions is occurring: 40% of the increase in cow numbers in New

[1] Typical figures for other developed countries are around 12%.

Zealand between 2007/8 and 2012/3 occurred in Canterbury (Livestock Improvement Corporation 2008) (DairyNZ 2013).

- The decreasing role of afforestation/deforestation in reducing New Zealand's net emissions: this is also significant in Canterbury where there has substantial conversion of forest blocks to pasture for dairying (12,700 ha between 2003 and 2012).

7.1.2 Projected Temperature Change

There has been a long-term increase in the average temperature for New Zealand of about 0.6 °C between 1920 and 2000. Figure 7.1 shows the annual average temperature from 1850 compared to the 1971–2000 average with blue bars showing the deviation below that average and the red bars showing above average deviations.

Figure 7.2 indicates the projected change in annual average temperature across New Zealand over 50 years (1990–2040). For Canterbury, this is 0.9 °C with a range across six scenarios of 0.2–1.9 °C. The projected increase is slightly higher in winter (1 °C), and lower in spring (0.7 °C).[2]

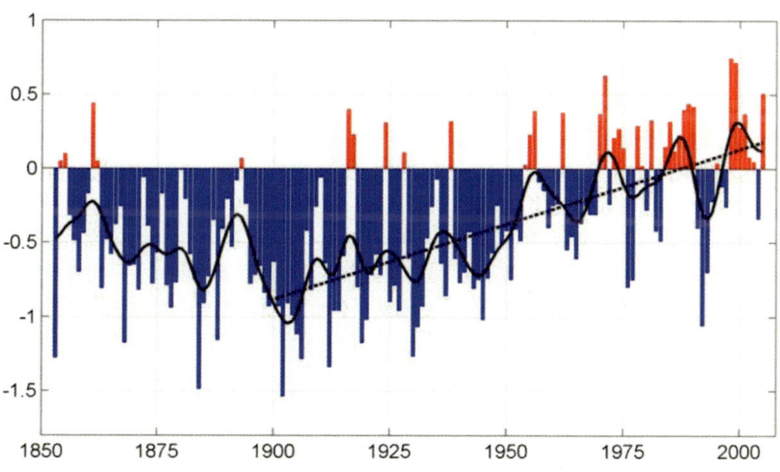

Fig. 7.1 New Zealand average surface temperature (°C) NIWA data (O'Donnell 2007)

[2] Temperature projections have been recently updated using the results of the IPCC Fifth Assessment Report (IPCC 2013). The ensemble average of projected changes in annual mean temperature between 1986–2005 and 2031–2050 for Canterbury vary from 0.7 °C to 1.0 °C for the four representative concentration pathways with a range from 0.4 °C to 1.6 °C. Projected increases for winter were slightly higher and spring slightly lower (Ministry for the Environment, 2016a, b, c).

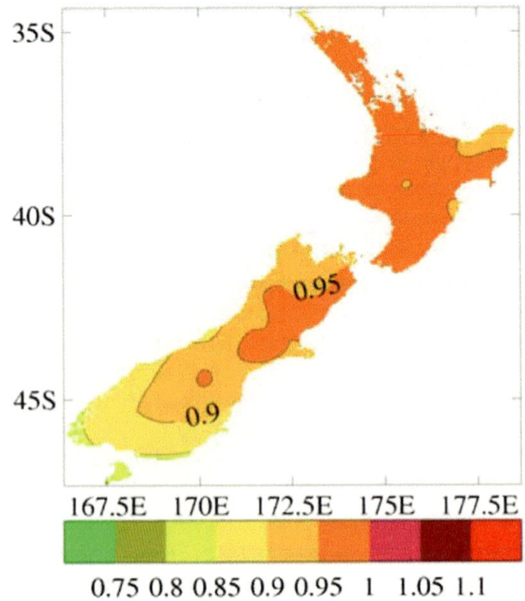

Fig. 7.2 Projected changes in annual mean temperatures (in °C) in 2040 relative to 1990 average over 12 climate models for A1B emission scenario (Ministry for the Environment 2008)

7.1.3 Projected Changes in Seasonal Rainfall

The projections for rainfall are more variable for the different climate models. There is also greater spatial and seasonal variability. Figure 7.3 shows the projected changes in summer and winter rainfall (in percentages) for 2040 relative to 1990. It is based on the average of the 12 climate models for the A1B emission scenario[3]. For the country, the general pattern is in summer for small increases on the east coast (2.5–5%) and marginally drier conditions on the west coast (0–2.5%). While in winter there are more significant changes projected with decreases on the east coast (7.5–10%) and increases on the west coast (5–12.5%).[4]

For Canterbury, lower winter rainfall on the Canterbury Plains means reduced aquifer recharge and lower flows in foothill rivers. However, for the major alpine rivers there is increased rainfall in the upper catchments in the Southern Alps.

---

[3] A1B scenario is one of the Special Report on Emission Scenarios (Nakicenovic and Swart 2000). It assumes rapid economic growth and global population that peaks mid-century and declines thereafter, rapid introduction of new and more efficient technologies and a balance of fossil and non-fossil energy sources. This assumes a doubling of global emissions from 1990 to 2050 and declining thereafter.

[4] Rainfall projections have been recently updated using the results of the IPCC Fifth Assessment Report (IPCC 2013). The ensemble average of projected changes in precipitation between 1986–2005 and 2031–2050 for Christchurch vary from 1 to 3% increase in summer and 0–4% decrease in winter, while for Tekapo in the Southern Alps it is 0–2% increase in summer and 6–11% increase in winter (Ministry for the Environment 2016c).

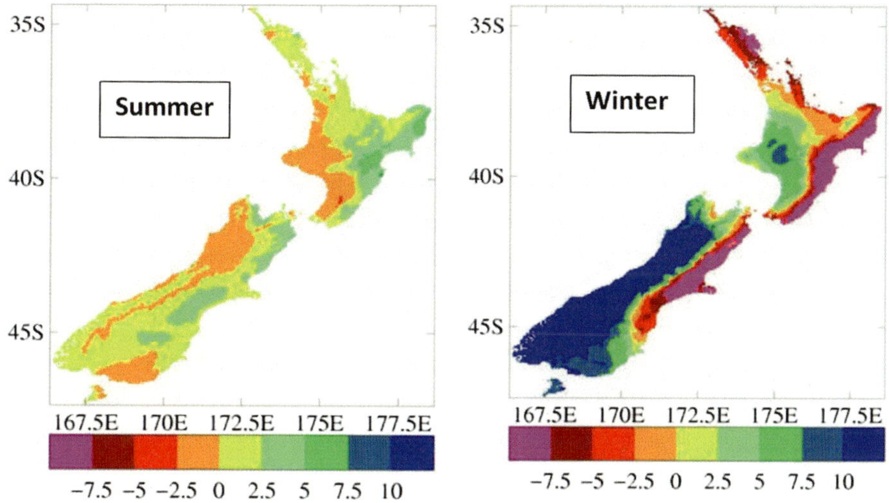

Fig. 7.3 Projected changes in seasonal mean rainfall (in %) for 2040 relative to 1990: Average over 12 climate models for A1B emission scenario (Ministry for the Environment 2008)

7.1.4 Projected Changes in Snowfall

Chinn investigated 127 glaciers in the Southern Alps (Chinn 1996). He found that on average glaciers had shortened by 38% and lost 25% in area. The upward shift of glacier mean elevation from 1890 to 1995 of 94 m is approximately equivalent to a temperature rise of 0.6 °C.

In the climate projections with higher winter temperatures, it is expected that snow cover will decrease and snowlines rise (Ministry for the Environment 2008). Figure 7.4 shows a projection of snow amount changes from the NIWA climate model for the A2 emissions scenario[5] for the 100 years from 1980–1999 to 2080–2099.

A decrease in winter snowfall and an earlier spring melt can cause marked changes in the annual pattern of river flow with higher flows in winter and early spring and lower flows in summer at the height of the irrigation season.

[5]A2 scenario assumes a heterogeneous world, increasing global population, regionally oriented economic development and slower technological change compared to other scenarios. This assumes a doubling of emissions from 1990 to 2040 and ongoing increases to 2100 (Nakicenovic and Swart 2000).

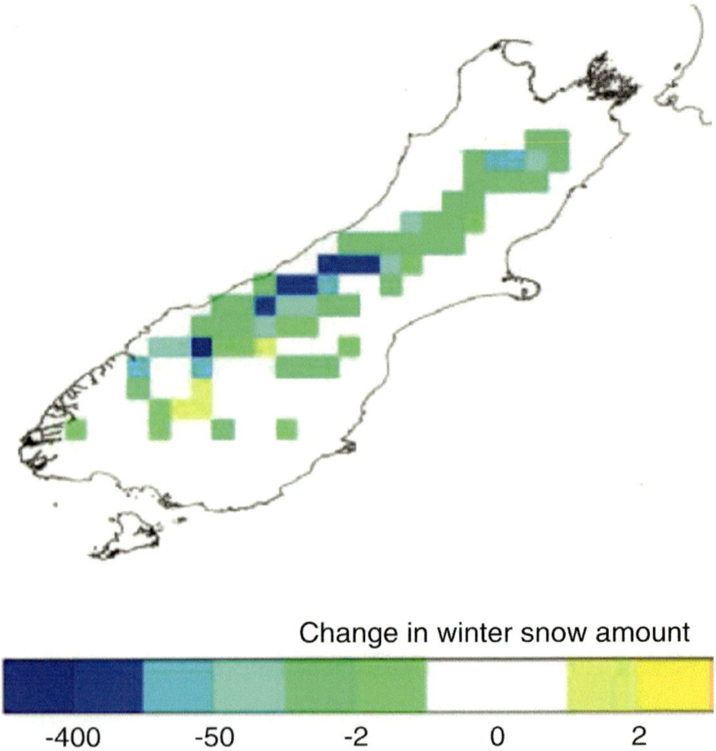

Fig. 7.4 Change in winter snow (in kg/m^2) between 1980–1999 and 2080–2099 under scenario A2 (Ministry for the Environment 2008)
Note: The contour intervals are not equally spaced. The snow amount is that lying on the ground averaged over the season. 1 kg/m^2 is equivalent to 1 mm rainfall.

7.1.5 Potential Evapotranspiration Deficit

Accumulated Potential Evapotranspiration Deficit (PED) is the amount of water that would need to be added to a crop over a year to prevent loss of production due to water shortage. PED can be used as an indicator of drought risk and as an indicator of irrigation demand. For unirrigated pastures, an increase in PED of 30 mm corresponds to approximately one week more of pasture moisture deficit and reduced grass growth (Mullen et al. 2005).

NIWA investigated four climate change projections from two global climate models (Hadley and CSIRO) downscaled to take account of New Zealand's local climate. This provided a range of projections from 25% ('low-medium') to 75% ('medium-high') of the projected global temperature range. 'Current' climate was based on data from 1972–2003, and projections were for '2030s' (2020–2049) and '2080s' (2070–2099).

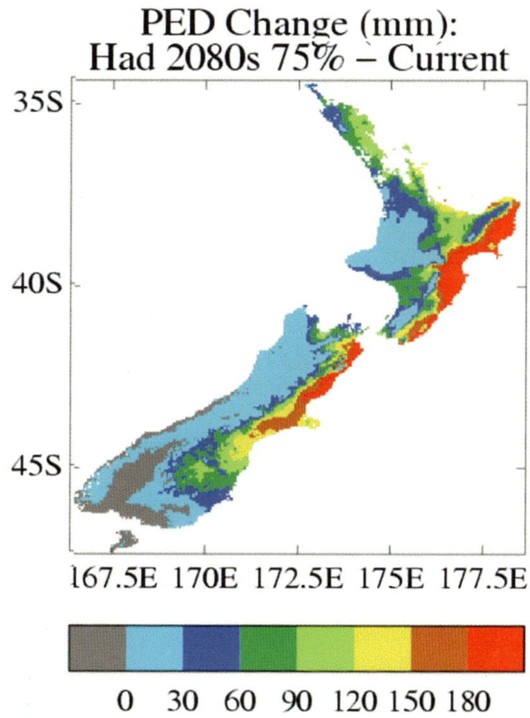

Fig. 7.5 Average change in annual accumulated PED (mm) between current climatology and projected climatology for 2080s using Hadley Model Scaled to IPCC 75% global warming (Mullen et al. 2005)

The modelling indicated an increase in drought in the eastern regions of New Zealand with the frequency of the current 1-in-20-year drought increasing between two and more than fourfold depending on the scenario. For Canterbury, which already has a high annual average PED of 322 mm (refer Sect. 7.3.1 above), climate projections indicate this will increase with some areas with PED increases of over 180 mm in 2080s with the medium-high projection (see Fig. 7.5).

7.1.6 Implications of Climate Change Projections for Water Management in Canterbury

The projections of climate change have significant implications for the management of freshwater in Canterbury. The most important changes are:

- The increase in PED which will generate increased irrigation demand
- The decrease in winter rainfall on the Canterbury Plains will reduce aquifer recharge and groundwater levels thereby reducing flows in groundwater-fed lowland streams
- The drier east coast in winter will lead to lower flows in foothill rivers
- The wetter west coast and warmer winters leading to reduced snow and increased winter flows but reduced summer flows in alpine rivers.

The implications will vary across the region. This is very much a nested system with global climate change downscaled to projected changes across New Zealand and the freshwater implications being catchment specific. The case study is presented of the changes projected for the Waimakariri catchment in central Canterbury, one of the region's alpine rivers. The critical variable is the reliability of supply to irrigators subject to maintaining environmental requirements for surface and groundwater.

7.1.7 Changes in Irrigation Reliability in the Waimakariri Catchment Under Climate Change Scenarios

The impact of a range of climate change scenarios on the irrigation reliability of the main irrigation scheme that extracts water from the Waimakariri River – the Waimakariri Irrigation Limited (WIL) –was undertaken (Srinivasan et al. 2011). WIL supports 18,000 ha of irrigated land. The investigations covered the three 20-year periods: 1980–99 ('1990 condition'), 2030–49 ('2040 scenario'), and 2080–99 ('2090 scenario'); and three climate scenarios: B1 (low emission), A1B (medium emissions), and A1FI (high emissions).

WIL is consented to abstract 10.5 m^3/s from the river during the irrigation season but the consent conditions mean the take can be restricted by the flow in the river. This is to protect the environmental flow requirements for the river. No abstraction is permitted when the river flow is below 41 m^3/s (full restriction); between 41 and 63 m^3/s a proportion of the consented take can be abstracted (partial restriction); and above 63 m^3/s the full consented take can be abstracted (no restriction). Previous studies concluded that low river flows limit irrigation supplies 11% of the time between September and December, and 48% of the time between January and April (Srinivasan and Duncan 2012).

Typical of Canterbury's alpine rivers, the Waimakariri has high rainfall in its headwaters in the Southern Alps (2000–5000 mm), moderate rainfall in the foothills (1000–2000 mm) and low rainfall on the plains (less than 1000 mm). The change in rainfall pattern projected by climate change was consistent across the scenarios (although varying in amount). The high rainfall upper catchment increased in rainfall particularly in the May–August period, there was a smaller increase in the foothills and the plains had a decrease in May–August. The projections for A1B scenario in 2040 is shown in Table 7.1.

The increased precipitation leads to increased mean annual flow, e.g. a 7% increase for the 2040 A1B scenario (Zammit and Woods 2011). However, the change in flow varies throughout the year. Figure 7.6 presents the average monthly flows for the different scenarios. There are large flow increases from May to September, but little change and even slight decreases between September and April (the southern hemisphere irrigation season). This is attributed to the increased temperature associated with climate change scenarios resulting in more rainfall and less snowfall and an earlier snowmelt.

Table 7.1 Projected precipitation in A1B 2040 scenario (in mm) and percentage change from 1990 condition (Srinivasan et al. 2011)

Period	Upper catchment (above 2000 mm)	Foothills (1000–2000 mm)	Plains (below 1000 mm)
Annual	3995 (+5%)	1310 (+4%)	799 (−1%)
Jan–Apr	1551 (+1%)	357 (+2%)	262 (+4%)
May–Aug	1301 (+11%)	478 (+5%)	274 (−6%)
Sept–Dec	1549 (+5%)	474 (+4%)	263 (<−1%)

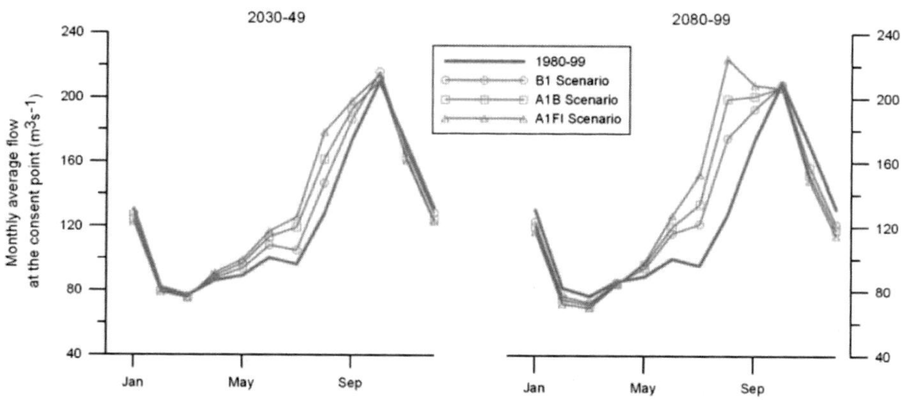

Fig. 7.6 Monthly average flows for scenarios (Srinivasan et al. 2011)

Table 7.2 Modelled water stored as snow (in mm): Average over 20-year period (Srinivasan et al. 2011)

Scenario	Annual average of maximum snow storage over 20-year period (mm)
1990 condition (1980–99)	155
2040 scenarios (2030–49)	
B1	134
A1B	109
A1FI	97
2090 scenarios (2080–99)	
B1	90
A1B	69
A1FI	40

The modelled amount of water stored as snow for the different scenarios is shown in Table 7.2. The table indicates the average of the maximum snow storage for the year over the 20-year period for each scenario. The table shows the decline from the 1990 condition where the annual average of the snow storage over the 20-year period (1980–99) of 155 mm to 109 mm in 2040 for A1B scenario (medium emissions) and

Table 7.3 Average duration (days) of longest continuous restriction over 20-year period (Srinivasan et al. 2011)

Scenario	Duration	Full restriction (100%)	Partial restriction (50–99%)	Partial restriction (1–49%)
1990 condition (1980–99)	27	6	11	10
2040 scenarios (2030–49)				
B1	28	6	13	9
A1B	29	7	13	9
A1FI	30	8	13	9
2090 scenarios (2080–99)				
B1	32	8	14	10
A1B	33	10	14	9
A1FI	34	12	14	8

a further decline to 69 mm in 2090. There are greater declines for the high emission scenarios (A1FI[6]) and lesser declines for the low emission scenarios (B1[7]).

So despite higher precipitation overall in the catchment with the reduction in snowmelt in spring there are more projected irrigation restrictions between September and December. One key summary indicator is average duration of the longest continuous restriction over the 20-year period. Table 7.3 shows the increased length of restriction with projected climate change and the greater proportion of full restrictions as part of the continuous restriction. For the 1990 condition the average duration is 27 days with 6 of those on full restriction. This can be compared to 29 days for the A1B 2040 scenario with 7 days of full restriction increasing to 33 days for A1B 2090 scenario with 10 of those days on full restriction.

7.2 Climate Change Response

7.2.1 New Zealand's Legislative Response to Climate Change

New Zealand's principal legislation relating to natural resources management is the Resource Management Act 1991 (RMA). The legislation was developed in the late 1980s prior to significance of climate change being appreciated. The RMA was

[6] A1FI scenario is one of the Special Report on Emission Scenarios (Nakicenovic and Swart 2000). It assumes rapid economic growth and global population that peaks mid-century and declines thereafter, rapid introduction of new and more efficient technologies with a technological emphasis on fossil-intensive sources.

[7] B1 scenario is one of the Special Report on Emission Scenarios (Nakicenovic and Swart 2000). It is based on a convergent world with the global population, that peaks in mid-century and declines thereafter, as in the A1 scenarios, but with rapid change in economic structures toward a service and information economy, with reductions in material intensity and the introduction of clean and resource-efficient technologies. The emphasis is on global solutions to economic, social and environmental sustainability, including improved equity, but without additional climate initiatives.

amended in 2004 to incorporate renewable energy and climate change provisions. There is also the Climate Change Response Act 2002 which provided the legal framework for New Zealand to meet its international obligations under the Kyoto Protocol. An amendment in 2008 established a greenhouse gas emissions trading scheme (ETS) which has been subject to further amendments.

The 2004 amendments to the RMA included the addition of "the effects of climate change" in Section 7 as one of the other matters that decision makers "shall have particular regard to" in exercising functions and powers under the RMA. However, it is noteworthy that climate change effects were listed in Section 7 (Other matters) rather than given status under Section 6 (Matters of national importance) which decision makers "shall recognize and provide for".

The 2004 amendments preclude regional councils from having regard to the effects of greenhouse gases on climate change (RMA Sections 70A and 104E). The intention was that climate change would be addressed as a national issue through a National Environmental Standard (NES): no NES has been promulgated. It was also stated in the purpose of the amendments that local authorities are to plan for the effects of climate change. However no legislative provisions have been made, rather guidance documents have been prepared (Ministry for the Environment 2008).

The Climate Change Response Act 2002 puts in place the legal framework to enable New Zealand to meet its international obligations under the United Nations Framework Convention on Climate Change (UNFCCC) and the Kyoto Protocol. The Act establishes a national agency and register to record and report greenhouse gas emissions to meet New Zealand's reporting obligations. It also provides the authority for the Minister for Finance to manage greenhouse gas emission units and to retire emission units equal to New Zealand's greenhouse gas emissions in the first commitment period (2008–12) under the Kyoto Protocol.

An amendment in 2008 established a greenhouse gas emissions trading scheme (ETS). The ETS was to cover all gases and all industries but with different entry times. Compliance for industries would require the surrender of a New Zealand emission unit or an international unit for each tonne of greenhouse gas emissions. New Zealand emission units were to be capped in number and were to be allocated by grandparenting (gifting) or auctioning. Trade-exposed industries were to receive a 90% free allocation of units to 2018 with phasing out by 2030. Forest owners with pre-1990 forests were to receive a fixed one-off free allocation of units. Post-1989 afforestation would earn credits while units would have to be purchased for deforestation.

With a change of government there were amendments to the ETS in 2009 (the Climate Change Response (Moderated Emissions Trading) Amendment Act 2009) and 2012 (the Climate Change Response (Emissions Trading and Other Matters) Amendment Act 2012). The current scheme has removed the cap on New Zealand emissions and unlimited importation of international units was permitted. Compliance now only requires one unit for every 2 tonnes of emissions. Trade-exposed, emission-intensive industries get free allocations based on production. The phase out of free allocations is now over a longer time period. There are no free allocations to industry that can pass on costs to consumers. There is an indefinite deferral of including agriculture emissions in the ETS.

The outcome of the ETS led to significant deforestation before the commencement of the commitment period in 2008 (Ministry for the Environment 2016b). The purchase of international units of dubious efficacy has removed the carbon price signal to motivate reduction in greenhouse gas emissions (Sustainability Council of New Zealand 2015). The free allocations to industry transfer the costs of compliance to the taxpayer (Sustainability Council of New Zealand 2015). The uncapped system with low carbon cost has led to growth in current and projected emissions.

7.2.2 Current and Projected Emissions Compared with Targets

Under the UNFCCC, the 1990 baseline for New Zealand's greenhouse gas emissions is 65.8 Mt. CO_2-e in terms of gross emissions and 36.9 Mt. CO_2-e in terms of net emissions (i.e. after deducting the contribution of land use and forestry as a net carbon sink of 28.9 Mt. CO_2-e) (Ministry for the Environment 2016a). Gross emissions for 2014 are calculated as 81.1 Mt. CO_2-e which is an increase of 23% above 1990 levels (Ministry for the Environment 2016a).

Projected gross emissions by 2030 represent a 42% increase above 1990 levels (Sustainability Council of New Zealand 2015). This can be compared with New Zealand Government's provisional gross emission target by 2030 of 30% below 2005 emissions (59.2 Mt. CO_2-e) which is equivalent to 10% below 1990 levels.

The science says that if we are serious about keeping warming to less than two degrees then we need about a 40% reduction from 1990 levels by 2030, 90% by 2050 and 100% by 2060 – and then negative emissions (removal of CO_2 from the atmosphere) for the rest of the century (Renwick 2015).

The ETS is the principal means that is currently in place to generate greenhouse gas reductions. An analysis of the ETS has estimated that the effect of the ETS will only be a 0.4% reduction in emissions by 2030 compared to government taking no action (Sustainability Council of New Zealand 2015).

Despite being the dominant source of New Zealand's greenhouse gas emissions, agricultural emissions are not part of the ETS and land use intensification, such as dairy conversions and forest clearance, are not subject to EIA evaluation under the RMA.[8] The agricultural sector is projected to provide 77% of the growth in emissions (Sustainability Council of New Zealand 2015). The Government's view was that the lack of mitigation options meant that agriculture should be kept out of the ETS (Editor 2011). However, there are mitigation measures and offsets available to address agricultural emissions.

[8] Greenhouse gas emissions have been specifically excluded from consideration by local government authorities in the consenting process (New Zealand's impact assessment process). This was on the basis that there would be a national approach through a National Environmental Standard (NES). However, no NES has been promulgated.

7.2.3 *Mitigation Approaches and Offsets Available for Agricultural Emissions*

In terms of greenhouse gas mitigation the most promising options for nitrous oxide have been identified as: nitrogen inhibitors that keep nitrogen in the less mobile ammonium form for longer, the use of herd shelters that can minimize the deposition of urine patches at high-risk times of the year, and, replacing nitrogen fertilizer inputs to boost pasture production with inputs of maize or cereal silage to reduce the amount of nitrogen ingested and excreted (De Klein et al. 2010).

For methane generation, the Parliamentary Commissioner for the Environment has identified potential methane reduction strategies that are in the research stages. These are selective breeding of low emission sheep and cattle, changing animal feed, and biotechnologies that target microbes in the rumen that produce methane (Parliamentary Commissioner for the Environment 2016).

7.2.3.1 Nitrification Inhibitors: Need to Consider Multiple Failure Pathways

The development of nitrification inhibitors is an excellent example of considering the nature of biophysical systems at the farm scale and identifying critical variables in an adaptive system and effective points of intervention to take management action. However, it also demonstrates the need to consider multiple failure pathways at larger scales as part of a nested adaptive system. While very effective at the farm scale, the presence of the active ingredient, dicyandiamide (DCD), in dairy exports has led to the suspension of production and use of nitrification inhibitors.

The concept of the nitrification inhibitors is designed to intervene in the nitrogen cycle in the soil profile. Nitrogen as ammonia percolates into the soil profile from urine patches, nitrogen fertilisers and nitrogen fixation by plants. The nitrification process converts the insoluble ammonia into nitrite and nitrate which leaches into the groundwater or volatilises as nitrogen gases, including nitrous oxide, into the atmosphere. The nitrification inhibitor, DCD, slows down the rate of nitrate production and thus reduces the nitrate leaching loss to groundwater and nitrous oxide loss to atmosphere. Figure 7.7 shows the nitrogen cycle in the soil profile and the point of intervention of the nitrification inhibitor in that cycle.

The effectiveness of the approach has been confirmed in experimental and field trials. Experiments involving two Canterbury soils and two North Island soils showed that the application of a fine particle suspension of DCD to grazed pasture was very effective in reducing nitrous oxide emissions with an average of 70% reduction (Di et al. 2007). Figure 7.8 shows the results of for a Canterbury Templeton soil indicating the daily flux of nitrous oxide for a urine rate of 1000 kgN per ha with and without DCD (as well as controls).

Lysimeter trials have indicated the effectiveness of DCD in reducing nitrate leaching losses. Trials on Templeton soils were found to reduce nitrate leaching

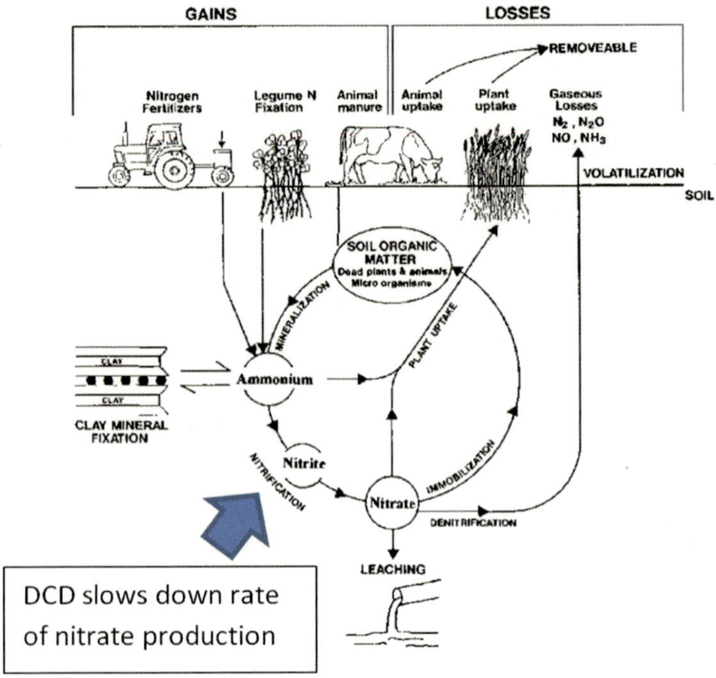

Fig. 7.7 Nitrification inhibitor: Point of intervention in nitrogen cycle in soil profile (McLaren and Cameron 1996)

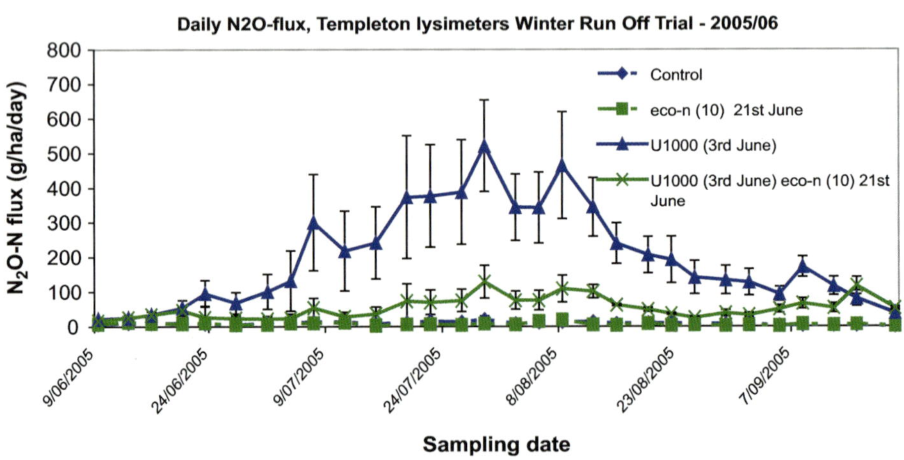

Fig. 7.8 Nitrous oxide flux reduction due to nitrification inhibitor (Di et al. 2007)

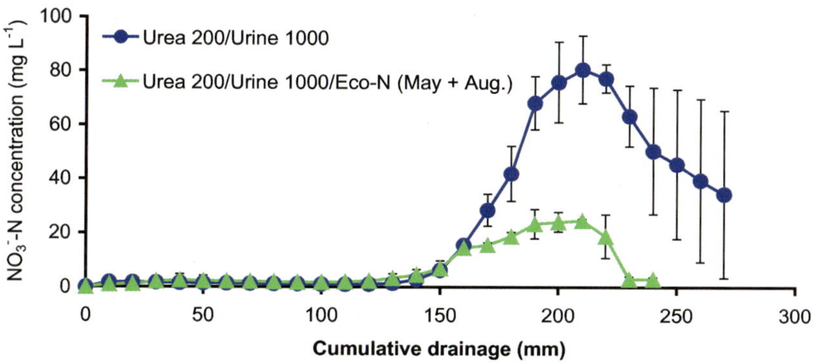

Fig. 7.9 Nitrate leaching reduction due to nitrification inhibitor (Di and Cameron 2004). Copyright The Royal Society of New Zealand; reprinted with permission of publisher Taylor & Francis

losses by 60% (Di and Cameron 2004). Figure 7.9 shows the results for irrigated pasture with urine only and irrigated pasture with application of DCD in May and August with respect to the nitrate concentration in drainage collected by the lysimeter.

However, in 2012, the US Drug and Food Administration added DCD to a list of substances to test for. Despite its use over many years, there has never been an international standard related to maximum residue levels in food products. Tests in New Zealand have shown the occasional presence of low levels of DCD coinciding with times of the year that the product is applied. Because zero detection of DCD cannot be guaranteed, producers of DCD like Ravensdown took the voluntary step to suspend its use (Rural News 24 Jan 2013).

So while nitrification inhibitors address crucial failure pathways of the nitrogen cycle for nitrous oxide emissions and nitrate leaching as well as increasing agronomic production, the potential health effects pathway and trade network failure pathway have not been adequately addressed.[9]

7.2.3.2 Offsets

Another strategy to address the increase in greenhouse gas emissions due to dairy conversions and forestry clearance in Canterbury (and elsewhere in New Zealand) is the requirement for offsets. The greenhouse gas emissions from dairy farms are variable: Ledgard examined 26 dairy farms in Rotorua and estimated an average of 9067 kgCO_2-e per ha with a range from 4504 to 12,198 kgCO_2-e per ha (Ledgard

[9] The example shows the importance of considering multiple failure pathways. DCD is effective in reducing nitrous emissions (contributing to addressing failure pathway 9 of climate change) and in reducing nitrate leaching (contributing to addressing failure pathway 2 of environmental impacts). However, there is no standard for DCD residue in food products (related to failure pathway 6 – the disease pathway) so when traces were found in milk products its use has been stopped (based on failure pathway 8 – collapse of trade network).

et al. 2010). Smeaton modelled a base dairy farm model of 9300 kgCO_2-e per ha compared to a sheep and beef farm of 3400 kgCO_2-e per ha (Smeaton et al. 2011). Thus, for a conversion from a sheep and beef farm to a dairy farm would require an offset of about 5900 kgCO_2-e per ha.

Mason and Ledgard are developing a calculator for making farms greenhouse neutral based on the number of hectares of radiata pine plantation which would be required for each 30 year period of farming, assuming a hectare of pine plantation can absorb 11,800 kgCO_2-e, allowing for harvesting (Mason and Ledgard 2013). Thus, for a dairy farm to be greenhouse gas neutral would require about 0.8 ha of pine plantation for each hectare of dairy farm, or, as an offset for a dairy conversion from a sheep and beef farm, about 0.5 ha of pine plantation for each hectare of dairy farm.

Another avenue for offsets is the generation of hydroelectricity as a component of irrigation storage (e.g. Opuha Dam) or tail race discharged (e.g. Rangitata Diversion Race (RDR)). The Highbank and Montalto power stations, associated with the 64,380 ha RDR scheme, generated 98 GWhr in 2011. In terms of fossil fuel emissions avoided (about 513 tCO_2-e per GWhr), the Highbank/Montalto generation would counterbalance about 5500 ha of dairy farm emissions (assuming 9.3 tCO_2-e per ha), or, offset the conversion of 8500 ha of sheep/beef farms to dairy farms (based on 5.9 tCO_2-e per ha differential emission rates).

However, the current provisions of the RMA require local authorities **not** to consider the discharge of greenhouse gas emissions. As shown by other jurisdictions, the environmental assessment process can be a very effective tool in reducing greenhouse gas emissions from projects. Box 7.1 sets out the policy for considering greenhouse gas emissions in environmental impact assessment in Western Australia and examples of its application resulting in improved design and adoption of offsets (Western Australia Environmental Protection Authority 2002).

Box 7.1: Use of Environmental Impact Assessment to Manage Greenhouse Gas Emissions of Development Proposals in Western Australia

With a heavy emphasis on resources and energy development, primarily for export, Western Australia is a major contributor of greenhouse gas emissions. The top ten emitters from mining and hydrocarbon projects emit an estimated 34.7 MtCO_2-e (6.5% of Australia's total emissions).

Environmental impact assessment (EIA) for new developments is primarily the responsibility of state governments. The Western Australian Environmental Protection Authority has recommended conditions aimed at reducing greenhouse gas emissions on development proposals using EIA since the late 1990s. Proponents of projects with significant greenhouse gas emissions are to (1) identify all greenhouse gas emission sources and calcu-

(continued)

Box 7.1 (continued)

late emissions in accordance with the National Greenhouse and Energy Reporting Act; (2) demonstrate that the proposal is designed and will be operated in a manner which maximises energy efficiency and minimises greenhouse gas emissions as far as practicable; and, (3) provide an analysis of greenhouse gas intensity (i.e. quantity of CO_2-e generated per tonne of product produced) and consider published benchmarked best practice for equivalent plants and equipment (Environmental Protection Authority 2015).

In the initial form of the EPA policy (Environmental Protection Authority 1998) the benchmarking was also against Australia's target from the 1997 Kyoto Climate Change Conference. Australia's 108% target represents a 25% reduction from "business-as-usual" predictions of greenhouse gas emissions for the year 2010 (which was 143% of 1990 levels). The expectation was that companies producing greenhouse gas emissions would go beyond a "no regrets" approach (i.e. implement more than cost neutral measures). Examples of the outcomes from the EIA processes for major mining and hydrocarbon projects are presented below:

- The Murrin Murrin Nickel-Cobalt project expansion was to increase to 250% of its original capacity. It involved the mining of a new ore body and transporting the ore to an expanded plant. The commitment in relation to reducing greenhouse gas emissions from the EIA process were: (1) the adoption of a recent development of nickel laterite processing – a no regrets measure achieving an estimated 10–25% reduction in greenhouse gas intensity; (2) indirect heating (rather than direct steam injection into the process); (3) rail transport of ore (rather than truck); and, (4) tree farming to offset emissions. This was estimated to achieve a 16–30% greenhouse gas emission reduction compared to 1990 business-as-usual (Environmental Protection Authority 1999).
- The Gas to Synthetic Hydrocarbons Plant on Burrup Peninsula was to process natural gas to produce 1240 tonnes per day of synthetic crude oil which can then be processed into specialty products such as lubricants and diesel fuel. The commitments in relation to reducing greenhouse gases from the EIA process were: (1) a 50% improvement in thermal efficiency compared to the pilot plant, (2) improved life cycle analysis of products, e.g. sulphur-free diesel fuel, and, (3) use of process steam for the state government's water desalination plant eliminating the need for an alternative fuel source (Environmental Protection Authority 2000).
- The Gorgon Gas Development Expansion on Barrow Island Nature Reserve was to expand liquefied natural gas production from 10 million

(continued)

Box 7.1 (continued)

tonnes per annum (MTPA) to 15 MTPA by the addition of a third gas processing train. The commitments to reduce greenhouse gas emissions from the EIA process were: (1) sequestering carbon dioxide emissions into a saline aquifer 2000 m beneath the ground, (2) LNG technology improvement, (3) use of sub-sea production system, and, (4) improved waste heat recovery. This reduced the greenhouse gas emission intensity from 0.89 tCO_2-e per tonne of LNG from the 1998 concept design to 0.35 tCO_2-e per tonne of LNG (Environmental Protection Authority 2009).

7.3 Drought and Its Management

Drought can be defined in biophysical terms. References are made to *meteorological drought*: "The state of the climate system that creates abnormally dry weather, prolonged enough for the lack of rainfall to cause serious hydrological imbalances" (Ministry of Agriculture and Forestry 2009). *Serious rainfall deficiency* is a period over 3 months or more where rainfall is below the lowest 10% of records but above 5%. *Severe rainfall deficiency* is a period over 3 months or more where rainfall is amongst the lowest 5% of records.

References are also made to *hydrological drought*: "a deficit of water in the landscape either in the groundwater reserves or in the surface hydrological system such as rivers, streams and lakes" (Ministry of Agriculture and Forestry 2009). A *significant drought* is defined as when the soil moisture deficit is greater than 110 mm, while a *severe drought* is when the soil moisture deficit is greater than 130 mm.

However, from a nested adaptive system perspective it is not only the biophysical system characteristics that are important but also the socio-economic response to drought conditions that is relevant. This is better reflected in the definition of *agronomic drought*: "a protracted period of deficient precipitation resulting in extensive damage to crop/pasture growth and production" (Ministry of Agriculture and Forestry 2009). Suitable indicators are grass growth rate, crop yields and livestock condition. Agronomic drought will often follow hydrological drought but can be influenced by management practices such as irrigation, supplementary feed and stocking rate practices.

The approach to drought adaptation presented here draws heavily from the study of farmer response in South Canterbury and North Otago (Burton and Peoples 2008). While the terminology used by Burton and Peoples is specific to agriculture, the concepts that they use are consistent with nested adaptive systems with consideration of responses at multiple levels (i.e. individual farmer, community affected by drought, government); over different time frames (i.e. tactical and strategic); and the interaction between socio-economic and biophysical systems (i.e. the adaptation

Table 7.4 Alignment of drought response approaches with sustainability approaches (Chapin et al. 2009 and Burton and Peoples 2008)

Sustainability approaches	Drought response approaches
Increase resilience	Ability to resist drought
Enhance adaptive capacity	Flexibility to deal with drought
Reduce vulnerability	Minimising the impact of drought
Enhance transformability	Adjustments to farm structure

strategy selected is dependent on the individual farmer, the nature of the drought, and, the economic conditions of the farming industry).

By combining the drought adaptation approaches from their literature review and survey research there is alignment with the general sustainability approaches identified by Chapin et al. (refer Table 4.2). From their survey research Burton and Peoples identified three drought response approaches:

- The first was the development of the farm to be *able to resist drought*: this is compatible with Chapin's *increase resilience approach.*
- The second was the development of a farming system that *provides flexibility to deal with drought*: this is compatible with Chapin's *adaptive capacity.*
- The third was *minimising the impact of drought*: this is compatible with Chapin's *reduce vulnerability.*

In their literature review they also identified approaches that were *adjustments to farm structure*, i.e. long-term crop or livestock changes, and, changes or investment in technology. This is compatible with Chapin's *enhance transformability.*

Table 7.4 summarises the alignment of the Burton/Peoples drought response approaches with Chapin's sustainability approaches. Specific strategies for each of these approaches are described below.

7.3.1 Ability of Farm to Resist Drought (Increase Resilience)

Burton and Peoples identify two main requirements to increase a farm's ability to resist drought: firstly, structure the farm for drought conditions, and secondly, build up reserves during good years. In terms of structuring the farm for drought conditions they identify the following strategies:

- Distribute land over different areas or climate zones, e.g. a lowland farmer having a high country block;
- Select vegetation and livestock to suit drought and farm, e.g. one farmer developed a merino-lucerne system;
- Plant shelter belts to reduce evaporation losses from drying winds; and
- Invest in irrigation and/or water storage to provide an alternative water supply.

In relation to building up reserves, they put forward two strategies:

- Ensure there is sufficient stored feed, e.g. provide for silage storage; and
- Maintain sufficient capital (or access to capital) for drought years.

7.3.2 Flexibility of Farm System to Deal with Drought (Enhance Adaptive Capacity)

The following strategies were found by Burton and Peoples to provide flexibility of the farm system to deal with droughts:

- Diversify production types on the farm to reduce the reliance on one farming system, e.g. a farmer adds cattle and deer to his stock to reduce his reliance on sheep;
- Develop a farming system with a decision point for drought response so that there is a planned approach to shift from the normal farming system to a drought system, e.g. farmers determine the time of year that a drought is likely to become apparent and plan to be able to destock at this point if required;
- Keep spare capacity in the farming system to allow flexibility: this usually means not overstocking so that overgrazing or loss of feed reserves does not occur. Depending on climate or market conditions there is then flexibility to make use of the spare capacity.

7.3.3 Minimising the Impact of Drought (Reduce Vulnerability)

During the drought, actions can be taken to reduce the effects of drought. Burton and Peoples findings included:

- The need to make decisions and take action early: delaying decisions to when a drought is well established can lead to "fire sale" situations for selling stock;
- Adjust stock grazing to drought conditions: this is both in terms of stock numbers and rate of rotation of grazing land;
- Buying in supplementary feed to offset loss of pasture;
- Agist stock outside of drought-affected areas.

7.3.4 Adjustments to Farm Structure (Enhance Transformability)

The literature review component of Burton and Peoples' investigation identified some long-term transformations from North American and Australian experiences with drought. These transformations included long term crop or stock changes, and, changes or investment in new technology.

One example is the transformation of the Oklahoma Dust Bowl from the 1930s. In the 1930s farmers were growing cotton which has a high water requirement and thus vulnerable to drought, and corn which was at the limits of its growing range and therefore sensitive to dryer than normal conditions. There has been a transformation to pastoral beef farming and growing soy beans in fertile lowlands. There has been recurrence of the meteorological conditions that occurred in the 1930s but not a recurrence of the dust bowl because of the changes in farm structure (McLeman et al. 2008).

In Australia, extensive pastoral farming is already in place so there is little scope for crop or livestock changes. However a study of the El Nino drought of 1991–5 (Stehlik 2003) identified one of the key management responses was the use of information technology to predict drought occurrences and to undertake farm budgeting.

7.3.5 Socio-economic Responses

The above discussion has focussed on the approaches relating to the biophysical aspects of the agricultural system. Burton and Peoples also identify economic and social responses for adapting to drought.

The following economic responses were identified:

- Raising capital to survive drought through selling stored feed to take advantage of higher prices or selling land;
- Off-farm employment for the farmer or spouse;
- Reducing household and farm expenditure;
- Increasing the family workload and reducing hired labour.

They also identified the following social responses to help cope with drought:

- Managing personal stress: living with drought creates psychological stress and can lead to increased suicide rates;
- Talking to other farmers and listening to what they are doing;
- Keeping in contact with the industry, such as the abattoir and bank manager;
- Using Rural Support Trusts that provide help during and after an adverse weather or environmental event;
- Using government assistance for financial and labour support and special recovery measures.

7.3.6 Political and Economic Context

Burton and Peoples highlight the significance of the political and economic context in relation to farmers' experience of drought in the 1980s and 1990s. In the language of nested adaptive systems, this is the *trade network failure pathway*.

Table 7.5 Summary of drought response strategies

Level	Socio-economic system (SES)	Linkages SES and BPS	Biophysical system (BPS)	Transformation
Broader region	Government response; Economic context	Prediction of drought events	Landholding diversification	Regional crop/ stock changes
Linkage to broader region	Industry support		Import feed; Agist stock	
Drought-affected area	Community networks	Rural support trusts	Farm systems suited to drought	Water storage
Linkage to Individual farmer	Relationships banker, abattoir; Off-farm job		Learning from others	
Individual farmer	Manage personal stress; Manage expenditure	Decision point for response	Agriculture diversity; Adjust stock levels; maintain spare capacity	Irrigation; On-farm storage

At the time of the drought in the 1980s, the economic situation facing farmers involved the removal of subsidies for their produce, inflation and interest rates that were high, low equity in land, and declining lamb prices. These conditions coupled with drought meant there was a significant impact on farmers.

In the same meteorological drought conditions in the 1990s, the economic context was quite different with low interest rates, higher equity in farms and recovery of lamb prices. As a result, farmers were in a better financial position and thus able to cope better with drought conditions.

7.3.7 *Summary of Sustainability Strategies*

The range of sustainability strategies identified by Burton and Peoples can be summarised in the classification framework developed in Chap. 4 for Chapin's general types of sustainability strategies (refer Table 4.3). Table 7.5 displays the drought response strategies in this classification framework.

7.4 A Resilience Approach to Flood Management

The Christchurch/Kaiapoi area is considered to be the largest economic asset in New Zealand at risk from flooding. Figure 7.10 shows the extent of the flood risk to the main urban areas of Christchurch and Kaiapoi from the Waimakariri River. Managing that risk is a significant issue for the sustainability of the city. Providing

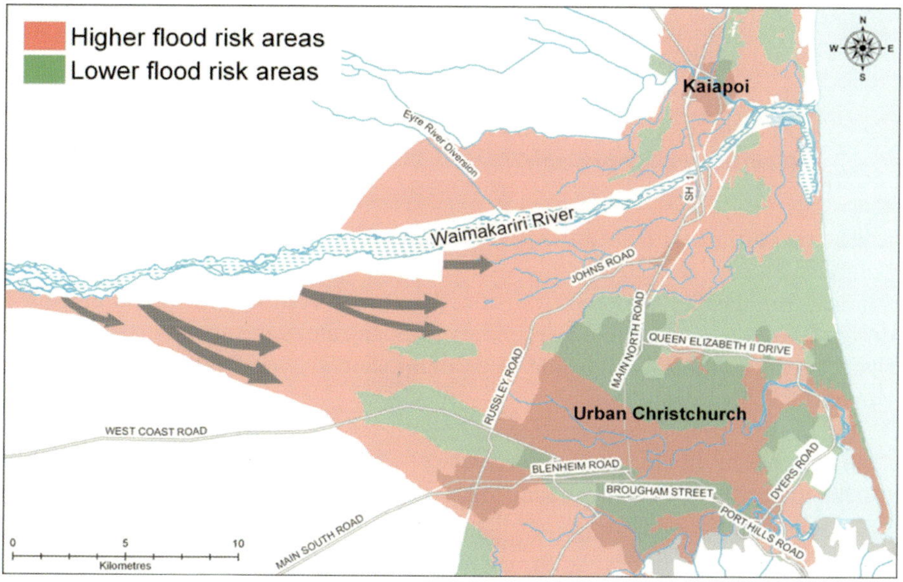

Fig. 7.10 Flood risk to the main urban areas of Christchurch (Environment Canterbury)

the adaptive capacity to manage the effects of a natural disaster on a socio-economic system is a type 3 sustainability issue (see Table 4.1).

7.4.1 Traditional Approach

The traditional approach to flood management is to provide stopbank protection for a flood of a specified return period. In the case of Christchurch, stopbanks were designed for a 1-in-500-year flood flow of 4730 cumecs (m^3/s). However, the traditional approach does not consider the risk of stopbank failure below the design flow. Recent North Island floods have experienced stopbank failure below design flow leading to flooding of "protected" areas and the inability of those floodwaters to return to the river because of stopbanks downstream. For the Waimakariri stopbanks there is considered to be a risk of breakouts from stopbank failure at 3300 cumecs which is 70% of the design flow.

Furthermore, the traditional approach does not provide adequate capacity for flood flows greater than the design flow. Climate change projections for the east coast of the South Island are indicating the occurrence of more extreme events. Also, there is the potential for braided rivers like the Waimakariri to transport large volumes of shingle and sediment reducing the existing channel capacity.[10] Thus a nominal design capacity is unlikely to be maintained in practice.

7.4.2 Resilience Approach

It has been estimated that in the next 30 years there is a 50% chance of stopbank failure and a 10% chance of urban area flooding. To address the issues of risk from a larger-than-design flood and containing breakouts from stopbank failure below a design flood, the provision of a secondary stopbank along the alignment of a natural river terrace on the southern side to accommodate a 1-in-10,000-year flow (6500 cumecs) has been designed.

The design concept is to contain and return breakout flow which involves flood storage between the primary and secondary stopbanks and returning that overflow to the main channel downstream. This also requires complementary work to strengthen and upgrade the stopbanks on the northern side of the river. The main proposed works are shown in Fig. 7.11 and the break and return scenarios are shown in Fig. 7.12.

There is also a number of mitigation measures incorporated in the design concept. These include:

- compensation for the potential damage from increased depth between the stopbanks
- rock lining in high velocity areas to reduce the risk of stopbank failure
- modifications to bridge embankments
- gravel removal from the channel to maintain channel capacity, and
- a flood warning and evacuation plan.

The need to shift from the traditional approach of designing for the risk associated with an infrequent event to a resilience approach that addresses the consequences of system failure is being recognised in other jurisdictions. As set out in Box 7.2, the dramatic failure of the levee system in New Orleans during Hurricane Katrina is a significant international example of the recognition of the need to manage for failure pathways rather than risk levels.

---

[10] Note that this is the natural geomorphological process that created the Canterbury Plains. Deposition of shingle and sediment in river channels gradually raises the elevation of the channel bed and reduces the flow capacity of the channel. A flood flow leads to overflow to what had become a flood plain at a lower elevation and the formation of a new channel. The process is then repeated eventually forming a shingle and sediment plain.

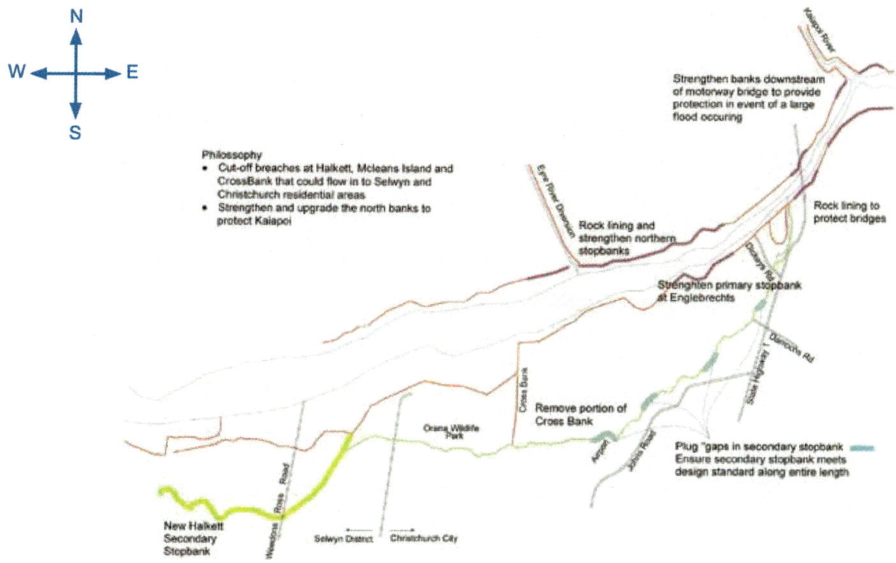

Fig. 7.11 Main works of flood management scheme for Christchurch (Environment Canterbury)

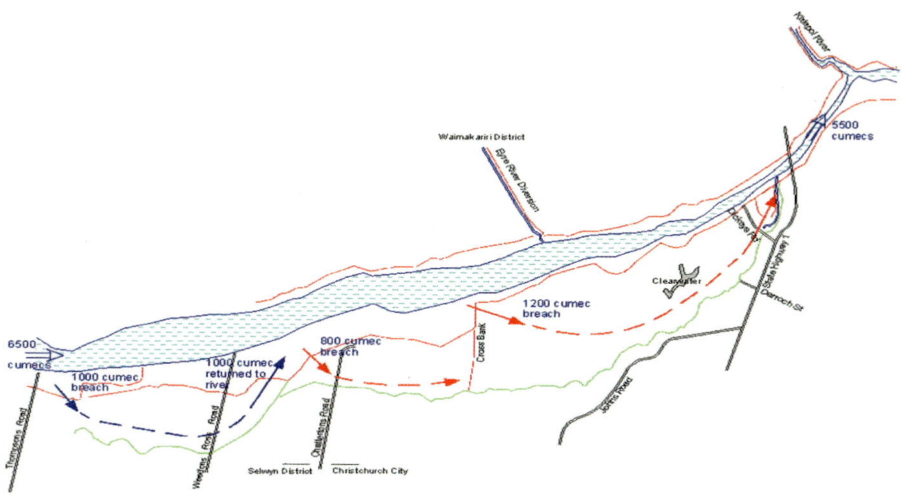

Fig. 7.12 Return of flood flows to the Waimakariri River (Environment Canterbury)

Box 7.2: Case Study of Hurricane Katrina

On 29 August 2005 Hurricane Katrina made landfall in southeast Louisiana. It caused severe destruction along the Gulf Coast from central Florida to Texas, much of it due to storm surge. The significant damage occurred in New Orleans which flooded primarily due to catastrophic failure of the levee system. Much of New Orleans is below sea level. There were around 1600 fatalities and further 400 missing presumed dead. Of the 73 neighbourhoods, 34 were completely inundated, 31 partially inundated, and only 8 did not flood. There were direct property losses of more than $20 billion with 78% attributed to residential property. Damage to the hurricane protection system was about $2 billion (US Army Corps of Engineers 2006b).

Katrina was a Category 5 storm (on the Saffir-Simpson scale) with up to 175 mph (285 km/h) until it was 170 miles (276 km) from landfall. At landfall wind speeds were at 127 mph (207 km/h) but the long path through the Gulf of Mexico built up surge and wave conditions larger than any previous storm.

There were 50 major breaches of the levee system with 46 of those due to overtopping and erosion and 4 were foundation failures of flood walls (I-walls) before water levels reached the top of the flood wall. Flooding covered about 80% of the New Orleans metropolitan area. About two thirds of the flooding was attributed to water flowing through the breaches and one third due to overtopping and the local rainfall (14 inches (360 mm) in 24 h) (US Army Corps of Engineers 2006b).

The levees and flood walls were designed for a specific hazard event (the Standard Project Hurricane – a fast moving Category 3 hurricane) representing the storm that would occur once in 200–300 years with winds up to 111–113 mph (180–184 km/h). This is based on the Saffir-Simpson scale which is a good predictor of wind damage but not such a good predictor of surge and wave generation potential of storms. Katrina exceeded the design criterion but the performance of the hurricane protection system was less than the design intent (US Army Corps of Engineers 2006b).

Some of the key design issues were as follows. Levees were built to an incorrect datum and no allowance had been made for subsidence: both factors meant some levees were below design level. The system had been designed in a piecemeal fashion resulting in inconsistent levels of protection. There had been overestimation of the shear strength at the levee toe because the shear strength at the centreline was assumed across the entire levee section. The design was based on 1 foot (0.3 m) waves while at least 4 foot (1.2 m) waves were experienced during Katrina. Overtopping by waves generated very high velocities over the crest and back sides of the levees leading to a high potential for scour and erosion. Examination of the levees that failed due to erosion

(continued)

Box 7.2 (continued)

determined that all were caused by the erosion of the crest and back face. The foundation failure of the flood walls was induced by a gap along the canal side of the flood wall. This failure mechanism was not considered in the original design of these structures (US Army Corps of Engineers 2006b).

In terms of nested adaptive systems, for the erosion failure the design criteria (Saffir-Simpson scale) focussed on the wrong critical variable and design did not address overtopping – the disturbance phase. For the floodwall failure, the relevant failure pathway was not considered.

The design also did not consider provisions for a larger-than-design event. Armouring of the crests and back sides were not included and would have reduced the risk of breaching when overtopping occurred i.e. improving the resilience of the system. Also pumping stations were not designed to operate in severe hurricane conditions. Pumping could have significantly reduced the duration of flooding (53 days) and in some areas the extent i.e. improving the restructuring of the system after disturbance.

The overarching lessons learnt as set out in the Interagency Performance Evaluation Task Force (IPET) report (US Army Corps of Engineers 2006b) are also interesting from a nested adaptive systems perspective. The stated lessons were:

- Resilience: referring to the ability to withstand catastrophic failure due to forces and conditions beyond those intended or estimated in the design;
- System performance: for any drainage basin the protection was as robust as the weakest component, e.g. some sections were incomplete, sections had different capabilities, subsidence had reduced capability over time;
- Risk and reliability: the traditional design approach as used in New Orleans was 'component-performance-based' i.e. using standards to define performance and relies on factors of safety to deal with uncertainty. What is needed is risk-based planning that is system-based requiring the entire system to be described in consistent terms and explicitly including uncertainty;
- Knowledge, technology and expertise: how does new knowledge from research get placed together to create knowledge for designers to avoid the lack of consideration of failure mechanisms.

In response to the failures in the New Orleans hurricane protection system highlighted by Katrina, the Corps initiated an Action Plan to address these shortcomings (US Army Corps of Engineers 2006a). These actions were grouped in four themes:

1. Comprehensive Systems Approach
   - Employ integrated comprehensive and systems-based approach

(continued)

Box 7.2 (continued)

- Employ adaptive planning and engineering systems
- Focus on sustainability

2. Risk-Informed Decision Making

- Employ risk-based concepts in planning, design, construction, operations and major maintenance
- Review and inspect completed works

3. Communication of Risk to the Public

- Effectively communicate risk
- Establish public involvement in risk reduction strategies

4. Professional and Technical Expertise

- Continuously reassess and update policy for program development, planning guidance, design and construction standards
- Employ dynamic independent review
- Assess and modify organisational behaviour
- Manage and enhance technical expertise and professionalism
- Invest in research.

This has led to a marked change in the Corps' approach to New Orleans and hurricanes. The Corps now refers to Hurricane and Storm Damage Risk Reduction System (HSDRRS) rather than 'hurricane protection system' indicating the existence of residual risk (US Army Corps of Engineers 2014). There is a recognition that no alternative can be formulated that will provide total protection to the planning area against all possible storms. They have adopted a strategy based on *multiple lines of defence* (Lopez 2006a, 2009).

As shown in Fig. 7.13, this strategy covers several geographical scales including the offshore coastal area (i.e. the offshore shelf, barrier islands, and sound), onshore coastal area (i.e. marsh landbridge and natural ridge), flood-related infrastructure (i.e. levees, flood gates, pump stations and highways), and floodplain components (buildings and evacuation).

Since Katrina in relation to flood-related infrastructure in New Orleans, there has been reconstruction of levees with armouring against erosion, floodwalls with improved foundation design (T-walls), outfall channels with closure structures, and, pump station repairs and provisions for pump stations to operate during hurricane conditions. These are to meet a one-in-a-hundred-year event rather than a particular storm category.

In addition to the reconstruction of structural measures, there is a programme to look at spatial scales beyond New Orleans and considering the entire Louisiana Coast – Louisiana Coastal Protection and Restoration

(continued)

Box 7.2 (continued)

(LACPR) (US Army Corps of Engineers 2009). The programme involves not only hurricane risk reduction but also coastal restoration. It also includes consideration of a Category 5 storm. The programme requires coordination and collaboration of Federal, State and local agencies.

Three categories of measures were considered:

- Coastal restoration measures – including land/marsh-building river diversions, freshwater distribution, mechanical marsh creation, barrier island/shoreline restoration, bank/shoreline ridge restoration;
- Structural measures – including surge reduction weirs, floodgates, levees, floodwalls and ring levees;
- Non-structural measures – including buy-out of properties or raising structures in place.

Rather than just cost efficiency as the evaluation criteria, measures were evaluated in terms of stakeholder input on preferences, direct and indirect environmental effects, effectiveness in reducing risk, project costs and the realities of future funding, as well as cost efficiency. The Corps is also developing a regional sediment budget with the objective of developing a Sediment Management Master Plan to provide a regional blueprint for the beneficial use of dredged material for habitat restoration (US Army Corps of Engineers 2009).

These changes represent a major shift in approach towards a multiple geographical scale approach to failure pathways and sustainability strategies. It also incorporates multiple criteria for evaluating measures rather than cost efficiency of meeting a particular design standard. It also considers greater-than-design events. It is a shift in the direction of a nested adaptive systems approach.

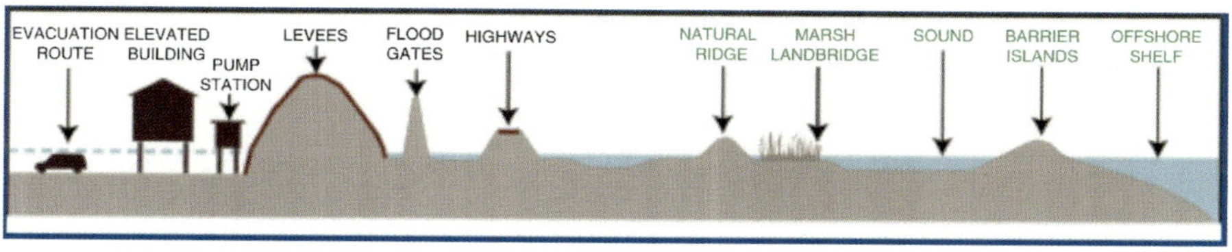

Fig. 7.13 Multiple lines of defence strategy (Lopez 2006b). Reprinted with permission from Lake Pontchartrain Basin Foundation

References

Burton, R., & Peoples, S. (2008). *Learning from past adaptations to extreme climatic events: A case study of drought*. Dunedin: AgResearch.

Chapin, F., Kofinas, G., & Folke, C. (2009). *Principles of ecosystem stewardship: Resilience-based natural resource management in a changing world*. New York: Springer.

Chinn, T. (1996). New Zealand glacier responses to climate change of the past century. *New Zealand Journal of Geology and Geophysics, 39*(3), 415–428.

DairyNZ. (2013). *New Zealand dairy statistics 2012–13*. Hamilton: DairyNZ.

De Klein, C., Monaghan, R., Ledgard, S., & Shepherd, M. (2010). *A system's perspective on the effectiveness of measures to mitigate the environmental impacts of nitrogen losses from pastoral dairy farming*. Paper presented at the proceedings of the 4th Australasian Dairy Science Symposium.

Di, H., & Cameron, K. (2004). Treating grazed pasture soil with a nitrification inhibitor, eco-n™, to decrease nitrate leaching in a deep sandy soil under spray irrigation – a lysimeter study. *New Zealand Journal of Agricultural Research, 47*(3), 351–361.

Di, H., Cameron, K., & Sherlock, K. (2007). Comparison of the effectiveness of a nitrification inhibitor, dicyandiamide, in reducing nitrous oxide emissions in four different soils under different climate and management conditions. *Soil Use and Management, 23*(1), 1–9.

Editor. (2011, September 20). *Farmers must share burden on emissions*. The Herald.

Environmental Protection Authority. (1998). *Guidance statement no. 12: Minimising greenhouse gas emissions*. Perth: Environmental Protection Authority.

Environmental Protection Authority. (1999). *Bulletin 931 Murrin Murrin Nickel-Cobalt Project Stage 2 expansion, 60km east of Leonora: Report and recommendations of the Environmental Protection Authority*. Perth: Environmental Protection Authority.

Environmental Protection Authority. (2000). *Bulletin 985 proposed gas to synthetic hydrocarbons plant, Burrup Peninsula, Western Australia: Report and recommendations of the Environmental Protection Authority*. Perth: Environmental Protection Authority.

Environmental Protection Authority. (2009). *Gorgon gas development revised and expanded proposal: Barrow Island nature reserve: Report and recommendations of the Environmental Protection Authority*. Perth: Environmental Protection Authority.

Environmental Protection Authority. (2015). *Environmental Protection Bulletin No. 24 greenhouse gas emissions and consideration of projected climate change impacts in the EIA process*. Perth: Environmental Protection Authority.

IPCC. (2013). The physical science basis: Contribution of Working Group 1 to the Fifth assessment report of the Intergovernmental Panel on Climate Change. Cambridge/New York.

Ledgard, S., Judge, A., Smeaton, D., & Boyes, M. (2010). *Greenhouse gas emissions from Rotorua dairy farms: Summary report*. Hamilton: AgResearch.

Livestock Improvement Corporation. (2008). *New Zealand dairy statistics 2007–08*. Hamilton: Livestock Improvement Corporation.

Lopez, J. (2006a). *The multiple lines of defense strategy to sustain coastal Louisiana*. Lake Pontchartrain Basin Foundation.

Lopez, J. (2006b). *Overview of multiple lines of defense strategy to sustain coastal Louisiana*. Lake Pontchartrain Basin Foundation & Coalition to Restore Coastal Louisiana. http://www.mlods.org/images/web_MLODS_overview.pdf

Lopez, J. (2009). The multiple lines of defense strategy to sustain coastal Louisiana. *Journal of Coastal Research* (Special Issue 54):186–197.

Manly, B. (2013). *2012 deforestaion survey*. Wellington: Ministry for Primary Industries.

Mason, E., & Ledgard, S. (2013). *Making your farm greenhouse gas neutral*. www.forestry.ac.nz/euan/carbonStockUnits.htm. Accessed 29 Dec 2013.

McLaren, R., & Cameron, K. (1996). *Soil science: Sustainable production and environmental protection* (2nd ed.). Melburne: Oxford University Press.

McLeman, R., Mayo, D., Strebeck, E., & Smit, B. (2008). Drought adaptation in rural eastern Oklahoma in the 1930s: Lessons for climate change adaptation. *Mitigation and Adaptation Strategies for Global Change, 13*, 379–400. https://doi.org/10.1007/s11027-007-9118-1.

Ministry for Primary Industries. (2013). *National exotic forest description as at 1 April 2012*. Wellington: Ministry for Primary Industries.

Ministry for the Environment. (2008). *Climate change effects and impacts assessment: A guidance manual for Local Goverment in New Zealand* (2nd ed.). Wellington: Ministry for the Environment.

Ministry for the Environment. (2013). *New Zealand's greenhouse gas inventory 1990–2011*. Wellington: Ministry for the Environment.

Ministry for the Environment. (2016a). *New Zealand's greenhouse gas inventory 1990–2014*. Ministry for the Environment. https://www.mfe.govt.nz/sites/default/files/media/Climate%20 Change/greenhouse-gas-inventory-snapshot-2016.pdf

Ministry for the Environment. (2016b). *The New Zealand emissions trading scheme evaluation 2016*. Wellington: Ministry for the Environment.

Ministry for the Environment. (2016c). *Climate change projections for New Zealand: Atmospheric projections based on simulations undertaken for the IPCC 5th assessment*. Wellington: Ministry for the Environment.

Ministry of Agriculture and Forestry. (2004). *A national exotic forest description as at 1 April 2003*. Wellington: Ministry of Agriculture and Forestry.

Ministry of Agriculture and Forestry. (2009). *Drought guide: Drought definition, recognition and assistance measures*. http://maxa.maf.govt.nz/mafnet/rural-nz/adverse-events/droughts/drought-guide.pdf

Ministry of Economic Development. (2013). *Energy in New Zealand*. Wellington: Ministry of Economic Development.

Mullen, B., Porteus, A., Wratt, D., & Hollis, M. (2005). *Changes in drought risk with climate change*. NIWA Client Report WLG 2005–23. Wellingtion: NIWA.

Nakicenovic, N., & Swart, R. (2000). *Special report on emission scenarios: A special report of Working Group III of International Panel on Climate Change*. Cambridge/New York: Cambridge University Press.

O'Donnell, L. (2007). *Climate change: An analysis of the policy considerations for climate change for the review of the Canterbury Regional Policy Statement*. Christchurch: Environment Canterbury.

Parliamentary Commissioner for the Environment. (2016). *Climate change and agriculture: Understanding the biological greenhouse gases*. Wellington: Parliamentary Commissioner for the Environment.

Renwick, J. (2015). *Updated: NZ climate target announced – Expert reaction*. http://www.sciencemediacentre.co.nz/2015/07/07/nz-climate-target-announced-expert-reaction/

Smeaton, D., Cox, T., Kerr, S., & Dynes, R. (2011). Relationships between farm productivity, profitability, N leaching and GHG emissions: A modelling approach. *Proceedings of the New Zealand Grassland Association, 73*, 57–62.

Srinivasan, M., & Duncan, M. (2012). Droughts and irrigation reliability: An investigation of a river-based scheme in New Zealand. *Journal of the Irrigation and Drainage Engineering, 138*(1), 80–89.

Srinivasan, M., Schmidt, J., Poyck, S., & Hreinsan, E. (2011). Irrigation reliability under climate change scenarios: A modelling investigation in a river-based irrigation scheme in New Zealand. *Journal of the American Water Resources Association, 47*(6), 1261–1274.

Stehlik, D. (2003). Australian drought as lived experience: Social and community impacts. In E. Butterill & M. Fisher (Eds.), *Beyond drought: People, policy and perspectives* (pp. 87–108). Collinwood: CSIRO Publishing.

Sustainability Council of New Zealand. (2015). *New Zealand's climate change: Targets, projections and liabilities*. http://www.sustainabilitynz.org/nzs-climate-change-targets-projections-and-liabilities/

US Army Corps of Engineers. (2006a). *Corps hurricane response: Task Force Hope Status Report, Sept 1 2006*. New Orleans: US Army Corps of Engineers Mississippi Valley Division.

US Army Corps of Engineers. (2006b). *Performance evaluation of the New Orleans and Southeast Louisiana hurricane protection system: Draft final report of the Interagency Performance Evaluation Task Force Vol 1 executive summary and overview*. New Orleans: US Army Corps of Engineers.

US Army Corps of Engineers. (2009). *Louisiana Coastal Protection and Restoration (LACPR): Final technical report*. New Orleans: USACE New Orleans District.

US Army Corps of Engineers. (2014). *Hurricane & storm damage risk reduction system*. New Orleans District: USACE http://www.mvn.usace.army.mil/Missions/HSDRRS.aspx. Accessed 14 Jan 2014.

Western Australia Environmental Protection Authority. (2002). *Guidance statement for mininmising greenhouse gas emissions*. Perth: WA EPA.

Zammit, C., & Woods, R. (2011). *Projected climate and river flow for the Waimakariri catchment for 2040s and 2090s*. Christchurch: NIWA.

Chapter 8
Socio-economic Failure Pathways

Abstract Unlike biophysical systems which follow natural processes, socio-economic systems are man-made. Institutional arrangements and management approaches are typically designed for a purpose and can be modified as circumstances change. In discussing socio-economic systems, not only are the current designs described but also the changing circumstances and innovations in institutional design concepts are considered. Circumstances have changed in Canterbury since the current institutional arrangements were designed in the 1980s.

Recommendations for changes in governance arrangements have been identified through the Canterbury Water Management Strategy and the Land and Water Forum. Furthermore, concepts of democracy have been developed, such as deliberative and monitory democracy, to address the shortcomings of representative democracy. A key change in Canterbury has been the shift from water as a relatively abundant resource to being a scarce resource, i.e. a common pool resource, for which self-governing communities have been demonstrated as the most effective governance arrangements. In other countries shifts from effects-based regulatory approaches to collaborative arrangements are occurring. Involvement of Māori in water governance and management is increasing leading to changes in institutional arrangements.

The shift from central government as the builder of irrigation infrastructure to reliance on the private sector has led a number of issues, such as, the contributions of infrastructure projects to restore ecological health and recreational amenity, and, the co-ordination of infrastructure for integrated water management.

The reliance on regulation to achieve compliance with environmental requirements has not been fully effective. Furthermore, voluntary environmental programmes have led to some improvements but have been inadequate to achieve environmental results. In Canterbury, this has led to the development of a spectrum of regulatory and voluntary approaches as well as the introduction of audited self-management to deliver the achievement of environmental outcomes.

At the individual level regulatory approaches are dependent on the effective exercise of authority. However, behavioural change at the individual level can also be achieved through motivation. Furthermore, the value of water is more than its utilitarian worth. There is a need for individual commitment to a water ethic to respect its significance to natural systems and other human values.

B.R. Jenkins, *Water Management in New Zealand's Canterbury Region*, Global Issues in Water Policy 19, https://doi.org/10.1007/978-94-024-1213-0_8

Keywords Institutional design • Governance arrangements • Compliance approaches • Infrastructure management

8.1 Governance Arrangements

8.1.1 Regional Councils

The governance arrangements for regional councils were established along the lines of a contemporary liberal democracy with representative government based on competitive elections, transparent decision processes and legal rights of appeal against council decisions. The council is supported by a bureaucracy where the council selected the chief executive who was responsible for employing staff to manage the council business. This is designed to provide a clear separation between governance and management. Regional councils were given rating powers to fund their operations subject to annual plans and public comment and hearing processes, and with annual performance and financial reporting against proposed outcomes and planned expenditure.[1]

In relation to water management, these governance arrangements are quite different from the catchment boards whose water allocation, water pollution, and flood and erosion management functions were incorporated in regional councils. Catchment boards were a mixture of elected and non-elected members with substantial government subsidies for on-farm assistance and annual reporting requirements to the relevant central government minister.

Much of the current debate in Canterbury is in relation to the return to an elected regional council from government-appointed commissioners. The first elections in 6 years were in October 2016 for a mixture of elected and appointed councillors.

However, the governance matters in relation to water need to be considered at a much broader scale than the composition of the regional council. The Canterbury Water Management Strategy recommended a nested governance structure for its implementation with:

- A National Tripartite Forum of central government, regional government and Ngāi Tahu to deal with national issues such as the Treaty of Waitangi, national policy instruments, and national strategic issues such as the integration of hydro generation and irrigation;
- A Regional Water Management Committee comprised of a representative of each Zone Committee, local and central government, Ngāi Tahu and other stakeholders to coordinate the development of a regional implementation programme

[1] Note formal performance reporting against community outcomes which are determined by a public process, was introduced in the Local Government Amendment Act 2002.

- Zone Management Committees comprised of community members, rūnanga[2] members and local authority members to coordinate zone implementation programmes (Canterbury Water 2009).

8.1.2 Land and Water Forum

The first report of the Land and Water Forum (Land and Water Forum 2010) also addressed the issue of water governance on a broader scale. The Forum identified a range of governance issues that needed to be addressed:

- Enabling iwi to give effect to the Treaty relationship with the Crown,
- Ensuring an intergenerational perspective in water management,
- Providing national direction and coordination,
- Identifying degraded waters in need of restoration,
- Overseeing the sustainable management of water,
- Facilitating essential water infrastructure,
- Enabling the efficient use of water,
- Recognising that management needs to be tailored to communities and catchments,
- Having sufficient technical and financial capacity to achieve integrated management,
- Having legitimacy and being efficient, effective, transparent and accountable, and
- Recognising iwi interests

To address these governance issues the main recommendations of the Forum were:

- Establishing a non-statutory National Land and Water Commission constituted on a co-governance basis to advise Ministers in relation to water management governance issues,
- Developing a National Land and Water Strategy on a collaborative basis to identify opportunities for enhancing cultural, economic, environmental and social values for water resources in an integrated way,
- Putting in place a National Policy Statement on Freshwater that establishes a framework of national objectives and policies for water,
- Improving central government agency contribution to national coordination and guidance, preparing national standards, providing technical and financial support to regional councils, and bringing together data on freshwater systems,
- Improving regional council performance through regional water strategies developed collaboratively, government appointees on regional council committees to

[2] Rūnanga is the representative body of whanau (family) traditional marae-based communities within their rohe (territory).

strengthen links to central government, iwi representation, comprehensive data management, and use of land use controls for diffuse sources, and
- Making the Parliamentary Commissioner for the Environment responsible for reporting on effectiveness of regional councils (biennially) and the system of land and water management (five yearly).

A major emphasis was placed on collaborative approaches to resolving water management issues. The Forum cited the following pre-requisites for a collaborative approach: leadership and facilitation, inclusiveness, open-mindedness by participants, capacity and resource, set time frames, and a final decision maker.

8.1.3 Theoretical Developments in Governance

There has also been the evolution in thinking about the nature of democracy, sometimes referred to as 'post-representative' democracy (Keane 2009). One evolutionary development is *deliberative democracy*[3] (Dryzek 2010; Hajer and Wagenar 2003). A system can be said to possess deliberative capacity to the degree it has structures to accommodate deliberation that is:

- Authentic, i.e. induces reflection upon preferences in a non-coercive fashion and is meaningful to those who do not share that viewpoint;
- Inclusive, i.e. provides the opportunity and ability for all affected interests to participate; and
- Consequential, i.e. the deliberations must make a difference in influencing collective outcomes.

The elements of a deliberative system are:

- Public space for free-ranging and wide-ranging communication,
- Empowered space for connecting the public discussions to institutions empowered to make decisions,
- Transmission of the public discussions to the decision-making institution,
- Accountability of the institutions to the public deliberations,
- The organisation design of the system to facilitate deliberation, and
- The degree to which these elements determine the content of collective decisions (Dryzek 2010).

Keane's review of the evolution of democracy leads him to the concept of *monitory democracy* in which power-monitoring and power-controlling devices have begun to extend sideways and downwards through the whole political order (Keane 2009). The new institutions of monitory democracy are defined by their commit-

[3] There is also *discursive democracy* which is a critical subcategory of deliberative democracy that emphasises the public discussion of differing perspectives (Dryzek 2010).

ment to strengthening the diversity and influence of citizens' voices and choices in decisions that affect their lives irrespective of the outcome of elections.

Some of the monitory mechanisms are focussed at the level of *citizens' inputs* to government or civil society institutions, e.g. citizens' juries. Other monitory mechanisms focus on the policy process and are designed to monitor *policy throughputs* e.g. independent public inquiries. Others are mechanisms that focus on scrutinising *policy outputs* e.g. performance monitoring institutions. Keane sees this evolution in democracy being made possible by what he calls *communicative abundance,* such as the availability of computerised media networks [4] (Keane 2009).

8.1.4 Institutional Design for Common Pool Resources

Ostrom reviewed many institutional arrangements for the governance of common pool resources,[5] like water. When common pool resources are scarce, there is the potential for the *tragedy of the commons.* Continued use of a scarce resource beyond its sustainability limit further degrades the resource and its availability to all. However, from the perspective of an individual user, there is no incentive for that individual to reduce the individual's use. Reduction of the individual's use incurs cost to the individual but benefits other users. Therefore each individual user is motivated to increase use. The cumulative effect is that all users wish to increase use thereby increasing the degradation of the resource.

Ostrom (Ostrom 1990) identified three governance prescriptions in relation to common pool resources:

- Centralised control (what she refers to as *Leviathan* based on Hobbes's term): government determines how the resource will be allocated and managed;
- Privatisation where access to the resource is governed by private property rights; and
- Self-governing communities where there is community determination of resource management requirements.

In her analysis, she found that survival of resources over long periods of time were associated with governance arrangements based on self-governing communities. However, not all self-governing communities were effective. She identified the following design principles for long-enduring common pool resource (CPR) institutions:

---

[4] Keane links *assembly democracy*, as practised in Ancient Greece, to the era dominated by the spoken word, *representative democracy* to the era of print culture, and *monitory democracy* to the growth of multi-media saturated societies (Keane 2009).

[5] Common pool resources are natural or human-made resource systems that are *non-excludable* i.e. it is difficult to exclude potential beneficiaries from accessing the resource, and *subtractable* i.e. use by one user subtracts from the available resource and reduces the availability of the resource to others.

- Clearly defined boundaries: individuals who have rights to withdraw resource units from the CPR must be clearly defined, as must the boundaries of the CPR itself;
- Congruence between appropriation and provision rules and local conditions: appropriation rules restricting time, place, technology, and/or quantity of resource units are related to local conditions and to provision rules requiring labour, material, and/or money;
- Collective choice arrangements: most individuals affected by the operational rules can participate in modifying the operational rules;
- Monitoring: the monitors who actively audit CPR conditions and appropriator behaviour are accountable to the appropriators or are the appropriators;
- Graduated sanctions: appropriators who violate operational rules are likely to be assessed with graduated sanctions (depending on the seriousness and context of the offence) by other appropriators, by officials accountable to these appropriators, or by both;
- Conflict resolution mechanisms: appropriators and their officials have rapid access to low cost local arenas to resolve conflicts among appropriators or between appropriators and officials;
- Minimal recognition of rights to organise: the rights of appropriators to devise their own institutions are not challenged by external government authorities;
- Nested enterprises for larger systems: appropriation, provision, monitoring, enforcement, conflict resolution, and governance activities are organised in multiple layers of nested enterprises.

These governance models have relevance in relation to the recent history of water resource management in Canterbury and New Zealand. Prior to the 1980s, centralised control was the dominant governance model where central government was a major actor in the provision of services in New Zealand. In relation to water management, the Ministry of Works was the designer and provider of water infrastructure. It built major hydro projects (such as the Waitaki scheme in South Canterbury) and irrigation projects (such as the Balmoral scheme in North Canterbury).

As discussed in Sect. 2.1, in the 1980s, there was a shift to a privatisation model where the role of government in relation to water management was limited to that of regulator. The Ministry of Works was disbanded. The Resource Management Act (RMA) was introduced. Regional councils were established. The role of government was to define the environmental constraints and leave it to the private sector to propose water resource development.

With water scarcity becoming a major issue in Canterbury, there is recognition of the need for a paradigm shift in water management in the region. Canterbury has chosen a strategy-driven collaborative governance approach based on Ostrom's self-governing communities model (Canterbury Water 2009). Following the recommendations of the Land & Water Forum, central government is also exploring collaborative governance approaches at the national level (Land and Water Forum 2010; Ministry for the Environment 2013).

8.1.5 Styles of Planning and Governance

Complementary to the evolution of democratic systems in relation to governance arrangements for policy making and implementation, consideration can also be given to the changes in the nature of planning systems and their implications for governance arrangements. Over the past 30 years since the creation of regional councils, the nature of water management in Canterbury has changed from managing a relatively abundant resource to one whose availability is constrained. Furthermore, the cumulative effects of water use on water quality and aquatic ecology are of concern. These changes have importance not only to water management but also to water governance.

As an abundant resource, it is possible to independently consider each applicant's proposal for water use on its merits and each applicant's interest in water use being relatively separate from other uses. RMA processes and their governance are consistent with these circumstances. However, for a constrained resource consideration needs to be given to the *interdependence* of proposed uses with other uses and to the *diversity* of interests of other users.

Innes and Booher have identified four main models of planning and policymaking in relation to the degree of interdependence of interests involved and the diversity of interests involved (Innes and Booher 2003). These comprise the technical bureaucratic model, the political influence model, the social movement model and the collaborative model. Each is useful under different conditions of diversity and interdependence among interests. A technical bureaucratic model works best where there is neither diversity of interests nor interdependence among interests. This is the assumed circumstance under RMA processes considering each proposal on its merits when water was in plentiful supply.

However, for multiple interests and multiple interdependencies, Innes and Booher argue that it is only a collaborative model that deals with diversity and interdependence because it needs to be inclusive and to explore interdependence in search of solutions.[6] From their real-world observations, the collaborative model allows for collectively beneficial solutions to complex and controversial problems in water management. Box 8.1 sets out their analysis of CALFED – an informal policy-making system engaging multiple agencies and stakeholders to address long standing water management issues in northern California.

Innes and Booher note that collaborative approaches do not fit readily into institutional arrangements for public agency decision making that exist in many countries. One of the obstacles that is most pervasive is the degree to which other models of policy making are firmly instituted in both practice and law. This is considered further with respect to Canterbury in Chap. 12.

---

[6] For the other planning models: the political influence model is most suited to high diversity but low interdependence of interests where the approach is co-opting players to buy into a common cause of action; the social movement model is suited to low diversity but high interdependence of interests where the approach is converting players to a vision and course of action.

Box 8.1: Governance for Resilience: CALFED as a Complex Adaptive Network for Resource Management (Booher and Innes 2010)

Federal and state agencies and nongovernmental stakeholders had been at odds over water management in northern California, centred around the Sacramento – San Joaquin Delta including the San Francisco Bay Estuary (– the Bay Delta) in relation to water quality, ecosystem quality, water supply reliability and levee system integrity. The governance system of interest group pressure on legislatures, hierarchical public agencies with narrow conflicting mandates, and adversarial legalism through the courts offered no opportunity for collective problem solving and solutions that would address the systemic issues.

CALFED began in 1994 as an interagency agreement among state and federal agencies (the Bay Delta Accord). It became a collaborative effort to manage the state's water system involving 25 state and federal agencies and more than 35 major stakeholder groups. The immediate stimulus for the agreement was the economic, environmental and political strains from stalemated decision making on water issues.

Booher and Innes identified several governance innovations associated with CALFED to address these challenges. These included:

- A distributed network structure: Four interlinked groups made up of agency staff and stakeholder representatives played a central role – one for co-ordination of water operations, a second for evaluating water supply alternatives, a third looking at the effects of water diversions on fisheries and a fourth as a co-ordinating team.
- Non-linear planning methods: Rather than traditional stepwise decision-making process, the collaborative interactions often went back and forth between such tasks as idea-generation and addressing implementation issues.
- Self-organising system behaviour: participants defined and redefined their tasks and scope along the way adapting to new problems and information. The system's behaviour was determined by their interactions and relationships not by their formal roles in their agencies.
- Collaborative interaction characteristics: the pattern of governance under CALFED varied significantly from the characteristics of earlier governance arrangements. Table 8.1 compares the different governance characteristics.

Under CALFED, collaboration largely replaced gridlock and litigation as a form of governance. A Programmatic Record of Decision was created as a strategic framework for decisions and incorporating linkages between projects. There was a shift from single-purpose projects to co-ordinated multi-

(continued)

Box 8.1 (continued)

purpose projects. Another change was to more local and regional initiatives rather than top-down decisions. CALFED was more open, transparent and inclusive of public involvement. It established an Independent Science Program and an interdisciplinary board to assess the science involved. CALFED also embraced the idea of adaptive management and learning to address uncertainty, change, conflict and complexity of the California water systems.

Booher and Innes suggest that a new process for governance of complex social-ecological systems is emerging. This involves:

- Structures involving interdependent network clusters rather than top-down hierarchies
- Distributed control as a source of direction rather than central control
- Open boundaries for involvement rather than closed boundaries
- Adaptability to changing goals
- Shared authority for decisions and actions rather than having a single authority
- System behaviour is determined by the interactions of participants rather than their formal roles
- The manager's role is as a mediator or facilitator rather than being directive as a programme controller
- Planning processes are adaptive rather than following step-wise procedures
- Success is measured by realisation of collective actions
- Deliberative democracy is conducive to collaboration rather than representative democracy.

Table 8.1 Governance characteristics before and after CALFED (Booher and Innes 2010)

Before CALFED	After CALFED
Grid-lock and litigation-driven processes	Collaborative processes
Project-by-project decisions	Comprehensive strategy with linkages and balancing requirements
Single-agency, single-purpose projects	Multiple purpose, interagency projects
Centralised decision making	Emphasis on local and regional solutions
Limited public involvement	Extensive public involvement and leadership
Internal agency science: no peer review	Independent science reviews
Mechanistic decision making based upon assumptions and mandates	Flexible, adaptive management and learning

8.1.6 Evolution of Environmental Governance in Other Countries

Governance in other countries has also evolved from frameworks based on effects-based management. Bleischwitz observed that as industry evolves its corporate governance of environmental issues then government needs to evolve as well (Bleischwitz 2007). Four stages of co-evolution have been identified as set out in Table 8.2. The first stage is the use of regulations to get industry to add on pollution reduction equipment. The second stage is the use of policy instruments and standards to achieve end-of-pipe pollution control in design of developments. The third stage is the use of integrated approaches with multiple government agencies working together on sustainability strategies and industry looking beyond their own operations to the management of the environmental integrity of its supply chain. The fourth stage is a shift to collaborative governance where government has a broader role of enabling society to achieve sustainability while industry is focussed on eco-efficient products and services.

The New Zealand governance arrangements are most closely aligned to the second stage of Bleischwitz's framework.

Australian Environmental Protection Agencies were established under effects-based legislation with the roles of setting environmental policy and standards, assessing the effects of proposals, monitoring and enforcement of compliance with environmental conditions, and, undertaking scientific investigations. Changes in Australian organisational arrangements have occurred since the 1980s when the New Zealand RMA was introduced. Rather than an emphasis on regulatory functions, there have been changes to multi-functional agencies. This reflects the shift of environmental management from an emphasis on mitigating adverse effects of economic development, to sustainable development where environmental considerations are one of the multiple objectives relating to the achievement of economic, social, cultural and environmental outcomes (Jenkins 2009).

Table 8.2 Co-evolution of corporate and political governance

	Political governance	Instruments	Corporate governance
1.	Environmental problem solving through regulation	Project mitigation pollution control	Environmental add-on cost to business
2.	Environmental policy through ministries	Environmental policies and standards	End-of-pipe pollution control
3.	Integrated management through agency coordination	Government sustainability strategies	Supply chain management
4.	Enabling of civil society	Collaborative governance	Eco-efficient products and services

Adapted from Bleischwitz (2007)

8.1.7 *Post-regulatory State*

There has also been a shift in thinking about the role of government from the 1980s. The concept then was that the role of government was to define through regulation the boundaries in which civil society could operate. Development would then occur by private development within the regulatory constraints set by government. This is consistent with the effects-based model of environmental management that underpins the Resource Management Act and the traditional role of EPAs.

There are three major trends away from this purely regulatory role of government (Scott 2005):

- Legal Theory of Autopoiesis

This approach relies on coupling of the legal system with other systems. One example is "audited self-management" which involves companies developing their own policies and plans to achieve environmental outcomes rather a reliance on prescribed rules and involves independent auditing of the outcomes to ensure compliance.

- Governmentality

This approach depends on social controls to complement legal controls. The concepts of self-governing communities and stakeholder-based management plans are examples of the application of this approach.

- Responsive Regulation

This approach is based on attempting to keep regulatory intervention to the minimum level necessary to achieve the desired outcomes, while retaining the capacity to intervene with more stringent measures. One example is the "enforcement pyramid" which involves using low level sanctions such as advice and warnings at the base of the pyramid and only escalating to more drastic remedies if those who are being regulated are unresponsive.

8.1.8 *Restorative Justice*

There are also alternative concepts being developed for achieving justice with respect to breaches of laws and regulations. This is the concept of *restorative justice*. This is based on a theory of justice that crime is an offence against an individual or community rather than the state. The victims of crime are active in the judicial process. Offenders are encouraged to take responsibility for their actions by apologising, making reparations or community service (Preston 2011; McElrea 2004).

A key part of the concept is that because crime hurts, justice should heal. The victim and surrounding community have been affected by the action of the offender and restoration is necessary. The offender's obligation is to make amends with both the victim and the involved community. There should be healing for the victim, offender and the community.

In the application of restorative justice to environmental issues, the victims of environmental crime can be the people affected by the adverse effects (e.g. impaired water quality), future generations (for degradation of a resource base), the environmental system, or, the environmental agency as a surrogate. The expectations of an offender are that they take responsibility for their conduct that caused the adverse effects and for the consequences of the adverse effects. Types of reparation include making good the environmental damage, paying the costs incurred by public authorities or affected parties, and, paying a monetary amount for an environmental purpose.

8.1.9 Governance and Indigenous People

In New Zealand, the principal governance document is the Treaty of Waitangi (Te Tiriti o Waitangi). Signed in 1840 by representatives of the British Crown and many Māori chiefs, Article 1 provides for the Crown's right to govern New Zealand, Article 2 guarantees "full, exclusive and undisturbed possession" (English version of the Treaty or "*tino rangatiratanga*" (Māori version of the Treaty) to Māori of their lands, estates, forest, fisheries and *taonga* (treasured things), while Article 3 affirms equal citizenship rights of all New Zealanders, including Māori (Waitangi Tribunal 2011).

Water is regarded as a taonga by Māori. While the Treaty gives the Crown the right to govern, in return it requires the Crown to protect *tino rangatiratanga* (full authority) of *iwi* (Māori tribes) in relation to *taonga katao* (all that they treasure) (Waitangi Tribunal 2011). The issue of ownership in the western sense is unclear. Rather, iwi have a responsibility as *kaitiaki* (cultural guardians) for safeguarding the *mauri* (life force) of water for the benefit of current and future generations. Iwi interests in freshwater reflect a diverse mix of cultural, customary and economic values (Brough 2010).

The RMA as the principal resource management statute makes the following provisions in relation to the Treaty. Section 6 identifies seven matters of national importance that shall be recognised by all persons exercising functions and powers in relation to resource management. One of these matters is "the relationship of Māori and their culture and traditions with their ancestral lands, sites, *waahi tapu* (sacred sites) and other taonga" (RMA s6(e)). Under Section 7 the Act identifies 11 other matters that persons exercising powers under it shall have particular regard to. One of these matters is *kaitiakitanga* (traditional guardianship – the active protection and responsibility for natural and physical resources by *tangata whenua* (people of the land) (RMA s7(a)). Under Section 8 persons exercising functions and powers under the Act shall take into account the principles of the Treaty of Waitangi. Furthermore, regional councils in preparing regional policy statements (RMA s61) and preparing regional plans (RMA s66) must take into account any relevant planning document recognised by an *iwi* (Māori tribe) authority (iwi management plan). There are also a variety of provisions requiring consultation with iwi.

While there is recognition of Māori interests in water it is the view of the Waitangi Tribunal[7] that current laws and policies do not support the kaitiaki relationships to the degree required by the Treaty. Māori are seeking decision making roles in the governance and management of water as a Treaty partner. They are seeking more than the consultation opportunities of the RMA. There are also concerns that iwi have limited capacity to engage in the consultation opportunities within the procedural timeframes of the RMA.

In Canterbury, there have been some recent agreements between the regional council and Ngāi Tahu (the iwi whose *rohe* (territory) includes the Canterbury region). One is the agreement to the restoration and rejuvenation of the *mauri* (life force) and ecosystem health of Te Waihora/Lake Ellesmere with the signing in 2011 of *Whakaora Te Waihora* – a restoration programme for the lake (Ministry for the Environment et al. 2011) followed by a co-governance agreement for the management of the catchment of the lake (Te Waihora Management Board et al. 2012). There has also been a regional relationship agreement – *Tuia* – between the regional council and the ten Papatipu Rūnanga[8] of the region (Environment Canterbury and Te Runanga o Ngāi Tahu 2012).

One of the challenges with a collaborative governance approach is how to integrate an approach to balance multiple stakeholder and community interests while working with one of the key interests as a Treaty partner. In the formulation of the Canterbury Water Management Strategy, kaitiakitanga was treated as a separate target to be achieved. As noted in Sect. 8.1 above, the proposed nested arrangements included a partnership of central government, regional government and Ngāi Tahu to address national issues including Treaty matters, a Regional Committee including Ngāi Tahu and other stakeholders, and, Zone Committees including rūnanga and community members. National level governance arrangements are still to be addressed. To accommodate the multiple stakeholder/Treaty partner concerns, the Region and Zone Committees were established with terms of reference requiring consensus as the basis for decision making. To address the potential for a veto by a stakeholder group, if consensus could not be reached then the committee would be disbanded and a new committee formed. This has not been necessary so far.

While co-governance and co-management arrangements are evolving, the rights and interests of Maori in New Zealand's freshwater management remain poorly defined and unresolved. In particular, the question of ownership of water is unresolved. The Crown claims that in common law no-one owns water, rather it is common property; it is only use rights that can be owned. The RMA is silent on the issue. There are other jurisdictions where water is seen as common property, such as

[7] The Waitangi Tribunal was established in 1975 by the Treaty of Waitangi Act 1975. The Tribunal is a permanent commission of inquiry charged with making recommendations on claims brought by Māori relating to actions or omissions of the Crown that potentially breach the promises made in the Treaty of Waitangi.

[8] Papatipu Rūnanga means those Rūnanga as recognised under the Te Rūnanga o Ngāi Tahu Act 1996 whose traditional territories are within the greater Canterbury region.

South Africa, that have incorporated the public trust doctrine in their water law (e.g. Republic of South Africa National Water Act 1998).

The public trust doctrine suggests that certain resources, like water, are the common, shared property of all citizens. The Crown holds these resources in trust for the common good and has a stewardship responsibility in perpetuity for these resources (Takacs 2008). Sax considered that public trusteeship created an obligation that could be enforceable against government (Sax 1970). The public trust doctrine has some parallels with the Māori concept of kaitiakitanga which is a responsibility for wise stewardship of resources.

However, by international standards, New Zealand is probably further advanced than other western countries with colonial backgrounds in addressing water governance issues and the rights of indigenous people. The situation most like New Zealand is probably Canada. In relation to Canterbury there are particular similarities to British Columbia which has adopted collaborative governance approaches to water management issues.

Canada has many treaties with its indigenous people and rewrote its constitution in 1982 to include protection of existing Aboriginal rights and title. Court cases have confirmed the need for provincial and federal governments to respect these rights and consult with First Nations where government actions can affect these rights (Low and Shaw 2012).

A workshop on First Nation communities and water challenges identified a number of key issues in relation to water governance, many of which have resonance in New Zealand. One was the lack of resources and capacity to the challenges they face. Another was the inadequacy of consultation by government and industry on projects affecting water in their territories. A third was the need for mutual respect. A fourth was the need for community water strategies reflecting their needs and vision. There was also identification of varying scales at which issues should be addressed. In addition, there was commonality in relation to some of the solutions sought, in particular, for First Nations to be active participants in water-related decision making, and, the importance of increased respect for and application of indigenous knowledge in water governance (von der Porten and de Loe 2010).

Case studies in of collaborative governance approaches from British Columbia also highlighted the distinction between Indigenous people as *nations* rather than one of many *stakeholders* in collaborative processes (von der Porten and de Loe 2013).

8.1.10 Overview of Evolving Practices

There are some interesting parallels in the evolution of changes in governance, institutional design for managing common pool resources, multi-functional agencies, and approaches to regulation. Hajer and Wagenaar discuss the shift from *government* to *governance* and from *institutions* to *networks* as a response to the *network society* in which we live (Hajer and Wagenar 2003). Innes and Booher see this as part of an international trend where communities, regions and even nations are

Table 8.3 Evolving governance practices

Governance practice	Practices of the 1990s	Evolving practices
Nature of democracy	Representative democracy	Deliberative and monitory democracy
Resource management	Assessment of environmental effects	Sustainability through self-managed communities
Regulation	Consents and enforcement	Post-regulation with audited self-management and restorative justice
Māori engagement	RMA processes	Co-governance and co-management arrangements

seeking collaborative ways to make policy as an alternative to confrontation, top-down decision-making, or, paralysis. It is the leading edge of new forms of governance and deliberation (Innes and Booher 2003).

Table 8.3 sets out the practices in place in the 1990s as the basis for resource management and practices now evolving to address the changing circumstances for water management in Canterbury.

In relation to the nature of democracy, there has been a shift from the reliance on an elected council to a collaborative form of governance with a high level of stakeholder and community engagement in strategic decision making consistent with deliberative democracy approaches.[9] In relation to controversial water management projects, there has been a shift from the reliance on the assessment of environmental effects and court processes toward community-based decision making consistent with self-managed communities.[10] In relation to regulation, there is a shift from enforcement of consent conditions to concepts of audited self-management and restorative justice consistent with post regulatory approaches.[11] In relation to Māori engagement there has been a shift from a reliance on RMA processes to evolution of co-governance and co-management arrangements.

Furthermore, it has been found that collaborative approaches are more likely to resolve conflicting views rather than representative or centrally directed approaches. The Canterbury experience has been that where people who have to live with the decision are involved in decision making, that it is more likely creative alternatives and mutual accommodation can be achieved.

The parallels of these evolving practices are not only related to the importance of stakeholder and community engagement in these practices, but also the increased role of achieving multiple community outcomes. In strategy development, this involved the use of the Regional Environment Report as an input to the Canterbury Water Management Strategy, and, the definition of outcome targets across 10 key areas of the strategy for the region. In the strategy implementation at the Zone level it involved Zone Committees determining how the targets would be met.

[9] Even the change to government-appointed commissioners at Environment Canterbury has not changed the shift to collaborative forms of governance.

[10] This is discussed further in Chap. 11.

[11] This is discussed further in Sect. 8.2.

It is not just the specification of multiple community outcomes but also the measurement of the achievement of those outcomes. This is consistent with monitory democracy and the auditing of self-management approaches. One of the challenges of collaborative approaches is ensuring the accountability of different agencies in a network of agencies for the contributions to community outcomes. This is discussed further in Chap. 11. A second challenge for collaborative approaches is the combining of engagement of multiple stakeholders with co-governance and co-management arrangements with iwi.

8.2 Infrastructure Management

8.2.1 Central Government Role

As noted in Chap. 2, central government changed its role in the late 1980s from water resource developer to water regulator when the Ministry of Works and Development was abolished and the Resource Management Act introduced. City and district councils retained their water supply, municipal wastewater treatment and stormwater management roles. Regional councils were given the flood management infrastructure responsibilities of catchment boards under the Soil Conservation and Rivers Control Act 1941.

From 1912 to 1987, the Ministry of Works and Development had the responsibility for the design, construction and operation of government-owned irrigation schemes as well as the responsibility for recommending annual water charges to the Minister for approval. Farmers frequently stated that the charges were set too high for farms to remain viable while officials believed the charges to be too low to recover capital costs associated with schemes (Farley 1994).

Of the economic reforms beginning in 1984, the primary focus was on the agricultural sector and virtually all agricultural subsidies were removed. Between 1988 and 1990, 49 government-owned irrigation schemes in New Zealand were sold to private irrigators. Very few of the schemes yielded a high sale price for government; many sold for $1 or less. Nearly $60 million of capital investment by government was unrecovered (Farley 1994).

Also, as part of the reforms the New Zealand Electricity Department was corporatized as the Electricity Corporation of New Zealand in 1987. Until then, New Zealand had a centrally run system of providers of generation, transmission, distribution and retailing. There were further reforms including in 1996 the establishment of a wholesale spot electricity market and generation assets split into several state-owned enterprises (Electricity Authority 2011). Subsequently all the electricity generation state-owned enterprises were fully or partially privatized.

For irrigation projects, this has led to the reliance on applicant proposals for water resource development. For irrigation, this has been mainly two types of proposals. One type is applications for bores to take groundwater from individual farmers. The second is applications from groups of farmers with irrigation scheme proposals. These are often with support of a corporate entity such as a power company where there is a small hydro-electric component and with territorial authority financial assistance.[12]

More recently there has been limited central government involvement through the Irrigation Acceleration Fund (refer Sect. 2.4.2). Central government allocated $35 million over 5 years to support the development of irrigation infrastructure proposals to the stage where they are "investment ready" (i.e. technically and commercially robust and demonstrating a high level of community support).

There has also been central government involvement through the establishment of the Crown Irrigation Investments Limited that acts on behalf of the New Zealand Government as a bridging investor for regional water infrastructure development (i.e. on a last-in first-out basis). Crown Irrigation has agreed terms for a $6.5 m secured, second ranking investment in Central Plains Water for a period of up to 5 years. Stage 1 is expected to cost between $120 m to $140 m with the expectation of $50 m in equity capital from farmer shareholders and other parties.

8.2.2 *Evaluation of Government Role in Water Infrastructure*

In a recent strategic review of national infrastructure (transport, telecommunications, energy, water and social infrastructure) each sector was analysed against guiding principles as "effective/could-be-further-developed/ineffective" (New Zealand Government 2011). The water infrastructure sector ranked poorly. For the three principles of (1) *investment analysis* (i.e. investment is well analysed and takes sufficient account of potential changes in demand), (2) *funding mechanisms* (i.e. maintaining a consistent and long-term commitment to infrastructure and utilizes a broad range of funding tools), and (3) *regulation* (i.e. regulation enables investment in infrastructure that is consistent with other principles and reduces lead times and uncertainty), the water infrastructure sector was ranked as "ineffective". It was considered that the water infrastructure sector "could be further developed" in relation to *resilience* (i.e. national infrastructure networks are able to deal with significant disruption and changing circumstances), in relation to *accountability and performance* (i.e. it is

[12] Examples include Opuha Dam which includes a 7 MW power station and started as a joint venture between Alpine Energy (the local electricity distribution company) and Opuha Water (a partnership of two irrigation cooperatives and private investors); Central Plains which started as a feasibility study between Christchurch City Council and Selwyn District Council and became a Trust with financial support from Dairy Holdings (a corporate dairy farmer); North Otago Irrigation which is a farmer cooperative with an initial shareholding by Meridian Energy (a power company operating the hydroelectric scheme on the Waitaki River, the water source for North Otago Irrigation) and with loans from Waitaki District Council; Barrhill Chertsey Irrigation scheme is a joint venture between Barrhill Chertsey Irrigation Limited, a cooperative company with nearly 20 farmer shareholders from within the mid Canterbury district and Electricity Ashburton Limited, the local cooperative lines company.

clear who is making decisions and on what basis and what outcomes were being sought), and in relation to *coordination* (i.e. infrastructure decisions are well coordinated across different providers and are sufficiently integrated with decisions about land use).

Strategic opportunities were seen for (a) better demand management practices and consistent performance criteria for water infrastructure; (b) the promotion of partnerships and activities within the sector; and (c) ensuring that water management assets contribute to improved social, economic, environmental and cultural well-being of communities (New Zealand Government 2011).

8.2.3 Canterbury Water Infrastructure

The CWMS identified the need for a Water Infrastructure and Services Entity to take on the role of designing, building, financing and operating the regional water storage and distribution system (Canterbury Water 2009). While the private sector can build individual projects, this does not provide for integrated water management. Examples of integrated water management that require the agreement of multiple actors include: (1) water use efficiency initiatives requiring reorganisation of consented allocations, such as, the use of surface water in the upper part of a groundwater catchment to enhance recharge (refer Sect. 6.1.3); (2) the use of managed aquifer recharge for high winter flows in alpine rivers to address adaptation to climate change projections; and (3) the conjunctive use of groundwater and surface water depending upon relative availability.

In their review of global threats to human water security and river biodiversity, Vorosmarty and his colleagues found that 80% of the world's population is exposed to high levels of threat to water security. Massive investment in water infrastructure and technology enables rich nations to offset high water stress levels. Less wealthy nations remain vulnerable. Rich nations tolerate relatively high levels of ambient stressors, then reduce their negative impacts by treating symptoms instead of underlying causes of incident threat. However, there is limited global investment in environmental protection and rehabilitation which means that stresses on biodiversity for many locations go unabated. Also some of the infrastructure investment, such as storage on major rivers and withdrawals from rivers can lead to biodiversity degradation (Vorosmarty et al. 2010).

Thus, to achieve the Canterbury Water Management Strategy of delivering programmes to restore ecological health and other targets such as recreational amenity (Sect. 3.2.4), there needs to be co-ordination of infrastructure investment and biodiversity restoration. This is unlikely to be achieved with a reliance on private sector investment. Experience in Canterbury has demonstrated that collaborative approaches to water resource development can lead to innovative proposals that address multiple objectives (Sect. 3.3.2).

Regional water resource development and regional water operation could also benefit from collaborative approaches. Table 8.4 shows the range of agencies and

Table 8.4 Agencies and institutions with water infrastructure roles

Infrastructure component	Agency or Institution[a]											
	CC/DC	RC	MoH	MfE	EPA	MPI	Treas	MoE	Hydro	Irrig	Ind	Farm
Storage												
Drinking	D, O[b]	P, R[b]	P, R	P	R							U
Industrial	D, O	P, R			R						D, O	
Irrigation	D	P, R[c]			R	P, D	P			D, O		D, O
Hydro	–	P, R		P[d]	R		P	P	D, O	D		
River diversion												
Irrigation	D	P, R				P, D	P			D, O		D, O
Industrial	D, O	P, R									D, O	
Stock water	D, O	P, R				P, R						D, O
Drinking	D, O	P, R	P, R	P								
Hydro	–								D, O	D, O		
Groundwater												
Drinking	D, O	P, R	P, R									
Industrial	D, O	P, R									D, O	
Irrigation	D, O	P, R				P, D				D, O		D, O
Distribution	D, O	P, R							D, O	D, O	D, O	D, O
Water treatment	D, O	R	P, R	P								
Wastewater	D, O	P, R									U	
Stormwater	D, O	P, R									U	U
Flood management	D, O	D, O		P					O			

[a]*CC/DC* City Council/District Council; *RC* Regional Council; *MoH* Ministry of Health; *MfE* Ministry for the Environment; *EPA* Environment Protection Authority; *MPI* Ministry for Primary Industry; *Treas* Treasury; *MoE* Ministry of Energy; *Hydro* hydro generator; *Irrig* irrigation company; *Ind* industry; *Farm* farmers

[b]*D* design of infrastructure; *O* operation of infrastructure; *P* policy related to infrastructure; *R* regulatory role; *U* user-provider

[c]Hawkes Bay Regional Council has recently taken on the role of developer of the Tukituki storage

[d]The main involvement of MfE in hydro is in relation to climate change

institutional interests with a design, operation, policy or regulatory role in water infrastructure in Canterbury. It would not appear practical to bring all of these interests into one organisation, nor is it necessary.

The private sector has demonstrated the ability to work together in the co-ordination of irrigation supply infrastructure. An example is provided in Box 8.2 which describes the network of agreements that have been put in place to enable Barrhill Chertsey Irrigation scheme to be developed and operate.

However, there are two key reasons that reliance on private sector networks alone is insufficient for sustainable water management. One is the contribution of projects to restore ecological health and recreational amenity; this requires government and community involvement. The second relates to the management of complex systems.

Box 8.2: Private Sector Agreements to Facilitate the Barrhill Chertsey Irrigation Scheme (BCIwater 2015)

Barrhill Chertsey Irrigation (BCI) stage one has built infrastructure to use up to eight cumecs from the Rakaia River at Highbank. TrustPower Limited (the owner of the existing Highbank power station) has installed intake, fish screening and pumping facilities to deliver water to the Rangitata Diversion Race (a headrace to an existing irrigation scheme) using some of the existing Highbank Power Station facilities.

Water swap arrangements managed by RDR Management Limited (the managers of the Rangitata Diversion Race) have allowed water to be delivered to a pipe network across the upper plains. The water is siphoned from the Rangitata Diversion Race to buffer ponds in the Highbank, Methven, Ashburton forks, Buccleugh and Mayfield areas. The initial distribution network includes five main pipelines with a total of 70 km of pipe.

Automation and modifications to the Rangitata Diversion Race, the buffer storage and distribution network is financed through a joint venture agreement between BCI and local electricity lines cooperative Electricity Ashburton.

A further three cumecs of water is being used through licence arrangements with Rooney Group Limited to operate Acton Scheme (another irrigation scheme) in the Rakaia and Pendarves area. Stage one is using the first of the consented BCI water take on arable, dairy and intensive pastoral farms.

Stage two of the BCI project will see the construction of a canal to deliver the remaining irrigation water to mid Canterbury by gravity. Significant electricity generation in conjunction with water currently under application by Ashburton Community Water Trust is planned. This stage will also likely include sizeable foothills storage to provide BCI and other irrigators in the district with improved reliability of supply.

The more complicated and delicately balanced systems grow, the more vulnerable they become to disturbances (Sect. 4.2.1).[13] Co-ordination of infrastructure for integrated water management could be achieved by establishing a network of infrastructure institutions that could incorporate other water users and stakeholders. Examples that have been developed in other countries are CALFED (Box 8.1) and the French basin authorities.

8.3 Compliance Approaches

One of the critical components of water resource management is where consent is granted for water take and use subject to conditions that there is compliance with those conditions. The Canterbury experience is one of resentment of regulation and even where there has been a long term programme of active enforcement with industry support, improvement in regulatory compliance has been slow and incomplete. Similarly, voluntary industry initiatives have encouraged performance improvement but have not achieved stated goals. The experience of dairy farm effluent management is used below to illustrate the limitations of regulatory and voluntary approaches. Gunningham has analysed the reasons for the failure of voluntary approaches (Gunningham 1995). This has led to the development of *audited self-management* as an alternative approach to achieving compliance. Its application is illustrated by a trial approach in the Te Ngawai catchment for water allocation.

8.3.1 Regulatory Approach to Compliance and Its Limitations

The regulatory approach to achieving compliance is to monitor performance with respect to consent conditions and plan rules developed under the RMA and take enforcement action against non-compliance. The desired outcome is to modify behaviour of actual and potential offenders by: educating resource users; promoting compliance with the RMA, plans and consents; using enforcement tools to obtain necessary action; and, promoting deterrence through appropriate penalties (Environment Canterbury 2010).

To evaluate the effectiveness of the regulatory approach, the Canterbury experience with regulating dairy effluent management is considered. This has been subject to intense regulatory efforts with all dairy farms being subject to annual inspections since the 2006/7 dairy season. This has been with increasing support from the dairy industry.

---

[13] This is considered to be one of the reasons for the failure of the sophisticated water management system in Angkor associated with the demise of the Khmer kingdom in the fourteenth century (Fletcher et al. 2008; Stone 2006).

The legal approach to managing water under the RMA generally follows the following steps:

1. Create an offence to use water
2. Create an authority to use water under certain conditions
3. Provide an enforcement mechanism in relation to the conditions
4. Create an authority to inspect compliance with those conditions
5. Provide a system of penalties that can be imposed for non-compliance
6. Establish an appeal process against the imposition of penalties.

In relation to dairy farm effluent management, s15 of the RMA places restrictions on discharges of contaminants into the environment unless allowed by a plan rule or a consent condition. Rules in a regional plan or conditions in an individual resource consent provide the basis for assessing compliance. Rules and conditions include requirements relating to dairy effluent storage, dairy effluent disposal, dairy effluent application rates to land, ponding of dairy effluent, and, avoidance of discharge to or near waterways.

Regional council officers make unannounced site inspections to determine compliance with requirements. The main compliance grades are: fully compliant, minor non-compliance, significant non-compliance, major non-compliance, not monitored or unable to be assessed, and enforcement action. Reports of the degree of compliance are sent to the consent holders highlighting issues encountered and remedial actions (with time frames for completion). Enforcement actions include formal warnings, issuing an abatement notice, fines and prosecution depending on the seriousness of the offence or the recurrence of offending.

Figure 8.1 shows the pattern of compliance for annual inspections from 2006–7 to 2012–3. There has been an improvement over 7 years in full compliance on initial inspection from 39.6 to 71.9% and a reduction in significant/major non-compliance from 17.7 to 6.8%. Follow-up inspections of those graded as having significant or major non-compliance achieved full compliance in 65% of cases. Enforcement action in 2012/3 included 9 abatement notices (requiring action to be taken or an inappropriate action to cease), 11 infringement notices (a punitive fine under the RMA), and 4 prosecutions (resulting in fines from $13,500 to $25,000).

However, it is not always possible to monitor conditions at the time of inspection. For example, for 272 farms (27%) in the 2012–3 season, it was not possible to assess compliance in relation to discharge of effluent not exceeding the water holding capacity of the soil or resulting in any ponding of the surface.

So even with a regular program of inspection, follow-up inspections and enforcement actions, regulatory enforcement is not fully effective. Furthermore, regulatory compliance does not encourage improved practice. Notwithstanding, inspectors found many examples of good management practices.

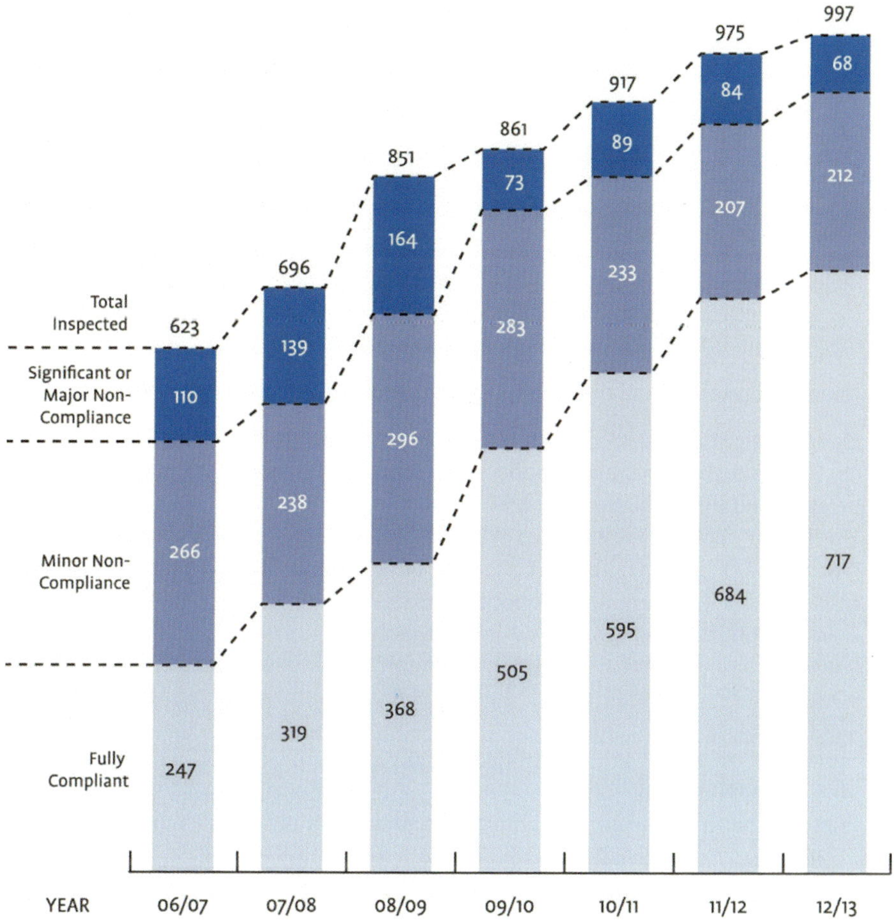

Fig. 8.1 Results of initial compliance inspections of dairy farm consents in Canterbury from 2006/7 to 2012/3 (Environment Canterbury 2013)

8.3.2 Voluntary Industry Initiatives

There has been an increasing willingness of industry to accept responsibility for environmental aspects of their operations. There is a growing shift in culture from reactive compliance with government regulations to voluntary proactive improvement of environmental performance.

A significant initiative of the dairy industry has been the *Dairying and Clean Streams Accord* established in 2003 (Fonterra Co-operative Groups et al. 2003). The Accord was between Fonterra (the dairy cooperative that manages most of New Zealand's milk production and processing), the regional councils and two central government ministries – Ministry for the Environment and the Ministry of

Agriculture and Forestry. There were two main purposes of the Accord. One was a statement of intent to promote sustainable dairy farming in New Zealand. The second was a focus on reducing the impacts of dairying on the quality of waterways.

The motivation for the Accord was because:

- Dairying is an important land use and industry for New Zealand,
- Ongoing intensification of dairy farming has increased the importance of addressing the impact on waterways,
- An industry based Accord sends a strong message to the public and consumers that environmental management is important, and
- If done well, industry self-management is more effective in achieving environmental outcomes than the reliance on rule-based regulation.

The Accord set a number of action priorities and performance targets:

- Dairy cattle are excluded from waterways that are deeper than a *red band* (an ankle height stripe on a gum boot) and wider than a stride: 50% of waterways to be excluded by 2007 and 90% by 2012.
- Bridges or culverts are in place where stock cross water courses: 50% to be in place by 2007 and 90% by 2012.
- Dairy farm effluent is appropriately treated and discharged: 100% compliance with consents and regional plans immediately.
- Nutrients are managed effectively to minimise losses to ground and surface waters: 100% of dairy farms to have in place systems to manage nutrient inputs and outputs by 2007.
- Significant wetlands are fenced and water regimes protected: 50% by 2005 and 90% by 2007.
- Fonterra and regional councils develop regional action plans by June 2004.

Fonterra also introduced initiatives to support the Accord (Fonterra 2013). One was focussed on improving effluent management systems. Farms that had been judged non-compliant by regional councils with respect to their consent conditions were visited by sustainable dairy advisors. Effluent improvement plans were put in place. Fonterra also stated that there would be payout deductions on suppliers with inadequate effluent infrastructure, however, this has not yet been implemented.

Fonterra also established a programme *Every Farm Every Year* in 2010. Farm assessors make 10,500 farm visits per year. Referrals can then be made to a sustainable dairy advisor team. A total of 3915 cases have been resolved between August 2010 and November 2013 (Fonterra 2013).

DairyNZ, the industry body for the dairy industry, also introduced initiatives in relation to dairy effluent management. It developed a farm dairy effluent code of practice (DairyNZ 2013). It introduced an effluent management module for the Agricultural Industry Training Organisation. It produced an engineering practice note for IPENZ (the Institute for Professional Engineers, New Zealand) on farm dairy effluent pond design and construction (IPENZ 2013). It also developed a short course on farm dairy effluent system design and management.

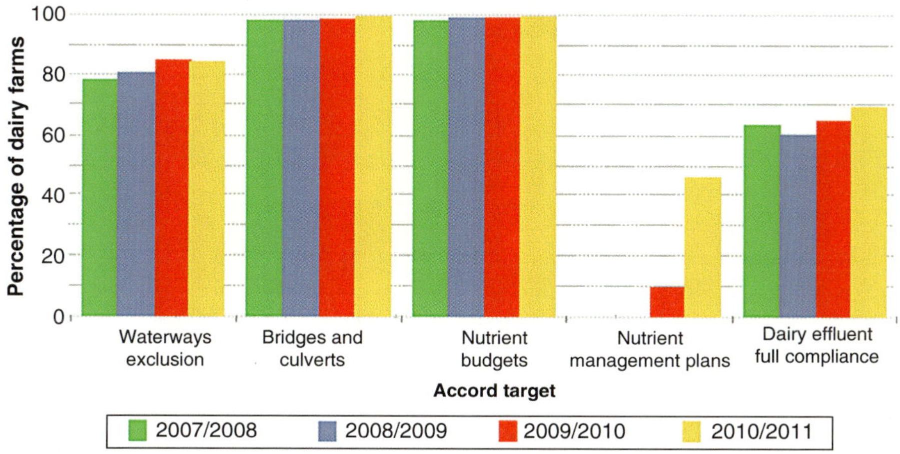

Fig. 8.2 Progress towards meeting the Dairying and Clean Streams Accord targets (2007/08–2010/11) (Fonterra et al. 2013)

The Accord partners produced a progress report on the achievement of the Dairying and Clean Streams Accord (Fonterra 2013). Figure 8.2 sets out the stated achievements of dairy farms in relation to the Accord targets. The results in some areas seem quite impressive with waterways exclusion achieved on 82% of farms and nearly 100% of farms with bridges and culverts, and 100% with nutrient budgets. In other areas performance was lower with nutrient management plans at 45% and effluent compliance at 69%.

However, catchment audits in Canterbury (Jones 2007) and a review by Fish & Game and Forest & Bird (Deans and Hackwell 2008) indicated that performance had been overstated. In Canterbury, the catchments of Rhodes and Petrie were audited. Waterways exclusion had only been achieved on 65% of farms. Not all cow crossing points of waterways had bridges or culverts. The nutrient budgets were found to be fertiliser plans. Effluent compliance was 65% (Jones 2007).

In the review by Fish & Game and Forest & Bird, their concern was that the Accord was not focussed on measurable improvements in water quality. The performance measures were focussed on elements of so-called *best practice.* The review noted that even in *best practice* catchments water quality had declined. Figure 8.3 shows *box and whisker* plots of bacterial levels in these catchments indicating the interquartile range of all sites exceeded the desired standard for the median of 110 *E. coli* per 100 mL.

The Review was also concerned that the exclusion of environmental interests from the development of the Accord led to "soft" targets. The Review noted that some targets were not actually measured, such as wetland protection, and some inappropriately measured with fertiliser plans claimed to be nutrient management systems. There was also difficulty in verifying the accuracy of self reporting. As a voluntary measure, the Accord failed to deal with serious non-compliance with its target requirements.

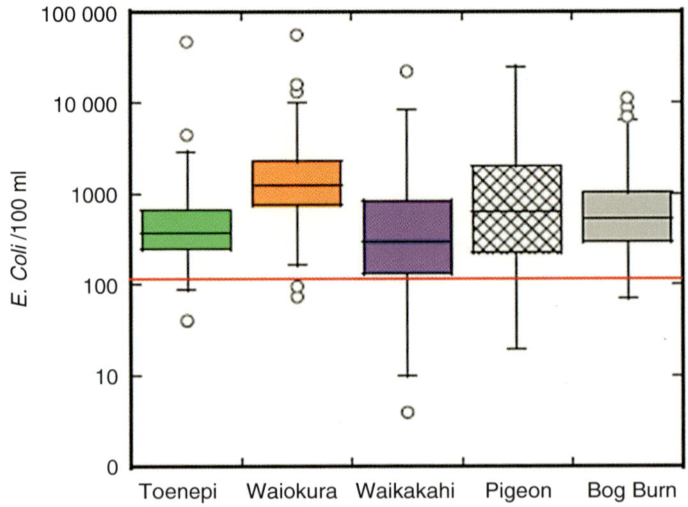

Fig. 8.3 Bacterial levels in best practice catchments (Wilcock et al. 2007)
Note: Box shows interquartile range of results; bar shows two standard deviations; and, circles show extreme values beyond two standard deviations. Horizontal line shows the median faecal coliform standard

The Review was concerned that there was only partial coverage of dairy farms because only consented farms were included. Farms with permitted activity status (i.e. they do not require consents but operate under rules set out in a regional plan) were not included in the Accord. Only farms with milking sheds were included; dairy run-off blocks (i.e. where cows graze but are not milked) were excluded from the Accord. The Review was also concerned that the Accord was used as an argument against the need for regulation when the Accord has been unsuccessful in reducing the impact of dairying on freshwater quality.

In summary, the Dairying and Clean Streams Accord has led to on-farm improvements and to industry initiatives to support actions on farms. However, the Accord targets were not met, audits indicated that self reporting was overstating achievements, and, overall water quality is still deteriorating.

8.3.3 Limitations of Voluntary Approaches

Gunningham undertook an analysis of *Responsible Care* – an international programme of the chemical industry for self-regulation (Gunningham 1995). The programme had two basic characteristics:

- The establishment of codes of practice to go *beyond compliance*

- The commitment to community participation and consultation: this included a National Community Advisory Panel and local community groups around manufacturing plants.

Gunningham identified three characteristics of voluntary self-regulation that impeded the effectiveness of the programme: (1) an *assurance problem*; (2) a *collective action problem*; and (3) a *credibility obstacle*. The assurance problem related to the competition between industries. Gunningham found that: "a firm does not withhold its contribution to a public good (e.g. a cleaner environment) based on a rational calculation of costs and benefits involved…but rather does so because it is unable to obtain the necessary assurance that other firms will contribute their fair share".

The collective action problem relates to the fact that self-regulatory schemes rely on regulation of the industry, by the industry, for the industry. Gunningham found that that the Chemical Industry Council did not validate the accuracy of the self assessments it received from firms, used moral pressure as the primary means of encouraging compliance, and, did not use the option of terminating membership of the voluntary scheme.

The credibility obstacle relates to the acceptance of the programme by the community. Gunningham found: "in terms of gaining public acceptance and credibility, Responsible Care is unconvincing. This is because it lacks effective strategies for monitoring and enforcement".

There are clear parallels with the shortcomings of the Dairying and Clean Streams Accord, indicating the limitations of relying on voluntary approaches to achieve satisfactory environmental and resource management.

8.3.4 The Concept of Audited Self-Management

In order to overcome these limitations of self-regulation but provide industry with an alternative to government regulation, the concept of audited self-management was introduced in Western Australia (Jenkins 1996). This concept involved the following approach:

- The environmental performance requirements were set by the regulator but industry was able to determine how to meet the requirements;
- Industry was required to have an environmental management system (EMS) with independent certification (by either the regulator or independent certifier);
- Industry was required to undertake measurements to demonstrate environmental performance requirements had been met with the measurements audited by an independent auditor (either the regulator or an independent auditor);
- The results of the measurements were to be publicly reported.

This approach was designed to address the shortcomings in self-regulation in the following way:

- The avoidance of soft performance targets but giving industry certainty of expectations: this was designed to address the collective action problem through the regulator setting the environmental performance requirements.
- Giving industry the flexibility to determine the practicality and cost-effectiveness in how the performance requirements are met – rather than specifying methods as occurs with many consent conditions.
- The use of the EMS to provide integrity in demonstrating that the proposed practices will achieve the desired environmental performance.
- The use of performance measurements with independent audit to avoid the reliance on self reporting and to address the assurance problem.
- The public reporting requirement to provide transparency and to overcome the credibility obstacle.

8.3.5 Applications of Audited Self-Management to Water Management in Canterbury

Two initial investigations into audited self-management (ASM) to test the application of the concept in Canterbury were undertaken. One investigation was to test stakeholder reaction to the concept in the Opihi catchment (Irrigation New Zealand 2008), while the second was the practical application to water extraction from the Te Ngawai River (Glubb and Miller 2006).

In the Opihi catchment some of the key findings were as follows. In terms of drivers for adopting ASM, water users believe that contributions to good water management can be made by tapping into the experience and energies of local communities. It was recognised that regional councils are constrained by resource limitations to increase the level of compliance monitoring. Central government recognised that more can be achieved by communities working in partnership with government rather than top-down regulation.

In order for ASM to be an effective alternative to compliance monitoring, the following attributes were considered necessary:

- The data for decision making needs to be robust;
- The measurement data and derived information needed to be widely accessible and in appropriate detail;
- Open and regular communication was needed between the system manager and those affected by decisions; and
- The governance arrangements needed to be democratic and the roles and responsibilities must be clearly defined and agreed.

To be effective in managing to water availability limits and for cumulative effects the need for decision support systems was recognised. There were different information requirements for different stakeholders. For example, irrigators needed information for water ordering, current and accumulated abstractions, whether flow restrictions on takes were in place, rainfall and evapotranspiration data and forecast; whereas the regional council needed information on consent compliance, flows to determine if restrictions on takes were required, the status of meters and measuring sites, and quality assurance of data and automated reporting.

The measurement requirements for managing water availability at the catchment level and the extent of cumulative effects is far more comprehensive than that for managing individual consents. This includes: real-time flow and use data, lake storage surface water levels, and, groundwater abstraction and pipeline flow data. To make effective use of the data there was a need seen for telemetered data, data sharing between the irrigation company and the regional council, a catchment simulation model for predictions and implications for water orders and rainfall forecast, a GIS-based system to locate flows and spatial information, and, an internet-based system for user access and integration with other stakeholder management systems.

A trial was undertaken in the Te Ngawai catchment to test water meters and develop a data logging system for the management of the total irrigation take from the river (Glubb and Miller 2006). This involved all consent holders in the catchment whose consent conditions were tied to the flow monitoring site at Cave. The Te Ngawai River is a foothill river that naturally reduces in flow in the late summer and autumn. In the irrigation season, when flows at Cave are below 500 L/s then takes are restricted to 50% of the consented rate. When flows at Cave are below 400 L/s then there is full restriction on irrigation takes (Glubb and Miller 2006).

The trial involved the establishment of a telemetry system and the design of an internet site. Real-time flow measurements of individual takes and river flows were taken. The website also provided climate station records and results of soil moisture probes. The individual takes results were combined to provide a measure of the total take from the river. It also enabled comparisons of the total take and the capacity of the flow in the river to provide for the total take.

A water user group was established amongst the irrigators. This enabled active management and comparing of allocated rates and volumes with the capacity to meet irrigation demand. The data management system was designed to provide alarms when consent conditions were exceeded, when river flow trigger levels for restriction on takes were reached, when allowable rates of take were exceeded, when daily or weekly volume limits were exceeded, or, when the combined rate of take was exceeded.

The key outcomes of the trial were that:

- The individual data allowed irrigators to manage water use and demonstrate compliance.
- The combined data allowed the water user group to share available water at times of restriction.
- The real-time data enabled irrigators to actively manage irrigation takes to maximise productivity while not placing the river on full restriction.
- The access to climate and soil moisture data facilitated water use efficiency.

The stakeholder reaction in the Opihi catchment and the outcomes of the Te Ngawai trial led to the incorporation of audited self-management in the CWMS, the Regional Policy Statement and the Land and Water Plan (refer Sect. 3.3.8).

8.3.6 Concluding Comments

The analysis has shown that there are failure pathways for both regulatory compliance and voluntary compliance with natural resource management requirements. Regulatory compliance can be ineffective when it is not possible to monitor requirements at the time of inspection and when there are economic, attitudinal or managerial barriers to compliance. Voluntary compliance is dependent for its effectiveness on there being a willingness to comply and can be ineffective because of assurance problems of competitors complying, collective action problem of industry not taking enforcement action against non-compliance, and credibility obstacles of community acceptance of industry self-regulation.

Two ways of addressing these failure pathways in Canterbury are (1) the development of a *regulatory spectrum* that provides a combination of regulatory and voluntary approaches, and (2) the exploration of *audited self-management.*

The regulatory spectrum is shown in Fig. 8.4 showing the span of activities from punitive and directive measures associated with major non-compliance and prohibited activities, through requesting and requiring in relation to minor non-compliance, to encouragement with respect to full compliance, best practice and innovation.

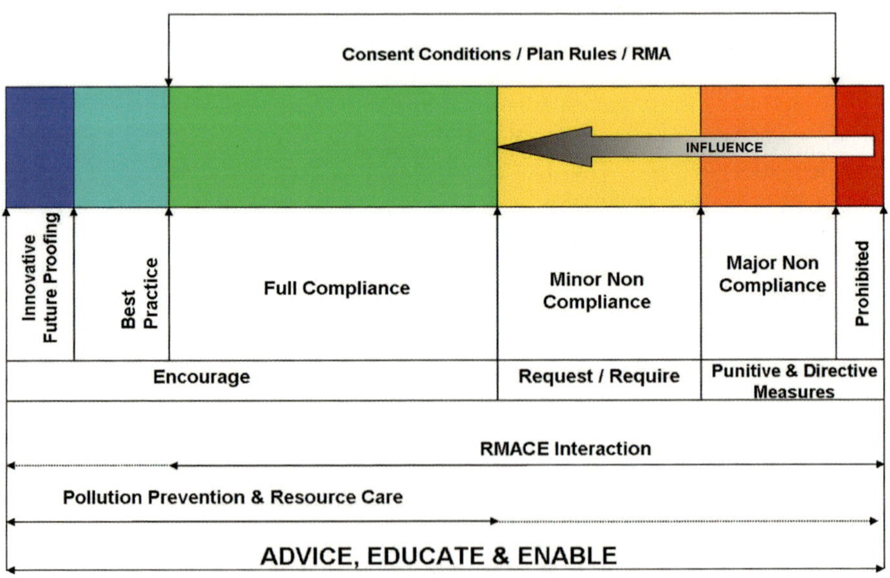

Fig. 8.4 Regulatory spectrum (Environment Canterbury)

Progress with audited self-management is dependent on developing the decision support systems in order to be able to measure both the cumulative effects of combined activities and the contribution of individuals to the cumulative effects. For water allocation, this requires metering of all takes which is progressively occurring throughout Canterbury and the determination of environmental flow requirements for all rivers which is nearing completion. For water quality, this requires setting cumulative effect limits and using measurement and modelling to determine individual contributions which are also progressing across the region (Sect. 3.3.8).

8.4 Individual Commitment

This section considers the individual level. The model for exercising authority for achieving compliance of Dornbusch and Scott is first described (Dornbusch and Scott 1975) and then applied to the regulatory enforcement function for regional councils. This is followed by barriers to compliance that have been identified in relation to Canterbury water management. The motivational model of Lawler and Porter for achieving behavioural change is described (Lawler and Porter 1967) and then applied to the collaborative approach of the Living Streams programme. The alternative to legal enforcement of restorative justice is outlined as a different way of achieving behavioural change. The section closes with the discussion of the need for a water ethic with an individual commitment to sustainability.

8.4.1 Exercising Authority and the Dornbusch and Scott Evaluation Process Model

Regulatory agencies have authority through legislation to enforce compliance with statutory requirements. Dornbusch and Scott have identified an *evaluation process* model that identifies the tasks required to exercise authority (Dornbusch and Scott 1975). This is a generalised model but can be used to describe the process experienced by individuals subject to regulation.

Dornbusch and Scott distinguish four components of the evaluation process for exercising authority:

- Allocating: assigning a defined task to be undertaken or the outcome to be achieved by an individual;
- Criteria Setting: defining the criteria to be used in evaluating the desired outcomes or performance requirements;
- Monitoring: the measuring of the performance and outcomes in relation to the set criteria; and
- Appraisal: assessing the performance and outcomes achieved.

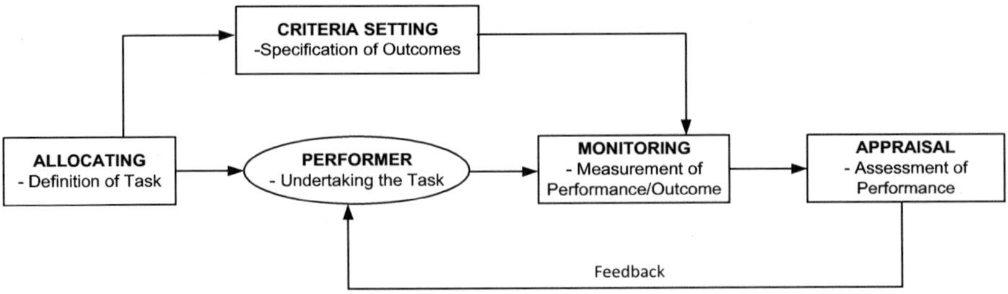

Fig. 8.5 Evaluation process for exercising authority (Dornbusch and Scott 1975)

These evaluation process components relate to the performer undertaking the requirements as shown in Fig. 8.5. The four components are required for an evaluation process to be effective. Authority is exercised when sanctions are used to reward or punish according to the appraisal of performance. In combination with the four components (i.e. allocating, criteria setting, monitoring, and appraisal) this provides a control system to achieve desired performance or outcomes.

8.4.2 Application of the Dornbusch and Scott Model to Regulatory Function

The evaluation process model for exercising authority can be applied to the regulatory enforcement function of the regional council under the RMA. In relation to *allocating,* the tasks to be undertaken are usually defined as plan rules or consent conditions. For the rules to be effective there needs to be *criteria setting*: the regulatory performance requirements need to have measurable criteria for monitoring compliance. *Monitoring* of performance is required and assessed for compliance (i.e. *appraisal*). Enforcement action can be taken against the person subject to regulation when non-compliance is found.

An example from dairy effluent management relates to the application of dairy effluent to land. In terms of allocating, the task is to manage the risk of nitrate-nitrogen leaching through the soil profile and contaminating groundwater. A condition is included in the dairy farm discharge consent to address this issue. In terms of criteria setting, to limit the loss of nitrate-nitrogen to groundwater there is an application limit of 200 kg/ha of nitrogen from dairy effluent per year. In terms of monitoring, the application limit requires a disposal area of approximately 3.25 ha for every 100 cows that are being milked. In terms of appraisal, six dairy farms were assessed as having application rates greater than 200 kg/ha/y (Burns 2013). To address this non-compliance the consent holder can either milk fewer cows or increase the effluent disposal area. Otherwise enforcement action would be taken.

8.4.3 Barriers to Compliance

Environment Canterbury has identified a number of barriers to compliance (Burns 2013). Some barriers are economic: firstly, where there has been inadequate capital investment in the effluent system to achieve compliance; secondly, where the focus is on return on investment with limited incentive to invest in infrastructure that does not increase production and financial return; and thirdly, where consent holders may not have the cash flow to invest in the changes to achieve compliance. Some barriers relate to knowledge and commitment: consent holders may not have an active role in farm management; farm management can be transitional with frequent movement of individual managers and sharemilkers; there may not be a culture of compliance; or, there may be literacy or language barriers. Other barriers relate to risk management: where there is no contingency planning for potential system failures.

In terms of the economics of improved environmental performance related to nitrate leaching, there have been two recent studies that have examined this issue. As part of the development of the Selwyn Waihora Zone Implementation Programme, financial modelling was undertaken of adopting nitrate leaching reduction actions for 18 farms representative of land use in the catchment (AgriBusiness Group 2012). In terms of the range of mitigation strategies considered, active water management and reducing stocking rates showed the greatest reduction in nitrate leaching. In terms of cost-effectiveness, DCD[14] use achieved 14% less nitrate leaching with improved cash position and total equity. Active water management (i.e. irrigation based on soil moisture demand) achieved 38% reduction at low cost. Reduced stocking rates (57% less nitrate leaching for 15% less cows on light soils) were achieved on improved cash position with reduced expenditure but with a reduction in total equity.

Smeaton et al. (2011) examined the relationship between farm productivity, profitability, nitrate leaching and greenhouse gas emissions for typical dairy and sheep/beef farms. They found that greenhouse gas emissions and nitrate leaching intensity were not significantly associated with profit and concluded that high profit per hectare can be achieved with a low greenhouse gas emission intensity or nitrate leaching intensity. They also found that the reasons why farmers have not already adopted improved environmental performance systems are complex but include: the requirement for a higher level of managerial skill, incompatibility with farm soil types or terrain, or, increased risk and capital cost to convert to a new system (Smeaton et al. 2011).

Thus, economics should not be a major barrier to achieving compliance or even to go beyond compliance.

There have also been factors identified that lead to full compliance and practices beyond compliance (Burns 2013). These include personal commitment: farmers take pride in their work; they have a connection to the land that is being farmed; and, there is a culture of compliance. Some are related to training: staff have completed agricultural training courses, and, there is support for the professional development

[14] DCD is dicyandiamide which is the active ingredient in nitrification inhibitors (Sect. 7.2.3)

of employees. Others relate to contingency management where risks have been identified and strategies put in place to avoid or manage the risks.

8.4.4 *Facilitation of Behaviour Through Motivation*

The factors leading to beyond compliance behaviour provide a second approach for regulatory agencies to achieve behavioural change for improved management of natural resources – an approach based on motivation. A theoretical model developed by Lawler and Porter (Lawler and Porter 1967) – the *expectancy theory of motivation* – is helpful in identifying a generalised model that can be tailored to natural resource issues.

Lawler and Porter proposed a multivariable model to explain the complex relationship between motivation, task performance and satisfaction. Figure 8.6 provides a diagrammatic representation of the model. Lawler and Porter state that effectiveness of task performance by individuals depends on the degree of motivation to put forth effort, the extent to which an individual possesses the relevant task abilities, and the extent to which an individual accurately perceives the role to be performed.

Motivation to put forth effort is a result of the value of the possible rewards to the individual and the individual's perceived probability that these rewards depend upon effort. The latter factor is the result of two other perceived probabilities, i.e. the probability that rewards depend upon performance and the probability that performance

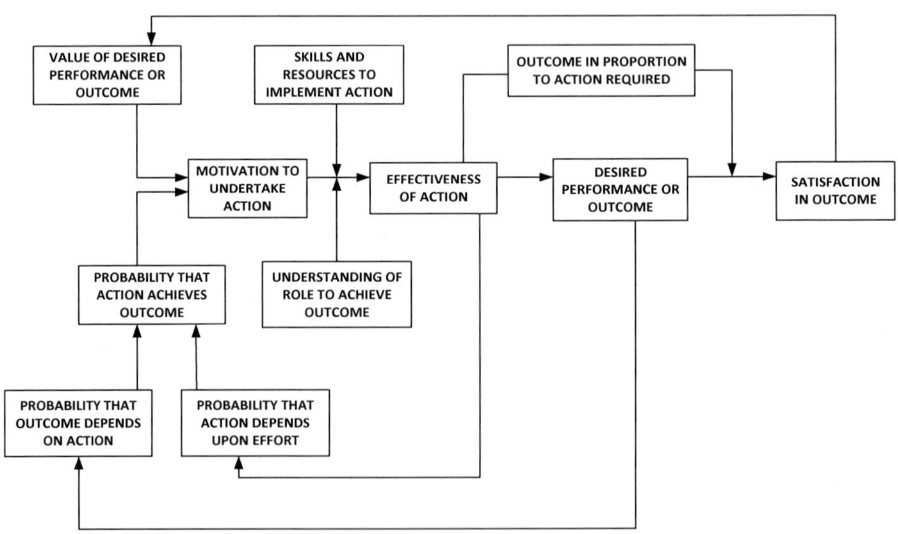

Fig. 8.6 Expectancy theory of motivation (Lawler and Porter 1967) (Reprinted with permission from Elsevier)

depends upon effort. Individual satisfaction depends upon the receipt of rewards that are perceived to be equitable for the level of effectiveness of task performance.

The model includes several feedback loops. One is from effectiveness of task performance to the probability that performance depends on effort; where there is evidence of the relationship between the effort put forth and performance, the perceived probability that performance depends on effort may change. Another feedback loop is from rewards to the probability that rewards depend on performance; where there is evidence on the relation between performance and rewards, the perceived probability that rewards depend on performance may change. A third feedback loop is between satisfaction and rewards.

With the model's emphasis on 'perceived probabilities' (or expectations in more common parlance), the model has been labeled 'the expectancy theory of motivation'.

8.4.5 Application of the Lawler and Porter Model to Improved Water Management

An example of the facilitation of behavior change by Environment Canterbury is the *Living Streams* programme which was designed to improve the health of rural waterways in partnership with local communities. The case of the Pahau River, a tributary of the Hurunui River, is considered below.

After the occurrence of algal blooms at the mouth of the Hurunui River, a catchment survey was undertaken to identify the main sources of nutrients in the catchment. The Pahau catchment was found to be the main contributor.

At a meeting with farmers in the catchment, once they accepted that the water quality outcome was a result of their performance, there was a willingness to take action, i.e. the link in the Lawler and Porter model between motivation and task performance. Farmers did not wish to be seen as "pariahs" in the community, i.e. the value of rewards for better water quality performance. However, farmers wished to be assured that actions would result in water quality improvement, i.e. the probability that performance depends upon effort.

Farmers voluntarily undertook a range of actions to improve water quality including: the control of stock access to rivers and the construction of reticulated stockwater supplies; fencing, and bridges and culverts at stock crossings of waterways; riparian planting for filtering runoff; improved effluent management; better scheduling of irrigation to reduce excess runoff; collection of excess irrigation runoff; and, a shift from inefficient border dyke irrigation to more efficient spray irrigation.

Rather than a regulatory role, the regional council undertook a facilitation role. This included: landholder and community engagement; provision of information; working with industry on management actions; investigation of the management of "wipe-off" water (i.e. excess irrigation runoff); and, monitoring of water quality.

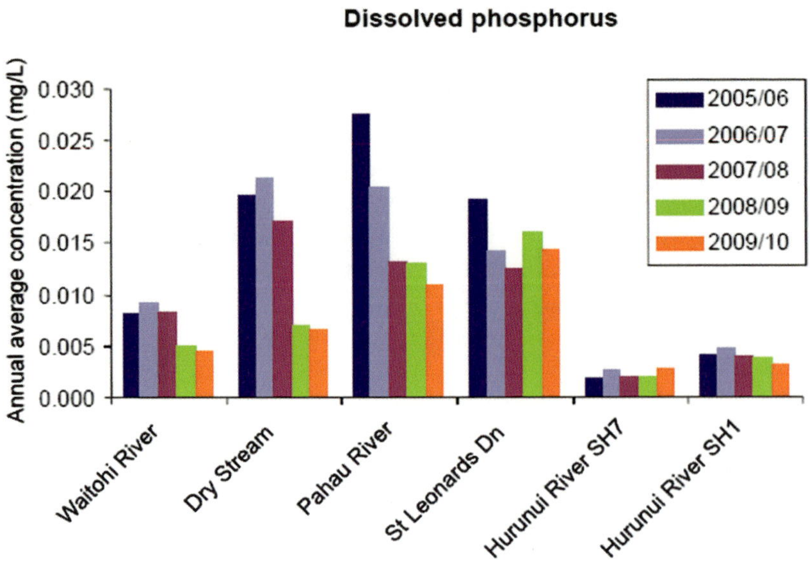

Fig. 8.7 Phosphorus reduction in the Pahau catchment (Environment Canterbury)

Figure 8.7 shows the reduction in concentration of dissolved phosphorus from the Pahau catchment after 5 years of the water quality improvement programme. The reduction indicates the success of this voluntary programme.

8.4.6 Alternative Approaches to Enforcement: Application of Restorative Justice

As well as using alternative approaches to regulation, Environment Canterbury has also been using alternatives to regulatory enforcement. Regulatory enforcement action for non-compliance is designed as punishment. The outcome for the offender is negative and there may be no beneficial outcome for the environment. The introduction of the option of environmental restorative justice in Canterbury has led to some constructive outcomes for offenders and the environment.

A precondition of the process is the admission of guilt to the offence which leads to the charges being withdrawn and the avoidance of an adversarial court process. The requirement for reparation is designed to generate a positive environmental outcome which is tailor-made to the situation. It provides an opportunity to rebuild the relationship of the regulator and the offender. It also provides an opportunity to restore the reputation of the offender. However, it is not a universal solution. There are concerns about favouritism, inconsistent application, the adequacy and appropriateness of reparation, and, the avoidance of the legal process.

One example of the application of restorative justice was in the case of a dairy effluent prosecution. A farm manager, who was having problems with effluent ponding due to the effluent irrigator operating too slowly, dug soakholes to eliminate ponding but directly contaminated the groundwater system. The owner (and consent holder) fired the farm manager who left New Zealand leaving the owner liable for prosecution. The owner took full responsibility for the offence and entered into restorative justice mediation. He agreed a reparation package with the following elements: increased effluent storage, an effluent control system with automatic cut-offs and text alerts of problems (the first use of this technology in the South Island), the installation of a lysimeter for measuring nitrate leaching (which was linked into a regional lysimeter monitoring programme), and, the conduct of a field day to demonstrate the new technology. The cost was greater than the likely fine from prosecution. The consent holder shifted from non-compliance to best practice in effluent management.

8.4.7 *Need for a Water Ethic*

Beyond the issues of compliance and motivation processes, and beyond the application of technology for improved performance and efficiency, there is something that is missing in terms of individual commitment. As Postel states "by and large society views water in a utilitarian fashion – as a *resource* valued only when it is extracted from nature and put to use" (Postel 2008). The RMA provides a framework for allocating rights in water. However, when water availability and the cumulative effects of water use are at sustainability limits, the life supporting role of water is at risk. There is a need to recognise our obligations and responsibilities in relation to natural systems dependent upon water. Postel talks about the *missing piece* in water management being a *water ethic* – "a guide to the right conduct in the face of complex decisions about natural systems that we do not and cannot fully understand. The essence of such an ethic is to make the protection of freshwater ecosystems a central goal in all that we do" (Postel 2008). This is an important link of the socio-economic systems back to the biophysical systems.

References

AgriBusiness Group. (2012). *Selwyn Te Waihora nutrient performance and financial analysis*. Christchurch: Environment Canterbury.

BCIwater. (2015). Background – Barrhill Chertsey Irrigation Ltd. www.bciwater.co.nz/background/. Accessed 7 Aug 2015.

Bleischwitz, R. (2007). *Corporate governance of sustainability: A co-evolutionary view on resource management*. Cheltenham: Edward Elgar.

Booher, D., & Innes, J. (2010). Governance for resilience: CALFED as a complex adaptive network for resource management. *Ecology and Society, 15*(3).

Brough, B. (2010). *Freshwater management*. http://www.iwichairs.maori.nz/Kaupapa/Freshwater/FreshwaterReport.pdf

Burns, M. (2013). *Canterbury region dairy report 2012–2013 season*. Christchurch: Environment Canterbury.

Canterbury Water. (2009). *Canterbury Water Management Strategy strategic framework*. Christchurch: Environment Canterbury.

DairyNZ. (2013). *Farm dairy effluent (FDE): A farmer's guide to building a new effluent storage pond*. Wellington: DairyNZ.

Deans, N., & Hackwell, K. (2008). *Dairying and declining water quality: Why has the "dairying and clean streams accord" not delivered cleaner streams?* Wellington: Fish & Game and Forest & Bird.

Dornbusch, S. M., & Scott, W. R. (1975). *Evaluation and the exercise of authority*. San Francisco: Jossey-Bass.

Dryzek, J. (2010). *Foundations and Frontiers of deliberative governance*. Oxford: Oxford University Press.

Electricity Authority. (2011). *Electricity in New Zealand*. Wellington: Electricity Authority.

Environment Canterbury. (2010). *Compliance monitoring and enforcement guidelines*. Christchurch: Environment Canterbury.

Environment Canterbury & Te Rūnanga o Ngāi Tahu. (2012). *Relationship agreement between Ngā Papatipu Rūnanga and Environment Canterbury*.

Farley, P. J. (1994). *Privatisation of irrigation schemes in New Zealand, Short report series on locally managed irrigation*. Columbo: International Irrigation Management Institute.

Fletcher, R., Penny, D., Evans, D., Pottier, C., Barbetti, M., Kummu, M., Lustig, T., & Authority for the Protection and Management of Angkor and the Region of Siem Reap Department of Monuments and Archaeological Team. (2008). The water management network of Angkor, Cambodia. *Antiquity, 82*, 658–670.

Fonterra. (2013). *Environmental sustainability on farm: Progress update November 2013*. Auckland: Fonterra Co-operative Group.

Fonterra Co-operative Groups, Regional Councils, Ministry for the Environment, & Ministry for Agriculture and Forestry. (2003). *Dairying and clean streams accord*. Wellington: Ministry for the Environment.

Fonterra, Local Government NZ & Ministry for the Environment, Ministry for Primary Industries. (2013). *The Dairying and clean streams accord: Snapshot of progress 2011/12*.

Glubb, R., & Miller, B. (2006). *Te Ngawai water use telemetry project*. Christchurch: Environment Canterbury.

Gunningham, N. (1995). Environment, Self-Regualtion, and the chemical industry: Assessing responsible care. *Law & Policy, 17*, 57–109.

Hajer, M., & Wagenar, H. (Eds.). (2003). *Deliberative policy analysis: Understanding governance in the network society*. Cambridge: Cambridge University Press.

Innes, J., & Booher, E. (2003). Collaborative policymaking: Governance through dialogue. In M. Hajer & H. Wagenaar (Eds.), *Deliberative policy analysis: Understanding governance in the network society*. Cambridge: Cambridge University Press.

IPENZ. (2013). *IPENZ practice note 21: Farm dairy effluent pond design and construction version 2*. Wellington: DairyNZ.

Irrigation New Zealand. (2008). *Decision support systems for audited self management*. Christchurch: Irrigation New Zealand.

Jenkins, B. R. (1996, December 5–6). *Best practice environmental regulation – The Western Australian approach*. Paper presented at the Environmental Management beyond 2000, Griffith University, Brisbane.

Jenkins, B. R. (2009, February 26). *Are regional councils past their use-by date?* Paper presented at the Environmental Defence Society conference – Reform in paradise: Threat or opportunity? Auckland.

Jones, R. (2007). *Dairying and Clean Streams Accord: Results of tier 2 monitoring within Rhodes Stream/Petries Creek catchment area*. Christchurch: Environment Canterbury.

Keane, J. (2009). *The life and death of democracy*. London: Simon & Schuster.

Land and Water Forum. (2010). *Report of the Land and Water Forum: A fresh start to fresh water*. Wellington: The Land and Water Trust.

Lawler, E., & Porter, L. (1967). Antecedent attitudes of effective managerial performance. *Organizational Behavior and Human Performance, 2*, 122–142.

Low, M., & Shaw, K. (2012). First nations rights and environmental governance: Lessons from the Great Bear Rainforest. *BC Studies, 172*(Winter 2011/12), 9–33.

McElrea, F. (2004, July 27). *The role of restorative justice in RMA prosecutions*. Paper presented at the Salmon Lecture to Resource Management Law Association.

Ministry for the Environment. (2013). *Proposed amendments to the National Policy Statement Freshwater 2011: A discussion document*. Wellington: Ministry for the Environment.

Ministry for the Environment, Canterbury Regional Council & Te Rūnanga o Ngāi Tahu. (2011). *Memorandum of understanding in relation to Whakaroa Te Waihora*.

New Zealand Government. (2011). *National infrastructure plan 2011*. Wellington: New Zealand Government.

Ostrom, E. (1990). *Governing the commons: The evolution of institutions for collective action*. New York: Cambridge University Press.

Postel, S. (2008). The missing piece: A water ethic. *American Prospect, 19*(6), 22–23.

Preston, B. (2011, March 22). *The use of restorative justice for environmental crime*. Paper presented at the EPA seminar on restorative environmental justice, Melbourne.

Sax, J. (1970). The public trust doctrine in natural resource law: Effective judicial intervention. *Michigan Law Review, 68*, 475–566.

Scott, C. (2005). Regulation in the age of governance: The rise of the post-regulatory state. In J. Jordana & D. Levi-Faur (Eds.), *The politics of regulation: Institutions and regulatory reforms for the age of governance*. Cheltenham: Edward Elgar Publishing.

Smeaton, D., Cox, T., Kerr, S., & Dynes, R. (2011). Relationships between farm productivity, profitability, N leaching and GHG emissions: A modelling approach. *Proceedings of the New Zealand Grassland Association, 73*, 57–62.

Stone, R. (2006). The end of Angkor. *Science, 311*(5766), 1364–1368.

Takacs, D. (2008). The public trust doctrine, environmental human rights, and the future of private property. *New York University Environmental Law Journal, 16*, 711.

Te Waihora Management Board, Te Rūnanga o Ngāi Tahu & Canterbury Regional Council. (2012). *Te Waihora Governance Agreement*.

von der Porten, S., & de Loe, R. (2010). *Water challenges and solutions in First Nations Communities*. Waterloo: Water Policy and Governance Group.

von der Porten, S., & de Loe, R. (2013). Collaborative approaches to governance for water and Indigenous peoples: A case study from British Columbia. *Geoforum, 50*, 149–160.

Vorosmarty, C., McIntyre, P., Gessner, M., Dudgeon, D., Prusevich, A., Green, P., Glidden, S., Bunn, S., Sullivan, C., Reidy Liermann, C., & Davies, P. (2010). Global threats to human security and river biodiversity. *Nature, 467*, 555–561.

Waitangi Tribunal. (2011). *Ko Aotearoa Tenei: A report into claims concerning New Zealand law and policy affecting Māori culture and identity*. Wellington: Legislation Direct.

Wilcock, R., Monaghan, R., Thorrold, B., Meredith, A., Betteridge, K., & Duncan, M. (2007). Land-water interactions in five contrasting dairy catchments: Issues and solutions. *Land Use and Water Resources Research, 7*, 2.1–2.10.

Chapter 9
Management of Waterborne Disease

Abstract Water quality impairment primarily through land use intensification contributes to waterborne disease through drinking water, contact recreation, contact with toxic algal blooms, and eating shellfish. Disease pathways are similar to other western countries but with the more rural character of New Zealand, pathogen contamination has a dominance of zoonotic agents such as Campylobacter. For the management approaches of these pathways New Zealand has adopted World Health Organization (WHO) recommendations with requirements for water safety plans for drinking water supplies, guidelines based on the Annapolis Protocol for contact recreation, an alert framework for cyanobacteria based on WHO guidelines, and, hazard analysis and critical control points for commercial shellfish harvesting. However, the status of drinking water supplies in Canterbury indicates large areas graded as unsatisfactory or unacceptable. Only 56% of monitored lake and river sites are graded as suitable for contact recreation. Eighty percent of river recreational sites have exceeded the alert criteria for toxic algal mats and nearly 30% have exceeded action criteria.

Based on nested adaptive systems analysis shortcomings in the management approaches are identified. Water safety plans for drinking water provide a sound basis for considering contaminant failure pathways, however, the approach does not address socio-economic issues like affordability. Recreational water quality management actions for unacceptable bacteriological or toxic algae levels are limited to public warnings that water bodies are unsuitable for contact recreation. This is similar for recreational harvesting of shellfish where a public health warning is the primary management action for phytoplankton levels exceeding health criteria. The hazard analysis and critical control points approach for commercial shellfish harvesting is much more comprehensive but management actions are focused on product contamination and do not address sources of contamination. The analysis highlights the need for management interventions to include proactive catchment management to prevent contamination. It also shows the need for sufficient organizational scale to achieve technical critical mass for water infrastructure management while maintaining local political control to manage the affordability of interventions.

Keywords Waterborne disease • Disease pathways • Management approaches

B.R. Jenkins, *Water Management in New Zealand's Canterbury Region*, Global Issues in Water Policy 19, https://doi.org/10.1007/978-94-024-1213-0_9

9.1 Disease Pathways

In relation to the potential waterborne disease pathways that affect human health, there appears to be four general types that need to be considered for Canterbury:

1. Ingestion of drinking water
2. Water contact recreation – microbiological ingestion
3. Contact with toxic algal blooms
4. Shellfish gathering from contaminated water

One indicator of significance of the type of disease is the number of notified cases. Notifications are based on medical practitioner and laboratory reports to the Medical Officer of Health so the data only records disease incidence where medical advice is sought. Table 9.1 sets out the notified disease rates for cases reported in 2014 (ESR 2015a) where exposure to water was identified as one of the risk factors. Where risk factors were identified, the percentage of cases related to water exposure pathways is also shown in the table. The table shows that campylobacteriosis is the most significant waterborne disease followed by giardiasis, cryptosporidiosis and salmonellosis in relation to consumption of untreated water and recreational water contact. There were also 18 cases of toxic shellfish poisoning notified (a rate of 0.4 per 100,000): all cases had eaten recreationally collected seafood.

In the summary of outbreaks[1] in New Zealand in 2014 (ESR 2015b) there were 820 outbreaks involving 14,235 cases associated with enteric disease. Waterborne transmission was attributed to 42 outbreaks and 131 cases as either the primary (28 outbreaks) or secondary mode (14 outbreaks). *Giardia* spp. was the most commonly reported pathogen (23 outbreaks and 77 cases) and then *Cryptosporidium* spp. (10 outbreaks with 26 cases) followed by *Campylobacter* spp. (4 outbreaks and 14 cases). Untreated drinking water supply was the most common contributing factor

Table 9.1 Notification rates for New Zealand in 2014 of diseases where exposure to water was reported as a risk factor (ESR 2015a)

		Exposure to risk factor (% reported)	
Notifiable disease	National notification rate (cases per 100,000)	Consumed untreated water	Recreational water contact
Campylobacteriosis	150.3	24.2	16.2
Giardiasis	37.9	33.3	37.6
Cryptosporidiosis	12.9	41.5	27.1
Salmonellosis	21.2	22.5	22.7
Yersinosis	15.1	16.9	19.7
Gastroenteritis	16.7	7.6	6.9
Shigellosis	2.8	12.0	16.1
VTEC/STEC	4.1	–	27.3

[1] Outbreak is defined as two or more cases (i.e. individuals affected) of a specific disease linked to a common cause (ESR 2015b).

(39 outbreaks and 125 cases) followed by inadequately treated water supply (9 outbreaks and 25 cases). Other contributory factors were treatment plant process failure (1 outbreak and 12 cases), source water quality inferior to normal (2 outbreaks, 9 cases) and contamination of post treatment water storage (1 outbreak and 2 cases).

Water quality was also a contributory factor where the environment was the mode of transmission. Exposure to recreational waters was implicated in 12 outbreaks and exposure to swimming pools in 11 outbreaks. In foodborne transmission, toxic shellfish poisoning was associated with 1 outbreak involving 13 cases (ESR 2015b).

9.1.1 *Ingestion of Drinking Water*

In 2006, Ball summarised that "There is ample evidence of waterborne disease outbreaks in New Zealand to indicate a significant risk of contracting gastro-intestinal disease (GID) from drinking-water that is untreated or inadequately treated. An average of 16.8 waterborne outbreaks (range from 6 to 27) occurs annually, affecting an average of 145 cases/year (range from 18 to 370). While the largest reported waterborne outbreak affected 3500 people (in Queenstown in 1984), the number of cases involved in most outbreaks is small, averaging nine cases per outbreak in 2001–2005, and is smaller than other countries for which data are available. This probably reflects the larger proportion of water supplies serving small communities in New Zealand compared to most other developed countries. This is consistent with the relatively poor compliance with the DWSNZ of the small community drinking-water supplies compared to that of the larger community supplies" (Ball 2006).[2]

Ball provides examples from Canterbury of disease occurrence from 1984 to 2005:

- 19 cases of campylobacteriosis in Ashburton in 1986 which was attributed to the delay after heavy rain in chlorinating the town water supply drawn from an infiltration gallery on the Ashburton River
- 42 cases of campylobacteriosis in1990 associated with a camp drawing water from springs on a farm with neither filtration or disinfection
- A cluster of campylobacteriosis cases in Waimate in 1992 attributed to the chlorinator being inoperative
- 6 diagnosed cases of campylobacteriosis in Fairlie in 1994 when turbidity associated with heavy rain reduced the efficacy of chlorination
- 33 cases of campylobacteriosis in Ashburton in 1996 associated with failure of the infiltration gallery after 3 days of heavy rain led to turbidity and faecal material from surrounding agricultural areas

[2] In August 2016, there was New Zealand's largest campylobacter outbreak with an estimated 5530 (or 39%) of Havelock North residents affected by gastroenteritis (Hawke's Bay District Health Board 2016). The contamination was found in a groundwater supply bore. An inquiry into the cause of the contamination is in progress.

- 59 cases of norovirus in Mt. Hutt in 1996 associated with a sewer discharge upstream of a water supply with filters removed and UV disinfection inoperative
- 5 confirmed and 18 possible cases of shigella in Banks Peninsula in 2004 with a break in a septic tank effluent pipe discharging above the intake of a spring-fed water supply.

Since 2005, there have been further disease outbreaks. In 2008, there was an outbreak of campylobacteriosis in Springston with 4 residents notified and a further 42 reporting symptoms (Medical Officers of Health 2008). The supply well was contaminated with faecal material possibly caused by agricultural activity in the area (Stuff.co.nz 12/03/2008). A new deeper well was commissioned.

In 2010, there was an outbreak of Cryptosporidium in Canterbury with 82 notifications in the January to March quarter compared to an average of 12 notifications for the previous 3 years. The disease is predominantly waterborne but no specific cause was identified (ESR 2010).

In 2012, there were 138 cases of confirmed or probable campylobacter which was attributed to faecal contamination of the Darfield water supply. The failure of the pump for the groundwater supply led to the use of infiltration galleries from the Waimakariri River (the town's supplemental source and former supply scheme). In mid-August heavy rainfall resulted in surface flooding and increased turbidity in the water supply. Monitoring indicated the presence of *E. coli*. The change in source led to reinstatement of chlorination but at the time of the contamination the chlorine cylinder was empty and the 'empty' alarm had been disconnected (Bartholomew et al. 2014). This resulted in an estimated community cost of between $0.7 million and $1.26 million (Sheerin et al. 2014).

One of the important issues for drinking water supply in Canterbury is source contamination associated with land use intensification either through leaching to groundwater, surface runoff or direct access of animals to streams (Davies-Colley et al. 2003). Groundwater is the dominant source of drinking water in Canterbury. Nitrate nitrogen and faecal contamination (as indicated by *E. coli* bacteria counts) are the most common health-related contaminants in Canterbury's groundwater (Environment Canterbury 2012). In the 2012 survey of groundwater monitoring wells for the region, 12% of wells exceeded health-based maximum acceptable levels (MAV) for *E. coli* and 11% of wells exceeded MAV for nitrate nitrogen. Table 9.2 sets out the exceedances for health and aesthetic criteria for drinking water from the 2012 survey.

Nitrate leaching losses were discussed in Sect. 3.1.6. The significant contribution of land use intensification, and in particular, the increased nitrate leakage to groundwater has been identified (Davies-Colley et al. 2003).

Close et al. (2008) investigated the effects of intensive dairying and border dyke irrigation on the leaching of *E. coli* and *Campylobacter* to shallow groundwater in the Waikakahi catchment in South Canterbury. *E. coli* were detected in 75% of samples and *Campylobacter* in 12%. An epidemiological assessment indicated a statistically significant increase in rate of campylobacteriosis compared to control groups in dairying areas without irrigation and the rest of the Canterbury region. *E.*

Table 9.2 Drinking water standard exceedances in Canterbury groundwater monitoring wells (Environment Canterbury 2012)

Health-based criteria	Maximum acceptable values	Exceedances in 289 wells monitored
E. coli	<I MPN/100 mL	36
Nitrate nitrogen	11.3 mg/L	33
Manganese	0.4 mg/L	11
Arsenic	0.01 mg/L	4
Aesthetic-based criteria	Guideline value	
Ammonia nitrogen	1.2 mg/L	6
Chloride	250 mg/L	0
Hardness (as $CaCO_3$)	200 mg/L	8
Iron	0.2 mg/L	19
Manganese	0.04 mg/L	26
pH	7.0–8.5	58
Sodium	200 mg/L	0
Sulphate	250 mg/L	1

coli were detected in all wells with concentrations ranging from <1 to 2400 MPN/100 ml[3] with an average of 40 MPN/100 ml and a median of 2 MPN/100 ml. The New Zealand drinking water standard is <1 MPN/100 ml.

There is a general trend throughout the data set of high concentrations of *E. coli* being detected in groundwater approximately 20–30 days after a coincidence of stocking and irrigation (or large rainfall event) in nearby paddocks. Concurrent trials on dairy farms using spray irrigation have identified very low levels of contamination in contrast to the areas irrigated by border dyke (Close et al. 2005). This highlights the impact of irrigation methods and poor irrigation efficiencies on leaching of microbes and contamination of groundwater.

9.1.2 Water Contact Recreation

Infections and illness due to recreational water contact are generally mild and therefore difficult to detect through routine surveillance systems (World Health Organization 2003). Available evidence from international studies suggests that the most frequent health outcome associated with exposure to faecally contaminated recreational water is enteric illness, such as gastroenteritis, through ingestion (Pruss 1988). Also, a cause-effect relationship between faecal or bather-derived pollution and acute febrile respiratory illness (AFRI) and general respiratory illness is biologically feasible through inhalation. A significant dose-response relationship between AFRI and faecal streptococci (as an indicator of contamination) has been reported (Fleisher et al. 1996). When compared with gastroenteritis, probabilities of

[3] MPN/100 ml refers to the most probable number of bacterial species per 100 ml of water.

Table 9.3 Guideline values of microbial quality of recreational waters (World Health Organization 2003)

Faecal contamination (95th percentile value of intestinal enterococci/100 mL)	Basis	Estimated risk per exposure
Category A ≤ 40	Below NOAEL (no observed adverse effect level)	<1% GI illness risk <0.3% AFRI illness risk
Category B 41–200	200/100 mL is above threshold of illness transmission	1–5% GI illness risk 0.3–1.9% AFRI illness risk
Category C 201–500	Substantial elevation in probability of adverse health outcomes	5–10% GI illness risk 1.9–3.9% AFRI illness risk
Category D ≥ 500	Significant risk of minor illness transmission	>10% GI illness risk >3.9% AFRI illness risk

contracting respiratory illness are generally lower and the threshold at which illness is observed is higher.

A review of international studies (Pruss 1988) evaluated the health risks associated with swimming in marine waters of poor microbiological quality. This included a study of health effects of bathing at selected New Zealand beaches (Mcbride et al. 1988). WHO derived guideline values for microbial water quality from these studies (Table 9.3). These microbiological assessment categories have been adopted in New Zealand (refer Sect. 9.2.2).

To examine the pathogen profiles for New Zealand freshwater conditions, a survey of microbial risk indicators and pathogens was carried out for 25 freshwater recreational and water supply sites (Till et al. 2008). It covered geographical areas across New Zealand and five categories of predominant environmental impact: forestry/undeveloped, municipal, dairy farming, sheep/pastoral farming, and birds. In contrast to studies in other countries which focused on human sewage, pathogen contamination in rural New Zealand can be expected to be dominated by zoonotic[4] agents such as *Campylobacter* and *Cryptosporidium.*

Of the pathogens tested, *Campylobacter* and adenoviruses were inferred to be most likely to cause human waterborne illness to recreational freshwater users. The critical value for *E. coli* as an indicator of increased *Campylobacter* infection is in the range 200–500 *E. coli* per 100 mL. Using the data from all sites an estimated 5% of notified *Campylobacter* cases in New Zealand could be attributable to freshwater contact recreation (Till et al. 2008). The results have been used to derive the national water quality guidelines for recreational freshwater in New Zealand (Table 9.4).

[4]Zoonotic refers to a disease that normally exists in animals but that can affect humans.

Table 9.4 Recreational site gradings for New Zealand freshwaters (Till et al. 2008)

Faecal contamination (95th percentile *E. coli*/100 mL)	Basis for derivation	Estimated risk of *Campylobacter* infection
Category A ≤130	NCRL (no calculated risk level)	<0.1% occurrence
Category B 131–260	Not too far above NCRL	0.1–1% occurrence
Category C 261–550	Substantial increase above background rate	1–5% occurrence
Category D >550	Significant risk of infection	>5% occurrence

9.1.3 Toxic Algal Blooms

Cyanobacteria (commonly known as blue-green algae) can form planktonic (suspended in the water column) blooms or dense benthic (attached to the substrate) mats. Some cyanobacteria include toxin-producing strains. Cyanotoxins have a broad range of toxicity mechanisms ranging from hepatotoxicity (toxic to the liver), neurotoxicity (toxic to nerves or nerve tissue), to dermatotoxicity (affects the skin) (Ministry for the Environment and Ministry of Health 2009). Studies have found elevated risks of gastrointestinal and dermatological symptoms associated with drinking and domestic use of water with raised cyanobacterial cell counts (El Saadi et al. 1995). The disease pathway is through the ingestion or inhalation of water during drinking or by contact during recreational activities.

Because of the multiple causes of gastrointestinal and dermatological symptoms, it is difficult to establish the dangers to health from algal exposure. An Australian survey investigated occurrence of diarrhea, vomiting, flu-like symptoms, skin rashes, mouth ulcers, fevers, and, eye or ear irritations of people engaged in water-based recreation. The study confirmed that symptom occurrence was associated with cyanobacterial exposure during recreational water-related activities. The study also found there was a significant trend of increasing symptom occurrence with increasing duration of water contact and with increasing cyanobacterial cell density, using a cut-off of 5000 cells per mL (Pilotto et al. 1997).

The health risks associated with benthic cyanobacteria are less well known. However investigations have revealed widespread distribution of toxic species linked to dog poisonings (Wood et al. 2007).

9.1.4 Seafood Contamination

One of the significant health pathways for contamination of seafood is from algal blooms. From a human health perspective, the critical issue is seafood contamination from biotoxins. Biotoxins generated by phytoplankton can cause toxic shellfish poisoning. Six forms have been found in New Zealand: paralytic shellfish poisoning (PSP), amnesic shellfish poisoning (ASP) and diarrhetic shellfish poisoning (DSP), neurotoxic shellfish poisoning (NSP), yessotoxin shellfish poisoning (YSP) and azaspiracid shellfish poison (AZP) (New Zealand Food and Safety Authority 2003).

Biotoxins can also affect marine ecological systems. Furthermore, high biomass-producing algal blooms can cause anoxia[5] affecting marine life.

Shellfish contaminated with sewage or other sources of bacteriological contamination may contain pathogens such as norovirus, hepatitis A, Shigella, Vibrio and Salmonella. These can cause dysentery, gastro-enteritis and other illnesses within a few hours or days and can have long term implications such as damage to blood, liver and immune system.

Marine biotoxin blooms are primarily natural events due to rapid growth of natural occurring phytoplankton in seawater. There are instances where nutrient loading from rivers can increase nutrients available for phytoplankton locally but upwelling nutrient supply is the usual driver of many bloom events.

For the east coast of the South Island of New Zealand, the Southland Current brings water from the south of New Zealand along the Canterbury Coast (Fig. 9.1). An example is the dinoflagellate bloom dominated by *Gymnodium* species in 1994. The bloom was first detected in oyster beds in Foveaux Strait (to the south of the South Island) in mid-January. The bloom gradually worked its way along the east coast with a "red tide" offshore at Timaru (in South Canterbury) causing widespread deaths of white-fronted terns, and reaching the Marlborough Sounds in the north-east of the South Island in May (Meduna 1994).

The mixed sand and gravel beaches from Oamaru north to Banks Peninsula provide little opportunity for the settlement of bivalve shellfish, paua or crayfish. In contrast, these species are found in the rocky reefs around Banks Peninsula, and the harbours and estuaries contain cockles and pipi. North of Banks Peninsula is a stretch of steep coarse-grained sandy beaches with beds of cockles present. From Amberley Beach north, the coast is comprised of exposed rocky reefs where crayfish and paua are the major non-commercially harvested species.

One survey indicated that the species most frequently targeted for non-commercial harvest are paua and crayfish with a lower number of trips targeting cockles, mussels and pipi. The trips targeting bivalve shellfish species in Canterbury comprised only 3.9% of the total trips in New Zealand (Fisher and Bradford 1998).

Canterbury contributes a small component of New Zealand's commercial aquaculture production with 3% of total production of Greenshell Mussels (New Zealand harvested product in 2011 was 101,311 tonnes (Aquaculture New Zealand 2015))

[5] Anoxia in waters refers to an absence of dissolved oxygen.

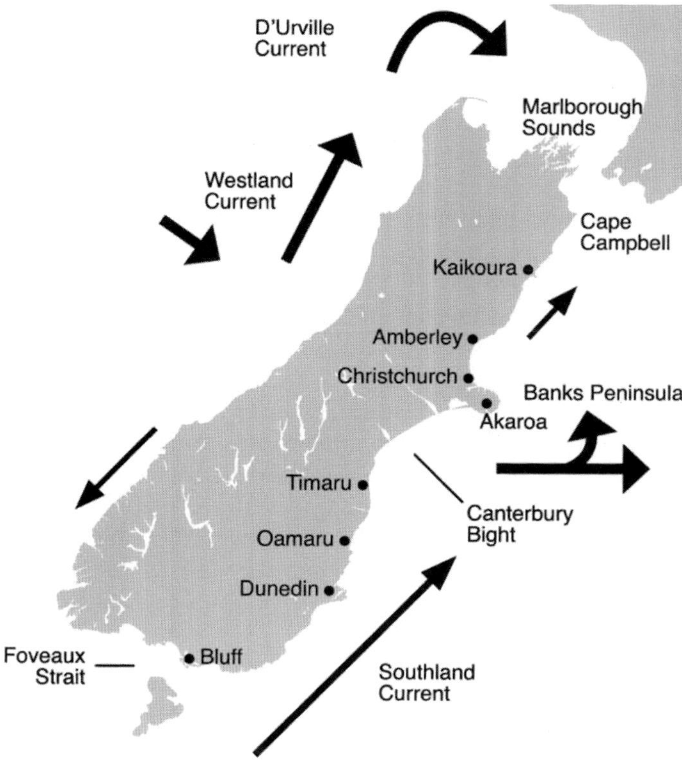

Fig. 9.1 Coastal current flows for South Island (Hay et al. 2000)

and 4% of total production of King Salmon (New Zealand harvested product in 2011 was 14,037 tonnes (Aquaculture New Zealand 2015)). Greenshell Mussel farms are sited in the northern bays and coastline of Banks Peninsula, while King Salmon farms are in Akaroa Harbour on the southern side of Banks Peninsula.

As the coastal current flows north up the Canterbury Bight it receives freshwater inputs from rivers, streams and drains as well as flows from Te Waihora/Lake Ellesmere and Wairewa/Lake Forsyth (Bolton-Ritchie 2008). Testing from the lakes indicate cyanobacteria concentrations above public health trigger levels. While data is scarce, modelled plume dynamics and available knowledge indicates that the lake plumes contribute little to toxic cyanobacteria in Banks Peninsula's coastal waters (Hadfield et al. 2012).

9.2 Management Approaches

9.2.1 Management of Drinking Water

The Health Act 1956 was amended by the Health (Drinking Water) Amendment Act in October 2007. Prior to the amendments, drinking water management was comprised of entirely voluntary elements. There were drinking water standards, a register of drinking water supplies, grading of those supplies, a drinking water database and annual reporting of microbiological and chemical quality, and, guidelines for drinking water quality management. Public health risk management plans were voluntary. The enactment of the 2007 Act Amendments resulted from a concern that the organization of New Zealand's drinking water supplies was not adequate to safeguard communities (Ministry of Health 2015). Only 71% of the population was served with water that complied with drinking water standards.

One of the significant changes in the 2007 Act Amendments was the requirement for public health risk management plans (now called water safety plans). These are the main action forcing mechanism in terms of improving drinking water standards.[6] The key requirements of water safety plans are:

- flow diagram of source, treatment and distribution;
- identification of risks/barriers to contamination;
- risk analysis considering causes of contamination, preventative measures, and corrective actions if measures fail;
- development of an improvement timetable to reduce risks;
- preparation of contingency plans for system failure;
- undertaking performance assessment by monitoring implementation and outcomes; and
- recording and communication requirements of performance (Ministry of Health 2005).

For drinking water, vulnerability can be assessed on the basis of whether any of the following four barriers to water contamination are missing:

- prevention of contaminants entering the source water;
- removal of particles from the water (where many of the germs reside);
- killing germs; and
- prevention of recontamination after treatment.

The assessment needs to cover the three aspects of water supply:

- the catchment and intake,
- treatment and storage, and
- distribution.

[6]This follows the World Health Organization approach (Davison et al. 2005).

Box 9.1 provides a summary of the Water Safety Plan (WSP) analysis for the Akaroa water supply in relation to water quality. It demonstrates the value of identifying risks in the catchment and treatment processes so that management interventions with respect to catchment protection and improved treatment processes can be developed.

Box 9.1: Akaroa Water Safety Plan

The Akaroa Water Safety Plan (Christchurch City Council 2014) provides the requirements of a WSP by describing the water supply scheme; identifying the *critical points* in the scheme; identifying the *barriers to contamination* in the scheme; providing a *risk assessment* for the water supply scheme; providing a detailed *improvement schedule* to mitigate the significant identified risks; providing the *contingency planning strategies* that will be adopted; and, discussing the *performance assessments* and the *auditing procedures.*

The risk assessment identified a number of significant risks in relation to the water supply system. The raw water was a combination of surface and groundwater. Two thirds of the surface water catchment was used for grazing livestock which have access to the streams: this was associated with increased turbidity after heavy rainfall. Also, there is a natural organic content in the water. The groundwater was from fractured basalt of volcanic origin with highly variable water quality which can have a low pH and elevated concentrations of iron, manganese and sometimes chloride.

There were two water treatment plants: L'Aube Hill and Aylmers Valley. The Aylmers Valley treatment process comprises a pre-treatment reservoir, dual media rapid sand filter and chlorination by injection of sodium hypochlorite. The pre-treatment and filtration processes are inadequate to reliably remove turbidity and do not meet the protozoa removal requirements (4 log removal). Furthermore, chlorination reacts with the organic content of the raw water and can produce organochlorides (trihalomethanes and haloacetic acids) that are health risks. In addition, chlorate is a byproduct of using sodium hypochlorite in the disinfection process. The treated water failed to meet the New Zealand drinking water standards for protozoa removal as well as the presence of organochlorides and chlorate are above acceptable levels.

To respond to these issues the improvement schedule which has now been implemented includes the identification of a catchment protection zone, purchase of a reserve to cover the catchment upstream of the water supply intakes and upgrading of the water treatment process. The process includes (1) potassium permanganate dosing to precipitate iron and manganese, (2) coagulation to gather up large molecules of natural organic matter, (3) membrane filtration with a pore size less than one micron to remove natural organic matter and protozoa, (4) disinfection with chlorine gas to avoid chlorate production, and (5) caustic dosing for pH adjustment.

In relation to catchment protection, current legislation allows for the protection of the quality and other aspects of the sources of drinking water supplies. Regional councils have responsibilities under the Resource Management Act (s30(1)(c)) to control land use in order to protect water quality in water supply catchments while Health legislation makes water suppliers responsible for the water supply from the point of abstraction to the consumer.

The National Environmental Standard for Sources for Drinking Water requires regional councils to ensure that effects on drinking water sources are considered when making decisions on resource consents and rules for regional plans.

Protection of drinking water is one of the targets in the Canterbury Water Management Strategy. Also the regional council Regional Plan that became operative in 2011 has a policy designed to avoid public water supply contamination by identifying protection zones around each community drinking water supply well or a specified distance from a surface water intake for a community water supply (Environment Canterbury 2011). Work is well advanced on defining these zones for Canterbury.

However, the implementation of Water Safety Plans and Catchment Protection Zones are still work-in-progress. Figure 9.2 shows a map of the status of drinking water supplies in Canterbury. There are large areas graded as unacceptable and unsatisfactory. Many areas are ungraded. Only a few areas like Christchurch, Ashburton, Waimate and Kaikoura have water supplies that are graded as satisfactory.

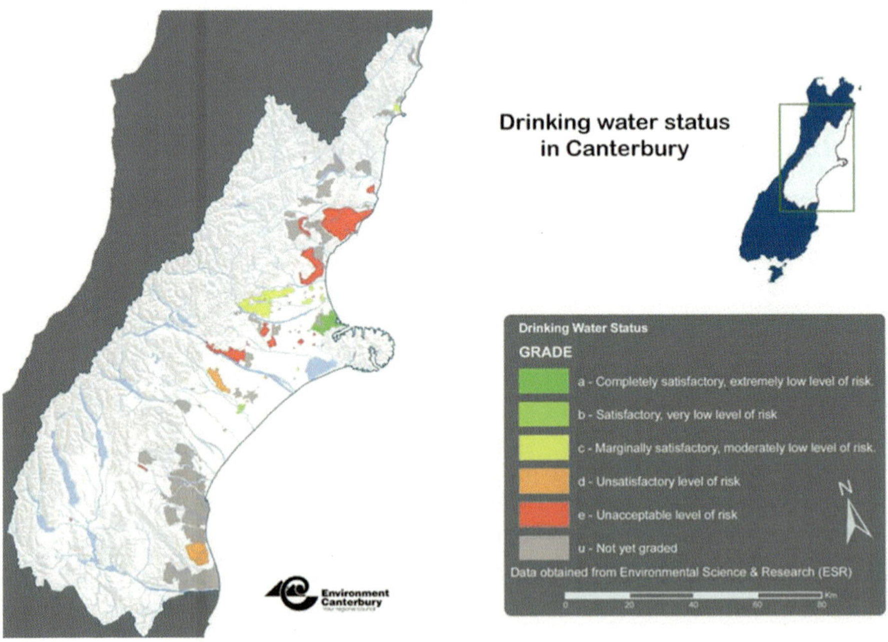

Fig. 9.2 Status of drinking water supplies in Canterbury (Environment Canterbury)

9.2.2 *Management of Contact Recreation Disease Pathway*

The vulnerability to water borne disease from water contact recreation can be assessed on the basis of the Suitability for Recreation Grading (SFRG) as set out in the Microbiological Water Quality Guidelines for Marine and Freshwater Recreational Areas (Ministry for the Environment 2002). These Guidelines incorporate a risk-based approach to monitoring water quality promoted by the World Health Organization in the Annapolis Protocol (World Health Organization 1999) and adapted for New Zealand circumstances. The guidelines use a combination of a qualitative risk grading of the catchment, supported by the direct measurement of appropriate faecal indicators to assess the suitability of a site for recreation. In addition, alert and action guideline levels are used for surveillance throughout the bathing season.

There are two components to providing a grading for a recreation site:

- the Sanitary Inspection Category (SIC), which generates a measure of the susceptibility of a water body to faecal contamination
- historical microbiological results, which generate a Microbiological Assessment Category (MAC), which provides a measurement of the actual water quality over time.

These two components are combined to give an overall Suitability for Recreation Grade (SFRG), which describes the general condition of a site at any given time, based on both risk and indicator bacteria counts. This grade helps to determine whether ongoing monitoring is required, and provides the basis for telling people whether or not the water is suitable for recreational use, from a public health perspective.

The assessment of SIC is based on a catchment checklist based on

- the wastewater discharges entering the water body and their degree of treatment, and,
- the stormwater discharges entering the water body and the land use associated with the contributing catchment and the density of bird and animal life.

For marine waters, the assessment of MAC is based on faecal enterococci levels and epidemiological research involving the direct measurement of gastrointestinal illness (GI) and acute febrile respiratory illness (AFRI). While for freshwater, MAC is based on *E. coli* concentrations based on quantitative risk assessment inferred from dose response relationships for Campylobacter which constitutes more than 50% of reported notifiable diseases in New Zealand.

The Guidelines identify the following management responses:

- an alert level based on sample results (exceedance of 140 enterococci per 100 mL for marine waters and 260 *E. coli* per 100 mL for freshwaters) where the response is to increase sampling to daily and a catchment assessment to identify possible sources

- an action level based on sample results (two consecutive samples greater than 280 enterococci per 100 mL for marine waters and one sample exceeding 550 *E. coli* per 100 mL for freshwaters) where the response is increased sampling, catchment assessment to identify source, public notification of health risks and discussion among local and health authorities on who is responsible for fixing the problem
- for high risk sites (those graded poor or very poor and usually having direct discharges) where the response is permanent health warning signs erected, and, the agencies responsible for infrastructure, land use affecting water quality, water quality monitoring and public health protection to discuss and document the remediation work required and the timeframes and budgets for achieving this.

The intent of the Annapolis Protocol was to replace the regulatory schemes using a pass/fail test based on percentage compliance with faecal indicator counts related to human waste, with a scheme that had a graded scale, incorporated the large number of factors that can influence site condition and encouraged the treatment of the sources of contamination to achieve a lower grading.

The gradings for monitored freshwater recreational sites in Canterbury range from "very poor" to "very good" (Robinson and Bolton-Ritchie 2014). Fifty-six percent of river and lake sites were graded as suitable for contact recreation (Environment Canterbury 2013). In the Sanitary Inspection Category assessment of freshwater recreation sites, the most common factor identified as having the primary influence on microbiological water quality of the monitored sites is whether the site is a focal point for run-off from an agricultural, urban or rural catchment. Poor and very poor gradings are mainly associated with urban stormwater, intensive agricultural land use, unrestricted stock access to waterways and density of birdlife.

The gradings for monitored marine recreational sites in Canterbury range from "poor" to "very good". The only poor gradings are from the Avon Heathcote Ihutai Estuary for sites near the mouths of the Avon Ōtākaro and Heathcote Rivers which receive stormwater and sewage overflows from reticulation infrastructure. For sites with "moderate" SIC gradings (the most adverse water quality grading for marine sites in Canterbury), the primary influences on microbiological water quality are urban stormwater (22 of 29 sites), bird density (14 of 29 sites) and high intensity agriculture (4 of 29 sites).

The main management response is one of making gradings public on the regional council website and placing warning signs at sites graded poor or very poor (which is considered unsuitable for contact recreation) and sites where water quality is temporarily degraded due to runoff from intense rainfall events.

For most sites the gradings have not changed over the past 5 years. However, there are examples where monitoring data has been used to initiate management action to address poor water quality.

One is the marine site, Corsair Bay in Lyttelton Harbour. Elevated faecal contamination levels detected during monitoring led to investigations by Christchurch City Council. These investigations identified leakage issues from a public toilet block and a private pump station that regularly overflowed. When both these issues were addressed the monitoring results showed a clear improvement (Bourke M, 2015, Corsair Bay, personal communication).

A second is a freshwater site, Coes Ford on the Selwyn River, a popular swimming site. However the site has been graded between "poor" and "very poor" for the past ten years based on faecal coliform monitoring (Robinson and Stevenson 2012). Another monitored site less than 4 km upstream at Chamberlains Ford is nearly always graded as "good". Silverstream (catchment area 1505 ha), a tributary of the Selwyn River which joins upstream of Coes Ford has been identified historically as a significant contributor of poor microbial water quality (Adamson and Main 1996).

The Silverstream Water Improvement Group was formed in 2002 as part of the regional council Living Streams programme. Land owners have been encouraged to exclude stock from waterways through fencing and stock crossings, and to establish vegetated riparian buffer strips. By 2006, 80% of the stream boundaries in the Silverstream catchment were fenced, increasing to 91% by 2013 with all but four stock crossings bridged (Glasgow 2013). While there is a long term downward trend in *E. coli* measurements at Coes Ford, the SFRG is still graded as "poor" (Robinson and Stevenson 2012).

9.2.3 Management of Cyanobacteria in Recreational Fresh Waters

New Zealand has established guidelines for the management of cyanobacteria in recreational fresh waters (Ministry for the Environment and Ministry of Health 2009). These guidelines are based on the multi-tiered approach recommended by the World Health Organization (World Health Organization 2003) and the Australian National Health and Medical Research Council (National Health and Medical Research Council 2008). New Zealand has a three-tier alert level framework for monitoring and management action. It is designed to provide a graduated response. There is a framework for planktonic cyanobacteria and one for benthic cyanobacteria which are summarized in Table 9.5.

The surveillance (green mode) is triggered when cyanobacteria are detected at low levels (for planktonic cyanobacteria up to 0.5 mm^3/L or 500 cells/mL; for benthic cyanobacteria up to 20% coverage). Ongoing monitoring is considered the appropriate management action.

The alert (amber mode) accommodates the transition to the action level. The cyanobacteria levels are between the low risk to health of the surveillance levels and the higher risk action levels (for planktonic cyanobacteria 0.5 to <1.8 mm^3/L for potentially toxic cyanobacteria, and, 0.5 to <10 mm^3/L for the combined total of all cyanobacteria; for benthic cyanobacteria 20 to 50% substrate coverage of potentially toxigenic cyanobacteria).

The management actions considered appropriate for the alert level are for the monitoring agency to notify the public health authority, to increase sampling

Table 9.5 Alert level framework for cyanobacteria (Ministry for the Environment and Ministry of Health 2009)

Alert level	Planktonic cyanobacteria		Benthic cyanobacteria	
	Monitoring situation	Management action	Monitoring situation	Management action
Surveillance (green mode)	Cell conc. less than 500 cells/mL.	Undertake weekly/fortnightly inspection and sampling spring to autumn.	Up to 20% substrate coverage of potentially toxigenic cyanobacteria.	Undertake fortnightly surveys spring to autumn
Alert (amber mode)	Biovolume toxic cyanobacteria 0.5 to <1.8 mm³/L. All cyanobacteria 0.5 to <10 mm³/L.	Notify public health unit. Increase sampling to weekly. Multiple sites inspected and sampled.	20–50% substrate coverage of potentially toxigenic cyanobacteria	Notify public health unit. Increase sampling to weekly. Consider increasing survey sites. Recommend public signage. Test for cyanotoxins.
Action (red mode)	Microcystins ≥12 µg/L or toxic cyanobacteria ≥1.8 mm³/L. All cyanobacteria ≥10 mm³/L 3 cyanobacteria scums present.	Continue monitoring. Notify public of health risk. If toxic taxa present test for cyanotoxins.	More than 50% substrate coverage of potentially toxigenic cyanobacteria. Up to 50% substrate coverage where potentially toxigenic cyanobacteria are detaching.	Immediately notify public health unit. Notify public of health risk. If potentially toxic taxa present test for cyanotoxins.

frequency and to consider inspecting and sampling from multiple sites. Additional actions for benthic cyanobacteria are to test for toxins and to alert the public through media releases and site signage.

The action (red mode) is when the guideline level designed to protect against health effects of repeated exposure to cyanobacterial toxins ingested during recreational activity is reached (for plankton cyanobacteria a concentration of 12 µg/L total microcystins or a biovolume estimate of 1.8 mm³/L of potentially toxigenic cyanobacteria); or, when there is an increased probability of respiratory, irritation and allergy symptoms from exposure to very high cell densities of cyanobacteria (for planktonic cyanobacteria 10 mm^3/L for total biovolume of all cyanobacteria, or, for benthic cyanobacteria 50% substrate coverage when mats commonly detach from the substrate and can accumulate on shorelines).

The main management action is public notification by public health units of the health risks by media releases and requesting city and district councils to erect signs at affected water bodies. Local doctors should be encouraged to report illness linked to water contact with cyanobacteria.

Testing for cyanotoxins is also recommended because of the persistence of cyanotoxins during post-boom decline. Lowered cell counts and biovolumes may not give a true indication of toxin levels. Alert levels should not be lowered until two successive results of biovolumes (for planktonic cyanobacteria) or substrate cover age (for benthic cyanobacteria) have been recorded.

Table 9.6 sets out the number of days between 2010/11 and 2015/16 where cyanobacteria levels at river-based recreational water quality sites were above 20% of substrate cover. Of the 35 river recreation sites that are monitored 28 (80%) have exceedances of the "alert" criterion of 20% substrate cover. Ten sites have recorded maximum values exceeding the "action" criterion of 50% substrate cover. The table indicates annual fluctuations in cyanobacteria cover, however the overall trend appears to be of increasing frequency of cyanobacteria cover particularly in Mid and South Canterbury.

In relation to lakes, three of the 15 lakes that are monitored for contact recreation water quality exceed the "action" criterion for planktonic cyanobacteria: these are the coastal lakes, Te Waihora/Lake Ellesmere and Wairewa/Lake Forsyth, and, the artificial lake, Lake Pegasus.[7] Table 9.7 sets out the results for the three lakes for the 2014/5 summer. The causes of the exceedances indicate different dominant plankton species; *Picocyanobacteria*[8] for Te Waihora/Lake Ellesmere which persisted above action levels from the summer of 2013/14 to the summer of 2015/16 (i.e. over two winters as well as the summer of 2014/15); Wairewa/Lake Forsyth had an initial bloom dominated by *Anabaena*[9] species in the early summer, followed by a *Nodularia*[10] dominated bloom in late summer indicating an increase in salinity. Pegasus Lake was subject to an *Anabaena* bloom.

[7] There are also two other lakes outside of the contact recreation water quality monitoring programmes that are monitored for cyanobacteria and where algal blooms are frequent: Lake Rotorua (near Kaikoura) and St Annes Lagoon (near Cheviot).

[8] *Picocyanobacteria* are cyanobacteria having dimensions less than 2 micrometres.

[9] *Anabaena* is a genus of filamentous cyanobacteria that exist as plankton and are known for nitrogen-fixing abilities.

[10] *Nodularia* is a genus of nitrogen-fixing cyanobacteria that occur mainly in brackish or saline waters.

Table 9.6 Number of days when cyanobacteria levels in rivers exceed 20% substrate cover

	Year					
Site	10/11	11/12	12/13	13/14	14/15	15/16
Kahutara R @ Peketa					24	64
Waiau R @ Waiau[a]		168				
Hurunui R @SH7						37
Hurunui R @ SH1		66	35	30		
Waipara R @ Stringers Br[a]	103	13				
Waipara R @ Teviotdale Br		38				
Ashley R @ Loburn Br[a]	18	68	28	25		
Ashley R @ SH1[a]	71		44	55	58	
Waimakariri R @ Reid's R		20				
Selwyn R @ Whitecliffs[a]	63		54	98	34	42
Selwyn R @ Glentunnel[a]	63		22	13		
Selwyn R @ Coes Ford						124
Ashburton R @ SH1		6			51	
Waihi R @ SH72				31	28	98
Waihi R @ Winchester				49		50
Temuka R @ Manse Br			27		29	109
Temuka R @ SH1				7	49	70
Opihi R @ Saleyards Br[a]	22	19			49	56
Opihi R @ SH1[a]	93	47	13		136	142
Opihi R @ Waipopo	14		56		16	124
TeNgawai R @ TeNgawai Br						28
Pareora R @ Evans Crossing						22
Pareora R @ Huts[a]	63	69	63	28		
Pareora R @ SH1[a]						36
Otaio R @ Gorge			19			
Waihao R @ Black Hole			77			
Waihao R @ Gum Tree Flat			35			
Waihao R @ Bradshaws Br					195	76
Hakataramea R @ SH82				52		
TOTAL	510	514	473	388	669	1078

Source: Environment Canterbury (2017) Dynes K personal communication
Note: Rivers are listed from north to south in the region. Where there are multiple sites on the same river the upstream sites are listed first
[a]Sites where greater than 50% substrate cover has been reported

Table 9.7 Plankton blooms in recreational lakes for summer 2014/15 (Bolton-Ritchie and Robinson 2016)

Lake	Bloom occurrence	Dominant species (mm^3/L)
Te Waihora/Lake Ellesmere	14/1 2014 to 18/2/2016	*Picocyanobacteria* > 1.8
Wairewa/Lake Forsyth	8/10/2014 to 13/11/2014	*Anabaena* > 1.8
	10/2/2015 to 14/4/2015	*Nodularia* > 1.8
Pegasus Lake	11/2/2015 to 10/4/2015	*Anabaena* > 1.8

9.2.4 *Shellfish Management*

Bivalve shellfish receive special consideration in food safety laws because:

- They are filter feeders and can accumulate hazardous levels of pathogens (bacteria, viruses and other illness-causing micro-organisms), biotoxins, and, chemical contaminants (such as heavy metals);
- There is no thermal process prior to sale to eliminate pathogens and shellfish are often eaten raw or lightly cooked.

Initially standards were developed to assure that shellfish exported to the United States were safe to eat. The management approach now incorporates all commercial shellfish for harvesting, processing and transport for local consumption and export as well as non-commercial shellfish food gathering.

The programme for non-commercial food gathering is based on monitoring for biotoxin levels at selected sites (by Ministry for Primary Industries) and for notification of cases of toxic shellfish poisoning (by District Health Boards). When biotoxins in shellfish samples have exceeded public warning limits, or when two or more cases of human illness have been notified, health authorities issue public warnings. There is weekly sampling of phytoplankton and fortnightly sampling of shellfish for PSP toxins for sites with a history of PSP contamination. Phytoplankton provides an early warning of toxicity in shellfish. Once phytoplankton triggers are exceeded shellfish samples are taken. Public warnings are issued when toxin levels detected in shellfish are above public warning levels (New Zealand Food and Safety Authority 2013).

For commercial marine farms, a much more comprehensive approach is taken. New Zealand has been developing and implementing hazard analysis and critical control points (HACCP) as a means of managing food safety concerns (Fishing Industry Inspection and Certification Council 1997). HACCP was originally developed by the Pillsbury Company in consultation with the US Army and US National Aeronautics and Space Administration to ensure food safety for astronauts.

The key steps in the HACCP process are:

- Conduct a hazard analysis – identify food safety hazards and preventative measures to control hazards
- Identify critical control points – identify points in the food manufacturing process where control can be applied
- Establish critical limits for each critical control point – determine maximum or minimum values for control purposes
- Establish critical control point monitoring requirements – identify monitoring procedures for control purposes
- Establish corrective actions – identify actions to be taken if the critical limit is not met
- Establish procedures for ensuring the HACCP system is working – establish verification requirements that the HACCP plan is working as intended

- Establish record keeping procedures – establish records to monitor plan implementation, verify the correct monitoring took place, and validate that the set limits control the hazard.

In relation to water quality management for shellfish harvesting, the key elements for New Zealand are as follows:

- Addressing the hazards of pathogens, biotoxins and chemical contaminants;
- Defining water quality of the shellfish growing area as the critical control point and using a "sanitary survey" for classifying growing areas;
- Setting water quality criteria for pathogen indicators (faecal coliforms in water and *E. coli* in shellfish) and marine biotoxins (toxic species of phytoplankton in water and marine biotoxin levels in shellfish);
- Establishing water and shellfish monitoring stations for routine monitoring of bacteriological quality, and, a marine biotoxin monitoring programme (as part of a marine biotoxin management plan) for each shellfish growing area;
- Establishing procedures for closure of harvesting areas; for making contaminated shellfish fit for human consumption through "relaying" (transferring shellfish to another growing area for contaminant reduction), "depuration" (immersion of shellfish in tanks of clean seawater to allow contaminants to be purged), or post-harvest treatment; and, for product recall;
- Undertaking annual reviews of the sanitary survey and marine biotoxin management plan as verification requirements; and,
- Defining procedures for surveys, monitoring, review and reporting (New Zealand Food and Safety Authority 2006).

The sanitary survey includes the following elements:

- Information on the growing area, its catchment and land use, the shellfish resources, harvest practices, and aquaculture-related facilities;
- The identification and evaluation of pollution sources that could affect the growing area;
- The hydrographic and meteorological characteristics (physiography, tides and currents, rainfall/runoff and river discharges, and winds);
- Water quality studies of contaminants and pollution events; and
- Analysis of the inter-relationships between the pollution sources, hydrographic and meteorological characteristics, and effects on the shellfish growing area.

The sanitary survey forms the basis for classifying the growing area. The classifications are:

- *Remote approved* where there is no human habitation or pollution sources in the catchment;
- *Approved* where the growing area is not subject to contamination that presents a health risk and meets acceptable standards, and, shellfish can be grown without relay, depuration or post-harvest treatment;

Table 9.8 Bacteriological standards for shellfish growing areas

	Open areas	Restricted areas
Faecal coliforms		
Median MPN	14/100 mL in water	88/100 mL in water
90% samples	43/100 mL in water	260/100 mL in water
E. coli		
Median MPN	230/100 mL in water	4600/100 g in shellfish
90% samples	700/100 mL in water	14,100/100 g in shellfish

- *Restricted* where there is a limited degree of pollution but shellfish can be made fit for human consumption by either relaying, depuration or post-harvest treatment;
- *Conditional* where the effects of pollution are identified and evaluated, and, where the growing area has a management plan for areas affected by pollution sources; and
- *Prohibited* where the growing area is adjacent a point source outfall, or pollution sources are unpredictable, or contamination levels are unacceptable.

The bacteriological standards related to open and restricted areas are set out in Table 9.8.

There is also a requirement for a marine biotoxin management plan for each growing area. The biotoxin management plan includes a monitoring programme for shellfish and phytoplankton, and the procedures that are to be followed when trigger levels are exceeded. Table 9.9 sets out the trigger levels for phytoplankton and the maximum permissible levels for marine biotoxins in shellfish. An Alert Level is defined when phytoplankton trigger levels are exceeded or biotoxins detected in shellfish but are below maximum permissible levels. Closure Level is defined when a biotoxin in shellfish exceeds a maximum permissible level. There are also trigger levels for the withdrawal of closure when biotoxins in shellfish are below the maximum permissible levels and for return to routine operations when biotoxins in shellfish have returned to background levels and phytoplankton levels are below trigger levels.

Box 9.2 summarises the application of this approach to an aquaculture farm in Menzies Bay on the northern coastline of Banks Peninsula.

9.3 Analysis of Approaches

9.3.1 Analysis of Approach to Drinking Water

The water safety plan approach considers the elements of an adaptive cycle: *exploitation* (use of the water for drinking purposes) in considering source, treatment and distribution; *accumulation* (of contaminants that pose a risk to public health) in

Table 9.9 Phytoplankton trigger and shellfish biotoxin levels

Phytoplankton action levels		Shellfish biotoxin maximum permissible levels	
Species	Level (cells/L)	Toxins	Level
Paralytic Shellfish Poison (PSP)			
Alexandrium minutum	100	Saxitoxins	0.8 mg/kg
Alexandrium ostenfeldii	100		
Alexandrium catenella	100		
Alexandrium tamarense	100		
Gymnodinium catenatum	100		
Amnesic Shellfish Poison (ASP)			
Pseudo-nitzchia australis	100,000	Domoic acid	20 mg/kg
Pseudo-nitzchia pungens			
Pseudo-nitzchia multiseries			
Pseudo-nitzchia turgidula	500,000		
Pseudo-nitzchia fraudulenta			
Pseudo-nitzchia delicatissima			
Pseudo-nitzchia pseudodelicatissima			
Pseudo-nitzchia multistriata			
Neurotoxic Shellfish Poison (NSP)			
Karenia brevis	1,000	Brevetoxins	20 MU/100 g or 0.8 mg/kg
Karenia/Karlodinium/Gymnodinium group	250,000		
Diarrhetic Shellfish Poison (DSP)			
Dinophysis acuta	500	Okadaic acid, dinophysistoins, pectenotoxins	0.16 mg/kg
Dinophysis acuminata	1,000		
Prorocentrum lima	500		
Yessotoxin Shellfish Poison (YSP)			
Gonyaulax cf. spinifera	100	YTX, 45 OH YTX, homo YYX, 45 OH-homo YTX	1 mg/kg
Protoceratium reticulatum	500		
Azaspiracid Shellfish Poison (AZP)			
		AZA1, AZA2, AZA3	0.16 mg/kg

Box 9.2: Management of an Aquaculture Farm in Menzies Bay

Menzies Bay is a relatively shallow (2 to 14 m deep) bay on the north coast of Banks Peninsula. Growing Area 1602 is in the bay. This is a commercial shellfish farming area that has been established to grow green-lipped mussels (*Perna canaliculus*) using the conventional long line technique. A sanitary survey was undertaken to define the quality of the shellfish growing area (Johnson 2003).

Menzies Stream with a catchment of 700 ha is the primary source of runoff discharged into Menzies Bay. There are also two ephemeral streams with a

(continued)

Box 9.2 (continued)

total catchment of 100 ha. Most of the catchment is in farmed pasture cover of predominantly sheep and some cattle with a relatively low stocking rate (2 stock units per hectare). No fertilizer has been applied for 30 years and pesticides and herbicides are not used. There are only 4 permanent residents and no reticulated services, no stormwater, no industrial wastes, and thus no point source discharges into Menzies Bay. Domestic stock are the major source of faecal coliforms in the catchment and faecal matter is likely to be flushed from the land into the bay when it rains.

Water quality studies confirmed peaks of faecal coliforms in streams associated with rainfall events with between 4 and 21% of stream results exceeding 1600 fc/100 mL. In the aquaculture farm area in Menzies Bay faecal coliform levels were observed to rise in response to rainfall but did not exceed 14 fc/100 mL and soon returned to background levels (<2 fc/100 mL). All sites complied with the water quality requirements of the standard IAIS 005.1 (New Zealand Food and Safety Authority 2003) of the faecal coliform median not greater than 14 fc/100 mL and not more than 10% of the data being greater than 43 fc/100 mL.

Shellfish flesh quality was also tested. On the sampling days, some faecal coliform levels in the flesh exceeded 300 per 100 g of flesh (the desirable level not to be exceeded in the standard (IAIS 005.1)). The faecal coliform levels in flesh did drop below this threshold within some days after rain.

Harvesting constraints for Growing Area 1602 Menzies Bay were set based on rainfall. Faecal contamination in runoff associated with rain events is the source of pollution but there are no streams with measured flows. There are no provisions for relay, depuration or wet storage for the aquaculture farm so closure is the management intervention to deal with water quality impairment.

The current requirements in the Banks Peninsula Management Plan (Simpson et al. 2012) which includes Growing Area 1602 Menzies Bay are based on rainfall recorded from the Menzies Bay rain gauge in the 24-h period 9 am to 9 am. If rainfall is less than 20 mm then no closure required. If rainfall is 20 mm or more but less than 35 mm then closure for 3 days. If rainfall is 35 mm or more but less than 50 mm then closure for 5 days. If rainfall is more than 50 mm then closure for 5 days. There are an estimated 30 days per annum necessitating closures in the area.

The Banks Peninsula Plan also incorporates a Marine Biotoxin Management Plan. Harvesting of shellfish from these areas is still at an early stage and year-round harvesting is not yet achievable. Sampling for biotoxin analysis of phytoplankton and shellfish occurs when the area is due to be harvested. Testing frequency is weekly for DSP, 2-weekly for ASP and 4-weekly for PSP.

(continued)

Box 9.2 (continued)

When phytoplankton are present above trigger levels for a biotoxin then the shellfish samples are analysed for the appropriate biotoxin. When shellfish biotoxin levels are above background levels but below maximum permissible levels then further investigation, usually re-sampling for the same biotoxin, is required. Precautionary measures may be taken when biotoxin levels are close to maximum permissible levels.

When biotoxin levels above the maximum permissible levels are detected, the growing area is closed immediately, and action is taken to ensure consignments of shellfish are not contaminated with biotoxins. When required, recall of product is put in place. The Public Health Authority investigates any suspected food poisonings in which marine biotoxins are implicated and attempts to trace the origins of the food to see if it could be from the growing area.

A testing regime for the implicated toxin and the closed area is developed for each closure. Re-opening of the growing area is based on 2 clear results below the maximum permissible level and the regulator is satisfied that no risk exists regarding that toxin.

considering risks and barriers to contamination; *release* (possible failure pathways by which contaminated water can cause water-borne disease) in considering risk analysis of contamination; and, *reorganisation* (measures, corrective actions and contingency plans) in considering improvements to reduce risks. The approach also addresses the key properties of *potential for resources,* i.e. contamination in this instance, *connectedness* through flow diagrams and contaminant pathways, and *resilience* of the drinking water system to contamination. The focus is on the impacts of water resource hazards on the socio-economic system, i.e. a type 3 sustainability issue[11]. The approach also incorporates an adaptive management approach in developing improvements, establishing contingency plans and monitoring performance and responding to the analysis of performance.

An example of the implications of the approach is the Cheviot water supply which has been on boil water notices for several years (Hurunui District Council 2011). The process of WSP preparation identified the bacterial contamination and turbidity of the source water (shallow groundwater connected to the Waiau River) and alternative treatment and supply options. However, a constraint on implementation is a type 4 sustainability issue (*capacity of the socio-economic system dependent on water to be maintained*) which is the affordability of the identified improvements. This has led to a political decision to delay the timing required for implementation. The requirements for a WSP are also limited to water supplies for

[11] In Chap. 4, four types of sustainability issues were identified: type 1 – capacity of the biophysical system to be maintained; type 2 – capacity of linkages of the socio-economic system to the biophysical system; type 3 – capacity of linkages of the biophysical system to the socio-economic system, and type 4 – capacity of the socio-economic system to be maintained.

communities greater than 500 even though the historical evidence indicates it is the small scale untreated or inadequately treated supplies that constitute a significant risk.

Similar situations to Cheviot occur across Canterbury as shown in Fig. 9.2 which shows the drinking water status for water supplies in Canterbury.

Although the disease failure pathway is at the individual level, the geographical scale considered is the aggregation of users of the Cheviot water supply system. There is also benefit in considering issues at the broader scale of the catchment. The bacterial contamination is a result of land use activities in the catchment of the Waiau River – a type 2 sustainability issue (cumulative effects of use on the water resource system). However, the turbidity is a type 1 issue at the catchment scale (contamination by natural source) from the contribution of glacial flour from erosion in the headwaters of the catchment.

Thus a resilience assessment of the four types of capacities for managing drinking water indicates: a sound approach for type 3 issues (analysing contamination risk pathways); inadequate approach for type 4 issues (affordability of actions to address contamination risk); the need to consider broader spatial scales (the catchment of the drinking water supply) and type 2 issues (cumulative effects of catchment land use); and, recognition of type 1 issues (glacial erosion causing turbidity) that cannot be managed at source. To achieve an acceptable drinking water supply, a broader sustainability strategy incorporating the outcomes of resilience assessment is needed.

The analysis highlights the issues of catchment contamination, scale of population served and affordability of management interventions which are considered further below (see Sect. 9.3.5).

9.3.2 *Analysis of Approach to Management of Contact Recreation*

The grading system for a recreation site provides a nested approach to failure pathways for water quality. The Sanitary Inspection Category provides a qualitative grading of contamination risk at the catchment scale for the catchment upstream of the site based on land use and pollution sources. The Microbiological Assessment Category provides a water quality assessment at the local scale based on bacteriological measurement.

However, the management interventions when acceptable thresholds for recreational use are exceeded of warnings through signage and public information are based on the transformation to a degraded state. While the intent of the Annapolis Protocol is to identify the sources of contamination, the New Zealand management guidelines do not focus on actions to improve water quality. As stated in the guidelines (page B3): "The guidelines do not specify that the cause of failure to meet the specified levels must be rectified. They merely require that the public is informed when beaches are not suitable for contact recreation."

There are examples of voluntary proactive approaches such as Corsair Bay and Silverstream catchment noted above. However, there is not a requirement or an authority for action to address degraded water quality.

9.3.3 Analysis of Approach to Management of Cyanobacteria

The approach to the identification of the presence of cyanobacteria at health-threatening levels is limited to the monitoring of sites for planktonic and benthic cyanobacteria. There is a graduated management response based on providing public warnings of the degraded state.

This management approach is less robust than the recreational water quality approach. The site assessment does not address upstream contaminant sources. The graded response is only a recognition of the degraded state and warnings to avoid contaminated areas.

9.3.4 Analysis of Approach to Commercial Shellfish Management

There are many similarities between components of the HACCP approach and a nested adaptive system approach although the language is different. The definition of failure pathways for nested adaptive systems (Sect. 5.1) is similar to undertaking a hazard analysis in the HACCP approach. The identification of critical variables and their thresholds for nested adaptive systems is similar to identifying critical control points and establishing critical limits for each critical control point in the HACCP approach. The development of management interventions for nested adaptive systems is similar to the establishment of corrective actions in the HACCP approach.

The main corrective actions for shellfish management (i.e. closure of harvesting areas, relaying to other growing areas, and clean water depuration to purge contamination, and, product recall) can be considered with respect to the resource outcomes for coupled socio-economic and biophysical systems, i.e. maintenance of the natural system, degraded state when a sustainability threshold is exceeded, or, transformation to a sustainable alternative (refer Sect. 4.2.8 and Fig. 4.8). Harvest closure and product recall reflect management of a degraded state, while relaying and depuration reflect transformation to an alternative state to reduce shellfish exposure to stress and facilitate recovery from contamination.

However, from a nested adaptive system perspective, there are other potential management interventions, in particular, the maintenance of the natural system by addressing the source of the contamination. The HACCP approach highlights the value of the sanitary survey and growing area classification in establishing the requirements for aquaculture management. Furthermore, the sanitary survey also indicates where management intervention could be effective in the upstream catchment or coastal area with respect to contamination sources.

The HACCP approach highlights the significant role of monitoring needed for sustainability management. One aspect is the monitoring of critical control points, e.g. bacteriological water quality and biotoxins in shellfish (refer Menzies Bay example in Box 9.2). Furthermore, there is monitoring of precursor events, e.g. monitoring of rainfall that generates contaminated runoff, and, phytoplankton biotoxin levels to provide advanced warning of hazards requiring corrective action.

It is interesting to note that the Action Plan has monitoring triggers for the build-up of contamination leading to closure if a threshold level for a biotoxin is exceeded, i.e. monitoring the front loop of the adaptive cycle (the exploitation and accumulation phases). Also, the Action Plan process has triggers for monitoring the recovery of the return of biotoxin levels to background levels, i.e. monitoring the back loop of the adaptive cycle (the release and reorganization phases). The HACCP process not only requires biophysical monitoring but also incorporates the need for monitoring the plan implementation (on the socio-economic side).

9.3.5 *Issues to Be Addressed*

The analysis of management approaches highlights a number of important issues in relation to sustainability for the management of health risks associated with waterborne disease. The first is in relation to proactive catchment management to prevent contamination. A second issue relates to the scale at which drinking water management is addressed. The third is in relation to the affordability of management interventions.

9.3.5.1 Proactive Catchment Management to Prevent Contamination

For the potential failure pathways for the management of waterborne disease there are distinct differences with respect to the incorporation of catchment management as a management intervention. For drinking water management, Water Safety Plans require a risk analysis for contamination associated with the water supply catchment and an improvement plan to manage risks through source protection or water treatment. As the Akaroa example (refer Box 9.1) indicates, the purchase of the upstream catchment to be able to control land use is incorporated in the improvement plan. In addition, the National Environmental Standard for Sources of Drinking Water requires regional councils to ensure the effects on drinking water sources are considered in decision making, and, the regional council is in the process of defining protection zones around groundwater wells and surface water intakes.

For the management of commercial aquaculture, the sanitary surveys quantify the risks associated with upstream contamination. However, the management intervention requirements only apply to the downstream activity (e.g. closure of growing areas for harvesting, and relaying or depuration of shellfish). Addressing the source of contamination is not a requirement.

For the management of recreational water quality, a qualitative assessment is made of sources of contamination in the catchment upstream of the recreational site. However, the requirements for management interventions are limited to public warnings of the unsuitability of the site for contact recreation and do not include rectification of the sources of contamination. The information on sources of contamination in the catchment are identified which can facilitate voluntary actions. However, with only a qualitative assessment of contamination, the extent on intervention required is unknown. As in the case of the Silverstream catchment, the voluntary actions have achieved an improved water quality at Coes Ford but not sufficient to meet contact recreation requirements.

For the management of cyanobacteria and non-commercial harvesting of shellfish, there is reliance on monitoring to identify the hazard. Thus the hazard is only identified after the monitoring has occurred. This can be compared to recreational water quality where the hazard has been identified in advance. With public warnings as the only management intervention, public health protection is dependent on informed public choice.

The Canterbury Water Management Strategy has targets to increase the percentage of population supplied with water that meets the New Zealand Drinking Water Standards for health-based determinants, to have active restoration programmes in place for mahinga kai waterways, and, to increase the percentage of lake and river sites used for contact recreation that meet recreational water quality guidelines. For these targets to be achieved then proactive catchment management approaches will be needed. The current national approaches are inadequate to support proactive catchment management activities. The approach to the development of sustainability strategies for catchments is addressed further in Chap. 10.

9.3.5.2 Appropriate Scale and Affordability for Management Interventions

The issues of appropriate scale and affordability of management interventions are now considered. At the national level, there is a relationship between scale of drinking water supply and compliance with drinking water standards. Table 9.10 sets out the degree of compliance for population supply zones from large (more than 10,000 people) to small (101 to 500 people) for the year 2013/14. The table shows declining compliance with smaller size. Large zones achieving 88.9% compliance while small zones only achieved 20.7% compliance.

It is also noteworthy that the scheduled improvements for Akaroa water supply, a minor and a small zone, (refer Box 9.1) occurred after the amalgamation of Banks Peninsula District Council with Christchurch City Council. Furthermore, Franklin

Table 9.10 Achievement against drinking-water standards 2013/14

Population supply zones	Percentage compliance	Population served in registered zones
Large zones (more than 10,000)	88.9	3,002,000
Medium zones (5001 to 10,000)	52.9	270,000
Minor zones (501 to 5000)	41.2	477,000
Small zones (101 to 500)	20.7	79,700
All zones	79.0	3,829,000

District Council water supplies which did not achieve compliance prior to amalgamation into Auckland Council in 2010 have been brought into compliance by Watercare, the Auckland Council CCO (Council Controlled Organisation) for water services (Watercare 2015). In addition the local sources of supply were constrained in being able to meet future demand from population growth: the groundwater from the Kaawa formation was reaching the sustainable limit of abstraction, and four of the six high demand streams within the water resource zones of Kingseat, Pukekohe, Ramarama and Waiuku were fully allocated (Franklin District Council 2007). To resolve the situation for Franklin District water was provided from Watercare's Waikato River pipeline to Auckland.

Abbott and Cohen have reviewed the studies on the economies of scale of water utilities. Their conclusions were that studies that looked at small water businesses generally found economies of scale could be achieved if they became larger. However, the findings also suggest that there is a critical level of output after which the scale economies will be exhausted. In the studies they reviewed this critical level ranged from 100,000 to 1,000,000 connections (Abbott and Cohen 2009).

Three general models of organizational arrangements to achieve increased technical scale and access to capital have been identified (Marques 2010). One is the English model which involves privatization of the ownership and operation of the assets and the establishment of an independent economic regulator to oversee the private operations. A second is the French model where the assets are publicly owned while the management and operation is undertaken by private entities under medium-to-long term concession contracts. Contracts are awarded according to a tendering or bidding process. Regulatory agencies are created to oversee the quality of outputs and deal with unforeseen circumstances. The third is the public operator model where there is state ownership and operation of the assets. For example, Scotland amalgamated its water authorities into one statutory authority (Scottish Water) with an independent economic regulator (Water Industry Commission) that sets the price of water.

A recent survey of studies comparing private versus public ownership found the available evidence inconclusive. Some studies showed that publicly owned water utilities perform better than private water utilities, others found the opposite, and some found insufficient evidence to make an assessment (Carvalho et al. 2012).

There have also been hostile reactions to the privatization of water especially when the cost to consumers has increased to recover the investment in improved water supply networks. Box 9.3 summarises the management of water supply in Cochabamba, Bolivia. The significant civil unrest in reaction to privatization led to

Box 9.3: Management of Water Supply in Cochabamba, Bolivia

In the 1990s Cochabamba water supply was performing poorly. Population was increasing and access to piped water decreased from 70% of the population to 40% (World Bank 2002). Privatisation of the city's water supply was seen as the way to improve the water supply network and get access to the capital required to invest in the needed infrastructure (Finnegan 2002).

Prior to privatization the water infrastructure of Cochabamba was managed by a state agency SEMAPA. There were also small water cooperatives and water carters who provided water to areas outside the SEMAPA network. Aguas del Tunari, a consortium led by International Water (a British firm owned by Bechtel) was the only bidder for the privatization contract. Aguas del Tunari were to take over the municipal network to provide drinking water to all of the people of Cochabamba (Nickson and Vargas 2002).

As a condition of the contract Aguas del Tunari had agreed to pay the $30 m of accumulated debt of SEMAPA and to finance the expansion and upgrading of the existing system (Finnegan 2002). Upon taking control the company raised water rates an average of 35% to about $20 per month (Blackwell 2002). This led to massive protests amongst the communities where family incomes were about $100 per month. The protests included a general strike, thousands involved in demonstrations, occupation of the central plaza and barricading of major highways. A state of siege was declared and there were violent clashes between demonstrators and law enforcement officers resulting in five deaths and mass arrests (Blackwell 2002).

The privatization contract was terminated. Water prices were returned to their pre-2000 levels with a group of community leaders running the restored state utility SEMAPA. SEMAPA has more than tripled the size of its service area since 2000, but at least 40% of the city's population, mainly in the southern part of the city, still lacks piped water. These residents are increasingly relying on traditional community-run water systems as an alternative (Achenberg 2013). In addition, these community systems have created an umbrella organization, ASICA-Sur (the Association of Community Water Systems of the South) to work with SEMAPA, to receive technical assistance and funding from international donor agencies. Rather than autonomous water governance, they are seeking a co-management solution (Marston 2014).

The water scarcity and funding limitations continue. SEMAPA does not currently have enough water in its network to supply the entire city, and financing of water infrastructure is still largely dependent on international donors. Additional supply is awaiting the completion of the Misicuni Dam. One vision is for an extended and improved public network to include the city's most marginal people. Another is the integration of the community-run systems as indivisible decision-making units through a co-management scheme with SEMAPA. SEMAPA would sell Misicuni water to the community-run systems which would control the distribution and pricing of water in their respective communities (Marston 2014).

the termination of the contract. However, Cochabamba still has problems with water availability and investment in water infrastructure. The situation in Cochabamba highlights the challenge in creating the scale to achieve the critical mass of technical expertise, the affordability of capital investment in the infrastructure, and, the autonomy of local decision making. It is also interesting to see the concept of co-management of local community systems with the state authority being put forward as an alternative to the expansion of the public operator model or privatization.

In this regard, there is an interesting variant of the Council Controlled Organisation model that has been developed for the Greater Wellington region. Capacity Infrastructure Services Limited was initially established as an organization owned by two council authorities (Wellington City Council and Hutt City Council) to manage water infrastructure distribution networks in both council areas. It has now grown to cover the greater Wellington region and is owned by all four local authorities and the regional council to provide management of water supply, wastewater treatment and stormwater management. It has been renamed Wellington Water. The assets are still owned by the individual councils. A representative from each council sits on the regional Wellington Water Committee that provides overall leadership and direction to the company. The institutional model retains local authority control and autonomy while achieving economies of scale[12], and technical critical mass for water management infrastructure management and operation. It also provides integration of the regional bulk water supply network (previously run by the regional council) with the local distribution networks (run by the local authorities).

To meet the dual requirements of affordability of management interventions and achieving an adequate scale for technical capability, a fourth model of organizational arrangements may be appropriate for the Canterbury region. This model has local control to ensure affordability but technical integration to ensure appropriate scale and cost effectiveness. The Wellington Water shared services approach and the Cochabamba co-management approach are examples of this fourth model.

References

Abbott, M., & Cohen, B. (2009). Productivity and efficiency in the water industry. *Utilities Policy, 17*, 233–244.

Achenberg, E. (2013). *From water wars to water scarcity: Bolivia's cautionary tale*. Harvard Review of Latin America, *XII*(2).

Adamson, B., & Main, M. (1996). *Potential sources of faecal coliform contamination at Coes Ford*. Christchurch: Environment Canterbury.

Aquaculture New Zealand. (2015). *Industry overview*. http://aquaculture.org.nz/industry/overview/

Ball, A. (2006). *Estimation of the burden of water-borne disease in New Zealand*. Wellington: Ministry of Health.

Bartholomew, N., Brunton, C., Mitchell, P., Williamson, J., & Gilpin, B. (2014). A waterborne outbreak of campylobacteriosis in the South Island of New Zealand due to the failure to implement a multi-barrier approach. *Journal of Water and Health, 12*(3), 555–563.

[12] There are about 140,000 connections for a population of about 400,000.

Blackwell, B. (2002). From coca to congress. *The Ecologist, 32*(9), 28–31.

Bolton-Ritchie, L. (2008). *Coastal water quality in selected bays of Banks Peninsula 2001–2007*. Christchurch: Environment Canterbury.

Bolton-Ritchie, L., & Robinson, K. (2016). *Water quality monitoring for contact recreation: Summary of the 2014/2015 season*. Christchurch: Environment Canterbury.

Carvalho, P., Macques, R., & Berg, S. (2012). A meta-regression analysis of benchmarking studies on water utilities market structure. *Utilities Policy, 21*, 40–49.

Christchurch City Council. (2014). *Akaroa water supply: Water Safety Plan 2014 revision (draft 2014)*. Christchurch: Christchurch City Council.

Close, M., Dann, R., Noonan, M., & Bright, J. (2005). *Transmission pathways from domestic livestock to water quality contamination of groundwater*. Christchurch: Institute of Environmental Science and Research.

Close, M., Dann, R., Ball, A., Pirie, R., Savill, M., & Smith, Z. (2008). Microbial groundwater quality and its health implications for a border-strip irrigated farm catchment, South Island New Zealand. *Journal of Water and Health, 6*(1), 83–98.

Davies-Colley, R., Cameron, K., Francis, G., Bidwell, V., Ball, A., & Pang, L. (2003). *Effects of rural land use on water quality*. Hamilton: NIWA.

Davison, A., Howard, G., Stevens, M., Callan, P., Fewtrell, L., Deere, D., & Bartram, J. (2005). *Water Safety Plans: Managing drinking-water quality from catchment to consumer*. Geneva: World Health Organization.

El Saadi, O., Esterman, A., Cameron, S., & Roder, D. (1995). Murray River water, raised cyanobacterial cell counts and gastrointestinal and dermatological symptoms. *Medical Journal of Australia, 162*, 18–24.

Environment Canterbury. (2011). *Natural Resources Regional Plan*. Christchurch: Environment Canterbury.

Environment Canterbury. (2012). *Annual groundwater quality survey 2012*. Christchurch: Environment Canterbury.

Environment Canterbury. (2013). *Canterbury Water Management Strategy targets: Progress report 2013*. Christchurch

ESR. (2010). Covering January–March 2010. *New Zealand Public Health Surveillance Report, 8*(2).

ESR. (2015a). *Surveillance report: Notifiable diseases in New Zealand 2014*. Porirua: The Institute of Environmental Science and Research.

ESR. (2015b). *Surveillance report: Summary of outbreaks in New Zealand 2014*. Wallaceville: The Institute of Environmental Science and Research.

Finnegan, W. (2002, April 8). Leasing the rain. *The New Yorker*.

Fisher, D., & Bradford, E. (1998). *National marine recreational fishing survey 1996: Catch and effort results by fishing zone*. Wellington: Ministry of Fisheries.

Fishing Industry Inspection and Certification Council. (1997). *A guide to Hazard Analysis Critical Control Point Systems in the seafood industry*. Wellington: MAF Regulatory Authority (Meat and Seafood).

Fleisher, J., Kay, D., Salmon, R., Jones, F., Wyer, M., & Godfree, A. (1996). Marine waters contaminated with domestic sewage: Nonenteric illnesses associated with bather exposure in the United Kingdom. *American Journal of Public Health, 86*(9), 1228–1234.

Franklin District Council. (2007). *Franklin District growth strategy 2051*. Pukekohe: Franklin District Council.

Glasgow, K. (2013). *Silverstream catchment report August 2013*. Christchurch: Environment Canterbury.

Hadfield, M., Fenwick, G., & Woods, C. (2012). *Effects of Lake Ellesmere (Te Waihora) openings on the coastal environment*. Christchurch: Environment Canterbury.

Hawke's Bay District Health Board. (2016). *Havelock North Campylobacter outbreak Update November 2016*. http://hawkesbay.health.nz/assets/News-and-Event-files/Gastro-update-flyer-Nov-A5.pdf

Hay, B., Grant, C., & McCoubrey, D.-J. (2000). *A review of the marine biotoxin monitoring programme for non-commercially harvested shellfish part 1: Technical report*. Wellington: Ministry of Health.

Hurunui District Council. (2011). *Drinking water standard compliance for Cheviot and a review of water treatment and source options, Cheviot Ward Committee report*. Amberley: Hurunui District Council.

Johnson, A. (2003). *1602 – Menzies Bay: Sanitary survey report*. Christchurch: Canterbury District Health Board.

Marques, R. (2010). *Regulation of water and wastewater services*. London: IWA Publishing.

Marston, A. (2014). The scale of informality: Community-run water Systems in Peri-Urban Cochabamba, Bolivia. *Water Alternatives, 7*(1), 72–88.

Mcbride, G., Salmond, C., Bandaranyake, D., Turner, S., Lewis, G., & Till, D. (1988). Health effects of marine bathing in New Zealand. *International Journal of Environmental Health Research, 8*(3), 173–189.

Medical Officers of Health. (2008, April 2008). Springston water supply outbreak. *Public Health Information Quarterly*, 1–2.

Meduna, V. (1994). Bloomin' hell: Toxic algal blooms have seen New Zealand shellfish put off-limits again, and probably not for the last time. *New Zealand Science Monthly, 5*(10), 6–7.

Ministry for the Environment. (2002). *Microbiological water quality guidelines for marine and freshwater recreational areas*. Wellington: Ministry for the Environment.

Ministry for the Environment and Ministry of Health. (2009). *New Zealand guidelines for cyanobacteria in recreational fresh waters – Interim guidelines*. Wellington: Ministry for the Environment.

Ministry of Health. (2005). *A framework on how to prepare and develop public health risk management plan for drinking-water supplies*. Wellington: Ministry of Health.

Ministry of Health. (2015). *Drinking water in New Zealand*. www.health.govt.nz/our-work/environmental-health/drinking-water/drinking-water-legislation. Accessed 26 Aug 2015.

National Health and Medical Research Council. (2008). *Guidelines for managing risks in recreational water*. Canberra: NHMRC.

New Zealand Food and Safety Authority. (2003). *New Zealand Fishing Industry Implementation Standards IAIS 005.1 Shellfish Quality Assurance Circular 1995 Issue 25*. New Zealand Food Safety Authority, Wellington.

New Zealand Food and Safety Authority. (2006). *Animal products (Specifications for Bivalve Molluscan Shellfish) notice 2006*. Wellington: Ministry for Agriculture and Forestry.

New Zealand Food and Safety Authority. (2013). *Non-commercial marine biotoxin monitoring programme: MPI VS Operating and Response Manaual*. Wellington: Ministry for Primary Industries.

Nickson, A., & Vargas, C. (2002). The limitations of water regulation: The failure of the Cochabamba concession in Bolivia. *Bulletin of Latin American Research, 21*(1).

Pilotto, L., Douglas, R., Burch, M., Cameron, S., Beers, M., Rouch, G., Robinson, P., Kirk, M., Cowie, C., Hardiman, S., Moore, C., & Attewell, R. (1997). Health effects of exposure to cyanobacteria (blue-green algae) during recreational water-related activities. *Australian and New Zealand Journal of Public Health, 21*(6), 562–566.

Pruss, A. (1988). Review of epidemiological studies on health effects from exposure to recreational water. *International Journal of Epidemiology, 27*, 1–9.

Robinson, K., & Bolton-Ritchie, L. (2014). *Water quality monitoring for contact recreation: Summary of the 2013–2014 season*. Christchurch: Environment Canterbury.

Robinson, K., & Stevenson, M. (2012). *Microbial water quality of the lower Selwyn River/Waikirikiri and Silverstream tributary*. Christchurch: Environment Canterbury.

Sheerin, I., Bartholomew, N., & Brunton, C. (2014). Estimated community costs of the outbreak of campylobacteriosis resulting from contamination of a public water supply in Darfield, New Zealand. *The New Zealand Medical Journal, 127*(1391), 13–20.

Simpson, D., Campbell, A., & O'Brien, G. (2012). *Banks Peninsula Management Plan: Growing areas 1601, 1602, 1604 and 1605*. Ministry for Primary Industries Wellington.
Stuff.co.nz. (12/03/2008). *Virus outbreak in Springston traced to water*. www.stuff.co.nz/the-press/news/national/312912/Virus-outbreak-in-Springston-traced-to-water
Till, D., Mcbride, G., & Ball, A. (2008). Large-scale microbiological study: Rationale, results and risks. *Journal of Water and Health, 6*(4), 443–460.
Watercare. (2015). *Franklin's water supply up to standard*. www.watercare.co.nz/about-watercare/news/Pages/Franklin's-water-supply-up-to-standard.aspx. Accessed 26 Aug 2015.
Wood, S., Selwood, A., Rueckert, A., Holland, P., Milne, J., & Smith, K. (2007). First report of homoanatoxin-a and associated dog neurotoxicosis in New Zealand. *Toxicon, 50*, 292–301.
World Bank. (2002). *Bolivia water management: A tale of three cities*. Washington, DC: Operations Evaluation Department World Bank.
World Health Organization. (1999). *Health-based monitoring of recreational waters: The feasibility of a new approach (the 'Annapolis Protocol')*. Geneva: WHO.
World Health Organization. (2003). *Guidelines for safe recreational water environments: Volume 1: Coastal and fresh waters*. Geneva: World Health Organization.

Chapter 10
Regional Level Socio-economic Failure Pathways

Abstract There are several potential socio-economic failure pathways at the regional level. One pathway relates to the contribution to trade networks: does the use of water and investment in infrastructure make an economic contribution? Economic analyses have been undertaken for the Canterbury Region using benefit-cost analysis with respect to infrastructure investment associated with the Canterbury Water Management Strategy, and at the national level using general equilibrium analysis for increased irrigation using developments from several regions but predominantly Canterbury. Analysis of economic externalities shows costs comparable with economic returns from benefit cost analysis. A significant international example is wheat production in Saudi Arabia with respect to this potential failure pathway. As a trade network failure, it is helpful to consider the concept of "virtual water" as trade does not occur directly in water but does occur in products that depend upon water for their production.

External intrusion is another failure pathway. The agricultural export capacity of New Zealand is attracting international interest in New Zealand farm land and food processing industries. There are competing interests between attracting investment in a capital constrained international economy, and maintaining local ownership of New Zealand assets, particularly productive farm land. As an island state, New Zealand is relatively protected from the foreign intrusion. A more complex international example of this failure pathway is water management in the River Jordan Basin. A summary of eight stages in water management shows the multiple facets of water conflict including boundary definition, transboundary planning, military clashes, diplomatic approaches, addressing resource depletion, peace agreements and joint approval processes.

A third failure pathway is the social effects of changing technology. This includes the social impact assessment of major water infrastructure projects. The Canterbury example of the impact of the Waianiwaniwa storage proposal on the affected land owners and the nearby town of Coalgate led to a rejection of the storage. A larger scale international example is the Three Gorges Project in China where millions of people were displaced. The shift from communal irrigation tanks to individual groundwater bores in the Deccan Plateau in India is used to illustrate the implications of changing water technology on the social structures for its management. The technology change led to a decline in community water management and maintenance institutions.

B.R. Jenkins, *Water Management in New Zealand's Canterbury Region*, Global Issues in Water Policy 19, https://doi.org/10.1007/978-94-024-1213-0_10

Keywords Economic contribution • External intrusion • Social effects of changing technology

10.1 Regional and National Economic Analysis

An important consideration in water resource management is whether the investment in infrastructure not only is financially viable but also is economically viable at the regional and national scale. Three economic analyses are discussed below. One is the benefit-cost analysis undertaken for storages under consideration in the Canterbury Water Management Strategy (Harris Consulting et al. 2010). The second is an analysis of externalities from the increase in dairying associated with irrigation (Tait and Cullen 2006). The third is a national analysis of the economic impact of increased irrigation in New Zealand using a dynamic Computable General Equilibrium Analysis (NZIER 2010a). As a significant component of New Zealand's agricultural production is for export there is the potential for trade network failure pathways. The trade is in virtual water, the water embedded in the traded products, rather than water being traded directly. An analysis of New Zealand's water footprint completes this section.

10.1.1 Regional Economic Analysis

The benefit-cost analyses focused primarily on the economic analysis of the water services infrastructure for storages at Tekapo, Coleridge, Lees Valley and Hurunui. The key elements of the model comprise: (1) the additional area that can be irrigated, (2) the capital costs of storage and irrigation infrastructure, (3) on-farm costs and benefits, (4) regional economics including output, value-added and GDP, and (5) employment on farm and in the regional economy.

Based on the assumptions incorporated in the model, the increase in irrigated area associated with the four storages was 236,000 ha. This requires a capital expenditure of $5.2 billion comprising storage ($0.56b), delivery infrastructure ($1.3b), and, on-farm infrastructure and farm system changes ($3.3b). Discounted cash flow analysis was undertaken at a discount rate of 8% (which was the Treasury guideline for government projects at the time of the analysis). This indicated a present value of benefits of $3.3b, a present value of costs of $2.5b, and, a net benefit of $0.8b.

The results were sensitive to capital costs, expected returns from intensification, and discount rate. Doubling the capital cost would lead a negative NPV (−$0.9b). A 20% reduction in returns reduces the net benefit to $0.2b. An increase in discount rate to 10% reduces the net benefit to $0.1b.

In relation to regional economic impact the analysis indicated:

- An increase in regional GDP of $0.4b per annum on farm and $1.7b per annum throughout the regional economy. (This can be compared to a regional GDP of $19.9b.)
- An increase in employment of 3000 jobs on farm and 17,000 throughout the regional economy. (This can be compared to regional employment of 238,000 full time equivalents.)
- An increase in household income of $0.8b per annum throughout the regional economy. (This can be compared to a regional household income of $11.8b.)

The regional economic analysis indicated that transferring water from Lake Tekapo to irrigate areas in South Canterbury was not economically viable. A more refined analysis (URS 2014) confirmed these findings. For both of the two possible concepts considered, the net income from irrigation would not cover the on-farm and off-farm capital and operating costs. In addition, there would be electricity generation losses because of the water diverted from hydroelectric power stations on the Waitaki River downstream of Lake Tekapo.

10.1.2 Externalities of Dairying Conversions

Tait and Cullen (2006) have provided an estimate of the negative externalities of a public good nature associated with dairy conversions in Canterbury. They identified annual external costs related to damage to water resources (surface water, groundwater and loss of angler values), damage to air resources (greenhouse gas emissions), damage to ecosystem biodiversity (shelter belt loss and sediment in surface water), and, damage to health (human pathogen-related diseases, and bovine TB). Based on 146,000 ha of irrigated dairy farming in Canterbury, Tait and Cullen calculated the external costs to be in the range $197–308 per hectare or $24–40 million per year.

Using this range of external costs for 236,000 ha expansion and an 8% discount then the net present value of the externalities is in the range $0.6–0.9b. This encompasses the estimated economic benefit from the benefit-cost analysis of $0.8b suggesting that the regional economic benefit is at best marginal.

10.1.3 National Economic Analysis

Dynamic Computable General Equilibrium (CGE) analysis is a more sophisticated framework for analyzing major investment decisions from a national perspective. It captures the inter-linkages between sectors as well as linkages to households (via the labor market), the government sector, capital markets and the global economy (via imports and exports). CGE is able to assess macroeconomic impacts on exchange rates, net foreign liabilities and the current account balance. These are particularly important when the extra production is exported. CGE models the dynamics of the economy including employment and wages response to labor demands, investment responses to rates of return and how New Zealand's net investment/savings imbalance increases or decreases net foreign liabilities. Benefit-cost analysis (BCA) is unable to address macroeconomic impacts and economy dynamics. Furthermore, BCA does not consider impacts on factor prices nor flow-on effects to other industries.

The analysis considered 14 proposed irrigation schemes that would irrigate 347,000 ha with 270,000 ha (78%) in Canterbury. The off-farm investment was estimated to be $2.7b and on-farm investment of $8.7b. This was based on offshore borrowing with interest on foreign debt paid at 7% per annum.

Based on the assumptions incorporated in the model, the analysis indicates that consumption[1] by 2035 is around $2b greater than if the irrigation schemes had not been implemented. There is a drop in consumption between 2025 and 2030 due to the repayment of offshore borrowing. Over 25 years, there is a net present consumption gain of $8b in GDP by 2035 which is about $2.1b (0.8%) higher than it would otherwise have been.

The majority of the gains in GDP come through agricultural production at the farm level projected to be about $1.5b in 2035. The off-farm sector expands in line with the farm sector (projected to be about $0.8b in 2035). However, for other sectors there is increased competition for resources and exchange rate appreciation which negatively affects other exporters. Despite some flow-on effects from farm supply industries and household expenditure industries there is a net loss to the rest of the economy (projected to be about $0.14b in 2035).

Sensitivity analysis shows that reducing the returns to irrigation by 20% has a relatively direct negative impact on the net present gains which fall by about $2b (25%). The price of milk solids would need to fall from $5.50/kg (which was assumed in the analysis) to $4.40/kg. It is noteworthy that after a rise in Fonterra's farmgate milk price to $8.40/kg in 2013/4, the forecast farmgate milk price in the first quarter of 2016 (8 March 2016) dropped to $3.90/kg which equates to a forecast farmer payout (milk price plus earnings) of $4.35–$4.45/kg (Fonterra 2016).

Because of the significance of irrigation to New Zealand's export economy, there is a potential trade network failure pathway for water management in Canterbury. An international example of a trade network failure pathway is Saudi Arabia which is set out in Box 10.1.

[1] Total consumption is considered the critical metric from a "NZ Inc." perspective. It measures the amount of income available to New Zealanders to spend on goods and services. This is GDP less repayment of debt and the opportunity cost of investment.

Box 10.1: Groundwater Depletion in Saudi Arabia – Failure Pathway Through Trade

In the 1980s Saudi Arabian government policies had a focus on food security and diversification of its production base to reduce reliance on petroleum for its national income. It had a policy of encouraging private investments in agriculture through subsidies, interest-free loans and development of water infrastructure. The withdrawal of groundwater mainly for irrigated wheat production led to a significant increase in agricultural water use from 6800 Mm^3 in 1980 to 21,000 Mm^3 in 2006 (FAO 2008). This compares with the estimated renewable groundwater of 2200 Mm^3. There has been a rapid decline in groundwater levels and predictions that the resources may not last more than about 25 years (FAO 2008).

By 1984 the kingdom was self-sufficient in wheat and after that became an exporter in wheat. In 1992 the kingdom authorized payments equivalent to $US2.1b to farmers for 1991s record wheat crop of 4 million tonnes. This amount could have been purchased on world markets for one fourth the price (Postel 1992). The root cause of the problem was the economic signals provided to farmers. Also the value of water was not priced in market transactions. This represents a trade network failure in water management.

A new governmental strategy has been developed and some of the subsidies and support programmes that contributed to depletion of groundwater resources have been discontinued or revised (FAO 2008). Key elements of the strategy to reduce water demand include: (1) Stopping the expansion of high water consuming crops and concentrating on high value added crops; (2) Stopping the distribution of agricultural land except in regions with sufficient renewable water resources; (3) Improving irrigation management, estimating crop requirements and encouraging tools like soil moisture probes for better irrigation scheduling; (4) Controlling water consumption through metering and water pricing for use greater than crop water requirements; (5) Supporting research on crop varieties that are resistant to drought, salinity and acid soils; and (6) Expanding the use of treated wastewater.

The strategy has been successful in reducing water demand. It was 14,700 Mm^3 in 2010 (Zaharani et al. 2011). However, this is still significantly more than the renewable groundwater resource.

10.1.4 Virtual Water

Trade network failure pathways in relation to water management are not the usual concept of trading as the trading does not occur in water directly but in commodities, primarily food, that have required water in their production.[2] It is known as trading in "virtual water" which refers to the hidden flow of water if food or other commodities are traded from one place to another. The virtual water content of a product can be defined as the volume of freshwater used to produce the product, measured at the place where the product was produced (Hoekstra et al. 2011).

Water footprints can be calculated for a process, a product, a consumer, group of consumers (e.g. a region or a country) or a producer (e.g. a public or private organisation). Water footprints are a measure of human appropriation of freshwater resources, and incorporate both direct and indirect water use of a consumer or producer. It has been found that 90% of water needed by individuals or the national economy is embedded in food consumption. Water footprints are measured in terms of water volumes consumed (evaporated or otherwise not returned) or polluted per unit of time (Hoekstra et al. 2011).

Three components are often distinguished: (1) green water footprint – the volume of rainwater evaporated or incorporated into a product, e.g. rain-fed agriculture; (2) blue water footprint – the volume of surface water or groundwater evaporated or incorporated into a product, e.g. irrigated agriculture; and (3) grey water footprint – the volume of polluted water due to the product.

To examine the trade in virtual water for a country, national water footprint accounts can be calculated. One account is the water footprint of national consumption. Another is the water footprint of national production.

The water footprint of national consumption is the total volume of freshwater used to produce the goods and services consumed by the inhabitants of the nation. This can be subdivided into: internal – the volume of water used from domestic sources; and external – the volume of water used in other nations to produce imported goods. Table 10.1 sets out the estimated water footprints for New Zealand consumption. Internal sources contribute 2627 Mm^3/year while external sources contribute 3578 Mm^3/year for a total water consumption footprint of 6205 Mm^3/year. At the time of the analysis (1996–2005), New Zealand's population was taken to be 3.906 million. This means a per capita consumption of 1590 m^3/year/person.

Table 10.1 New Zealand water footprint for consumption (Mm^3/year) (Mekonnen and Hoekstra 2011)

	Green	Blue	Grey	Total
Internal	1864	353	410	2627
External	2812	228	538	3578
Total	4676	581	948	6205

[2] There is a direct overseas trade in bottled water from New Zealand of about 9 million litres per year. This issue has recently become highly contentious. There is significant opposition for a public resource to be used for private profit without payment of a royalty nor any benefit to local communities (NZ Herald 14 March 2017).

Table 10.2 New Zealand water footprint for production (Mm^3/year) (Mekonnen and Hoekstra 2011)

	Green	Blue	Grey	Total
Crops	3426	809	296	4521
Grazing	11,748			11,748
Animal		378		378
Industrial		10	38	48
Domestic		102	288	390
Total	15,174	1298	622	17,094

The water footprint of national production is the total volume of freshwater consumed or polluted within a nation due to activities in different sectors of the economy, i.e. agriculture, industry, domestic. Table 10.2 sets out the estimated water footprints for New Zealand production. Rainfed agriculture is the dominant component – 15,174 Mm^3/year (89%) of the total 17,094 Mm^3/year. The per capita water footprint of national production is 4380 m^3/year/person.

Comparing New Zealand production (4380 m^3/year/person) with New Zealand consumption (1590 m^3/year/person) indicates that New Zealand is a net exporter of virtual water at 2790 m^3/year/person. While the primary sector of the New Zealand economy accounts for 7.5% of GDP, it contributes over 50% of New Zealand's export earnings (Treasury 2015). Water is therefore a major factor in New Zealand's export economy. Expansion of trade is strongly influenced by the availability of water and by the improved productivity of water. It also means that there is a significant opportunity cost associated with the inefficient use of water when water is at availability limits.

10.2 External Intrusion

Two main types of external intrusion are considered. One is in relation to access to water while the other is in relation to access to land (and its associated water). New Zealand is fortunate as an island country as it avoids transboundary conflict in relation to water catchments and land boundaries. It has also set up its primary water management agencies (regional councils) based on catchments.

International examples provide greater insight into external intrusion as a potential failure pathway. Box 10.2 summarises water conflict in the Jordan River Basin where access to water has been a significant component in the conflict between Israel and its Arab neighbours. Box 10.3 summarises international foreign investment in farmland where intrusion is purchasing land rather than invasion of land or armed conflict. The relevance of these summaries to Canterbury's and New Zealand's circumstances is considered below.

Box 10.2: Water Conflict in the Jordan River Basin
Characteristics Leading to Conflict

Gleick identifies the following characteristics that make water likely to be a source of strategic rivalry: (1) the degree of scarcity; (2) the extent to which the water is shared by more than one region or state; (3) the relative power of the basin states; and, (4) the ease of access to alternative freshwater sources (Gleick 1993). He cites the Middle East as "perhaps the clearest example of a region where fresh water supplies have had strategic implications".

Scarcity is an issue in the Jordan Basin. Allan estimates that given the current population (in 2002) of the Jordan Basin, the region would need about 15 billion m^3/year of water to be self-sufficient. There is less than 3 billion m^3/year available with an additional 1–2 billion m^3 of soil water. This annual deficit of 10–12 billion m^3/year has existed since the 1950s (Allan 2002).

The water resources of the Jordan basin are shared. The three principal sources of the northern River Jordan are: (1) the Hasbani River (annual flow about 250 Mm^3) with its source 30 km north of the Lebanon-Israel border, with the Hazbaya springs in Lebanon and Wazzini spring which was in Syrian territory until the Six Day War in 1967; (2) the Banias Springs (annual flow about 125 Mm^3) in the Golan Heights inside Syria until 1967; and (3) Dan Spring (annual flow about 250 Mm^3) within the borders of Israel. The Jordan River (annual flow about 1300 Mm^3) flows south to the Sea of Galilee and is then joined by the River Yarmuk (with sources in Syria and Jordan) before discharging into the Dead Sea (Medzini 2001). In addition to the Jordan River there are two aquifers that are transboundary water resources: the Coastal Aquifer which serves Israel, Gaza and Egypt, and, the Mountain Aquifer which serves the West Bank (Selby 2013). Refer Fig. 10.1.

The relative power and the power relationships in the Middle East are extremely complex. Medzini highlights not only the power relationships between the states in the River Jordan drainage basin but also the power struggle between the eastern and western blocs and their relevance to water management in the Jordan Basin (Medzini 2001). Selby identifies the power imbalance between Israel and the Palestinian Authority as a major factor in the negotiations in relation to water (Selby 2013).

Alternative freshwater resources are limited. The MENA (Middle East North Africa) Region, of which the Jordan Basin is a part, is the driest and most water scarce region in the world. MENA has about 0.7% of the world's available freshwater resources but has 5% of the world's population (Ju'ub undated). The Nile Basin to the west of the Jordan Basin is highly contested as is the Tigris-Euphrates Basin to the east. To the north is the Latini River (annual flow about 410 Mm^3/year) in Lebanon which has been the target of proposed Syrian, Jordanian and Israeli water solutions (Kiser 2000).

(continued)

Box 10.2 (continued)

Stages in Israel's Approach to Water Management

Israel's approach to water management has evolved through a series of stages. The first stage was the definition of boundaries of Israel following the Balfour Declaration of 1917 granting Jews a national home in Palestine. In his detailed review of historical documents, Medzini found that the Zionist movement sought borders that would offer security and an economic future based on land and water for agriculture and the creation of hydroelectric power for the future state. The Zionist movement sought borders which included the sources of the River Jordan and the southern part of the River Latini. The border agreement signed between Britain and France in 1923 was a compromise. The River Latini remained in Lebanon, the Banias Springs in Syria and the Dan Springs in Palestine (Medzini 2001).

The second stage was in the 1930s and 1940s when independent water resource development studies were undertaken: two influential ones were the Transjordanian study and the Zionist study. The Transjordanian study concluded that the naturally available water resources were not sufficient to sustain a Jewish homeland. The Zionist study concluded similarly but noted that by diverting water from the Jordan River basin for support of agricultural and residential development a Jewish state with 4 million new immigrants would be sustainable. The tensions in Palestine exploded into an all-out Palestinian revolt in 1936 that lasted for 6 months demanding a halt to Jewish immigration and to the sale of land to the new immigrants (Haddadin 2006).

The third stage (1951–1953) occurred after Israel was established in 1948 and 1948 Arab-Israeli war. Water projects became a cause for military clashes as Israel attempted to implement projects such as the National Water Carrier project to divert water from the Jordan River, the intake of which was originally located in a demilitarised zone between Israel and Syria (Medzini 2001).

Integrated basin management through US diplomacy was attempted as a fourth stage in 1954–1957 using the model of the Tennessee Valley Authority: what is referred to as the Johnston Plan (the US envoy who mediated the plan). A negotiated plan was eventually agreed by technical experts from the riparian states but was not ratified by the Israeli Knesset or the Arab League. However Jordan and Israel agreed to operate within their allocations defined in the Johnston Plan, and with financial support from the US, two major projects were completed: the Israeli National Water Carrier Project and the Jordanian East Ghor Main Canal (Haddadin 2006).

Arab states took action in 1964 in an attempt to prevent Israel from completing the National Water Carrier Project leading to a fifth stage. Arab states agreed to divert sources of the Jordan River to deny availability of this water to Israel. Periodic armed conflict lasted until the Six Days War in 1967. The outcome of the war was that Israel gained control of the Banias Springs, the

(continued)

Box 10.2 (continued)

Wazzini Springs and the north bank of the Yarmuk River. Israel was able to prevent the construction of the dam across the Jordan River planned by the Arab states (Medzini 2001).

However conventional freshwater sources were insufficient to meet Israel's increasing demand for water. A sixth stage was Israel looking at alternatives to the development of freshwater resources because aquifers were depleting and the Dead Sea declining in volume. The invention of drip irrigation and the adoption of improved irrigation techniques reduced agricultural demand. As a result the average requirement for water decreased from 8700 m^3/ha in 1975 to 5500 m^3/ha in 1995 (ISCID undated). The use of treated wastewater for agriculture enabled additional freshwater to be made available for urban use. More than 80% of household wastewater is recycled amounting to 400 Mm^3/year. This constitutes about 40% of water use in agriculture (Brenner 2012; Rabinovitch 2010). Israel also established a policy of importing virtual water, such as importing rather than growing grain. By 2000, grain imports to Israel (including Palestine) and Jordan exceeded 5 million tonnes annually. If all freshwater in those three territories had been used for grain production only 3 million tonnes could be grown (Allan 2002). Israel's water footprint for consumption is estimated to be 2303 m^3/year/person while its water footprint for production is only 654 m^3/year/person. This equates to importing 1649 m^3/year/person in virtual water: for a population of 6.134 million this is equivalent to 10,100 Mm^3/year (Mekonnen and Hoekstra 2011).

The seventh stage in water management relates to the Oslo Accords (a set of interim agreements between Israel and the Palestinian Liberation Organisation mediated by Norway and signed in 1993) and the Israel-Jordan peace treaty (a comprehensive treaty that settled relations between the two countries signed in 1994). While the Oslo Accords did not reach a comprehensive peace agreement, elements of the Accords remain. It was a step towards Palestinian self-government and established several water management institutions including the Palestinian Water Authority, an Israeli-Palestinian Joint Water Committee and Joint Supervision and Enforcement Teams for the Mountain Aquifer associated with the West Bank. Annex II of the Israel-Jordan peace treaty includes allocation agreements for the Yarmuk and Jordan Rivers, cooperation commitments to find additional water and build storage, water quality protection provisions, and groundwater management provisions. In addition a Jordan-Israel Joint Water Committee was established (Shamir 1997; Berland 2000). The agreements involved Israel providing more water both to Palestinian settlements and to Jordan. Israel was exchanging water for peace. Manna states in relation to the Israel-Jordan peace treaty: "Jordan and Israel looked for different outcomes while signing the treaty. For Jordan it meant receiving a sufficient amount of water from Israel; for Israel it meant receiving Jordanian recognition of it as a state" (Manna 2006).

(continued)

Box 10.2 (continued)

The eighth stage was the construction of desalination plants. With increasing internal demand from population and industrial growth as well as commitments to neighboring states additional supply was achieved through desalination of seawater and brackish water. Construction was accelerated during a drought from 1999 to 2002 (Tenne 2010). While restrictions were applied during this drought period, they were not sufficient to prevent overexploitation of aquifers and the Sea of Galilee (Fischhendler 2008). There were also concerns with climate trends that were consistent with future projections of climate change: these were increasing summer temperatures and decreasing rainfall in the catchment of the Sea of Galilee (Givati 2012). A desalination capacity of 750 Mm^3/year is sought by 2020: this represents 75% of the domestic water demand (Tenne 2011). There are four plants currently operating with the plant at Sorek being the largest in the world.

Multiple Potential Failure Pathways

The Israeli situation is an interesting example of a country having to deal with multiple potential failure pathways. The initial stages show the challenges with the foreign intrusion pathway: the first stage of boundary definition leading to transboundary catchment and aquifer planning difficulties (second stage) and then military clashes over water projects (third stage). Diplomatic approaches to integrated basin management while not getting political ratification did facilitate some projects to be initiated (fourth stage). However, project implementation led to further armed conflict (fifth stage).

Ongoing increases in water use led to a cumulative depletion of natural resources pathway with aquifers declining and Dead Sea levels dropping. The sixth stage of strategies for addressing natural resource depletion opened up other potential failure pathways. Irrigation improvements potentially open a marginal productivity pathway (i.e. investing in improvements without achieving a commensurate return). Nevertheless Israeli management of the complexity has led to improved water efficiency and increased agricultural productivity: in the past 40 years water consumption by agriculture has declined but output of field crops per unit of water has grown sevenfold (Kislev 2011). Wastewater recycling opens up a disease failure pathway. In 1970 use of untreated wastewater for irrigation led to a cholera outbreak in Jerusalem (Fattel et al. 1986). However, Israel is now recognized as a world leader in wastewater treatment and recycling. Israel reuses about 80% of its wastewater mostly for agriculture (Wintgens and Hochstrat 2006). The reliance on virtual water opens up a vulnerability to a trade network failure pathway. Israel produces less than half the food needed to feed its population (Kislev 2011).

The strategy for dealing with the foreign intrusion failure pathway was the development of peace agreements (seventh stage). This has not been possible

(continued)

Box 10.2 (continued)
with all riparian states. Water was an important component for both the Jordanian and Palestinian agreements. Implementation of the agreements has been challenging. For the Israel-Jordan peace treaty not all of the objectives have been fulfilled (Susskind and Islam 2012). However Berland identifies three examples of implementation problems where disputes which could have threatened the treaty were resolved (Berland 2000). One of the key issues of the treaty was finding an additional 50 Mm^3/year of water for Jordan; this has only recently been resolved with the signing of an agreement between Israel, Jordan and the Palestinian Authority for a desalination plant at Aqaba on the Red Sea in Jordan. The initial development is proposed to have a capacity of 65–85 Mm^3/year with the brine discharge of 120–130 Mm^3/year being sent by pipeline to the Dead Sea. The desalinated water will go to Eilat in Israel and Aqaba in Jordan. The agreement includes the release of 50 Mm^3/year of water from the Sea of Galilee to Jordan, and, Israel providing 20–30 Mm^3/year to Palestinians on the West Bank (Hosking 2015; Future Directions 2015).

In contrast, the Israeli-Palestinian Joint Water Committee has not delivered mutually beneficial outcomes. All water projects for the West Bank area require approval by the Joint Water Committee on the basis of consensus between the two parties. While nearly all Israeli projects have been approved there has been rejection of many Palestinian proposals particularly for new production wells. Seventeen years on from the Oslo Accords, new Palestinian wells are providing only 13 Mm^3/year. This is much less than the 20.5 Mm^3/year from wells promised for the 5-year interim period and well below the 70–80 Mm^3/year defined in the Accord for Palestinian future needs. Between 1995 and 2010 for Palestinian West Bank water supplies the gross per capita supply has fallen from 105 to 72 m^3/year. This has been attributed to the power imbalance between the two parties (Selby 2013).

The investment in desalination plants (eighth stage) does provide a high level of security for urban water in Israel. However, it does not address the continuing cumulative effects on aquifer depletion and Dead Sea decline. For the Western Basin of the Mountain Aquifer in the period 1970–2006 average annual outflows reached 434 Mm^3 while average recharge from rain amounted to 385 Mm^3. The injection of 15 Mm^3 of water into the aquifer has made up for part of the over extraction but leaves a deficit of 34 Mm^3 (UN-ESCWA and BGR 2013). Groundwater abstraction in Gaza has reached 180 Mm^3/year whereas the natural aquifer recharge is estimated to be only 55 Mm^3/year. Consequently the aquifer is being depleted and suffers from seawater intrusion (AquaPedia 2014). Flow in the Jordan River which is the main source of water to the Dead Sea has reduced from about 1500 Mm^3/year to less than 150 Mm^3/year. Between 1976 and 2009 the Dead Sea dropped more than 20 m and is declining at about 1 m/year. Even the brine discharge from the proposed Aqaba desalination plant only represents about 10% of the shortfall in flow to the Dead Sea (Tahal 2010).

Fig. 10.1 Map of Jordan River Basin (adapted from UNEP/DEWA/GRID 2016). Permission from UNEP http://www.grid.unep.ch/products/4_maps/jordan.gif Jordan River Basin 2001-10

Box 10.3: International Foreign Investment in Farmland

One of the effects of the 2007/8 food crisis due to a spike in the price of agricultural commodities was an increase in transnational acquisition of farmland (Anseeuw et al. 2012; von Braun and Meinzen-Dick 2009). Compiling information from a variety of sources the Land Matrix Project (Anseeuw et al. 2012) identified reports of 1217 agricultural land deals amounting to 83.2 million ha of land in developing countries equivalent to 1.7% of the world's agricultural area. The reports of land deals peaked in 2009.

The analysis of international investment in farmland indicated that the main countries targeted were in Africa (Sudan, Ethiopia, Mozambique, Zambia, Tanzania, Madagascar and Congo) and South-East Asia (Philippines, Indonesia and Laos). The following characteristics of the targeted countries were identified as: (1) a focus on the poorest countries that are less involved in world food exchange, (2) weak land tenure but with relatively high levels of investor protection, and, (3) a high prevalence of hunger (Anseeuw et al. 2012).

The land targeted by investors was where there was a large yield gap – a difference between potentially achievable yields and current crop production. In addition the targeted land was where additional inputs (namely water, fertilisers, seeds, infrastructure and knowhow) may create greater yields. Furthermore, the focus was on land with high accessibility in close proximity to a major city.

The main countries investing in farmland were emerging countries (China, Brazil, and South Africa) and the global north (USA, Europe). The main characteristics identified of investor countries were that they were (1) net importers of food, (2) had growing populations and consumption, (3) had a high demand for food, biofuels, and raw materials (such as palm oil, rubber), and (4) were water constrained. It was estimated that the increase in water consumption in the targeted countries was 12.7%, whereas it was expected that there would be a positive effect on the freshwater balance in the investors' countries of origin.

A range of implications were identified for targeted countries. There was a loss of production for the targeted countries with agricultural output going to the origin country of investors. There was displacement and eviction of current landholders. There were disputes over who owns land and who has the rights to sell the land. While compensation was being paid, it was considered low compared to international prices. There was employment generation. However, some of the employment is job replacement for farmers who have lost land access. Infrastructure improvements were made in relation to project infrastructure, access to markets, and, health and education facilities (Anseeuw et al. 2012).

Von Braun and Meizen-Dick advocate for a code of conduct to address the threats of foreign investment in land to create opportunities in the countries

(continued)

Box 10.3 (continued)
from that investment. The key elements of the code include: (1) transparency in negotiations and free, prior and informed consent of existing landholders; (2) respect for existing land rights including customary and common property rights; (3) sharing of benefits so that the local community benefit from foreign investment; (4) environmental sustainability to ensure water, soil, biodiversity and other impacts are addressed; and, (5) adherence to national trade policies where domestic supplies have priority when food security is at risk.

10.2.1 Conflict in Relation to Access to Water

In Box 10.2 a series of stages are identified in the Israeli approach to water management. The initial stages of the Israeli situation focus on boundary definition to ensure access to water. The closest parallel in the Canterbury situation is the dispute over the southern boundary of the regional council between Canterbury and Otago associated with the Waitaki River. Large braided rivers like the Waitaki were a major barrier to movement and often formed the provincial and district council boundaries. The Waitaki River was the historical boundary between North Otago and South Canterbury (Brooking 1998).

In establishing the regional council boundaries in 1989 most of the Waitaki catchment was incorporated in the Canterbury Region leaving the Waitaki District straddling two regions. There have been efforts by the Waitaki community to become part of Otago again (Ansley 1999). The boundary has added administrative complexity to water projects that cross the boundary, e.g. the Irrigation North Otago scheme, which extracts water from the Waitaki River (in Canterbury) to irrigate areas in the Kakanui catchment south of the Waitaki catchment boundary (in Otago), needed water take consents from the Canterbury Regional Council and water use consents from the Otago Regional Council. While it has led to verbal clashes and political debate, it has not led to military clashes as in the Jordan River Basin.

A striking difference between Canterbury and Israel is the degree of water scarcity and the institutional response. Israel is an importer of virtual water whereas Canterbury and New Zealand are exporters of virtual water. Canterbury is only starting to address the issue of water use efficiency, whereas Israel is a world leader in irrigation efficiency, wastewater recycling and desalination. An associated contrast is the approach to water infrastructure with a strong central government direction in Israel compared to the heavy reliance of the private sector in New Zealand.

Both Israel and Canterbury have turned to collaborative processes to resolve conflict. For Israel this is to manage external conflict whereas for Canterbury it is internal conflict. The Israeli-Palestinian experience also highlights a potential threat to effective collaboration – that of a power imbalance between the parties to collaboration effort. This is a concern being raised in Canterbury.

10.2.2 Conflict in Relation to Access to Land

External intrusion as a potential failure pathway associated with conflict relating to land access in New Zealand is primarily through foreign investment. As set out in Box 10.3, the 2007/8 food crisis led to an increase in transnational acquisition of farmland particularly by countries that were importers of food with growing populations and were water constrained. The countries targeted were those with large yield gaps and where additional inputs including water could increase yields. To address issues such as landholder displacement, land rights disputes and compensation, a code of conduct for foreign investment has been proposed. As described below foreign investment in New Zealand agribusiness is occurring. However, New Zealand is a relatively efficient exporter of agricultural products so the reasons for investment are different from other global investments and appear to be related to control of the supply chain to improve food security for investor countries. Foreign investment is also subject to the Overseas Investment Act which has provisions related to the concerns in the recommended code of conduct.

10.2.2.1 Foreign Investment in New Zealand

An analysis of foreign direct investment in New Zealand has been undertaken (KPMG 2015). The total of overseas investment applications approved over the two-year period 2013–2014 was approximately $14.2 billion. Agribusiness accounted for approximately 11% of the total investment. The dairy sector was the dominant source with 31% of the agribusiness sector in milk processing and 20% in dairy farms. Wine (12%) and horticulture (5%) were other sectors involving irrigation. However, it was expected that investment in the dairy sector will decline with the decline in milk prices globally. Although this may be offset by speculative buying of farms in the event forced sales occur in this sector. China and Hong Kong represent 49% of the investment in agribusiness with the majority of this being the results on investment in the dairy sector. Other regional sources of investment in agribusiness were Europe (14%), North America (8%) and Singapore (7%). In the 5-year period (2010–2014), foreign direct investment in land acquisition amounted to approximately 595,000 ha which is 5% of New Zealand's agricultural and forestry land area. Most of the land acquisition was in forestry (56%) and sheep and beef farms (30%) with smaller areas in dairy (12%), wine (1.2%) and horticulture (0.3%). Origin of investment was North America (48%) mainly in forestry, Europe (18%) and China/Hong Kong (13%) (KPMG 2015).

10.2.2.2 Overseas Investment Act

The purpose of the Overseas Investment Act is to "acknowledge that it is a privilege for overseas persons to own or control sensitive assets by (a) requiring overseas investment in those assets, before being made, to meet criteria for consent, and (b) imposing conditions on those overseas investments" (Overseas Investment Act 2005). Investment in land that is considered "sensitive" includes non-urban land greater than 5 ha (e.g. farmland) as well as land designated for conservation, recreation, heritage or historic purposes, or specified islands, and, lake beds, foreshore or seabed. Criteria for consent for overseas investment in farmland relate to whether the overseas person (a) has business acumen relevant to the investment, (b) has demonstrated financial commitment to overseas investment, (c) is of good character, and (d) is not ineligible for a visa; whether the investment will benefit New Zealand in a way that is substantial and identifiable; and, whether the farmland has been offered for acquisition on the open market.

Factors for assessing benefit of overseas investment in sensitive land cover economic issues (such as job creation, new technology, increased export, added market competition, additional investment and increased processing), environmental issues (such as conservation, protecting indigenous vegetation and habitat, fisheries and pest control), and social/cultural issues (such as historical and cultural significance, heritage covenants, and public access).

The criteria in the Overseas Investment Act address many of the issues identified in the proposed code of conduct of von Braun and Meizen-Dick (refer Box 10.3).

The requirement that farmland has been offered for sale on the open market relates to free, prior and informed consent of existing landholders. The requirement for substantial and identifiable benefits relates to the local community benefitting from foreign investment. The factors for assessing investments concerning environmental issues relate to environmental sustainability while the factors concerning social/cultural issues relate to customary and common property rights. A notable difference is in relation to trade policy, where New Zealand as a major food exporter is seeking increased export rather than prioritizing domestic supplies when food security is at risk.

Applications are assessed by the Overseas Investment Office (OIO) and advice is provided on how applications should be determined. The decision is made at Ministerial level. The issue is politically controversial with interest groups like *Save Our Farms* and *Campaign Against Foreign Control of Aotearoa* seeking a moratorium on farm sales to foreigners (CAFCA 2016) while other interests have supported foreign investment (NZIER 2010b). Two recent decisions that generated significant public debate were the applications to purchase Crafar Farms and the application to purchase Lochinver Station.

Crafar Farms, a group of 16 farms, was New Zealand's largest family-owned dairy farm business. It was put into receivership in October 2009. Crafar Farms was involved in multiple prosecutions for pollution offences and poor animal welfare. An application by Natural Dairy (NZ) Holdings Limited (a Hong Kong based company) to acquire Crafar Farms was declined by Ministers based on the OIO recommendation (New Zealand Government 2010). The recommendation to decline was based on the view that the good character of the director was affected because she was facing charges and because of her conduct in bankruptcy proceedings (LINZ 2010). Two Principals of the company were later convicted of money laundering (Field 2014).

However a subsequent application by Milk New Zealand Holdings Limited (a subsidiary of the Shanghai Pengxin Group) was granted consent (New Zealand Government 2012). Conditions to provide substantial and identifiable benefits to New Zealand included; investing a minimum of $14 m to improve economic and environmental sustainability of the farms, protecting two Māori pa sites; improving walking access to a forest park and waterfall; establishing an on-farm training facility; and assisting Landcorp (a state-owned enterprise of the New Zealand government) to extend its business in China. The Ministers' decision was appealed by a rival bidder to the High Court which required a re-evaluation of the consent compared to the offer of a New Zealand consortium. After further OIO advice the Ministers approved the consent (New Zealand Government 2012). Based on the improvements achieved on the farms, Milk New Zealand received the supreme award for the 2015 BNZ New Zealand Chinese Business Awards (Fox 2015).

Lochinver Station is a 13,687 ha sheep and beef farm near Lake Taupo. An overseas investment application was received from Pure 100 Farm Limited (a local subsidiary of Shanghai Pengxin Group). The OIO recommendation was that the consent be granted. The OIO noted the creation of contractors' jobs, increased exports from the conversion of forestry and wilding pine to dairy, and processing of the forestry land felled. The OIO listed other benefits that the deal offered including a financial contribution to the local Rangitaiki School, the agreement to sell lake and river beds to the Crown, conservation measures to protect waterways and trout, provisions for walking access, and, preservation of historic sites including a submerged waka. However Ministers declined the application (New Zealand Government 2015). While the Ministers acknowledged that the sale could provide benefit to New Zealand they considered that the benefits are not likely to be substantial and identifiable.

Lochinver was then sold to a privately owned New Zealand farming group Rimanui Farms Ltd (Rural News Group 2015a). Shanghai Pengxin initially sought a judicial review of the Government's decision to decline its application to purchase Lochinver Station (Rural News Group 2015b) but this has now been withdrawn (Tipa 2016). Daking NZ Farm Group (a company controlled by the Shanghai Pengxin Group) then withdrew its proposed acquisition of farmland in Northland citing the Lochinver decision (Peterson and Wilson 2016).

The discussion of the case studies demonstrates that as a potential failure pathway relating to external intrusion the relationship between retaining local ownership and accessing beneficial foreign investment is finely balanced.

10.3 Social Effects of Changing Technology

This section considers the potential failure pathway associated with the introduction of new technology that changes relationships within society. A recent Canterbury example of this issue was the proposed storage in the Waianiwaniwa Valley as part of the Central Plains Water Scheme. The social impact on the people and the community of the Waianiwaniwa Valley was a significant factor in the RMA Hearing Commissioners recommendation that the applicant withdraws the dam and reservoir from its proposal.

The potential scale of social changes associated with major storages is illustrated by the example of the Three Gorges Dam on the Yangtze River in China. The social implications of this project are summarized in Box 10.4.

Box 10.4: Social Implications of the Three Gorges Project

The Three Gorges Dam is 181 m high and 2335 m long. It is the largest hydroelectric generation project in the world with an installed capacity of 22,500 MW. The reservoir surface area is 1045 km^2. The project has submerged 13 cities, 140 towns and 1350 villages and displaced 1.3 million people. By the end of 2008 expenditure had reached more than 140 billion yuan[3] of which 65 billion yuan was spent on relocating affected residents (from Chinese sources quoted in Wikipedia).

The decision to build the dam was controversial. It was not until the early 1990s after the Tiananmen incident and after a national debate that the government took the authoritarian decision to build the dam (Lin 2007). There had been continuing suppression of dissenting viewpoints on environmental impacts and social issues relating to the Three Gorges Project including restrictions on public information and debate, extending to arrests of political activists opposed to dam construction (Human Rights Watch 1995).

Priority had been given to building the dam to provide electricity, flood control and navigation. However inadequate attention has been paid to the problems of the people affected by reservoir inundation (Heming et al. 2001).

The resettlement process for the Three Gorges project was modified from earlier approaches that had been considered by Chinese authorities to be unsuccessful. Resettlement funding was included in the overall project budget and development resettlement schemes were incorporated to address the daily subsistence problems of the resettled population. However, there were specific challenges associated with the Three Gorges Project. One was the lack of available farmland for the resettled rural population. A second was the linkage

(continued)

[3] One Chinese yuan equals $NZ 0.20. Expenditure on the project is equivalent to $NZ 28 billion with $NZ 13 billion spent on relocation.

Box 10.4 (continued)
between resettlement and environmental capacity; in particular, opening new land for agricultural purposes resulted in erosion. A third was that funding which had been earmarked for resettlement and reconstruction of infrastructure was embezzled or spent elsewhere. A fourth was public participation was limited to the village level concerning selection of resettlement sites, reallocation of land resources and dispute settlement between migrants and their hosts: public participation in earlier project decision making processes was next to nothing. A fifth was the lack of law protecting the rights and interests of the people displaced (Heggelund 2003).

In addition to previous resettlement approaches of settling migrants in nearby areas on land to be farmed and allowing migrants to move to and live with relatives in urban areas, a third strategy of moving migrants far away was introduced for the Three Gorges Project. Local resettlement was replaced by a combination of local and distant resettlement. Regions below the dam that benefitted from cheap electricity or improved flood control were selected for sharing the task of resettling people displaced by the Three Gorges Project (Heming et al. 2001).

Evaluations of the resettlement strategies have found issues with each of the three approaches. Resettlement of farmers was from fertile flat land in the valley floor to higher ground that was less productive and smaller in size, leading to a decline in income. Host communities were reluctant to give up rights to fertile land. Resettlement to urban areas included arrangements for jobs with industry. However, migrants were not always hired or often lost their new urban jobs due to the surplus of labor in the urban industrial sector and the low educational level of the displaced people. For those relocated far away from their place of origin there were difficulties in rebuilding livelihoods, difficulties integrating into the host community, loss of social networks, and difficulties of coping with a strange and new environment (Heming et al. 2001).

As the government only carried out limited consultation with the people affected concerning the issues of compensation and resettlement, the displaced people had no option but to become involved in protests to express their concerns. Protesters were arrested or dispersed by police and petitions of complaints were ignored by officials (Lin 2007). The consequences of resettlement have been economic impoverishment, social instability and environmental degradation (Heming et al. 2001).

However, there can also be more subtle and indirect social changes associated with the introduction of new technology. The widespread adoption of tube wells in India for irrigation displacing the system of water tanks has led to significant social changes. The management of water tanks was supported by local community institutions for the sustainable and equitable management of water resources. With the change to the individual management of tube wells, the local community institutions have declined and groundwater drawdown is threatening the sustainable management of groundwater. The situation is described in Box 10.5.

Box 10.5: India: a Shift in Technology Leads to Social Change (Chandrakanth and Romm 1990)

Irrigation tanks (small reservoirs) have a long history in the Deccan Plateau of India. More than half of the tanks were built by Hoysala rulers between the eleventh and thirteenth centuries. Tanks were constructed by village communities, private individuals and the State. The tanks were designed for drought storage. Nearly every valley would contain a series of tanks with overflow from an upstream tank flowing into the next downstream tank. The tank storages also recharged groundwater. Maintenance was needed for desilting and repair. The silt was used by farmers to improve land fertility.

Institutions governing construction and maintenance were largely religious in nature. Construction and maintenance of irrigation tanks were of fundamental importance to the prosperity of society, and were considered to be one of the seven meritorious acts a person could perform in a lifetime. Temples were a major land owner. Temples maintained village tanks when they breached. They provided funds to villagers to maintain tanks and leased land to farmers to encourage tank construction. Farmers who did not maintain their tanks would lose their right to two thirds of the land leased to them in favour of farmers who maintained tanks at their own expense. Villages had committees for the supervision of tanks. Their role was to invest endowments received for silt removal and repair.

The advent of the Green Revolution in the 1960s and the introduction of new technology, in particular, tube wells with electric pumps, rural electrification and chemical fertilisers, led to significant social change. Irrigation from wells increased from 6.6 million ha in the 1950s to 33.6 million ha in 2002–2003 (Mukherjee 2007).

Groundwater rights prevailing in India could be characterized as a version of the English doctrine of absolute right under which landowners have an absolute right to water and their land and are not constrained in the volume that they can abstract. In addition, credit incentives and subsidized electricity costs encouraged groundwater extraction and use (Mukherjee 2007).

Groundwater development brought considerable economic growth and diversification in rural areas. However, it has also brought degradation of the resource base. In recent years it is threatening livelihoods and long term availability of groundwater. In Karnataka, 20% of wells go dry every year (Mukherjee 2007). State controls on groundwater have been fraught with resistance from farmers.

(continued)

Box 10.5 (continued)

Use of tanks has reduced. The customary obligations of collective efforts in maintaining the tanks have declined and the conditions of the tanks have deteriorated. The social system at the village scale that supported sustainable water management has been replaced by a system of individual rights with State controls which have been ineffective in groundwater management.

10.3.1 Social and Community Considerations in the Waianiwaniwa Storage Proposal

The Waianiwaniwa Valley storage proposal was part of the Central Plains Water Scheme consent application that was evaluated by RMA Hearing Commissioners. The Central Plains Scheme included in the consent application was to irrigate up to 60,000 ha of farm land using water diverted from the Waimakariri and Rakaia Rivers (Fig. 10.2). A storage in the Waianiwaniwa Valley was designed to receive diverted water that could be made available for the irrigation scheme when diversion of flows from the Waimakariri and Rakaia were restricted during periods of low flow. The provision of storage increases the reliability of supply to irrigators.

The storage involves an earth dam, about 2 km in length and 55 m in height in the Waianiwaniwa Valley immediately upstream of the village of Coalgate (population 264). The dam would have created a reservoir flooding about 12 km^2 of the valley directly affecting 29 properties and 15 households (Taylor Baines in association with Fitzgerald Applied Sociology and People & Places 2007).

In a separate Minute, the Hearing Commissioners concluded that while the dam and reservoir "may promote the economic wellbeing of the wider Canterbury and national community, they would not promote the *social, economic and cultural wellbeing* of the Waianiwaniwa and Coalgate communities" (Milne et al. 2009).

The Commissioners' conclusions were influenced by the close proximity of the proposed dam to Coalgate – less than half a kilometer from the edge of the township. Specific impacts of concern in relation to the Coalgate community were: (1) adverse effects associated with noise, dust and general activity during construction; (2) the increase in heavy vehicle traffic during construction; (3) the adverse effects on property values; (4) visual impact of the dam; (5) residents' concerns about the risk of dam failure; (6) the ongoing uncertainty about whether the dam would be built; and (7) the potential for division in the community between opponents and supporters of the dam.

For the community in the Waianiwaniwa Valley, the Hearing Commissioners noted that property purchase and compensation would likely address most if not all economic impacts within the valley but would not address the social impacts. They recognized a strong sense of attachment by many landowners and their families to the properties and that such attachments could not be adequately addressed by com-

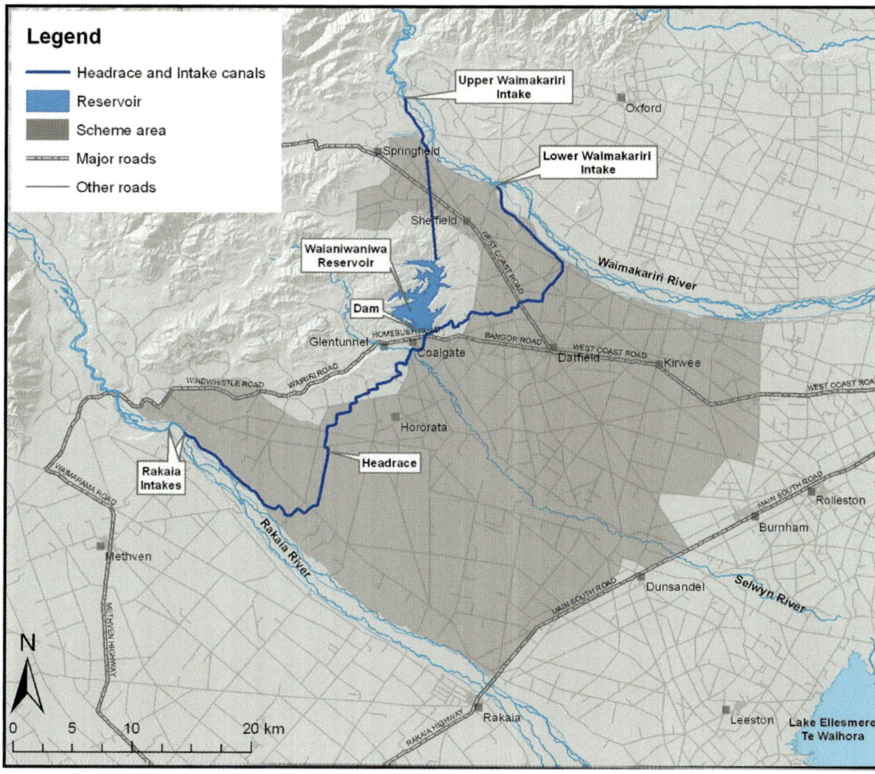

Fig. 10.2 Main elements of Central Plains Water Scheme consent application (Taylor Baines in Association with Fitzgerald Applied Sociology and People & Places 2007)

pensation. The Hearing Commissioners also considered that there was a perceived unfairness of one set of farmers taking other farm land for private benefit.

The Hearing Commissioners accepted that the Central Plains Scheme would probably result in significant net economic benefits at the regional level and to some extent at the national level. Furthermore, it was accepted that the Waianiwaniwa Valley storage would improve irrigation reliability. However in adopting "an overall balancing approach" on the total effects of the storage in the Waianiwaniwa Valley, the Hearing Commissioners concluded that approving the storage would not provide sustainable management and in particular the dam and reservoir would not enable the social and cultural well-being of affected people and affected communities (Milne et al. 2009).

10.3.2 International Examples

10.3.2.1 Three Gorges Project

In addition to the dramatic difference in scale, there are a number of contrasts between the Waianiwaniwa Valley storage and the Three Gorges Project. There are also some similarities in relation to achieving societal sustainability. One of the key differences is the ability in the Waianiwaniwa Valley storage for affected people to be able to express dissent and be heard prior to decisions being made on the project. Another key difference is the distinction between private benefit and public benefit associated with the projects. There is an unfairness for private interests to benefit at the expense of others. Notwithstanding, even where projects are for public benefit there is an obligation on those who benefit (or governments on their behalf) to provide social justice for those adversely affected and promote the social well-being of affected parties, both the people displaced and the communities absorbing the displaced people. As well as adequate financial compensation, there are societal impacts that cannot be adequately mitigated by financial compensation alone.

10.3.2.2 Water Management in the Deccan Plateau

As well as the direct effects of technology leading to social displacement and social change, there is also the effects of technology on the social structure for its management. As discussed in Box 10.5, the communal social structures that maintained drought storages in the Deccan Plateau for centuries were weakened with the shift to individually managed groundwater pumping systems under a system of landowner water rights and ineffective State controls.

References

Allan, J. A. (2002). Hydro-peace in the middle east: Why no water wars? A case study of the Jordan River Basin. *SAIS Review, XXII*(2), 255–272.

Anseeuw, W., Boche, M., Breu, T., Giger, M., Lay, J., Messerli, P., & Nolte, K. (2012). *Transnational land deals for agriculture in the global south: Analytical report based on the Land Matrix database*. Bern/Montpellier/Hamburg: CDE/CIRAD/GIGA.

Ansley, B. (1999). *The borderers, Listener* (Vol. 168). Auckland: Bauer Media Group.

AquaPedia. (2014). *Coastal Aquifer (Israel, Palestine)*. https://aquapedia.waterdiplomacy.org/wiki/index.php?title=Coastal_Aquifer_(Israel,_Palestine). Accessed 1 Apr 2016.

Berland, A. (2000). *The water component of the peace process between the Israelis and the Palestinians*. MA thesis, Tufts University, Medford.

Brenner, A. (2012). Limitations and challenges of wastewater reuse in Israel. In F. Quercia & D. Vidojevic (Eds.), *Clean soil and safe water, NATO Science for Peace and Security Series C: Environmental Security* (pp. 3–9). Dordrecht: Springer.

Brooking, T. (1998). *An old and early dispute, rapid resolution, small adjustment, a permanent solution and its recent overthrow: The historical background to the Otago/Canterbury Regional Council boundary dispute*. Dunedin: Otago Regional Council.

CAFCA. (2016). *Foreign control – Key facts*. Campaign Against Foreign Control of Aotearoa. http://canterbury.cyberplace.co.nz/community/CAFCA/key-facts.html. Accessed 18 Apr 2016.

Chandrakanth, M., & Romm, J. (1990). Groundwater depletion in India – Institutional management regimes. *Natural Resources Journal, 30*(Summer), 485–501.

FAO. (2008). *Irrigation in the Middle East in figures – AQUASTAT Survey 2008*. Rome: Food and Agriculture Organization of the United Nations.

Fattel, B., Yekutiel, P., & Shuval, H. (1986). Cholera outbreak in Jerusalem 1970 revisited: The case for transmission by wastewater irrigated vegetables. In J. Goldsmith (Ed.), *Environmental epidemiology: Epidemiological investigation of community environmental disease*. Baco Raton: CRC Press.

Field, M. (2014, August 26). *Money laundering conviction for Natural Dairy duo*. NZ Farmer.

Fischhendler, I. (2008). Institutional conditions for IWRM: The Israeli Case. *Groundwater, 46*(1), 91–102.

Fonterra. (2016). *Farmgate milk prices*. http://www2.fonterra.com/our-financials/farmgate-milk-prices. Accessed 10 Mar 2016.

Fox, A. (2015, April 17). *Train wreck remodelled as award winner*. NZ Farmer.

Future Directions. (2015). Sharing *Water: Israel and Jordan finalise agreement for Red Sea – Dead Sea pipeline*. http://www.futuredirections.org.au/publication/sharing-water-israel-and-jordan-finalise-agreement-for-red-sea-dead-sea-pipeline/. Accessed 1 Apr 2016.

Givati, A. (2012). *Climate trends in Israel and effects on water resources*. Tel Aviv: Water Authority.

Gleick, P. H. (1993). Water and conflict: Fresh water resources and international security. *International Security, 18*(1), 79–112.

Haddadin, M. (2006). *Water resources in Jordan: Evolving policies for development, the environment and conflict resolution*. Washington, DC: Resources for the Future.

Harris Consulting, Aqualinc Research, Agribusiness Group, & Butcher Partners. (2010). *Economic modelling: A tool to assist evaluation of the Canterbury Water Management Strategy and trial scenario 1 assessment*. Christchurch: Canterbury Water.

Heggelund, G. (2003). The Three Gorges Dam: Taming the waters of the Yangtze – Creating social instability. *NIASnytt, 2*, 12–14.

Heming, L., Waley, P., & Rees, P. (2001). Reservoir resettlement in China: Past experience and the Three Gorges Dam. *The Geographical Journal, 167*(3), 195–212.

Hoekstra, A., Chapagain, A., Aldaya, M., & Mekonnen, M. (2011). *The water footprint assessment manual: Setting the global standard*. London: Earthscan.

Hosking, J. (2015). Red Sea-Dead Sea Project: Too little, too late? *Venture*. http://www.venturemagazine.me/2015/06/red-sea-dead-sea-project-too-little-too-late/. Accessed 1 Apr 2016.

Human Rights Watch. (1995). *The Three Gorges Dam in China: Forced resettlement, suppression of dissent and labor rights concerns* (Vol. 7, Issue. 2). Human Rights Watch Publications.

ISCID. (undated). *Israel*. Israel National Committee of International Committee on Irrigation and Drainage. www.icid.org/v_israel.pdf. Accessed 26 Mar 2016.

Ju'ub, G. (undated). *Water conflicts in the Middle East between the present and the future*. http://www.alazhar.edu.ps/arabic/ewi/researches/gazapaper-new/water.pdf. Accessed 23 Mar 2016.

Kiser, S. (2000). *Water: The hydraulic parameter of conflict in the Middle East and North Africa*. Colorado: Institute for National Security Studies.

Kislev, Y. (2011). *The water economy of Israel*. New York: Taub Center.

KPMG. (2015). *Foreign direct investment in New Zealand: Trends and insights*. Auckland: KPMG.

Lin, T.-C. (2007). Environmental NGOs and the anti-dam movements in China: A social movement with Chinese characteristics. *Issues & Studies, 43*(4), 149–184.

LINZ. (2010). *Ministers decline Natural Dairy application Crafar farms*. http://www.linz.govt.nz/news/2010-12/ministers-decline-natural-dairy-application-crafar-farms. Accessed 18 Apr 2016.

Manna, M. (2006). *Water and the treaty of peace between Israel and Jordan*. Bristol: Center for Macro Projects and Diplomacy Roger Williams University.
Medzini, A. (2001). *The River Jordan: The struggle for frontiers and water 1920–1967*. London: SOAS Water Issues Group.
Mekonnen, M., & Hoekstra, A. (2011). *National water footprint accounts: The green, blue and grey water footprint of production and consumption*. Delft: UNESCO-IHE.
Milne, P., Fenemor, A., O'Callaghan, R., & Nixon, B. (2009). *Minute 10 of Commissioners: Preliminary recommendation and summary of conclusions and reasoning in relation to the proposed Waianiwaniwa reservoir and dam and the upper Waimakariri intake*. Christchurch: Environment Canterbury.
Mukherjee, S. (2007). *Groundwater for agricultural use in India: An institutional perspective* (Vol. 187). Bangalore: Institute for Social and Economic Change.
New Zealand Government. (2010). *Ministers turn down Natural Dairy application to buy Crafar farms*. https://www.beehive.govt.nz/release/ministers-turn-down-natural-dairy-application-buy-crafar-farms. Accessed 18 Apr 2016.
New Zealand Government. (2012). *Ministers approve Crafar farms bid*. https://www.beehive.govt.nz/release/ministers-approve-crafar-farms-bid. Accessed 18 Apr 2016.
New Zealand Government. (2015). *Ministers decline overseas purchase of Lochinver Station*. https://www.beehive.govt.nz/release/ministers-decline-overseas-purchase-lochinver-station. Accessed 18 Apr 2016.
NZIER. (2010a). *The economic impact of increased irrigation: A dynamic Computable General Equilibrium analysis of increased irrigation in New Zealand*. Wellington: Ministry of Agriculture and Forestry.
NZIER. (2010b). *In defence of foreign investment*. New Zealand Institute of Economic Research. http://nzier.org.nz/publication/in-defence-of-foreign-investment. Accessed 18 Apr 2016.
Overseas Investment Act. (2005). New Zealand.
Peterson, A., & Wilson, K. (2016, January 22). *Ministers blocked the overseas purchase of Lochinver Station. What does this mean for investors?* NZ Winegrower .
Postel, S. (1992). *Last Oasis: Facing water scarcity*. New York: W.W. Norton.
Rabinovitch, A. (2010). Arid Israel recycles waste water on a grand scale. *Reuters*. http://www.reuters.com/article/us-climate-israel-idUSTRE6AD1CG20101114
Rural News Group. (2015a, November 6). Lochinver stays in NZ ownership. *Rural News*.
Rural News Group. (2015b, October 15). Shanghai Pengxin seeks judicial review. *Dairy News*.
Selby, J. (2013). Cooperation, domination and colonisation: The Israeli-Palestinian Joint Water Committee. *Water Alternatives, 6*(1), 1–24.
Shamir, U. (1997). Water agreements between Israel and its neigbours. *Yale University Middle Eastern Natural Environments Bulletin, 103*, 274–296.
Susskind, L., & Islam, S. (2012). Water diplomacy: Creating value and building trust in transboundary water negotiations. *Science & Diplomacy, 1*(3).
Tahal. (2010). *Dead Sea study: Best available data report revised*. Red Sea – Dead Sea Water Conveyance Study Program. Tahal Group, Or Yehuda.
Tait, P., & Cullen, R. (2006, February 8–10). *Some external costs of dairying*. Paper presented at the Australian Agricultural and Resource Economics Society Conference, Sydney.
Taylor Baines in Association with Fitzgerald Applied Sociology and People & Places. (2007). *Central Plains Water Enhancement Scheme: Social Impact Assessment*. Christchurch: Taylor Baines.
Tenne, A. (2010). *Sea water desalination in Israel: Planning, coping with difficulties, and economic aspects of long-term risks*. Tel Aviv: Water Authority.
Tenne, A. (2011, October). *The Master Plan for Desalination in Israel 2020*. Paper presented at the OECD, Paris.
Tipa, P. (2016, April 1). Shanghai Pengxin withdraws judicial review. *Dairy News*.
Treasury. (2015). *New Zealand economic and financial overview 2015*. Wellington: New Zealand Government.

UNEP/DEWA/GRID. (2016). *Map of Jordan River Basin.* http://www.grid.unep.ch/products/4_Maps/jordanb.gif

UN-ESCWA and BGR. (2013). Western aquifer basin. In: *Inventory of shared water resources in Western Asia.* Beirut: United Nations Economic and Social Commission for Western Asia and Bundesanstalt für Geowissenschaften und Rohstoffe.

URS. (2014). *Tekapo transfer review 2014 with water quality and cultural aspects.* Christchurch: Environment Canterbury.

von Braun, J., & Meinzen-Dick, R. (2009). *"Land Grabbing" by foreign investors in developing countries: Risks and opportunities.* Washington, DC: International Food Policy Research Institute.

Wintgens, T., & Hochstrat, R. (2006). *Report on integrated water reuse concepts. AQUAREC Integrated Concepts for Reuse of Upgraded Wastewater* (Vol. D19). Brussels: European Commission.

Zaharani, K., Al-Shayaa, M., & Baig, M. (2011). Water conservation in the Kingdom of Saudi Arabia for better environment: Implications for extension and education. *Bulgarian Journal of Agricultural Science, 17*(3), 389–395.

Part III
Sustainability Methods

Chapter 11
Sustainability Assessments

Abstract Sustainability assessments have two key components: vulnerability assessment which is the analysis of potential failure pathways for a natural resource system that could threaten its sustainability; and sustainability strategy formulation which addresses the critical variables associated with potential failure pathways so that the system stays within the resilience thresholds and avoids the change in structure and function of the system.

A sustainability assessment of eutrophication from land use intensification in the catchments of six New Zealand lakes highlights that there are different water quality pathways with different critical variables for each lake for what is ostensibly the same issue of nutrient enrichment. This means the management interventions have to be tailor made for each lake. The assessment also shows that the current levels of intervention are insufficient to achieve the desired water quality objectives.

A sustainability assessment of the declining area of the Aral Sea not only identifies failure pathways where system thresholds have been dramatically exceeded, but also demonstrates the need to consider the linkages between socio-economic systems and biophysical systems at three spatial scales to understand why deterioration of social and ecological components dependent on the Aral Sea are likely to persist.

Keywords Adaptive cycles • Failure pathways • Critical variables • Management interventions • Adequacy of interventions

11.1 Sustainability Assessment Framework

In Sect. 4.2 the elements of an approach to achieving sustainability were identified. The sustainability assessment framework is based on these elements.

For defining the resource management issue as a nested adaptive system, the elements are:

- The adaptive cycle of exploitation, accumulation, disturbance/release, and reorganisation (refer Sect. 4.2.1);
- Socio-ecological systems as linked adaptive cycles with a biophysical system linked to a socio-economic system (refer Sect. 4.2.2); and,

B.R. Jenkins, *Water Management in New Zealand's Canterbury Region*, Global Issues in Water Policy 19, https://doi.org/10.1007/978-94-024-1213-0_11

- The nesting of adaptive cycles to link systems operating at different spatial and time scales (refer Sect. 4.2.3).

For vulnerability assessment, the elements are:

- The definition of failure pathways which can cause nested adaptive socio-ecological systems to collapse (refer Sects. 4.2.4 and 4.2.5); and,
- The identification of critical variables and their thresholds for these failure pathways (refer Sect. 4.2.7).

For formulating sustainability strategies, the elements are:

- The potential management interventions and the adequacy of these interventions to address the vulnerability of the adaptive systems (refer Sect. 4.2.8); and
- The identification and implementation of sustainability strategies for maintaining or transforming the nested adaptive system (refer Sect. 4.2.9).

Examples of defining resource management issues as nested adaptive systems were provided in Sect. 5.1 for irrigation from the Waimakariri River, Christchurch City water supply and gravel extraction from the Waimakariri River.

The remainder of this section describes the concept of vulnerability assessment (Sect. 11.1.1) and the concept of sustainability strategies (Sect. 11.1.2).

11.1.1 Vulnerability Assessment

In defining the phases of a failure pathway for vulnerability assessment consideration is given to the analysis of disasters. While the subject matter is much broader than water resource management, the patterns of failure and recovery are a useful starting point. Hewitt has analysed many types of failures and identified a failure and recovery sequence (Hewitt 1997). He has defined eight phases in disaster and recovery beginning with "preconditions leading up to disaster" (two phases), the "disaster" (four phases), and "recovery and reconstruction" (two phases).

The first precondition phase Hewitt describes as "everyday life". These are the arrangements in place prior to a failure. These include the risks associated with a particular lifestyle or approach, the routine safety measures, the social construction of vulnerability, and the degree of emergency preparedness.

The second precondition phase Hewitt calls "premonitory developments". There is an incubation period for potential disasters often associated with an erosion of safety measures, heightened vulnerability, and where warning signs are misread or ignored.

The disaster component of the sequence begins with a "triggering event or threshold" (phase 3). This is the threat period at the beginning of a crisis or impending failure. There may or may not be time to allow for warnings or evacuations.

Phase 4 is the "impact and collapse" phase. This is the disaster proper with concentrated death or destruction. This is followed by a "secondary and tertiary damages" phase (phase 5). This comprises the impact on the survivors and post-impact hazards. Phase 6 is the arrival of "outside emergency aid". This is the organised response providing rescue, relief, evacuation and shelter provision.

The disaster sequence is followed by recovery and reconstruction. Phase 7 is the "clean up and emergency communities" phase. This includes the establishment of relief camps, emergency housing, clearance of wreckage, on the biophysical side; and, blame and reconstruction debates, disaster reports, evaluations and inquiries, on the socio-economic side.

The final phase is "reconstruction and restoration" (phase 8). This is the re-establishment of everyday life and incorporates disaster-related development and hazard reduction, as well as private and recurring communal grief.

These phases can be related to an adaptive cycle (Fig. 11.1). The exploitation phase of the adaptive cycle is the everyday life phase of the disaster and recovery sequence. The accumulation phase of the adaptive cycle is the premonitory development phase with heightened vulnerability. The disturbance phase is the triggering event with the associated release represented by the impact and collapse, and, secondary and tertiary damages. The reorganisation phase begins with clean up and emergency communities in the initial stages of reorganisation and the contribution of outside emergency aid. There are then three potential outcomes (consistent with the three general types of outcomes identified in Sect. 4.2.9 and Fig. 4.8 of maintenance of the system, active transformation to a new state, or transformation to a degraded state). One is the maintenance of the original system, i.e. restoration in the disaster failure and recovery sequence (8a in Fig. 11.1). A second is the active trans-

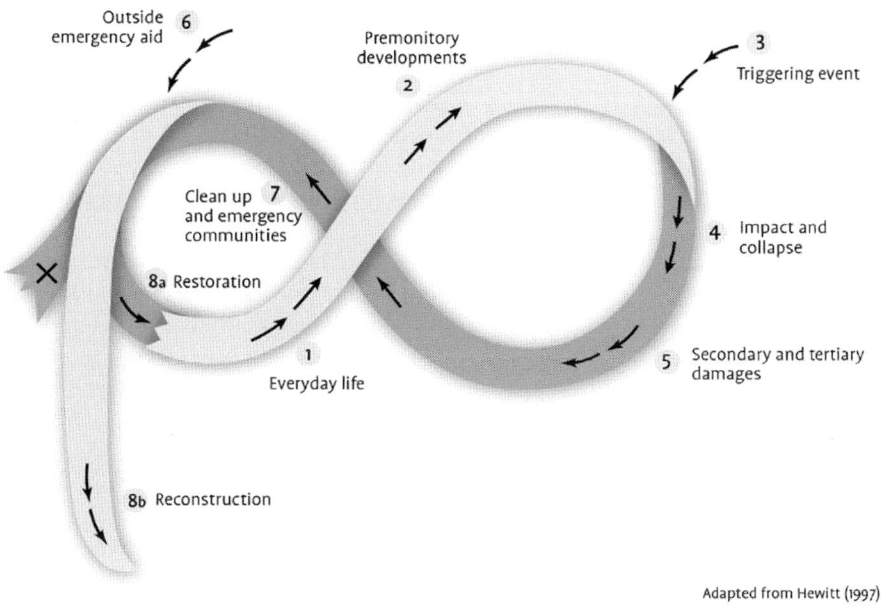

Fig. 11.1 Disaster failure and recovery sequence as an adaptive cycle

formation to a new system, i.e. reconstruction in the disaster failure and recovery sequence (8b in Fig. 11.1). The third is the unintended transformation to a degraded state if restoration or reconstruction does not occur (X in Fig. 11.1).

While Hewitt's terminology relates more to the consequences for human society of industrial and natural disasters, the concepts can be readily translated into consequences for natural resource systems of human-induced or natural failure pathways. Some water resource management examples of the disaster failure and recovery sequence are described below.

In relation to "everyday life", examples that represent preconditions for failure pathways are inefficient irrigation practices that lead to water availability failures, and, land use practices resulting in nitrate leaching that leads to water quality failures. An example of "premonitory developments" is soil moisture measurement. The need for soil moisture measurement is often ignored and even where soil moisture measurement is in place it is often not used to manage irrigation applications. This misreading or ignoring of soil moisture measurement can lead to either excessive irrigation resulting in using more water than necessary or insufficient irrigation resulting in production loss.

Examples of 'triggering events" that provide advice of impending failure include forecast of flood events (in the short term) and projections of climate change (in the long term). The effects of drought such as crop loss or lack of feed and water for stock are examples of "impact and collapse". An example of "secondary and tertiary damages" is the failure of primary stopbanks when the design flood is exceeded leading to flooding of adjacent land uses. For "outside emergency aid", examples include interbasin transfers for watersheds with shortages in water supply, and, oxygen injections to offset dissolved oxygen depletion for lakes with thermal stratification.

The introduction of boil water notices in Cheviot for the town's drinking water supply is an example of "clean up and emergency communities" response to a contaminated water source. The re-establishment of macrophytes and revegetation of native riparian species around Te Waihora/Lake Ellesmere is an example of "restoration", while the use of constructed wetlands for sediment retention is an example of "reconstruction".

The above discussion illustrates the relevance of considering failure pathways in the context of adaptive cycles. It also highlights the potential for external linkages (e.g. triggering events and outside aid) to other socio-economic or biophysical systems.

As noted in Sect. 5.1, for the analysis of potential failure pathways there are critical variables relevant to a system's resilience, i.e. the capacity of a system to absorb disturbance and still retain its basic function and structure. In looking at biophysical system vulnerability for the climate variability failure pathway, a critical variable for water availability for agricultural protection is potential evaporation deficit (refer Sect. 3.1.1). While for environmental degradation in relation to river flow, the critical variables are minimum, flushing and flood flows (refer Sect. 3.1.4). For socio-economic system vulnerability in relation to water availability, a critical variable is reliability of supply (refer Sect. 5.1.1).

Examples of thresholds for these critical variables have been provided above in relation to the management of irrigation from the Waimakariri River for the Waimakariri Irrigation Scheme. For potential evaporation deficit, using irrigation to maintain the threshold of soil moisture at 50% of field capacity was used to define "ideal" irrigation to offset evaporation losses in excess of rainfall. In relation to flow regimes, total restrictions on irrigation takes are imposed when the Waimakariri River flow is below 46 m^3/s in order to protect low flows in the river (refer Sect. 5.1.1). In relation to reliability of supply, irrigation scheme storage was introduced in the Waimakariri Irrigation Scheme to raise reliability of supply from 1 year in 42 to 23 years in 42 (refer Sect. 5.1.1).

11.1.2 Sustainability Strategies

Critical variables and associated thresholds become the targets for potential management interventions to prevent failure pathways from degrading natural resource systems. Sustainability strategies are combinations of management interventions to maintain the structure and function of the natural system, or, to transform a natural resource system to a new state which is more resilient to the drivers of change. A key focus is on the predictions of the outcomes of management interventions to achieve a sustainable natural resources system.

One example is the strategic investigations undertaken for the Canterbury Water Management Strategy for nitrate leaching from increased land use intensification (Sect. 3.2.5). Modelling of nitrate leaching from existing land use was carried out which correlated well with the field monitoring of nitrate concentrations in groundwater: this indicates some areas of shallow groundwater (0–50 m) exceeding the drinking water standard (Fig. 3.11). The investigations found that that if all potentially irrigable land was irrigated with current land use practices then a substantial proportion of the groundwater in the Canterbury Plains would exceed the nitrate drinking water standard (Fig. 3.11). Modelling also indicated that a 20% reduction of nitrate leaching associated with full intensification would reduce nitrate levels to be comparable to the existing land use (at the time of the analysis – 2009). To achieve nitrate levels in groundwater not exceeding the standard would require further reductions in nitrate leaching.

A second example is the approach to nutrient management in the Hurunui catchment. Elevated nutrient levels and faecal contamination have been highlighted as water quality issues in the tributaries and the lower Hurunui River (Hayward 2009b; Ausseil 2010). Three work streams were adopted (Enfocus 2011). One was a policy work stream that provided an overall framework for the project. A second was a science work stream that modelled the effects of land use intensification on nutrient levels and a range of possible thresholds for defining nutrient load limits, e.g. chlorophyll biomass, nitrate toxicity (Norton and Kelly 2010). The modelling was undertaken for five future use scenarios[1] (Lilburne et al. 2011). This information on nutrient modelling together with economic modelling of the five scenarios was input to the third work stream – the community work stream. This multi-stakeholder group undertook a deliberation process to consider future development scenarios and agree nutrient load limits.

A key task in developing sustainability strategies is the identification of possible management interventions. The process of intervention identification can be achieved by considering the interventions that are possible on the adaptive cycle phases of the nested adaptive system for the failure pathway. An example is provided below of Lake Waitawa, a small (16 ha), shallow (<7 m) coastal lake on the Kapiti Coast of the North Island of New Zealand. It has a catchment of 278 ha with 94% pastoral cover. It receives treated wastewater from a recreational camp. One of the potential failure pathways is from wastewater disposal (Jenkins 2015).

For wastewater process and disposal into the lake, the adaptive cycle phases are as follows:

- Exploitation (wastewater): the generation of greywater and wastewater from the camp
- Accumulation (wastewater): the build-up of wastewater and sludge in the treatment pond
- Release (wastewater): overflow from pond into lake via wetland
- Exploitation (lake): discharge of wastewater into lake
- Accumulation (lake): build-up of nutrients and pathogens in the lake
- Release (lake): algal growth in lake
- Reorganisation: lake reorganisation is dependent on reorganisation of wastewater treatment and disposal.

These adaptive cycle phases are shown in Table 11.1 together with possible management interventions.

A number of interventions have been identified which would improve lake water quality. At the exploitation (wastewater) phase, there could be a shift from flush to composting toilets. At the accumulation (wastewater) phase, the treatment pond could be desludged and the UV treatment increased by moving the trees that shade the pond. At the release (wastewater) phase, treatment of the overflow through the wetland could be increased through formalising wetland treatment. At the exploitation (lake) phase, rather than discharging to the lake, the alternative of land-based effluent disposal could be implemented. At the reorganisation (phase), the level of wastewater treatment could be increased to improve the quality of the effluent.

[1]The five scenarios were: (1) current land use; (2) business as usual intensification in line with historic trends; (3) extensive irrigation – full irrigation of suitable land; (4) conservative – all productive land converted to forestry; and, (5) 1990–5 water quality – land use change and mitigation to achieve water quality of the early 1990s.

Table 11.1 Wastewater adaptive cycle phases and potential interventions for Lake Waitawa (Jenkins 2015)

Adaptive cycle phases	Intervention
Exploitation (wastewater) Generation of greywater/wastewater	Composting toilets
Accumulation (wastewater) Build-up of wastewater and sludge in pond	De-sludge ponds Tree removal to improve UV treatment
Release (wastewater) Overflow to lake via wetland	Wetland treatment
Exploitation (lake) Discharge of wastewater to lake	Land-based disposal
Accumulation (lake) Build-up of nutrients and pathogens	
Release (lake) Algal growth in water	
Reorganisation (wastewater)	Improved wastewater treatment

11.2 Sustainability Assessment of Six New Zealand Lakes

This section describes the application of the sustainability analysis framework to the six New Zealand lakes (Jenkins 2016).[2] All six lakes are subject to nutrient enrichment with concerns relating to lake eutrophication. The first step in the application of the framework is defining the resource management issue as a nested adaptive system. This is followed by the vulnerability assessment of the lakes with respect to lake eutrophication. The failure pathways and critical variables are considered for each of the lakes: Lake Brunner, Lake Taupo, Lake Rotorua, Lake Omapere, Te Waihora/Lake Ellesmere and Waituna Lagoon. (The lake locations are shown in Fig. 11.2.) Then the differences in critical variables for the different lake systems are compared. The different critical variables and different failure pathways mean there is a need for different management interventions to achieve sustainable water quality. Next, the current approaches being taken for each of the lakes are outlined and an assessment made of whether the level of management intervention is likely to achieve sustainable water quality.

11.2.1 Lake Eutrophication as a Nested Adaptive System

Lake eutrophication is an example of an adaptive cycle. Land use intensification (i.e. exploitation phase) can lead to a build-up of nutrients in the lake water column or sediments (i.e. accumulation phase). Sufficient increase of nutrients can change

[2] Text for Sect. 11.2 drawn from Jenkins (2016) reprinted by permissions of the publisher (Taylor & Francis).

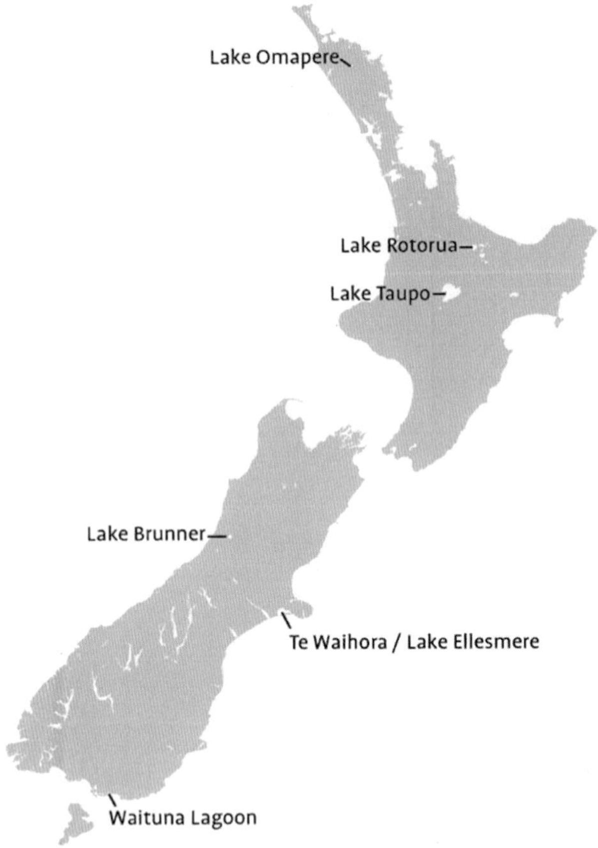

Fig. 11.2 Location of lakes (Jenkins 2016). Reprinted by permission of the publisher (Taylor & Francis)

the trophic state of the lake (i.e. disturbance phase) leading to increased algal blooms (i.e. release component of the disturbance phase). The lake restructures (i.e. reorganisation phase): this can be algal die-off and loss from the lake (i.e. recovery of the original system), or to long term decline in water quality (i.e. an alternative degraded system).

Describing eutrophication of lakes as an adaptive cycle involves consideration of at least two geographic scales: the lake and the catchment upstream of the lake. The dominant cause of lake eutrophication is the increase in nutrient-intensive land uses in the catchment typically associated with agriculture. One linkage between the catchment and the lake is through the accumulation of nutrient levels in soils leading to a release through soil erosion and runoff to nutrient accumulation in the lake downstream. The accumulation of nutrients in the lake and the lake sediments leads to the disturbance of eutrophication in the lake. Furthermore, for reorganisation of the degraded lake back to a higher water quality level usually requires reorganisa-

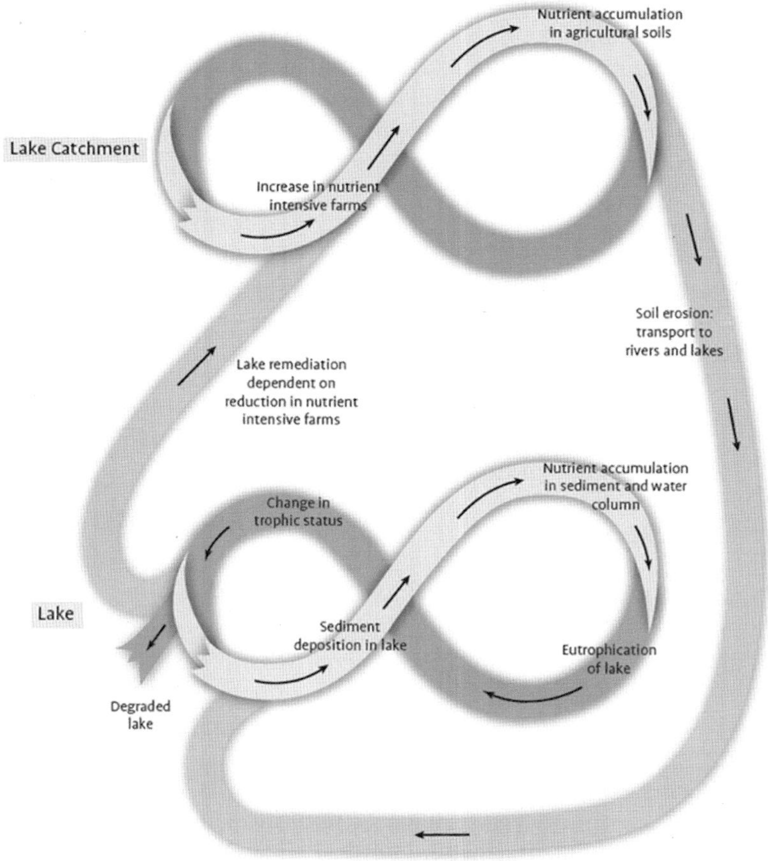

Fig. 11.3 Lake and catchment as nested adaptive cycles (Jenkins 2016). Reprinted by permission of the publisher (Taylor & Francis)

tion in the catchment through a reduction in nutrient intensity of land use in the catchment. It is a nested system as depicted in Fig. 11.3 with the phases in the adaptive cycle as follows:

- Exploitation of catchment: increase in nutrient-intensive land uses
- Accumulation in catchment: nutrient build-up in agricultural soils
- Release in catchment: soil erosion and sediment transport of nutrients to rivers and lake
- Accumulation in lake: nutrient build-up in lake sediments and water column
- Disturbance in lake: eutrophication in lake
- Reorganisation in lake: degraded lake unless reduction in nutrient-intensive land uses
- Reorganisation in catchment: reduction in nutrient-intensive farms.

There can also be other linkages between the catchment and the lake, such as nutrient-laden catchment runoff entering the lake and nutrient leakage to groundwater from catchment land use entering the lake via the groundwater system.

There can also be different time scales. For example, if the dominant nutrient input from a catchment to the lake is via groundwater then there can be a considerable time delay between land use change and equilibrium with nutrient input to the lake.

In lake eutrophication, critical variables include the nutrient loads being received from the upstream catchments. Thresholds are the tipping points for critical variables that can change the state or function of a socio-ecological system. An example is the lowering of dissolved oxygen of the bottom waters of a lake to the point that triggers the release of nutrients from lake bed sediments.

11.2.2 Failure Pathway Analysis of the Lakes

11.2.2.1 Lake Brunner

Lake Brunner is a large (41 km^2), deep lake (maximum depth 109 m) on the west coast of the South Island. The lake is oligotrophic and algal productivity is very strongly limited by the availability of phosphorus (Rutherford et al. 2008). Land use intensification from dairying has led to concerns that water quality will decline. Mean total phosphorus concentrations have increased from 5.1 mg/m^3 in the early 1990s to 6.1 mg/m^3 in the mid-2000s with comparable increases in Chlorophyll a (1.1 mg/m^3 increasing to 1.4 mg/m^3).

The lake has a long residence time (1.14 years) which leads to relatively high retention of nutrients in the lake (estimated at 52%). While nutrient storage reduces sensitivity to phosphorus loading this is only while there is well oxygenated bottom water. When lakes with long residence times have moved beyond nutrient loadings resulting in rates of algal productivity that cause anoxia in bottom waters, it is very difficult to restore them to a state with acceptable water quality (Verburg 2009).

11.2.2.2 Lake Taupo

Lake Taupo is New Zealand's largest lake (622 km^2) with a maximum depth of 60 m in the central North Island. The lake is oligotrophic (TLI = 2.1; mean 2005–9). Bioassays indicate that the lake is sensitive to nitrogen, and, algal growth in the lake increases in response to more nitrogen. Under pre-agricultural development the nitrogen load is estimated to be 650 t/year and it is currently estimated to be 1360 t/year. It takes a long time for the effect of intensifying land use to be seen in the lake. This is because the groundwater that carries much of the nitrogen from the land can take many decades to reach the lake. Studies estimate that between 20 and 80% of the current additional load of nitrate is yet to come before equilibrium is reached with current land use (Vant and Smith 2004).

There is a seasonal pattern of algal biomass with August typically the peak month. Both the annual mean and maximum values for Chlorophyll a in the upper 10 m of the lake have increased since regular monitoring in 1994 (average 0.5 mg/m^3, winter maximum 1.1 mg/m^3) to peaks in 2003 (average 1.1 mg/m^3, winter maximum 3.0 mg/m^3). Values have remained around this level since then or slightly declined.

11.2.2.3 Lake Rotorua

Lake Rotorua is a moderate sized lake (80.6 km^2) with a maximum depth of 45 m. The lake is eutrophic (TLI 4.6 – three yearly average to 2011). Water quality deterioration was observed between 1978/83. There has been substantial algal bloom activity (Bay of Plenty Regional Council 2012b). Sources of nutrients comprise nutrients already in the lake and sediments, nutrients entering the lake, and, nutrients in groundwater yet to reach the lake. The current nitrogen load entering the lake is estimated to be 755 tN/year, with nearly 80% from pastoral farming. A target load has been set at 435 tN/year (for a TLI of 4.2). The phosphorus load is estimated to be 40 tP/year with a sustainable load of 37 tP/year (Bay of Plenty Regional Council 2012a).

Lake Rotorua has about three periods of stratification of 10 days duration each year when oxygen concentrations become zero in the bottom waters. Each stratification event releases the equivalent of about one-third to one-half of the annual incoming load from the bottom sediments (Hamilton 2011).

11.2.2.4 Lake Omapere

Lake Omapere covers an area of 12 km^2 with a catchment of 17 km^2. It is shallow with an average depth of 2 m. It discharges to the Utakura River into Hokianga Harbour in Northland. The lake has been affected by land use change and other human actions, including pastoral farming, lowering the lake, drainage of wetlands, causing a decline in water quality (TLI = 6.1 2005–9 mean). Oxygen weed was introduced in the 1970s covering much of the lake and collapsed in 1985 causing severe blue-green algal blooms. The lake switches between algal-dominated and macrophyte-dominated states with periods of low Chlorophyll a (around 10 mg/m^3 1994–7 and 2008–12) with periods of high Chlorophyll a (maximum of 250 mg/m^3 1998–2008).

Nutrient loads in the lake are high (e.g. peak in-lake values in 2005–6 of 2100 mg/m^3 TN and 290 mg/m^3 TP). The TN:TP ratio for the lake varies substantially over time suggesting that there is no consistent nutrient limitation. In times of high suspended sediment, light can be the limiting factor. There is an estimated 500 t of TN and 50 t of TP in the top 2 cm of lake bed sediments which is frequently re-suspended into the water column through wind and wave action. Despite recent efforts at riparian management, there are still high surface water inputs to the lake with phosphorus peaks after rainfall and nitrogen in stream baseflows with concentrations exceeding lake concentrations. Dissolved oxygen levels are generally above 6 mg/L considered suitable for fish (Northland Regional Council 2007) (Gray 2012).

11.2.2.5 Te Waihora/Lake Ellesmere

Te Waihora/Lake Ellesmere is a large shallow brackish coastal lake. At a typical water depth of 1 m, the lake occupies approximately 200 km^2, about 7% of the catchment of 2560 km^2 (Taylor 1996). The lake is artificially opened to the sea to manage lake level, originally to manage the flooding of farm land, then to include water levels for bird habitat, and, more recently for timing of openings to facilitate migration of long fin eels.

Lake sediment evidence indicates a decline in water quality from the time of forest clearance for agriculture 150 years ago (Kitto 2010). The lake is hypertrophic with high levels of nitrogen (TN average around 2000 mg/m^3) and phosphorus (TP average around 200 mg/m^3). However, phytoplankton growth is light limited (Chlorophyll a typically 60–90 mg/m^3) due to high suspended sediment levels from wind-driven re-suspension of bed sediments and sediment inputs from stream inflows and lakeshore erosion. The lake does not regularly undergo severe oxygen depletion or toxic algae blooms (Hayward 2009a).

The predominant source of nutrients is from tributary streams to the lake (N – 98%, P – 90%). However, these are groundwater-fed streams, so from a catchment perspective infiltration from land use to groundwater is a critical failure pathway. The current N load to the lake is estimated to be 2650 tN/year. The equilibrium load for the 2011 land use is estimated to be 4100 tN/year, accounting for the time lag in travel of groundwater to the lake (Selwyn Waihora Zone Committee 2013). Water quality monitoring shows an overall reduction in nutrient loadings to the lake. However, this is driven primarily by reduction in flows over the past 10 years rather than a reduction in instream nutrient concentrations.

The brackish nature of the lake (typically 4–6 ppt) results from sea water inflows during lake openings and wave overtopping the gravel bar separating the lake from the ocean. Reduced freshwater inflows to the lake from increased surface and ground water abstraction has had the perverse effect of *reducing* salinity in the lake because there are fewer openings to the sea. This can be the nature of feedback loops in nested adaptive systems.

11.2.2.6 Waituna Lagoon

Waituna Lagoon is a coastal lake with a gravel barrier separating it from the ocean (similar to Te Waihora/Lake Ellesmere). The lagoon and wetland complex covers approximately 36 km^2 (Thompson and Ryder 2003). The lake area is about 16 km^2 which reduces to 7.2 km^2 after lake opening. It has a maximum depth of about 3.3 m reducing to 1.6 m on opening (Hamill 2011). As noted by the Lagoon Technical

Group: "through land development of the catchment over the past century (e.g. clearance of wetlands, drainage enhancement and fertiliser inputs) and an opening regime managed for farm drainage, the lagoon is now experiencing a number of ecological problems. This includes a decline in abundance of *Ruppia* (seagrass) that is central to the lake's ecological functioning, increased abundance of nuisance filamentous algae and reduced oxygenation of bed nutrients" (Lagoon Technical Group 2011).

The lake is eutrophic (TLI mean 2005–9 is 4.4). Estimated nutrient inputs from surface water in the catchment have increased from 179 t/year in 1995 to 433 t/year in 2009 for TN, and from 9.7 t/year (1995) to 21 t/year (2009) for TP (Hamill 2011). Nutrients from groundwater have been estimated to be 28–48 t/year TN and 1.4–2.4 t/year TP (Rissman et al. 2012). Based on DIN:TP ratios Waituna Lagoon is probably P limited (Schallenberg et al. 2010). However surface water data from 2001–2010 suggests that both N and P could be limiting at different times (Lagoon Technical Group 2011).

Sedimentation rate monitoring shows elevated rates (2.5–3.0 mm/year) of fine sediment deposition in the localised areas since c1960 to the present (Lagoon Technical Group 2011). Water clarity as measured by Secchi disc of about 1.2 m is not sufficient for light to reach the bottom of the deeper parts of the lagoon (Hamill 2011). Sediment anoxia has become widespread throughout the lagoon since 2007 (Lagoon Technical Group 2011).

Waituna Lagoon is considered to have a high risk of 'flipping' from its high value clear-water seagrass (*Ruppia*) dominated state to a highly undesirable turbid algal dominated (phytoplankton/epiphyte) state due to excessive inputs of nitrogen, phosphorus and sediment (Lagoon Technical Group 2011). The reduction of macrophytes to below certain thresholds of biomass or percentage bottom cover facilitates wind-induced sediment resuspension which further reduces water clarity often initiating a collapse of macrophyte communities in shallow lakes (Schallenberg and Sorrell 2009).

Lake openings facilitate nutrient flushing. However, this also increases salinity which adversely affects *Ruppia* growth. Timing of lagoon openings needs to ensure closure prior to the main *Ruppia* growing and germination period (spring-summer) (Lagoon Technical Group 2011).

11.2.3 Overview of Critical Variables

Table 11.2 summarises key critical variables for the six lakes under consideration in relation to the main failure pathways with to water quality. The key feature of the table is that although all lakes are subject to nutrient enrichment, the critical variables for the water quality failure pathways are different.

In terms of the catchment, the type of nutrient varies (N, P or both) and the hydrology pathway (groundwater, surface water, groundwater-fed tributaries) also varies. This is related to both the nature of the catchment and the nature of the lake. Sediment can also be a critical variable.

Table 11.2 Critical variables on water quality failure pathways for the lakes (Jenkins 2016)

Lake	Catchment variables	Lake variables
Lake Brunner	P loading in surface runoff	DO in bottom waters
Lake Taupo	N loading in groundwater	Algal blooms in lake
Lake Rotorua	N and P loading from surface and ground water	Algal blooms in lake DO in bottom waters Remobilisation of N and P from sediments
Lake Omapere	N (baseflow) and P (peaks after rainfall) loading from surface water	Toxic algal blooms Algae macrophyte switching High N and P sediment load mobilised by wind DO levels for mussels
Te Waihora/Lake Ellesmere	N and P loading in groundwater-fed tributary streams Sediment from streams	Lake level and lake openings Wind resuspension of sediments High N, P and sediment loads Loss of macrophytes
Waituna Lagoon	N and P loading from surface water with groundwater contribution Sediment from streams	Risk of flipping from macrophytes to algae Lake openings Wind resuspension

In terms of the lakes, there are different trigger points for change of state or function including DO in bottom waters, algal blooms in upper waters, remobilisation of nutrients in sediments from deoxygenation (in deeper lakes) and from wind resuspension (in shallow lakes), lake openings and levels (in coastal lakes) and macrophyte-algae switching.

What are also different between the lakes from a nested adaptive system perspective are the different downstream effects of nutrient enrichment of the lakes. For Lake Brunner and Lake Taupo the impact is on the water quality of their outlet rivers, the Arnold and Waikato Rivers respectively. For Lake Rotorua, the nutrients discharge into Lake Rotoiti: the Ohau Diversion Channel has been built to direct the discharge to bypass Lake Rotoiti to the Kaituna River (the outlet of Lake Rotoiti). For Lake Omapere, the nutrients, algae and sediments have had devastating effects on the water quality, aquatic ecology and food sources of the Utakura River and Hokianga Harbour. For Te Waihora/Lake Ellesmere and Waituna Lagoon the lake outlet is to the sea creating a freshwater plume: this is important for diadromous fish and eel passage and may be the source of offshore algal blooms.

11.2.4 Current Management Interventions for Each Lake

11.2.4.1 Lake Brunner

In 2004, the West Coast Regional Council notified a Proposed Water Management Plan. The Plan identified the Lake Brunner catchment as a Special Management Area (West Coast Regional Council 2010) with the objective of improving water quality to reach an average water clarity of 5.3 m by 2020 based on achieving the water quality levels that existed in 2004.

Farm plans had been developed with farmers in the catchment in 2005. There was some fencing and bridging work undertaken but not all participated. Lake nutrients continued to increase (Horrox 2009).

Based on trend analysis of total phosphorus, NIWA research indicated that if the present land uses in the catchment (intensive dairy farming) continued to develop at the same rate using the same land use practices, the transition of the lake from oligotrophic to mesotrophic state would occur in 2040 (Verberg et al. 2013).

The Regional Plan contains rules tightening dairy effluent management requirements, agricultural land development, controls on access to riparian margins, controls on phosphorus applications (limited to 2005–10 rates), and phosphorus water solubility. There are also non-regulatory measures for farm environmental plans with involvement of the NZ Landcare Trust and Dairy NZ.

However, specific phosphorus yields in the catchment are high – 2.4 kgP/ha/year compared to 1 kgP/ha/year for typical farms (Rutherford et al. 2008). Also recent water quality monitoring shows an increase in hypolimnetic oxygen consumption (a critical variable for the lake) with the DO level in bottom waters continuing to decline and down to its lowest recorded level value of 5 mg/L (45% saturation) (West Coast Regional Council 2011).

A partnership was formed with the regional council, central government and Westland Milk products to address farm management. Also, the Lake Brunner Community Catchment Care Group was formed with support from the New Zealand Landcare Trust and the regional council to enhance public land around the lake. The regional council and Westland Milk Products funded the preparation of a second round of farm plans in the catchment. In 2013, $200,000 funding was provided from the Fresh Start for Freshwater Fund[3] to help with fencing and riparian planting along waterways by farmers and the community group (Stuart 2016).

In recent years (2001–2015), declining trends in oxygen depletion rates at 100 m were less pronounced. Nitrate levels continue to increase but the critical variable for Lake Brunner, DRP, showed no significant change. Chlorophyll *a* and clarity show improvements and a slight decline was apparent for the TLI in 2013 (TLI = 2.77) and 2014 (TLI = 2.8) but increased in 2015 (TLI = 3.1) (Land Air Water Aotearoa 2016). Tributary water quality levels show increases in turbidity and bacterial contamination as well as increases in nitrogen (West Coast Regional Council 2016).

[3] Refer Sect. 2.4.2

The increased level of intervention has reduced the rate of decline in lake water quality but the continuing decline in tributary water quality indicates that further remediation is needed to achieve the Plan objective of a TLI of 2.8 by 2020.

11.2.4.2 Lake Taupo

The key management intervention in the catchment to return Lake Taupo to 2001 water quality levels by 2080 is the establishment of nitrogen discharge allowances (NDAs) for the manageable nitrogen loads generated by land use intensification. A target has been set for a 20% reduction of current nitrogen leaching by 2018 (Waikato Regional Council 2011).

The Lake Taupo Protection Trust has the task of permanent removal of this 20% of the manageable nitrogen (186 tN/year of 930 tN/year). The Trust is to achieve this through land use change to lower nitrogen-leaching land uses. There is an $81.5 m public fund with contributions from central, regional and district government. The fund is managed by the Trust and can apply the fund to purchase nitrogen either by purchasing land or NDAs (Lake Taupo Protection Trust 2013).

The target has been achieved ahead of schedule but with 90% of the reduction through purchases by the Trust rather than farmers trading discharge allowances. The price in 2012 was around $NZ300 per kg (OECD 2015). The target has been achieved by taxpayer/ratepayer funds rather than polluter pays.

Because of the time lag of groundwater from past land uses, the goal of achieving improved water quality has been set for 2080. However, indications are that a 20% reduction in nitrogen leaching from current land use will be insufficient to reduce the catchment load to meet 2001 water quality levels. Scientific estimates for the exact percentage of the load to come range from 30% to 41% of the current manageable load with other estimates as high as 80% (Hadfield et al. 2007). These higher estimates mean greater reductions in nitrogen (and greater cost in purchasing land and NDAs) would be needed to meet the water quality target.

11.2.4.3 Lake Rotorua

For management interventions in Lake Rotorua, the Rotorua Lakes Strategy Group (Te Arawa Lakes Trust, Rotorua District Council and Bay of Plenty Regional Council) prepared a Lakes Rotorua and Rotoiti Action Plan with an Action Plan Working Group of community and stakeholder organisations (Environment Bay of Plenty et al. 2009). The Action Plan is based on limiting catchment nitrogen loads to 435 t/year and phosphorus loads to 37 t/year. Catchment inputs are currently 556 t/year for nitrogen and 39 t/year for phosphorus. With groundwater lag effects, it is estimated that the nitrogen load in equilibrium with current land use is 746 t/year. In addition, there are nutrients retained in lake bed sediments estimated to release 360 t/year of nitrogen and 36 t/year of phosphorus.

Management interventions in the Action Plan to achieve the reductions in nitrogen are heavily dependent on land management improvements in the catchment with reductions of 30 t/year through adoption of best management practices (by 2012), 56 t/year through adoption of new but known technology (by 2019), and, 84 t/year by innovative (i.e. yet to be developed) new technology (by 2029). Another major potential contributor to nitrogen reduction was the diversion of Hamurana Stream from the lake to the Ohau channel downstream of the lake with an estimated reduction of 53–92 t/year. Phosphorus reduction was focussed on lake bed sediment treatment (25 t/year) and flocculation of phosphorus in tributary streams (4 t/year) (Environment Bay of Plenty et al. 2009).

Current implementation programmes are looking to a 320 t/year reduction from the catchment with 270 t/year from changes in pastoral land use and 50 t/year from engineering solutions. This would require a 51% reduction in relation to current nitrogen losses from pastoral land use. Farm interests have estimated that the farm costs amounted to $88 m and would result in a loss of farm value of $35 m (Omundsen 2013).

The 2008 Regional Plan (Rule 11) put in place nutrient benchmarks on pastoral properties mainly based on 2004/5 land use. There was to be no net increase of the export of nitrogen and phosphorus, or the increase had to be offset on the property or within the same lake catchment. The Proposed Regional Policy Statement includes the cap of 435 t/year for nitrogen inputs to the lake. This cap was appealed by Federated Farmers in part because it did not address the economic consequences on land owners and imposed unachievable targets on the community. It also sought a collaborative approach to addressing water quality issues (Federated Farmers of New Zealand 2012).

A Funding Deed was agreed between central, regional and district government for lake water quality improvements in the region with $45.5 m allocated to Lake Rotorua. An implementation schedule based on the Action Plan has been developed with annual reports on progress. The most recent progress report indicated that success of flocculation of phosphorus in tributary streams was achieving greater phosphorus removal than predicted. It also noted that progress on targeted nutrient reduction from land use was behind schedule and land use negotiations were on hold until an integrated rules and incentive fund were developed. The report also indicated on improvement in TLI for Lake Rotorua (down to 4.06) – attributed to in-lake interventions and favourable climatic conditions. However nitrogen and phosphorus levels continued to increase and clarity decline (Bay of Plenty Regional Council et al. 2013).

An agreement has been reached between Federated Farmers, the Lake Rotorua Primary Producers' Collective and the regional council (the Oturoa Agreement) in February 2013 to meet the nitrogen cap by 2032 with 70% of the required reduction reached by 2022 (the original RPS deadline). Further negotiations through the Rotorua Lakes Stakeholder Advisory Group have led to an agreement to achieve the pastoral land use reductions for nitrogen of 270 t/year through a combination of rules (140 t/year), incentives (100 t/year) and gorse removal (30 t/year). This would require dairy farms to reduce nitrogen leaching to 35 kgN/ha/year and drystock farms to reduce nitrogen leaching to 13 kgN/ha/year. This compares with current estimated averages of 54 kgN/ha/year for dairy and 15.7 kgN/ha/year for drystock farming (Omundsen 2013).

While the proposed management interventions have uncertainties because they are dependent on new technologies, uncertain costs, modelled approaches, and, implementation of major land use and land management changes, they represent a significant programme to achieve sustainable water quality objectives.

11.2.4.4 Lake Omapere

Lake Omapere was vested to the Lake Omapere Trust in 1955. The Trust represents the Ngāi Puki-nui-toui, the local iwi for the region. Recent management interventions for the lake include the introduction of grass carp (2000 and 2002) to address oxygen weed that at times covered the entire lake (e.g. in 1999) and a $0.6 m restoration and management project (2003–2010) administered by the Lake Omapere Project Management Group. The project included the voluntary adoption of farm plans, fencing and riparian planting and a voluntary lake strategy (Northland Regional Council 2013).

Gray summarises the Lake Omapere water quality situation as follows. Lake water quality has improved since 2007 but this is most likely as a result of a natural phenomenon in the lake with the state switching between algal dominated and macrophyte dominated, rather than due to the project and restoration efforts. The available data suggests lake sediments still contain high nutrient levels, which provide an internal nutrient source through wind resuspending sediment into the water column. The data also suggests that external inputs into the lake have not improved. Nutrient levels in catchment streams and drains are still high. Mussel numbers are stable in the lake and as they can filter algae from the water column, they are likely to be one of the main reasons for the lake improving (Gray 2012).

11.2.4.5 Waituna Lagoon

For the management of Waituna Lagoon, the regional council has formed a Waituna Partners Group of government agencies, industry groups and community groups to manage the lagoon in a sustainable way (Environment Southland 2013). The Lagoon Technical Group of scientists has been formed to advise on water quality and lagoon processes. There is a Lake Waituna Control Association that manages lake openings and a Waituna Liaison Committee for stakeholder engagement.

There have been drainage enhancement and rock protection works undertaken to reduce sediment loads in tributaries. Lake openings are managed for local flooding and nutrient flushing. The dairy industry has prepared a Sustainable Milk Production Plan for farmers to identify on-farm actions to achieve key environmental outcomes and supported a dairy farmer initiative for a Waituna Lagoon and Catchment Action

Plan. There have been winter grazing trials to reduce overland flow and sediment losses (Environment Southland 2013).

Plan Change 13 to the Region Water Plan proposes to require consents for new dairy farms and the regional council is developing a regional strategy with community involvement called Water and Land 2020 in response to the National Policy Statement requirements.

However, as Scanes summarises, in the last 10 years the ecological condition of Waituna has been in rapid decline. It has changed from a high value seagrass (*Ruppia*) dominated state to a more degraded condition with nuisance epiphyte and algal blooms. Consequent sediment anoxia is causing additional stress to the keystone *Ruppia* species (Scanes 2012).

Current expert opinion is that unless urgent intervention occurs, the lagoon will almost certainly undergo a rapid change in state to an even more degraded phytoplankton-dominated system (e.g. with algal blooms) which would endanger the *Ruppia* community and change the fundamental values and character of the lagoon (Lagoon Technical Group 2011).

Scanes' analysis of catchment loads from similar lagoons in New South Wales suggests that the loads required to maintain a moderate environmental quality (some eutrophic symptoms but still supporting healthy seagrass and fish communities) would be a total nitrogen load of 9 tN/km^2/year and a total phosphorus load of 0.57 tP/km^2/year. This represents a 52% reduction in TN load and a 23% reduction in TP load over 2010 conditions (Scanes 2012).

11.2.4.6 Te Waihora/Lake Ellesmere

In relation to the management interventions for Te Waihora/Lake Ellesmere the most significant is the rehabilitation programme for the lake – Whakaora Te Waihora – which is a $12 m programme supported by central and regional government, Ngāi Tahu and Fonterra (Te Waihora Co-Governance Group 2011). There have also been changes in the lake opening regime to incorporate long fin eel passage as well as wader habitat and managing flooding of surrounding farmland.

Ngāi Tahu is the owner of the lake bed under the Ngāi Tahu Settlement Act. There is a Joint Management Plan with the Department of Conservation for the lake bed (Department of Conservation and Te Runanga o Ngāi Tahu 2005). There is also a governance agreement between Ngāi Tahu and the regional council. Local organisational arrangements include the Lake Settlers Association that are involved in lake opening decisions and the Waihora Ellesmere Trust representing community interests. With the Canterbury Water Management Strategy, there is the Region Committee developing a Regional Implementation Programme and the Selwyn Waihora Zone Committee developing a Zone Implementation Programme (ZIP).

In the catchment, the Canterbury Water Management Strategy provides the regional framework for addressing land use intensification and water quality issues while the Selwyn Waihora ZIP provides recommendations for the management of the lake catchment which is to receive statutory backing through the Regional Land and Water Plan.

Investigations for the ZIP have estimated the current nitrogen load from the catchment to the lake to be 2650 tN/year with a time lag for further groundwater input to achieve equilibrium with current (2011) land use of 4100 tN/year. With the Central Plains Project and other intensification that can occur in the catchment, the nitrogen load which is estimated to reach 5600 tN/year, while the intensification would add an estimated $300 m in regional GDP (Selwyn Waihora Zone Committee 2013).

The Zone Committee has proposed the adoption of a solutions package to achieve farm management improvements that are 12½% better than "good management practice". This is designed to achieve a reduction in catchment nitrogen load to 4800 tN/y. This improved practice is estimated to reduce regional GDP by $30 m. However aspirations to achieve a macrophyte lake would need to reduce the nitrogen catchment load to 800 tN/year as well as other measures (Selwyn Waihora Zone Committee 2013).

11.2.5 Adequacy of Management Interventions

Table 11.3 summarises the management interventions in relation to physical activities, regulatory activities and organisational arrangements for each of the lakes and their catchments.

The analysis shows that while management interventions are occurring to reduce nutrients loads in each catchment they are insufficient to achieve sustainable water quality outcomes for Lake Brunner, Lake Taupo, Lake Omapere, Waituna Lagoon and Te Waihora/Lake Ellesmere. For Lake Rotorua an ambitious nutrient target has been set and agreed with key stakeholders to achieve sustainable water quality but there are uncertainties with developing new technologies, costs, reliance on modelling and ability to implement. There is a need for increased interventions and activities to support interventions in order to achieve sustainable water quality in the lake systems.

What is also evident is that new approaches are emerging that could provide the basis to improve water quality. There are a variety of technical innovations such as farm management plans, improvements in land management, riparian plantings, constructed wetlands, lake bed treatments, scientific investigations, water quality models and improved monitoring. There are also indications of incorporating Māori resource stewardship concepts into management approaches, in particular, the restoration of *mauri* for Lake Omapere and *mahinga kai* for Te Waihora/Lake Ellesmere.

An important component of the technical innovations for management intervention decision making is the use of modelling. Much of the initial modelling related to the replication of the lake response to increased nutrients (i.e. failure pathways). However, there is an increasing use of modelling to predict the outcomes of management interventions (i.e. sustainability strategies). This includes the nutrient load reductions from changing land management practices, the transport of nutrient accumulations in groundwater, the tipping points for managing system resilience, and, the effects of rehabilitation programmes. In addition to biophysical modelling

Table 11.3 Summary table of current management interventions (Jenkins 2016)

Lake	Physical activities	Regulatory activities	Organisational arrangements
Brunner	Two rounds of farm plans	Revised Regional Plan P application limits	Government/industry partnership Community Group with NZ Landcare Trust help Dairy NZ farm plans
Appraisal: Actions have reduced the rate of decline but further remediation needed			
Taupo	188 tN of trades with trust	Nitrogen discharge allowances Cap and trade market in NDAs 20% target reduction in N	Lake Taupo Protection Trust ($81.5 m) "Partnerships of Innovation"
Appraisal: N load greater than 20% in groundwater Long term return to 2001 levels by 2080 needs further intervention			
Rotorua	Lake bed treatment P flocculation in streams *Best management practices*[a] *New technologies* *Land use change to low N activities*	Nutrient benchmarks Reduce N: 755–435 t/year Deed of funding agreement ($45.5 m) *Investigate nutrient trading*	Te Arawa Lakes Trust Rotorua Lakes Strategy Group Stakeholder Advisory Group Lakes WQ Society Chair in lake management Oturoa agreement
Appraisal: Significant programme underway to achieve catchment nutrient reduction and lake sediment source reduction: uncertainties with new technologies, costs, models and implementation			
Omapere	Carp introduction to address oxygen weed Voluntary farm plans Fencing & riparian planting	Voluntary lake strategy	Lake Omapere Trust Lake Omapere project management Group
Appraisal: Recent water quality improvements due to macrophyte/algae cycling Tributary nutrient levels still exceed lake nutrient levels Inadequate actions to achieve sustainability			
Waituna	Drainage works Sustainable milk production plans Winter grazing trial *Constructed wetlands*	*Region Plan Change: consents for dairy farms* *Water and Land 2020*	Waituna Partners Group Lagoon Technical Group Waituna Liaison Committee Lake Waituna Control Association

(continued)

Table 11.3 (continued)

Lake	Physical activities	Regulatory activities	Organisational arrangements
Appraisal:			
Load of 19 tN/km²/year to be reduced to 9			
Load of 0.74 tP/km²/year to be reduced to 0.57			
Lake openings to flush nutrients limited by salinity effects on seagrass			
Inadequate actions to achieve sustainability			
Te Waihora	Whakaora Te Waihora Change in lake opening regime	Canterbury water management strategy Zone implementation programme Regional land and water plan	Region Committee Zone Committee Ngāi Tahu/ECan agreement Lake Settlers Association Waihora Ellesmere Trust
Appraisal:			
Current N load 2650 t/year, 2011 land use equilibrium load 4100 t/year			
Central Plains and other intensification: 5600 t/year			
Proposed solution package: 4800 t/year ($30 m reduction in regional GDP)			
Inadequate action to achieve macrophyte lake of 800 t/year			

[a]Items in italics are interventions in progress

there has been the financial modelling of the implications of changing farm management practices and the modelling of change to regional economies.

There are also regulatory changes such as catchment nutrient limits, nutrient discharge allowances, and consent controls, as well as non-statutory strategies.

Of particular interest is the importance of collaborative organisational arrangements. This is evident in a variety of forms:

- Intergovernmental partnerships
- Iwi-government governance and management agreements
- Strategy groups among key stakeholders
- Stakeholder advisory groups
- Community engagement mechanisms
- Region and Zone implementation committees
- Technical advisory groups
- Funding Trusts and financial Deeds of Agreement.

A proactive approach to the institutional design of organisational arrangements was an intentional component of the Canterbury Water Management Strategy based on the work of Ostrom for the design of collaborative approaches to the management of 'common pool resources' like water. It is evident from the Lake Rotorua experience that collaborative approaches have also been a reactive approach to provide a creative alternative to the stalemate generated by the adversarial legal processes of the Resource Management Act.

The sustainability framework based on nested adaptive systems has provided an insightful basis for analysing the issue of nutrient enrichment for lakes in New Zealand. The concepts of failure pathways, critical variables, their thresholds, and the adequacy of management interventions provide a deeper understanding than the frameworks currently in use to address this issue. The analysis also highlights the importance of integrating institutional and economic design into the management of biophysical systems. This is considered further in the next section.

11.3 Management Interventions to Achieve Sustainability

11.3.1 An Intervention Framework

This pattern of emerging approaches for achieving sustainable management of water in New Zealand can be placed in the framework of a socio-economic adaptive cycle:

- The use of human and economic resources for stakeholder, cultural and community engagement to consider how to collaboratively address the issue of sustainability of our water resources as well as the investment of technical resources to understand the issues and financial resources to undertake actions (i.e. exploitation phase).
- The accumulation of knowledge, social, cultural and economic capital to develop integrated approaches to sustainable strategies (i.e. accumulation phase).
- The formulation of new approaches to water management that change existing practices (i.e. disturbance/release phase).
- The development of new institutional arrangements to implement the new approaches to water management (i.e. reorganisation phase).

This ***has the potential*** to lead to the adoption of management interventions to the biophysical system of water resources that achieve sustainability. Similarly, the failure to adopt appropriate management interventions ***will*** lead to ongoing degradation of our water resources as the six case studies demonstrate.

This cycle can be shown diagrammatically as a lissajous figure for the socio-economic system (Fig. 11.4).

Management interventions in the biophysical system of water resources can occur at each of the phases of the biophysical adaptive cycle:

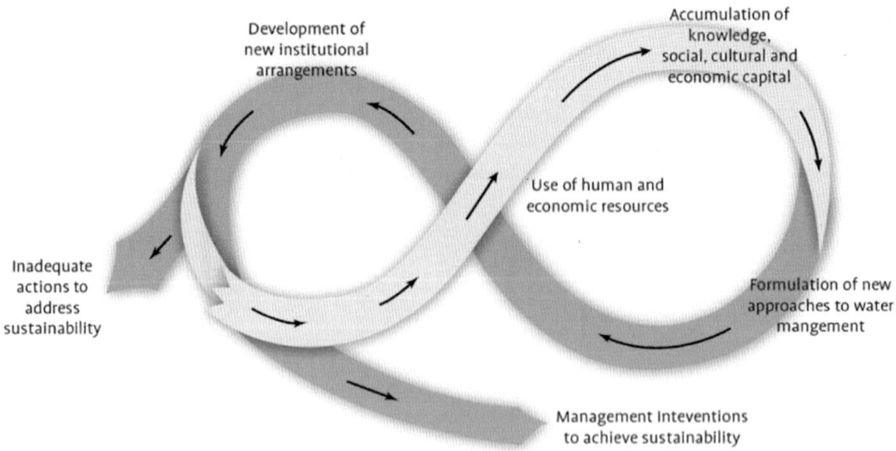

Fig. 11.4 Socio-economic system framework to develop management interventions to achieve sustainability

- Reducing the pressure on the resource ("reducing vulnerability" in Chapin's terms) in the exploitation phase: an example is the reduction in catchment nitrogen loads on a lake;
- Addressing the legacy issues of accumulated changes in the past ("enhancing adaptive capacity" in Chapin's terms) in the accumulation phase: an example is the lake bed treatment to reduce remobilisation of phosphorus;
- Increasing the resilience of the system to accommodate disturbance ("increasing resilience" in Chapin's terms) in the disturbance/release phase: an example is lake aeration to prevent stratification;
- Rehabilitating the adverse effects of the system ("enhance transformability" in Chapin's terms) in the reorganisation phase: an example is the reestablishment of macrophytes in a lake.

This can be shown diagrammatically as a lissajous figure for interactions in the different phases of the biophysical system (Fig. 11.5).

The socio-economic system (Fig. 11.4) and the biophysical system (Fig. 11.5) can be linked to show an overall framework for management intervention pathways to achieve sustainability (Fig. 11.6).

11.3.2 Overview Comments

The overall findings from the sustainability assessments of the six lakes were that:

- All require reductions in land use intensification in their catchments to achieve sustainable water quality;

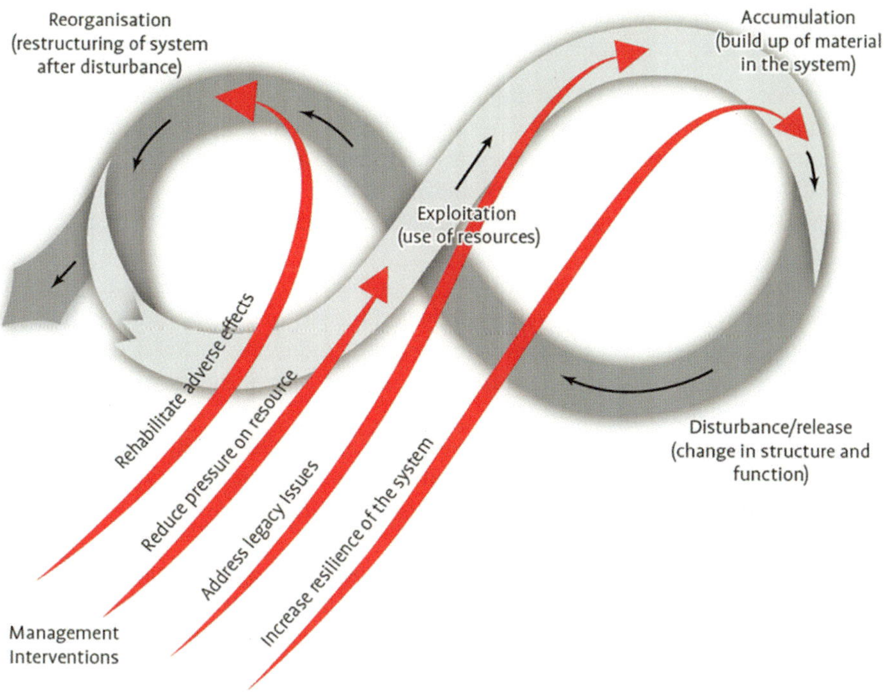

Fig. 11.5 Management interventions for each phase of the biophysical system

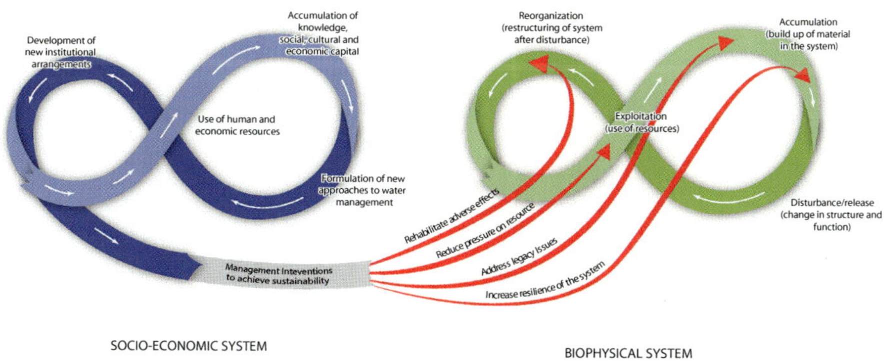

Fig. 11.6 Management intervention pathways to achieve sustainability (Jenkins 2016). Reprinted by permission of the publisher (Taylor & Francis)

- There is evidence of positive innovations in water management; and
- There is a need for greater management interventions.

The analysis also showed the importance of collaborative arrangements between government, community, industry, iwi and technical experts. This is consistent with the Local Government Act approach to sustainable development of government/community/industry partnerships to achieve community outcomes. It contrasts with the RMA emphasis on economic and social use of resources subject to acceptable effects.

The analysis highlights the limitations of New Zealand's current system of effects-based management. If we have reached the sustainability limits of resource use, then further land use intensification requires reduction in impacts in ***current*** land use. A system of impact assessment based on ***new*** projects will not achieve sustainable management.

The analysis also highlights the need to consider sustainability issues as nested socio-ecological systems. Maintaining our ***natural capital*** also requires building our ***social, cultural and economic capital*** to develop management interventions that are implemented to achieve sustainability.

However socio-economic drivers can also impede biophysical sustainability. Box 11.1 describes the decline of the Aral Sea. This is the result of the diversion of river inflow for irrigation to benefit 31 million people in the Aral Sea Basin. The shrinking of the Aral Sea has left a salinized remnant that has decimated the fishery, turned wetland ecosystems into desert ecology, and impaired the health and livelihoods of the population of 4 million people around the Aral Sea. Understanding the linkages between the socio-economic systems and the biophysical systems at three spatial scales is needed to understand why the degradation of the Aral Sea is likely to persist.

Box 11.1: The Decline of the Aral Sea

The Aral Sea is an internal drainage basin receiving water from the Amudarya and Syrdarya Rivers in Central Asia (Fig. 11.7). The middle reaches of the Amudarya and Syrdarya Rivers were identified by the Soviet Union for achieving self-sufficiency in cotton production. Commencing in the 1960s, major diversions were made to increase the irrigation of land for cotton growing. The area of irrigated land in the Aral Sea basin increased from 4,111,000 hectares in 1960, to 6,127,000 hectares in 1980 and to 7,403,000 hectares in 1990. From 893,000 metric tonnes in 1940, annual cotton production increased to 3,982,000 metric tonnes in 1966–70, 4,894,000 metric tonnes in 1971–75, and peaking at 5,666,000 metric tonnes in 1977 when all lands that could be allocated to cotton were in use. (Zonn 1999).

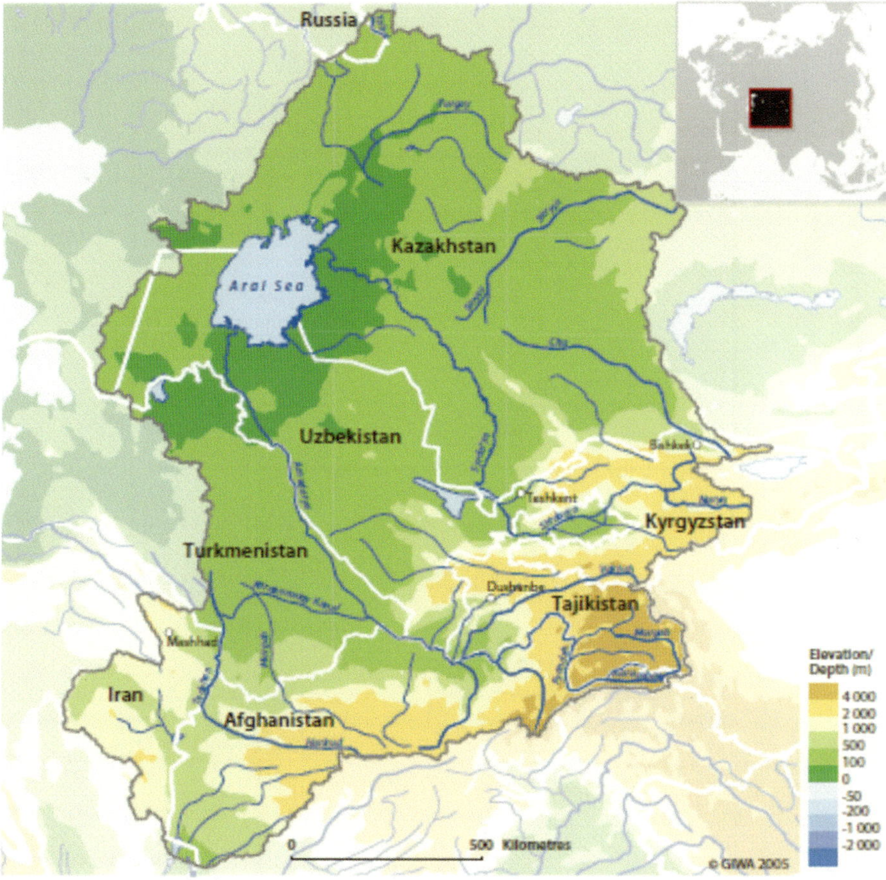

Fig. 11.7 Aral Sea basin (Severskiy et al. 2005). Reprinted with permission from UNEP

In terms of nested adaptive systems, at the geographical scale of the Aral Sea basin, there was "exploitation" of resources through the increased use of water to irrigate land for cotton leading to "accumulation" in the form of increased cotton production (Fig. 11.8).

This diversion of water from the Amudarya and Syrdarya Rivers has an impact on the flow of water entering the Aral Sea. From an annual average flow into the top of the Amudarya River of 47.0 km^3 during 1931–60, there was a 10.9 km^3 decrease in 1961–70, a 25.7 km^3 decrease during 1971–80, and 38.1 km^3 decrease for 1981–90. For the Syrdarya River with an annual average flow of 15.0 km^3 during 1931–60, there were decreases of 6.24 km^3 in 1961–70, 11.1 km^3 in 1971–80, and 12.9 km^3 in 1981–90 (Tsytsenko and Sumarokova 1999).

Past irrigation had been confined to deltaic and littoral zones. River flow reductions by irrigation diversions had been offset by reductions in natural evaporation, transpiration and infiltration to groundwater. In addition, net use was decreased by

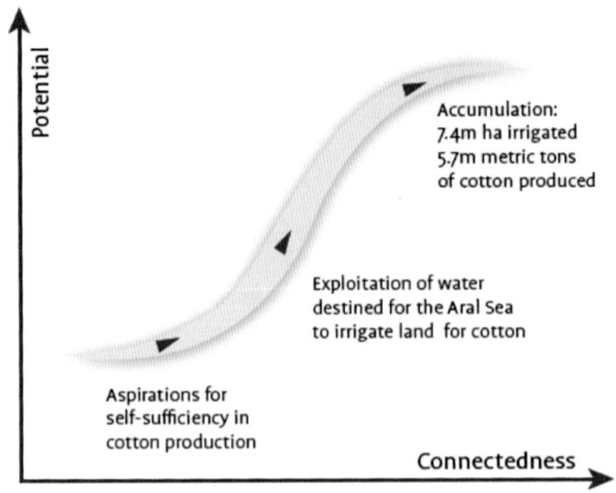

Fig. 11.8 Exploitation and accumulation phases of increased irrigation for cotton in the Aral Sea basin

Table 11.4 Aral Sea decline in inflow, volume, area and depth (Bortnik 1999)

		End of period			
Period	Average annual inflow to the sea (km^3)	Lake volume (km^3)	Lake area (km^2)	Lake level (m above sea level)	Average annual change in depth (cm)
1951–60	58.4	1083	67,100	53.4	+6.7
1961–70	43.3	951	60,200	51.2	−21.8
1971–80	16.7	628	50,800	45.4	−57.6
1981–89	4.2	329	36,500	38.6	−76.3
1994					
Small sea		20	3,000	39.5	
Large sea		256	30,100	36.9	

irrigation return flows. This level and pattern of irrigation retained a sustainable flow into the Aral Sea. However the irrigation expansion into desert areas exceeded these offsets and irrigation drainage ended in terminal lakes where it evaporated (Micklin 2010).

These significant inflow reductions lead to (negative) accumulation effects of reduced lake volume, lake area and lake depth (Table 11.4). Average annual lake inflow decreased from 58.4 km^3 in the decade 1951–60 to 4.2 km^3 in 1981–9. This resulted in a decline in lake volume from 1083 km^3 (1951–60) to 329 km^3 (1981–9), a decline in lake area from 67,100 km^2 (1951–60) to 36,500 km^2 (1981–9), and, a decline in lake depth of 14.8 m. Continuing lake level decline led to the separation into two lakes: the "small sea" associated with the Syrdarya flows, and the "large sea" associated with the Amudarya (Fig. 11.9).

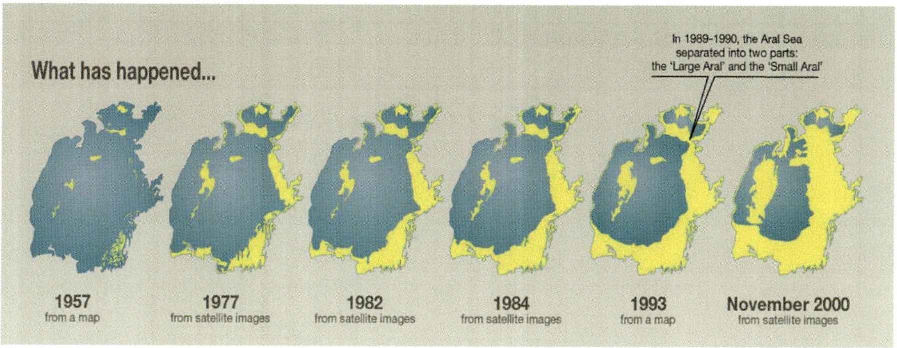

Fig. 11.9 Decline in Lake Area of the Aral Sea 1957–2000 (Rekacewicz 2005). Permission from GRID-Arendal http://www.grida.no/resources/5615

There was also an increase in contaminants entering the sea from highly mineralised irrigation return flows that also contained pesticides. Although the sea has no outlet, prior to 1960, the salt budget was relatively stable. From 1911 to 1960 the mean salinity only changed from 9.6 to 10.3‰. By 1970 it reached 11.6‰. By 1980 it increased to 17‰, and by the end of the 1980s it was 27‰. After the sea separated into two lakes the large sea continued to increase approaching 40‰ by 1995 while the small sea was in the range 24–27‰ (Bortnik 1999).

The declining lake levels led to decline in groundwater commencing at the margins of the lake such as the river deltas. In 1960–65 the depth to groundwater in the deltas fluctuated around 0.5 m. By 1980 depth to groundwater was more than 5 m, while by 1985 it was more than 10 m (Novikova 1999).

The connections between the changes at the scale of the Aral Sea Basin and the Aral Sea itself are shown in Fig. 11.10.

The changes in water quantity and quality in the Aral Sea had consequences for biophysical systems dependent of the lake. The decline in groundwater levels led to degradation of the water-dependent ecosystems. The change in salinity affected the fish populations in the lake. The drying of the former sea bed left the soil and dried salt exposed to wind erosion resulting in dust storms that were visible by cosmonauts in space.

The vulnerability of groundwater-dependent ecosystems to changes in groundwater level are shown in Table 11.5, indicating threshold values for critical levels for ecosystem maintenance. The adaptive cycle for this change is shown in Fig. 11.11. With the groundwater table below 10 m, there is a shift to new types of desert ecology. Initially there is a shift to halophytic vegetation on solonchak soils (alkaline, highly saline soils). With desalination of the soils there is then a development of a takyr process (cracked mud salt flat) and the introduction of desert species.

There are a series of thresholds in relation to the fish population associated with increased salinity. The first significant change was noted in the mid-1960s with increased salinity (11–14‰) at the fish spawning areas at the margins of the Aral Sea with drastic effects on the development of fish roe, larvae and fry. By 1971 when average salinity for the Aral Sea exceeded 12‰, there was a decline in adult

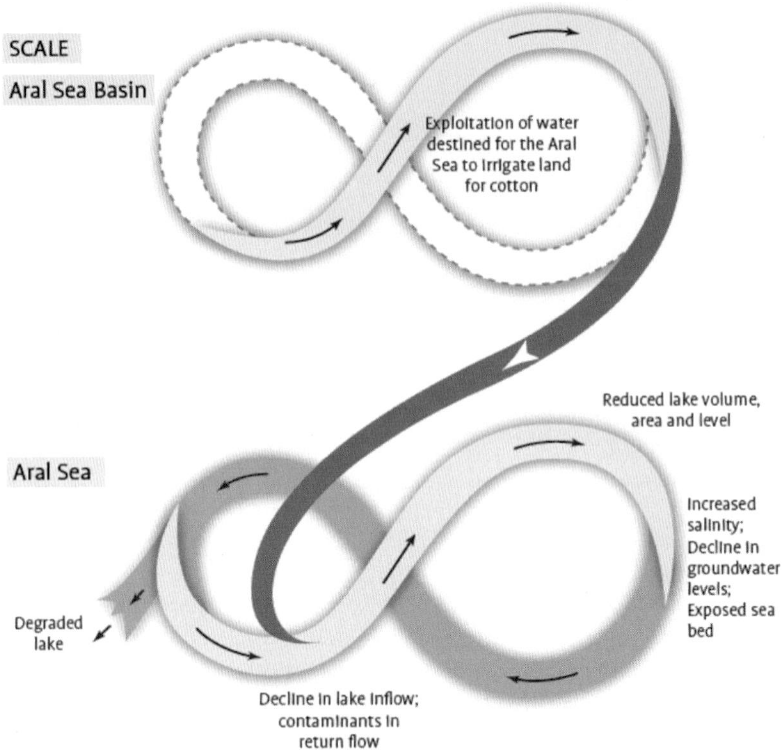

Fig. 11.10 Connection of the Aral Sea basin scale to the Aral Sea scale

Table 11.5 Threshold values for groundwater levels in delta ecosystems (Novikova 1999)

Ecosystem type	Critical groundwater level (metres below ground level)
Meadow-boggy ecosystems	0.5 ± 0.5
Meadow ecosystems	0.5–1.5
Shrub	0.5–5
Solonchak ecosystems	0–5
Wood ecosystems	0.5–10 (15)

freshwater fish. When salinity exceeded 14‰ in 1975 reproductive processes for freshwater species were affected. In the late 1980s average salinity was in the range 22–28‰ brackish species were lost. A further threshold could be reached in the Large Aral Sea when salinity reaches 50–55‰ when species of marine origin would not survive. It is interesting to note that attempts to introduce additional fish species since the 1920s have also had effects on the fish ecology. One side effect has been the presence of species with greater salinity tolerance than species native to the Aral Sea. Instead of the original 40 species, by the 1990s, there are now only five species, four of which had been introduced (Aladin 1999).

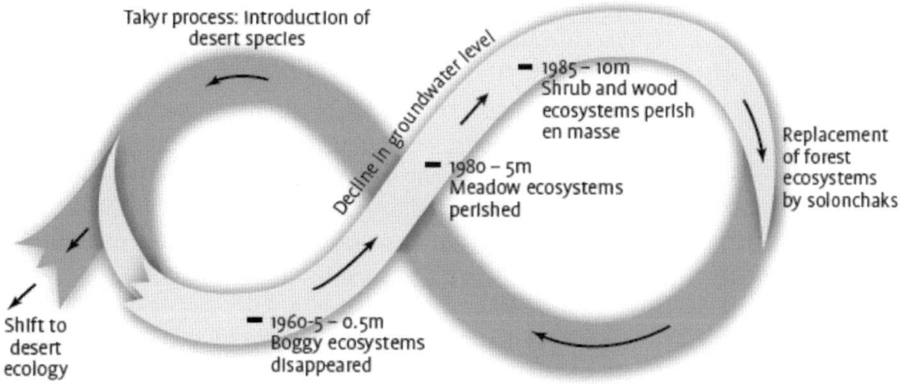

Fig. 11.11 Adaptive cycle for shift from delta ecosystems to desert ecosystems

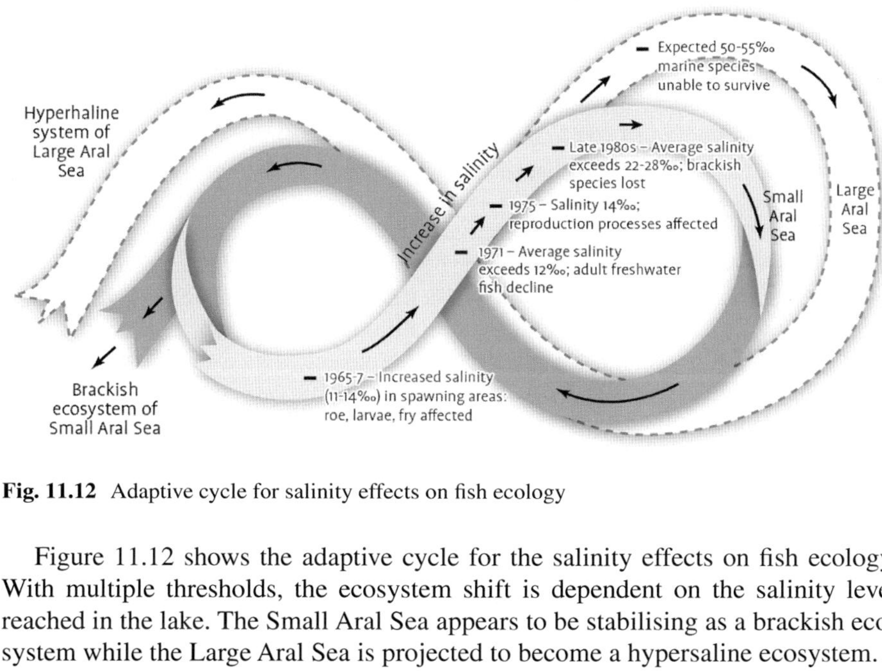

Fig. 11.12 Adaptive cycle for salinity effects on fish ecology

Figure 11.12 shows the adaptive cycle for the salinity effects on fish ecology. With multiple thresholds, the ecosystem shift is dependent on the salinity level reached in the lake. The Small Aral Sea appears to be stabilising as a brackish ecosystem while the Large Aral Sea is projected to become a hypersaline ecosystem.

The changes to the biophysical systems associated with the Aral Sea had implications for the socio-economic system associated with the Aral Sea. The change in fish ecology led to a significant decline in fish catch and abandonment of the fishing industry. There was also a major impact on human health. This led to migration from the Aral Sea.

Table 11.6 Decline in commercial fish catch from the Aral Sea (Letolle and Mainguet 1993)

Year	1960	1965	1970	1975	1980	1985	1990
Catch (metric tons)	43,430	31,040	17,460	2940	0	0	0

Table 11.6 sets out the sharp decline in commercial fish catch from 43,430 metric tons in 1960. The former sea ports of Muynak and Aralsk are now stranded because of the receding shoreline. For a few years fish were transported to Muynak for processing. The commercial fishing industry has collapsed with consequential unemployment and economic decline for Muynak and Aralsk (Glantz 1999).

There were significant health effects at the time of the decline of the Aral Sea. There were increases in water-borne diseases such as typhoid and viral hepatitis. There were coincident with increased mineralisation and bacterial pollution of surface water sources and of drinking water. There was also an increase in infectious non-enteric morbidity as well. Increased chemical pollution was associated with agricultural use of pesticides and fertilisers, and increased salt from wind-blown dust from the dried lake surface. Residual pesticides were measured in breast milk. In addition, there was a decline in nutrition in the communities around the Aral Sea. This has been attributed to the decline in fish in people's diets, a decline in food availability due to loss of meadows and desertification, and, the conversion of land use from food growing to cotton growing (Elpiner 1999).

According to World Bank background studies, about 4 million people living around the Aral Sea and delta areas have lost their livelihoods from the decline of the fishing industry and the encroachment of saline soils. However at the geographical scale of the Aral Sea Basin there have been benefits to about 31 million people, mainly in Uzbekistan and Turkmenistan from the increased irrigation (Barghouti 2006).

Figure 11.13 shows the links between the biophysical and socio-economic systems at their different spatial scales. On the biophysical side the increased water diversions at the Basin scale result in a decline lake inflow, reduced lake size and increased salinity at the scale of the Aral Sea. This leads to a decline in fish stocks in the fishery dependent upon the lake. The decline in the fishery links to the socio-economic system with a decline in fish catch. At the scale of the Aral Sea community this then leads to a collapse in the Aral Sea economy. At the Basin scale, there is a decline in the economy due to the decline in the Aral Sea fishing industry. However, this decline is more than offset by increase in the Basin economy from increased cotton production and other productivity gains from irrigated agriculture.

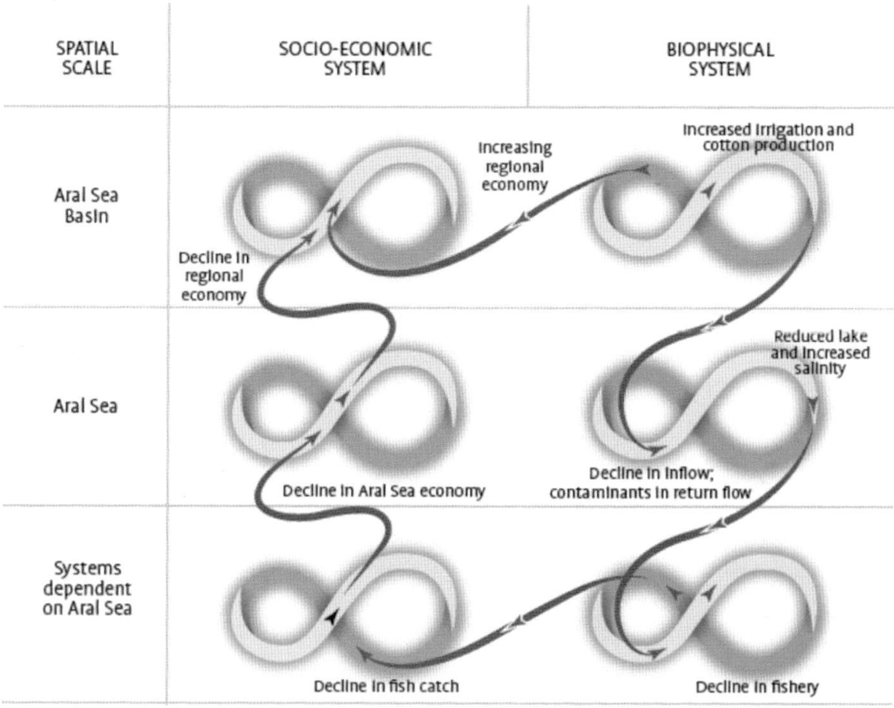

Fig. 11.13 Links of biophysical and socio-economic systems for Aral Sea at multiple spatial scales

References

Aladin, N. V. (1999). Creeping changes in biological communities in the Aral Sea. In M. H. Glantz (Ed.), *Creeping environmental problems and sustainable development in the Aral Sea basin*. Cambridge: Cambridge University Press.

Ausseil, O. (2010). *Hurunui River – Influence of the middle reaches on water quality of the Lower Hurunui River (2005–2008)*. Christchurch: Environment Canterbury.

Barghouti, S. (2006). *Case study of the Aral Sea water and environmental management project*. Washington, DC: The World Bank.

Bay of Plenty Regional Council. (2012a). *Improving water quality in Lake Rotorua: Information on the way land is used*. Bay of Plenty Regional Council Strategic Policy Publication 2012/03, Whakatane.

Bay of Plenty Regional Council. (2012b). Lake water quality. www.boprc.govt.nz/environment/water/rotorualakes/lakewaterquality. Accessed 8 Dec 2013.

Bay of Plenty Regional Council, Roturua District Council, Te Arawa Lakes Trust. (2013). Rotorua Te Arawa Lakes programme annual report 2012–2013.

Bortnik, V. N. (1999). Alteration of water level and salinity of the Aral Sea. In M. H. Glantz (Ed.), *Creeping environmental problems and sustainable development in the Aral Sea basin*. Cambridge: Cambridge University Press.

Department of Conservation, Te Runanga o Ngāi Tahu. (2005). *Te Waihora joint management plan: Mahere Tukutahi o Te Waihora*. Christchurch: Te Runanga o Ngāi Tahu and Department of Conservation.

Elpiner, L. I. (1999). Public health in the Aral Sea coastal region and the dynamics of changes in the ecological situation. In M. H. Glantz (Ed.), *Creeping Environmental Problems and Sustainable Development in the Aral Sea Basin*. Cambridge: Cambridge University Press.

Enfocus. (2011). *Nutrient management in the Hurunui: A case study in identifying options and opportunities*. Christchurch: Environment Canterbury.

Environment Bay of Plenty, Roturua District Council, Te Arawa Lakes Trust. (2009). *Lakes Rotorua and Rotoiti Action Plan*. Whakatane: Environment Bay of Plenty.

Environment Southland. (2013). Waituna Lagoon. http://www.es.govt.nz/environment/land/wetlands/waituna-lagoon/. Accessed 29 Nov 2013.

Federated Farmers of New Zealand. (2012). Notice of appeal to Environment Court against the decision on the Proposed Regional Policy Statement – Coast Environment, Water Quality and Land Use.

Glantz, M. H. (1999). Sustainable development and creeping environmental problems in the Aral Sea region. In M. H. Glantz (Ed.), *Creeping environmental problems and sustainable development in the Aral Sea basin*. Cambridge: Cambridge University Press.

Gray, T. (2012). *Review of the Lake Omapere Restoration and Management Project*. Whangarei: Northland Regional Council.

Hadfield, J., Sylvester, W., Vant, W., & White, P. (2007). *Statement of agreed matters between technical experts for the Appellants and Respondents*. Before the Environment Court in the matter of Appeals on Proposed Variation 5 to the Waikato Region Plan – Lake Taupo Catchment, Hamilton.

Hamill, K. (2011). *Southland Water 2010: Our ecosystems technical report for lakes and lagoons*. Invercargill: Environment Southland.

Hamilton, D. (2011, April 7–8). *Fix the lakes: Where we are what can be done*? Paper presented at the Rotorua Lakes symposium 2011, Rotorua.

Hayward, S. (2009a). Water quality in the Ellesmere catchment. In K. Hughey & K. Taylor (Eds.), *Te Waihora/Lake Ellesmere: State of the lake and future management*. Christchurch: EOS Ecology.

Hayward, S. (2009b). *Water quality issues of the Amuri Basin streams and the Hurunui River: Summary – March 2009*. Christchurch: Environment Canterbury.

Hewitt, K. (1997). *Regions of risk: A geographical introduction to disasters*. Harlow: Longman.

Horrox, J. (2009). West Coast Regional Council – Water monitoring in Lake Brunner. In West Coast Regional Council (Ed.), *Lake Brunner project update*.

Jenkins, B. R. (2015). *Sustainability analysis of five lakes in the Greater Wellington Region*. Waterways Centre for Freshwater Management Report 2015–005, Christchurch.

Jenkins, B. R. (2016). Sustainability analysis of the management approach for six New Zealand lakes. *Lake and Reservoir Management, 32*(2), 101–115. https://doi.org/10.1080/10402381.2015.1135491. http://www.tandfonline.com.

Kitto, S. (2010). *The environmental history of Te Waihora – Lake Ellesmere*. MSc thesis, University of Canterbury, Christchurch.

Lagoon Technical Group. (2011). *Interim recommedations to reduce the risk of Waituna Lagoon flipping to an algal-dominated state*. Invercargill: Environment Southland.

Lake Taupo Protection Trust. (2013). *The Trust*. http://www.laketaupoprotectiontrust.org.nz/page/lake_19.php. Accessed 28 Nov 2013.

Land Air Water Aotearoa. (2016). *Lake Brunner*. https://www.lawa.org.nz/explore-data/west-coast-region/lakes/lake-brunner/

Letolle, R., & Mainguet, M. (1993). *Aral*. Paris: Springer.

Lilburne, L., Elliott, S., Bidwell, V., Shankar, U., Kelly, D., & Hanson, C. (2011). *Hurunui catchment-scale land use and water quality modelling report*. Christchurch: Environment Canterbury.

Micklin, P. (2010). The past, present, and future Aral Sea. *Lakes and Reservoirs: Research and Management, 15*, 193–213.

Northland Regional Council. (2007). *Lake Omapere water quality report 2006*. Whangarei: Northland Regional Council.

Northland Regional Council. (2013). *Lake Omapere Restoration*. http://www.nrc.govt.nz/Your-Council/Council-Projects/Lake-Omapere-Restoration-Project/. Accessed 29 Nov 2013.

Norton, N., & Kelly, D. (2010). *Current nutrient loads and options for nutrient load limits for a case study catchment: Hurunui catchment*. Christchurch: Environment Canterbury.

Novikova, N. M. (1999). Priaralye ecosystems and creeping environmental changes in the Aral Sea. In M. H. Glantz (Ed.), *Creeping environmental problems and sustainable development in the Aral Sea Basin*. Cambridge: Cambridge University Press.

OECD. (2015). *The Lake Taupo nitrogen market in New Zealand: Lessons in environmental policy reform*. Paris: OECD.

Omundsen, S. (2013). *Framework for allocation and incentives in the Lake Rotorua catchment*. Report to Bay of Plenty Regional Council, Strategy, Policy and Planning Committee, Whakatane.

Rekacewicz, P. (2005). *Aral Sea: Trends and scenarios*. UNEP/GRID-Arendal. https://www.grida.no/resources/5793

Rissman, C., Wilson, K., & Hughes, B. (2012). *Waituna catchment groundwater resource*. Invercargill: Environment Southland.

Rutherford, K., Chague-Goff, C., & McKerchar, A. (2008). *Nutrient load estimates for Lake Brunner*. NIWA Client Report HAM2008-060, Hamilton.

Scanes, P. (2012). *Nutrient loads to protect environmental values in Waituna Lagoon, Southland NZ*. Invercargill: Environment Southland.

Schallenberg, M., & Sorrell, B. (2009). Regime shifts between clear and turbid waters in New Zealand lakes: Environmental correlates and implications for management. *New Zealand Journal of Marine and Freshwater Research, 43*, 701–712.

Schallenberg, M., Larned, S., Hayward, S., & Arbuckle, C. (2010). Contrasting effects of managed opening regimes on water quality in two intermittently closed and open coastal lakes. *Estuarine, Coastal and Shelf Science, 86*(4), 587–597.

Selwyn Waihora Zone Committee. (2013). *Addendum to Selwyn Waihora Zone Implementation Programme (21 July version)*. Christchurch: Environment Canterbury.

Severskiy, I., Chervanyou, I., Ponomarenko, Y., Novikova, N. M., Miagkov, S. V., Rautalahti, E., & Daler, D. (2005). *Global International Water Assessment: Regional assessment 24 Aral Sea*. University of Kalmar on behalf of United Nations Environment Programme.

Stuart, B. (2016). *Review of the Lake Brunner Project 2015*. Hamilton: NZ Landcare Trust.

Taylor, K. J. W. (1996). *The natural resources of Lake Ellesmere (Te Waihora) and its catchment*. Christchurch: Canterbury Regional Council.

Te Waihora Co-Governance Group. (2011). *Whakaora Te Waihora: Joint cultural and ecological restoration plan*. Christchurch: Environment Canterbury.

Thompson, R., & Ryder, G. (2003). *Waituna Lagoon: Summary of existing knowledge and identification of knowledge gaps*. Science for Conservation 215. Department of Conservation, Wellington.

Tsytsenko, K. V., & Sumarokova, V. V. (1999). Change of the rivers' flow in the Aral Sea Basin (in connection with the problem of quantitative assessment and consideration of environmental after-effects). In M. H. Glantz (Ed.), *Creeping environmental problems and sustainable development in the Aral Sea Basin*. Cambridge: Cambridge University Press.

Vant, B., & Smith, P. (2004). *Nutrient concentrations and water ages in 11 streams flowing into Lake Taupo*. Environment Waikato Technical Report 2002/18R, Hamilton.

Verberg, P., Horrox, J., Chaney, E., Rutherford, J., Quinn, J., Wilcock, R., & Howard-Williams, C. (2013). Effects of nutrient loading on the trophic state of Lake Brunner. *Marine and Freshwater Research, 46*, 436–446.

Verburg, P. (2009). *Effects of nutrient loading in Lake Brunner*. NIWA Client Report HAM2009-137, Hamilton.

Waikato Regional Council. (2011). *Waikato Regional Plan Variation No. 5 – Lake Taupo Catchment*. Hamilton: Waikato Regional Council.

West Coast Regional Council. (2010). *Proposed Regional Land and Water Plan*. Greymouth: West Coast Regional Council.

West Coast Regional Council. (2011). *Lake Brunner water quality: December 2011*. Greymouth: West Coast Regional Council.

West Coast Regional Council. (2016). *Lake Brunner water quality update: December 2015*. http://www.wcrc.govt.nz/Documents/Environmental%20Management/Council%20Monitoring%20Reports/Lake%20Brunner%20WQ%20update%2023%20Dec%202015%20.pdf

Zonn, I. S. (1999). The impact of political ideology on creeping environmental changes in the Aral Sea Basin. In M. H. Glantz (Ed.), *Creeping environmental problems and sustainable development in the Aral Sea Basin*. Cambridge: Cambridge University Press.

Chapter 12
Decision Making for Sustainability

Abstract The traditional approaches of planner-led technical decision making and process-led legal decision-making had not been effective in resolving water management issues in Canterbury. Technical processes did not address the diverse range of stakeholder concerns while legal processes led to unresolved conflicts. Collaborative governance approaches were adopted which relied on community engagement to develop the Canterbury Water Management strategy (CWMS).

Strategic Choice was adopted as a decision-making methodology to deal with multi-criteria, multi-stakeholder decision-making characterized by urgency and uncertainty within a collaborative approach. Rather than relying on the analysis of environmental effects for assessing options, a sustainability appraisal framework was adopted for evaluating economic, environmental, social and cultural criteria for both outcome achievement and negative impacts.

When resource availability is constrained and cumulative effects of development exceed environmental quality limits, there are significant shortcomings in current effects-based approaches to water resource decision making. Further resource extraction or water use for any new development, even with minor adverse effects will exacerbate an already unsatisfactory situation. There is a need for outcome based approaches to keep overall resource extraction and impacts of use within sustainability limits.

Keywords Community engagement processes • Strategic choice • Sustainability appraisal

12.1 Multi-stakeholder Decision Processes

12.1.1 Types of Decision Making Processes

The two dominant paradigms for decision-making in relation to public planning decisions are planner-led technical decision-making and process-led legal decision-making.

An example of the planner-led technical decision-making is set out in the American Planning Association's *Planning Theory for Practitioners* (Brooks 2002).

B.R. Jenkins, *Water Management in New Zealand's Canterbury Region*, Global Issues in Water Policy 19, https://doi.org/10.1007/978-94-024-1213-0_12

In this model the planner defines the problem, then considers alternatives and analyses them, with choices made for 'best-fit' considering feedback from the client group, and then the planner designs and implements courses of action (McAllister 1982).

An example of the process-led legal decision-making is Schedule 1 of the *Resource Management Act 1991* (NZ) (*RMA*) in New Zealand.[1] This defines a process of plan notification, calling for submissions, allowing for further submissions on the submissions received, establishing a hearing panel to hear submitters, decision-making by the hearing panel, and appeals to the Environment Court and further appeals to higher courts.

Another process-led legal approach comes from the UK Government *Planning Portal* website which sets out the process of how and when planning applications are decided and the options available once a decision is made (Department for Communities and Local Government and UK Government 2014). The process includes checking alignment with local planning instruments, consulting on the proposal, developing recommendations to address issues raised, reporting to decision makers, decision notification and grounds of appeal.

Public planning decisions involve both planner-led technical aspects and process-led legal aspects with one usually more dominant than the other in different circumstances. This leads to either technical planners or the legal authority having the dominant role in decision-making.

However, for many planning decisions which involve multiple stakeholders with different values, the planner is unlikely to be the most appropriate person to identify all of the issues to be addressed, the most appropriate way of addressing those issues, how the different interests can be reconciled, and, the appropriate weighting of multiple decision criteria. It is the stakeholders themselves that are better qualified to perform these tasks (Jenkins 2013a). Thus planner-led technical decision-making is unlikely to be appropriate.

Furthermore, in terms of process, planning authority processes to assess planning proposals typically involve consultation with affected parties, but the planning authority (or planning tribunal or court) is the arbiter between proponents and those affected (both supporters and opponents). However, these formal legal processes are unlikely to achieve reconciliation between proponents and affected parties. Indeed there is a tendency for these processes to be adversarial (Lomax et al. 2010). Thus process-led legal decision-making is unlikely to be appropriate.

Alternative approaches to designing decision-making processes which directly involve the range of stakeholders affected by planning decisions are being proposed. These include collective choice arrangements described by Ostrom (1990) and deliberative democracy described by Dryzek (2010) as discussed in Sect. 8.1.

[1] Refer Sect. 2.2 and in particular Table 2.1.

However, there are few examples of how to design and incorporate such decision-making approaches into public planning processes.[2]

Collaborative governance approaches involve multi-stakeholder decision processes. There is not a "one-size-fits-all" formula; rather there is a framework of matters to be considered. This includes the context involving the process design, the linkage to decision making, the identification of issues, stakeholder identification, facilitation back-up and funding. It includes framing of the process in terms of group composition, goals and agenda. Inputs comprise stakeholder preparations, agreed rules and procedures, mechanisms to address power gaps between participants, and, capacity building of stakeholders. Dialogue during the process needs established communication channels, facilitation and rapporteuring leading to decision making and closure of issues. Outputs need to be documented and facilitate the implementation of action plans; the ongoing stakeholder processes need to have an impact on official decision making and continue to relate to non-participating stakeholders and the general public. Throughout the process there needs to be "metacommunication", i.e. time for reflection, reassessment and feedback (Hemmati 2002).

One example that is consistent with this framework was the Regional Council's Living Streams Programme (Jenkins 2009a). The programme was designed to improve land management practices that have an impact on water quality and to maintain the health of waterways. The programme comprised three stages of (1) investigation: collecting information about the catchment and stream health in order to identify the issues that need to be addressed in a catchment report; (2) involvement: the results of the investigations are presented to landowners and community groups to review the catchment report, consider options and develop an action plan to meet community goals for the catchment; and (3) implementation: council officers working alongside landowners to encourage the voluntary undertaking of improvements, to monitor water quality changes, to inform the community of progress, and to reassess the actions required. Funding support was available through an Environmental Enhancement Fund (Environment Canterbury 2009b).

Table 12.1 sets out how the design of the Living Streams Programme relates to the framework of the multi-stakeholder decision processes. Sect. 8.4.5 described the outcome of the Pahau catchment from the Living Streams Programme. There was a significant reduction in pollutant load from the voluntary actions leading to improved water quality.

---

[2] One example is CALFED – the Californian water planning and management process (Booher and Innes 2010) – refer Sect. 8.1.5 and Box 8.1. A second example is the NeWater Project where seven international river basins were selected as case studies: see Jaroslav Mysiak et al. (eds), *The Adaptive Water Resource Management Handbook* (Earthscan, 2010).

Table 12.1 Living Streams programme as a multi-stakeholder decision process

Multi-Stakeholder Decision Process Framework[a]	Living Streams Programme[b]
Context	
Process design	Three stages: Investigation, involvement and implementation
Linkage to decision making	Voluntary actions by landowners on their land
Identification of issues	Through catchment report preparation
Stakeholder identification	Land owners and community groups in catchments with degraded water quality
Facilitation backup	Resource care officers at regional council
Funding	Landowner contribution with support from environment enhancement fund
Framing	
Group composition	All landowners and communities in catchment
Goals	Improve land use practice affecting water quality
Agenda	Set by catchment report
Inputs	
Stakeholder preparations	Involve landowners and community groups in action plan development
Agreed rules and procedures	Approach defined in living streams handbook
Power gaps	All participants treated equally
Capacity building	Extension role of resource care officers
Dialogue	
Communication channels	Community meetings
Facilitation	Facilitation by resource care officers
Rapporteuring	Documentation of agreements in action plan
Decision making	Voluntary agreements with landowners
Closure	Monitoring of water quality improvements
Outputs	
Documentation	Catchment report and action plan
Action plan implementation	Agreed actions undertaken
Ongoing stakeholder processes	Community meetings and on-site facilitation
Impact on official decision making	Programme part of regional council activities
Throughout the process	
Meta-communication	Monitoring of effects of implementation and reassessment of actions required
Non-participating interests	Reports public, open invitation to participate
Relating to general public	Reports public, meetings open to public

[a]Adapted from Hemmati (2002)
[b]Extracted from Jenkins (2009a)

12.1.2 Decision Making for Canterbury Water Management

12.1.2.1 Sustainability Limits Reached

As set out in Sect. 3.1.3, there has been extremely rapid growth in water allocation and use in Canterbury in the last 20 years. This is predominantly associated with the expansion of dairying and the demand for irrigation water to improve pasture growth and thereby dairy production. New Zealand has the highest growth rate in irrigation of any OECD country[3] with most of that growth occurring in Canterbury which has 70% of New Zealand's irrigated land.

This has led to the sustainability limits of water availability in Canterbury being reached for the current means of abstraction for both surface water and groundwater.[4] It has also led to cumulative effects from land and water use which have compromised water quality and freshwater ecosystems (Jenkins 2009b).

12.1.2.2 Strategic Investigations

The first stage of strategic investigations into water management was commenced after droughts in the late 1990s indicated issues with water availability based on an analysis of future demand and supply (refer Sect. 3.2.2) (Morgan et al. 2002). This planner-led technical investigation indicated that, under low flow conditions in rivers from which irrigation water is taken, current peak demand cannot be met by current abstraction methods. Most irrigation abstraction was based on run-of-river schemes relying on direct withdrawals from rivers. On an annual basis, however, water is available to meet future demand but would require major water storages for water to be available in the irrigation season. This finding led to a second stage of strategic investigations to identify sites for major water storage options for the region with respect to their hydrologic feasibility (refer Sect. 3.2.3) (Dark et al. 2008). This was also a planner-led technical investigation by consultants.

A third stage was designed as the evaluation phase of the potential storage options by a 20-person multi-stakeholder group (refer Sect. 3.2.4). This group was supported by subregional groups (for north, mid, and south Canterbury) who provided input to the evaluation. The multi-stakeholder group used a sustainability framework for comparing storage options. However, the multi-stakeholder group also expressed concerns about broader water management issues, and the multi-stakeholder group

[3] OECD indicates a 90 per cent increase in irrigated area in NZ from 1990–92 to 2001–03 compared to the OECD average of 6% (OECD 2008).

[4] Irrigation was primarily by run-of-river withdrawals (i.e., direct takes from rivers) and groundwater pumping from unconfined aquifers. For run-of-river withdrawals the sustainability limit is reached when available flow in the river above environmental flow requirements cannot meet the total water allocations granted to water users. For unconfined groundwater extraction, the sustainability limit is reached when the volume extracted by all users exceeds the allocation limit based on aquifer recharge for the groundwater basin.

recommended that before strategic water storage and water management decisions are made, rigorous scientific and public consideration is required of:

- the impacts of land use intensification and its effects on water quality;
- mitigation and management systems for water quality; and
- methods for maintaining or improving flow variability and low flows in major rivers (Whitehouse et al. 2008).

This led to a fourth stage – the development of the Canterbury Water Management Strategy (CWMS) (refer Sect. 3.2.5). It is the community engagement process as part of the CWMS that is the subject of this chapter.

12.1.2.3 Statutory Processes

While the strategic investigations (which were planner-led technical studies) were underway, a number of statutory processes (that is, process-led legal processes) were also in progress. One was the development of a Natural Resources Regional Plan for Canterbury (Environment Canterbury 2011). Regional plans are statutory instruments prepared by regional councils under the Resource Management Act (RMA) that state the objectives for the region for natural resources, the policies to implement the objectives, and the rules to implement the policies. Also in progress were also applications for Water Conservation Orders (WCOs) by environmental interests.[5] WCOs provide for the preservation of water bodies in their natural state, and the protection of environmental values, and of characteristics of 'outstanding significance' of water bodies according to Māori values (*tikanga Māori*). There were also a series of controversial irrigation and hydro generation projects initiated by development interests that were progressing through resource consent processes.[6]

All of these statutory processes were based on procedures specified in the RMA: Schedule 1 for the preparation of a regional plan, part 6 for resource consents and part 9 for WCOs. Community involvement in these processes was as affected parties or submitters (typically in opposition). The legalistic nature of the processes created an adversarial decision-making system. Information was in the form of evidence to hearing processes in an interrogative process and legal judgement formed the basis of decisions, usually in the Environment Court. Processes were protracted and acrimonious (Stevens 2003; TVOne News (online) 2009; Fernando and Werellagame 2008).

---

[5] One for the Rangitata River and later one for the Hurunui River.

[6] Major projects included Project Aqua, Central Plains Water and the Hurunui Water Project.

12.1.2.4 Paradigm Shift

The planner-led technical process with its limited focus on major storage to address the single issue of water availability had been found by the stakeholder response to be insufficient to address the broader water management issues, such as water quality and environmental flows. The process-led legal processes had been unable to effectively resolve the conflicting perspectives for water development in the region.

A different paradigm was needed (refer Sect. 3.2.1). Work on collaborative approaches such as the Living Streams Programme had been initiated in 2004 by the Canterbury Regional Council at the scale of tributary catchments with considerable success at resolving water management issues at this scale (Gunningham 2008). It was clear that the Canterbury water management issues needed to be considered at multiple scales. For Canterbury, there were at least four spatial scales:

- the regional level where the key issues are water availability and land use intensification;
- the catchment level at which the sustainability levels of water use, the cumulative impacts of water use, and, the reliability of supply for irrigation are the main issues;
- the subcatchment level, where environmental flow requirements in river reaches and the management of stream water quality and riparian margins are the main issues; and
- the property level, where the land use practices that influence water quantity and quality are defined.

An overall strategic framework was formulated based on Ostrom's self-managed community approach to governing common pool resources[7] and Gunderson and Holling's concept of nested adaptive systems for managing natural resources.[8] One of the key elements of Ostrom's design principles for managing common pool resources, such as water, is the 'collective choice arrangements'. The community engagement process was based on Ostrom's concept of collective choice arrangements that was undertaken to facilitate collective decision-making for the Canterbury Water Management Strategy (CWMS), that is, the fourth stage of the strategic process).

---

[7] Ostrom (1990) considers institutional designs for common pool resources such as water. She compares different governance models: government direction, privatisation and self-managed communities. Her research indicates self-managed communities as the most enduring form of institutional design for common pool resources and identifies institutional design principles for self-managed communities. Refer Sect. 8.1.4.

[8] Gunderson and Holling (Gunderson and Holling 2002) provide an operational framework for defining sustainability on the basis of an 'adaptive cycle' of exploitation of resources, accumulation of material, disturbances to the system and its potential to reorganise after disturbance. Adaptive cycles are 'nested'—operate at different spatial and time scales—but are linked. Refer Sect. 4.2.1.

12.1.3 Community Engagement Process for the Canterbury Water Management Strategy

The CWMS involved the following major activities:

- stakeholder and community engagement on the development of strategic options,
- definition of the strategic options for regional water management,
- community consultation on their option preferences,
- strategic investigations of outcomes, and
- sustainability appraisal of options.

From Stage 2 of the strategic investigations, the strategy process had been reporting to the Canterbury Mayoral Forum (which comprised all the mayors of the district and city councils, the chair of the regional council and their chief executives). A Steering Group under the auspices of the Mayoral Forum provided oversight of the process which was managed by the Regional Council. For Stage 4, the membership and role of the Steering Group was expanded. It comprised representatives of local and regional government, tangata whenua[9] as well as farming, environmental, industry and recreational interests. It was empowered to make recommendations to the Mayoral Forum.

12.1.3.1 Stakeholder and Community Engagement Process

The community engagement process in Stage 4 had activities based on seven milestones:

1. Release and announcement of the process
2. Definition of the process
3. Identification of the breadth and uses and benefits
4. Public reporting of uses and benefits
5. Achievement of depth and sophistication of strategies and substrategies
6. Public engagement on strategy options
7. Implementation and update.

Milestone 1 (Release and Announcement) involved the release of the Stage 3 report and the commencement of the CWMS (Stage 4). A clear separation was sought by the Steering Group for the shift from the technical work that had been focused on storage to a broader concept of water resource management in order to highlight the change in emphasis.

Milestone 2 (Defining the Process) had the objective of defining an effective and credible process for the development of a strategy for integrated water management in contrast to the more limited scope of water availability and storage which had

[9] *Tangata whenua* means 'people of the land' in Māori and refers to people with a traditional connection to the land.

been the focus of the earlier work. Key activities undertaken to achieve the milestone were:

- face-to-face briefings and discussions with stakeholders,
- email exchanges with more remote stakeholders, and
- approval of the Steering Group of the process.

One of the key principles of Ostrom is for community to be able to influence the design of the collective decision-making process.

Milestone 3 (Identifying Breadth of Uses and Benefits) had the objective of ensuring a broad identification of uses and benefits of all stakeholders for incorporation in a water management strategy. This was an extensive activity involving 11 stakeholder engagement meetings throughout the region and seven meetings on specific topics. It used a collaborative governance software package Open Strategies as a framework for engagement and as a web-based recording system of stakeholder views on the uses and benefits of water in the Canterbury region (Driver 2014).

Milestone 4 (Public Reporting of Uses and Benefits) involved documenting the outputs of the stakeholder engagement process, and making the outputs publicly available and open to public scrutiny. A web site (Canterburywater.org.nz) was established for all reports associated with the development of the water management strategy. Also, Open Strategies had the facility for stakeholder groups to review how their views had been recorded and to modify the wording of how their views had been expressed.

Milestone 5 (Achieving Depth and Sophistication of Strategies and Substrategies) had the objective of focusing stakeholders on defining and prioritising sub-strategies and projects to provide for the uses and benefits identified at Milestone 3 (or as modified from Milestone 4). The original intention was to use the facility in Open Strategies which links Projects and their Results to achieving Uses and Benefits (PRUB). However, it was found that there were thousands of linkages to be considered and the available technical data had limitations in quantifying the PRUB linkages. For community engagement a less detailed and higher level approach was needed which could cope with greater information uncertainty. The Strategic Choice framework was chosen to identify strategic options for regional water management (Friend and Hickling 2005) (Canterbury Water 2008).

Milestone 6 (Public Engagement on Strategy Options) had the objective of presenting to the public the strategic options for regional water management that had emerged from the stakeholder engagement process. This was designed to ensure that the public was fully aware that this was an opportunity to have significant input into the water management strategy.

Milestone 7 (Implementation and Update) had the objective of incorporating the Canterbury Water Management Strategy into statutory documents, creating methods for updating the strategy, and developing projects to implement the strategy.

The stakeholder and public engagement tasks – Milestones 3–7 – are discussed in more detail below.

12.1.4 Identification of Uses and Benefits

Based on the work on defining the process from Milestone 2 discussions, there was support for stakeholder group meetings at locations across the region. Eleven locations were identified.[10] There was also interest in addressing specific topics: economics (relating to energy and tourism); *tangata whenua* (*Ngāi Tahu* and *rūnanga*)[11]; youth; water quality and drinking water; and land use intensification.

Facilitated workshops were conducted by the regional council using the web-based collaborative governance tool *Open Strategies* to document the input from the variety of workshops in a coherent framework. The concept of *Open Strategies* is to enable multi-stakeholder groups to define multiple projects that can contribute to the range of benefits sought by the multiple interests. The *Open Strategies* framework links projects to the results achieved by those projects; the results are linked to uses of these results to members of the community; and the uses are linked to benefits to the community.

Milestone 3 was to define the breadth of uses and benefits that the stakeholders sought from water management in Canterbury. The purpose of the workshops was for stakeholders to identify their uses and benefits. Community input to this process was extensive. Summaries of the output of the workshops were displayed on the *Canterbury Water* website.

The workshops also led communities to identify values associated with water that were at a higher level than benefits from water use by different stakeholders.

The process of identifying uses and benefits using the facilitated workshops was pivotal as the starting point in defining for the CWMS a vision statement, a definition of priorities and principles to underpin the strategy, and, ten target areas for the strategy that projects and actions which form the strategy are designed to achieve.

12.1.5 Public Consultation and Reporting on Principles and Uses and Benefits

From the outcomes of the stakeholder workshops there were more than 4500 stakeholder comments. Using thematic analysis *Open Strategies* identified 11 themes based on stakeholder values that could underpin a water management strategy for Canterbury. It also summarised the range of uses and benefits identified in the stakeholder workshops. This was received by the Steering Group and public feedback was sought on this information (Canterbury Water 2008).

---

[10] From north to south these locations were Hurunui, Rangiora, Central Christchurch, Akaroa, Darfield, Rakaia-Methven, Ashburton-Hinds, Timaru, Fairlie, Waimate, and Omarama.

[11] *Ngāi Tahu* is the Māori tribe whose *rohe* (tribal territory) includes the Canterbury Region. *Rūnanga* are Māori groupings centred on the *whānau* (family) and *hapū* (sub-tribe) of *marae* (tribal meeting place) based communities.

From the Steering Group review of the public feedback the 11 themes were modified into ten fundamental principles: sustainability; *kaitiakitanga*[12]; instream values; region-wide (in terms of input and statutory adoption); non-abstractive uses (for example, food gathering and swimming); efficient and effective water and land management; drinking water; maintenance of essential character of waterways; public access to waterways; and stock exclusion from waterways. There was also a wide-ranging specification of uses and benefits under general categories of economic, environmental, cultural and social.

The public input confirmed the list of uses and benefits and suggested some changes to the fundamental principles.

12.1.6 Achievement of Depth and Sophistication of Strategies

12.1.6.1 Strategic Choice

An important component of strategy formulation is selecting a framework designed for the type of decision situation. *Open Strategies* required information on the links between projects, results, uses and benefits. While some of this information had been generated, there wasn't sufficient information to make effective use of the *Open Strategies* framework for the development of alternative regional water management strategies. The development of strategic options was facilitated by the use of *Strategic Choice* (Friend and Hickling 2005). This approach for option development and selection arose from experience of decision-making in environments where inter-organisational collaboration was essential to successful service delivery (Midgley 2000). As described further in Sect. 12.2 below *Strategic Choice* is more a method of problem structuring rather than problem solving. It is designed for finding solutions to complex problems where there is incomplete information, many interconnecting issues, uncertainty about possible effects of options, and multiple interests with conflicting objectives.

Rather than the steps in the planner-led technical decision process of 'define problem/develop alternatives/evaluate alternatives/make decision', *Strategic Choice* considers multiple problems are to be addressed and comprises four modes of 'shaping, designing, comparing and choosing' in order to deliver a 'commitment package' including multiple decision outputs. The commitment package includes early actions, explorations in response to uncertainty, and arrangements for deferred decisions.

A workshop of the Steering Group and technical support group addressed the shaping and design modes. This led to four strategic options for regional water management which were subject to sustainability appraisal workshop (comparing mode) and a choosing mode involving community consultation, public hearings and stake-

[12] *Kaitiakitanga* means the exercise of guardianship by the *tangata whenua* of an area in accordance with *tikanga Māori* (Māori customs) in relation to natural and physical resources.

holder engagement. The Strategic Framework document (Canterbury Water 2009) set out the commitment package.

The shaping mode involves defining *key decision areas* which were derived by the uses and benefits from Milestone 4. It also involves identifying *links between decision areas* as well as *other decision areas* that could be affected by courses of action to achieve uses and benefits. In addition, this mode includes defining priority decision areas in terms of urgency and importance (referred to as *problem focus*). A key input for these tasks was the Canterbury Regional Environment Report (Environment Canterbury 2008) which had analysed the resources, processes and outcomes with respect to water management in Canterbury.

The designing mode involves identifying possible options for each decision area and then the incompatibilities between options in order to develop a working short-list of possible strategic options. There had been two 'camps' prior to this stage in strategy development. One favoured further storage as the priority. The other opposed storage and wanted the adverse effects of water use and land use intensification addressed before any further storage development was contemplated. The workshop identified another option of improved water use efficiency which would make additional water available and reduce the water quality contamination from excess runoff or groundwater leakage but would involve reconfiguration of existing consents.

12.1.6.2 Strategic Options

From the *Strategic Choice* workshop, four strategic options for regional water management were agreed by the Steering Group for public consultation:

- Option A: Business as usual (base case) – current RMA approach that was effects-based and applicant-driven;
- Option B: Advance environmental protection first then infrastructure development – set limits, initiate restoration and improve efficiency;
- Option C: Reconfigure consents and infrastructure to improve reliability and enhance the environment – redistribution for integrated water management; and
- Option D: Advance infrastructure development with environmental repair and protection – storage incorporating environmental mitigation.

12.1.7 Public Engagement on Strategy Options

12.1.7.1 Public Consultation on Options

All households in Canterbury (about 150,000) had delivered to them a booklet describing the fundamental principles and the four strategic options for regional water management (Environment Canterbury 2009a). Over 1000 submissions were received and more than 100 were heard at public hearings conducted by the Steering Group members. The booklet also included a request for feedback on the preferred option. From the responses it was clear that there was little support for Option A (Business as usual). Option D (Storage-led strategy) and Option B (Environment-led strategy) were the most favoured. However, Option C (Efficiency-led strategy) received considerable first preference support and was the dominant second preference (refer Sect. 3.2.5).

12.1.7.2 Sustainability Appraisal

The four options were subject to a sustainability appraisal by the Steering Group and an Officials Group (technical advisors) using the framework developed by Sadler et al. (2008) to reflect New Zealand institutional arrangements. The framework is founded on four pillars of sustainability (social, economic, environmental and cultural) which correspond to the four well beings of the Local Government Act.

The appraisal process involved an intensive month-long period of identifying sets of social, economic, cultural and environmental capital assets that are involved in regional water management and selecting assessment criteria to reflect these assets. In an application workshop involving community representatives and technical specialists over 2 days (Russell and Ward 2010), participants reviewed evaluation criteria and scale descriptions for the four groups of capital assets on a 5 point scale (from −2 strong negative impact to +2 strong positive impact with the neutral position 0 representing the status quo). Once the evaluation criteria had been amended, each group was asked to identify points on the five-point scale that represented an acceptable minimum position (quadruple bottom line) and a desirable objective position (quadruple top line).

The four options were then scored against the amended evaluation criteria. Some of the key findings of this appraisal were as follows:

- Option A (business as usual) was below the sustainability bottom line on nearly all criteria;
- Option B (environment-led) scored well on environmental criteria but is below the bottom line on economic criteria;
- Option C (efficiency-led) scored above the bottom line on nearly all criteria; and
- Option D (storage-led) scored well on economic criteria but is below the bottom line on environmental criteria.

When considered at the sub-regional level, the workshop participants considered that combinations of options B, C and D were most likely to achieve sustainability at the sub-regional level. The sustainability appraisal approach and its application to the CWMS are described in Sect. 12.3 below.

12.1.8 Strategic Framework and Implementation Programme

The *Canterbury Water Management Strategy: Strategic Framework* document (Canterbury Water 2009) prepared under guidance from the Steering Group was released by the Canterbury Mayoral Forum in November 2009. The document provided a vision and principles for the CWMS. The vision statement of what success would look like for the desired outcome of the CWMS is '[t]o enable present and future generations to gain the greatest social, economic, recreational and cultural benefits from our water resources within an environmentally sustainable framework.'

First order priorities for water management were identified in the Strategic Framework document as: 'environment, customary use, community supplies and stock water'. Second order priorities were: 'irrigation, renewable electricity generation, recreation and amenity'.

Primary principles for defining the basis for water management were identified as: 'sustainable management, regional approach, and tangata whenua'. Supporting principles were: 'natural character, indigenous biodiversity, access, quality drinking water, recreational opportunities, and community and commercial use' (Canterbury Water 2009).

It also summarised the key challenges facing the Canterbury region, and the outcomes of the CWMS process with respect to regional water management options and their sustainability assessment. CWMS has been designed to deliver on a set of targets in the following areas:

- drinking water
- irrigated land area
- energy security and efficiency
- ecosystem health/biodiversity
- water use efficiency
- *Kaitiakitanga*
- regional and national economic growth
- natural character of braided rivers
- recreational and amenity opportunities
- environmental limits.

The *CWMS: Strategic Framework* document also provided the approach for developing the implementation programme for the strategy and the issues to be covered by those programmes. It continues the nested approach to collaborative governance with its multi-stakeholder Regional Water Management Committee[13] to address regional issues, and 10 Zone Committees of community members and

[13] 'The Regional Committee is a committee of the Canterbury Regional Council and includes representatives from Environment Canterbury, the territorial authorities, rūnanga and Ngāi Tahu, as well as one member from each of the zone committees' (Environment Canterbury 2015e).

rūnanga representatives[14] to facilitate community-driven implementation programmes to meet the CWMS targets. A Water Executive unit, as part of the Canterbury Regional Council, was established to facilitate the delivery of the implementation programmes. In addition, the *CWMS: Strategic Framework* document indicated how these programmes would be given statutory backing through regional policy statements and regional plans.

12.2 Strategic Choice

12.2.1 Decision Characteristics for Strategy Development

In the development of the CWMS an important consideration was the appropriate methodology for strategy development. There were particular characteristics of the situation which led to the selection of "Strategic Choice" (Friend and Hickling 2005) as the strategic methodology.

One characteristic was that there was not one criterion for decision making, rather there were multiple criteria that needed to be achieved. This meant that strategy development was not suited to Operational Research style optimisation.

A second characteristic was that there were multiple stakeholders with polarised views of what should happen. This meant that multi-criteria analysis would result in winners and losers depending upon the weight given to particular criteria. There was uncertainty about the guiding values for making decisions.

The third characteristic was that despite considerable investigation work having been undertaken, there was incomplete information available for predictive modelling for all issues. The ability for a rational, comprehensive decision-making process using quantitative systems analysis was constrained by uncertainty in information.

A fourth characteristic was the urgency for decision making. Consent applications for taking more water were increasing in number as the realisation that water availability had become the key constraint to improving agricultural productivity. Agriculture is the major export earner for New Zealand.

A fifth characteristic was the need for collaboration among organisations and among different interests. This was needed at government level where regional councils had the prime responsibility for water management while district and city councils had the prime responsibility for land management. Collaboration was also needed between commercial and community interests. Making water available for

[14] Zone committees comprise: (1) One member appointed by the Regional Council who is an elected member; (2) One member appointed by each Territorial Authority operating within the Zone Boundary who is also an elected member; (3) One member from each Rūnanga that falls within the Zone, or two representatives where only one Rūnanga falls within the Zone; (4) Between 4 and 7 community members from a range of backgrounds and interests (Canterbury Water undated).

irrigation meant there was less water available for instream uses like recreation, and there were increased environmental effects from the increased takes and use of water. There was uncertainty because decisions about making water available were connected to other decision problems.

As noted above (Sect. 12.1.6), Strategic Choice is a method of problem structuring (as distinct from problem solving) designed to facilitate strategy development for situations like water management with these characteristics (Midgley 2000).

The concept of Strategic Choice is described below. This is followed by a description of the "shaping" and "designing" modes of the Strategic Choice process which led to the development of four strategic options for water management in Canterbury. The "comparing" and "choosing" modes of the process are then described. This led to the preparation of a strategic framework document (Canterbury Water 2009) in the form of a "commitment package". This is the final output of the Strategic Choice process which includes proposals for immediate actions, steps to deal with the remaining areas of uncertainty, and processes about how deferred choices should be made. The commitment package and the progress of its implementation complete this section.

12.2.2 Concept of Strategic Choice[15]

Strategic Choice was designed to address complex problem situations. Simple problems can be solved by *designing* possible courses of actions and then *comparing* them in terms of their ability to solve the problem. For complex problems there are multiple problem inputs. In *Strategic Choice* before embarking on the *designing mode* of decision making, it is appropriate to undertake a *shaping mode* where the connections between the multiple problems to be addressed are considered. With multiple, connected problems the *comparing mode* is unlikely to find an agreed course of action that best suits all problems. In *Strategic Choice,* a *choosing mode* is added that identifies proposed commitments to action progressively through time.

In the *shaping mode,* the first fundamental element is the identification of *decision areas.* These are problem situations where people see an opportunity to choose between different courses of action.

The next fundamental element is to identify the connections between the different *decision areas.* These are referred to as *decision links* which is the relationship between two *decision areas.* Linked *decision areas* need to be considered jointly rather than separately.

---

[15]The material for this section is drawn from Planning under Pressure: the Strategic Choice Approach (Friend and Hickling 2005). Terms with specific meanings in Strategic Choice are shown in italics. Where alternative language was used in the process the Strategic Choice equivalent is also indicated.

The third fundamental element is the mapping of *decision links* across multiple *decision areas*. For the Canterbury exercise these linkages were identified in a series of tables.[16]

The final fundamental element of the *shaping mode* is focusing on selected clusters of *decision areas*. A cluster of *decision areas* is referred to as a *problem focus*. They are priority decision areas in terms of urgency and importance.

Moving into the *designing mode* the first fundamental element is the development of *decision options*. The term *decision option* refers to a course of action to address a problem in a *decision area*.

The next fundamental element is testing the compatibility of options in interconnected *decision areas* (i.e. where there are *decision links*). For the Canterbury exercise the compatibility was identified through a series of tables.[17] From the compatibility analysis, combinations of compatible options from different decision areas can be identified.[18]

The final fundamental element of the *designing mode* is the creation of *decision schemes*. A *decision scheme* is a combination of options from multiple *decision areas* that are compatible.

In the *comparing mode* for complex problems there are multiple criteria for assessing the merits of different *decision schemes*. However, there is not always a defined scale of measurement, so the more general concept of *comparison area* is used in *Strategic Choice* as the first fundamental element of the *comparing mode*.

Once a set of *comparison areas* has been selected, it can be used for a *relative assessment* of *decision schemes* under consideration. *Relative assessment* is the second fundamental element of the *comparing mode*.

Based on the *relative assessment* it is possible to judge the comparative advantage between alternatives. *Advantage comparison* is the third fundamental element of the *comparing mode*.

The final fundamental element of the *comparing mode* is the development of a *working shortlist* of the few *decision schemes* considered feasible after the *advantage comparison*.

The *choosing mode* focuses on the actions to be taken. The first fundamental element is defining *uncertainty areas*. These can be uncertainties about guiding values, uncertainties needing further investigation, or uncertainties about coordination with related decisions.

The second fundamental element in the *choosing mode* is determining whether the level of uncertainty is acceptable for making decisions or whether an exploratory action is warranted to resolve uncertainty. Any course of action designed to alter the current state of doubt is called an *exploratory option*.

---

[16] *Strategic Choice* (Friend and Hickling 2005) use a *decision graph* for the mapping of *decision links*.

[17] *Strategic Choice* (Friend and Hickling 2005) uses an *option bar* – a matrix that displays two options from different *decision areas* indicating compatibility or incompatibility.

[18] *Strategic Choice* (Friend and Hickling 2005) uses an *option graph* to graphically display compatibility between *decision options* from different *decision areas*.

Consideration of timing of decisions is the third fundamental element of the *choosing mode.* Commitment to a specific option is referred to as an *action scheme.* Commitments to courses of action may also be deferred to later in the process.

The final output of the *Strategic Choice* process (and the fourth fundamental element of the *choosing mode*) is the concept of the *commitment package.* This is conceived as an assemblage of immediate actions (*action schemes*), *exploratory options*, deferred choices, and a set of understandings about the way in which any deferred choices should be addressed.

Figure 12.1 sets out the sequence of fundamental elements for the four decision modes. While the solid arrows indicate the general sequence of decision making steps, the dashed arrows indicate that the process is unlikely to be linear with the potential for feedback throughout the process and the potential for multiple cycles. The concept of *Strategic Choice* is reflecting the need with complex problems to learn to work with cyclicity rather than linearity, subjectivity rather than objectivity, uncertainty rather than certainty, and selectivity rather than comprehensiveness.

12.2.3 Shaping and Designing Modes

12.2.3.1 Preparation for the Strategic Choice Workshop

The two-day workshop of members of the stakeholder Steering Group supported by the Officials Group to provide technical assistance was the pivotal exercise in defining strategic options for future water management in Canterbury.

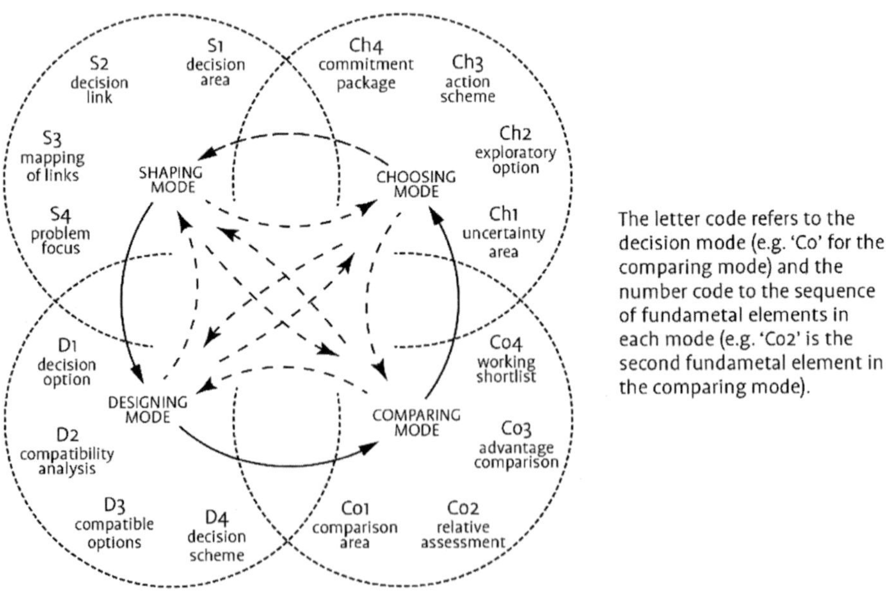

Fig. 12.1 The strategic choice process (Adapted from Friend and Hickling 2005)

Prior to the workshop the Officials Group had compiled a package of material for the Steering Group members including summaries and key documents in 15 areas.[19] For the workshop the major *decision areas* (i.e. the problem situations where people see the opportunity to choose between different courses of action) were suggested to be:

- Environmental flows
- Water quality
- Land use
- Allocation
- Demand management
- Infrastructure
- Biodiversity.

In addition, the topics of funding and governance were seen as *decision areas* of a different nature.

For each of the *decision areas* the Steering Group was provided with a topic description, the fundamental principles relevant to the topic, the commonalities across all options, draft options for addressing the topic, potential measurement criteria, and, uncertainties and risks. For example, the topic description for infrastructure was: the scale and type of new or re-engineered infrastructure for the storage and distribution of water, including concepts such as piped rather than channel distribution of irrigation water. Relevant fundamental principles for infrastructure were that: non-abstractive and abstractive uses of equal importance; efficient and effective water and land management; access to high quality drinking water; and, integration of water quality and quantity.

The draft options for "infrastructure" are set out in Table 12.2. They represent improving water availability at different scales: on-farm, catchment, alpine river storage, and, multiple catchment solutions involving inter-catchment transfers.

There were a number of elements that were considered common to all infrastructure options. One was that the development and redevelopment of infrastructure is integrated with water allocation, water use and the management of effects of use. A second was that the stretch target[20] is the "optimisation" of the size, configuration, cost and impacts. The third was that all options are integrated with the provision of environmental flows and the management of biodiversity, land use and the effects on water quality and on society.

Potential measurement criteria for infrastructure options comprised: availability and reliability of water for abstractive uses; regional economy impacts; infrastructure costs; viability and vibrancy of rural communities associated with water avail-

[19] These areas were: (1) tangata whenua (indigenous people) values, (2) environmental information on water quality and water quantity, (3) economic well-being, (4) social and cultural well-being, (5) water quality, (6) biodiversity, (7) physical nature/form of water bodies, (8) availability and reliability of water for extractive use, (9) non-extractive uses – recreation, tourism and cultural uses, (10) environmental flows and water levels, (11) control/regulation of land use, (12) demand management, efficiency, redistribution and property-level infrastructure, (13) infrastructure (storage options), (14) allocation systems, (15) governance.

[20] Stretch targets define a high and difficult level of success as an ambitious long-term goal.

Table 12.2 Options for infrastructure

Option	Description
1. Focus on on-farm or property scale infrastructure	Property-scale storage and distribution that enables better use of an individual's allocation. This would involve leaving individuals to develop and fund their own infrastructure development around existing community infrastructure.
2. Focus on catchment or local scale infrastructure	Catchment or local scale infrastructure that enables better use of a catchment and individuals' allocations. This would involve improving the capability, utilisation and operation of current infrastructure as a focus, and facilitating the development of new infrastructure to "fill the gaps".
3. Focus on new storage of alpine river water	This would focus on the provision of new large-scale storage as the primary solution to water supply needs.
4. Develop a multiple catchment solution	This would focus on developing a "plumbing system" for the provision of all of Canterbury's water needs. It would be integrated with demand management and allocation.

ability; recreational opportunities; opportunities to exercise kaitiakitanga.[21] Uncertainties and risks for infrastructure were seen as the economic implications of various supply reliabilities for irrigation; and, incorporating future needs.

This material was provided to the Steering Group as background to the workshop.

12.2.3.2 Strategic Choice Workshop

For the two-day workshop, the first day focused on information provision, while the second day focused on drafting alternative strategies. The formal workshop commenced with an explanation of the *Strategic Choice* concept and the format of the workshop. The contents of the background material package were also explained.

The initial workshop discussion was on what the *decision areas* were, the definition of scenarios (*decision schemes*), *decision options*, and, what the final strategy might look like (*commitment package*). *Decision links* (i.e. interconnections between decision areas) were also identified, for example:

- water availability connected to land use intensification
- land use intensification connected to water quality impairment
- extraction for water availability linked to adverse effects on biodiversity, recreation, natural character and kaitiakitanga.

The definition of *decision links* led to the development of *decision options*. The example relating to infrastructure is set out in Table 12.3 for the options identified in Table 12.2 plus the null option of no storage.

[21] Kaitiakitanga means the exercise of guardianship by the tāngata whenua (indigenous people) of an area in accordance with tikanga Māori (Māori customs) in relation to natural and physical resources; and includes the ethic of stewardship.

Table 12.3 Decision options for infrastructure

Decision links	Options				
Water availability	No storage	On-farm storage	Alpine storage	Water use efficiency	Integrated water management
Land use intensification	No further intensification	Limited intensification	Increased intensification depending on volume/location	Increased intensification	Increased intensification
Water quality	No change	Water quality worse	Dependent on land use practices; water quality worse	Reduced leakage; water quality improves	Water quality improves
Biodiversity	Decrease or no change	Decrease or no change	Storage effects depend on type/location	No change	Biodiversity opportunities
Recreation	No change	No change	Lake opportunity increase; rivers depend on flow regime	No change	Uncertain
Change to existing consents	No change	No change	Some change	Major change	Very significant change

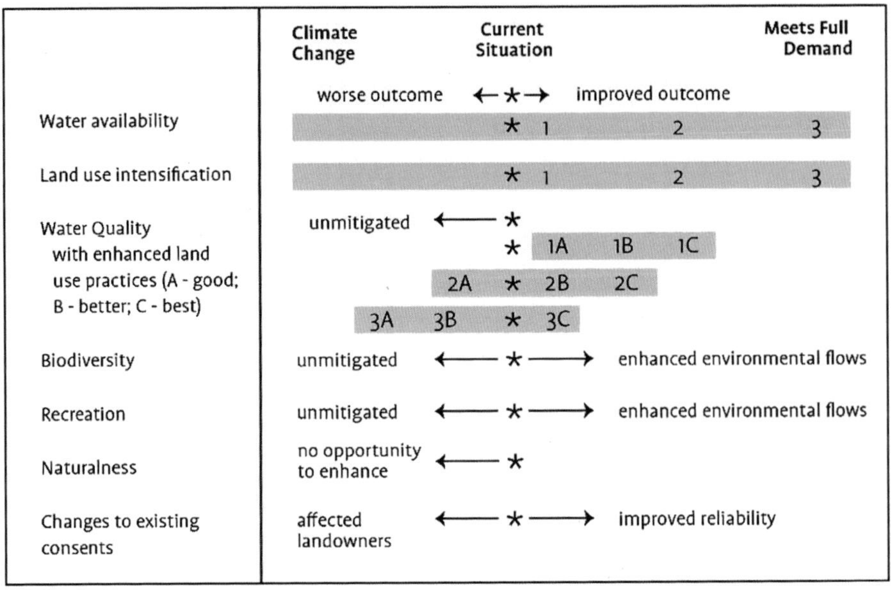

Fig. 12.2 Alpine storage decision scheme (Environment Canterbury)

The next stage in the workshop was the development of *decision schemes.* These are combinations of *decision options* drawn from each of the *decision areas* which represent compatible combinations. The links to other *decision areas* were explored to develop *decision schemes* to address all *decision areas*. For example, in developing the alpine storage option a table of the compatible actions was identified (Fig. 12.2). The key *problem focus* of alpine storage is increasing water availability for further land use intensification. The figure indicates the notional increases in water availability (labelled 1, 2 and 3) between current availability (indicated by an asterisk) and meeting full demand (i.e. all potentially irrigable areas having access to available water). Note that the projected effects of climate change would lead to a reduction in water availability (refer Sect. 7.1). Land use intensification also shows three notional increases (also labelled 1, 2 and 3) associated with increased water availability.

For water quality, increased land use intensification would lead to a worse outcome if unmitigated. Enhancements of land use practices would be an option to offset increased land use intensification. Different levels of enhancement are notionally shown in Fig. 12.2: A – good practice; B – better practice; and C – best practice. The impact on water quality then depends upon the combined effect of land use intensification and land use practices, e.g. 1A is a combination of a small increase in water availability coupled with a shift to "good" land use practices. Note that strategic investigations had determined that for full intensification of the Canterbury plains all farms (existing and new) would need to shift to best practice (i.e. option 3C) to avoid further deterioration of water quality (Bidwell et al. 2009).

Decision areas of biodiversity and recreation would result in worse outcomes from increasing water availability for irrigation. Improvements on the current situation would require enhancements of environmental flows. For the *decision area* of naturalness there is no opportunity to enhance the outcome with the introduction of alpine storage.

The workshop identified an additional *decision area* of "changes to existing consents". For some of the increased water availability options the current consents would need to be changed. This would be viewed negatively by current consent holders. Potentially offsetting this negative effect is the opportunities for enhanced reliability of supply.

There had been two diametrically opposed views about water strategy for Canterbury prior to this stage in strategy development. One view favoured further storage as the priority. The other view opposed storage and wanted the adverse effects of water use and land use intensification addressed and a moratorium on storage. The *Strategic Choice* workshop examined these two views in the development of alternative *decision schemes*.

The *designing mode* led to more accommodating versions of these two strategies. For the storage strategy, environmental mitigation was added to advancing infrastructure development. The moratorium on further development was modified to advancing environmental protection first before proceeding with any further development.

The workshop identified another option of improved water use efficiency which would make additional water available and reduce the water quality contamination from excess runoff or groundwater leakage but would involve reconfiguration of consents.

From the *Strategic Choice* workshop, four *decision schemes* were agreed by the Steering Group for public consultation:

- Option A: Business as usual (base case) – current RMA approach that was effects-based and applicant-driven;
- Option B: Advance environmental protection first then infrastructure development – set limits, initiate restoration and improve efficiency;
- Option C: Reconfigure consents and infrastructure to improve reliability and enhance the environment – redistribution for integrated water management;
- Option D: Advance infrastructure development with environmental repair and protection – storage incorporating environmental mitigation.

The workshop also considered the interfaces between the shaping, designing and comparing modes. The work with the public engagement on uses and benefits combined with the fundamental principles agreed by the Steering Group provided the basis for the Officials Group to construct *decision areas* for consideration by the Steering Group for the *Strategic Choice* workshop. In the *Strategic Choice* workshop, the Steering Group identified four strategic options (*decision schemes* in the language of *Strategic Choice*). There was discussion of the *comparing mode* involving the assessment of the strategic options against sustainability criteria and public consultation on the strategic options through public meetings, public submissions on a strategic options document, and public hearings.

12.2.4 Comparing Mode

As described above in relation to public engagement on strategic options (Sect. 12.1.7), the comparing mode consisted of two components. One was the public consultation on options. This indicated that continuing with the current RMA approach was the least preferred option. However, it did not indicate a preferred strategy. The second component was the sustainability appraisal undertaken by the multi-stakeholder Steering Group. The four options were scored (*relative assessment* in the language of *Strategic Choice*) against evaluation criteria agreed by the Steering group. One of the key findings of this appraisal (*advantage comparison* in the language of *Strategic* Choice) was that the business-as-usual option based on the RMA was unsatisfactory in terms of achieving the sustainability bottom line on nearly all criteria. The efficiency-led option (Option C) could meet nearly all criteria but required changes to existing uses to be implemented. At the sub-regional level, the workshop participants considered that combinations of Options B (environment-led), Option C (efficiency-led) and Option D (storage-led) were most likely to achieve sustainability.

The comparing mode identified the need for change and the direction of change but not a specific blueprint for implementation.

12.2.5 Choosing Mode, Commitment Package and Implementation

The Strategic Framework document (Canterbury Water 2009) reflected the output of the S*trategic Choice* approach and was framed in the form of a *commitment package*. In terms of output, the document had redefined the strategic water management issues facing Canterbury from the initial focus on water availability to a focus on integrated water management with ten target areas from the fundamental principles to be addressed including water availability.

The document outlined the key challenges facing the region. It also redefined the strategic vision of what success would look like with a collaborative community-driven approach rather than an adversarial regulatory-driven approach under the Resource Management Act. The document provided clarity about the vision and principles underpinning future water management. The results of the process followed in strategy development and the strategic investigations undertaken were summarised in the document. In particular it highlighted the issue of water quality impairment from land use intensification unless land use practices were substantially improved.

There were immediate *action schemes* identified both in terms of on-the-ground actions, such as the "Immediate Steps biodiversity programme" and the establishment of a "water executive" to coordinate implementation, and in terms of process such as the formation of zone and regional committees to develop zone implementation programmes and a regional implementation programme.

There were further *exploratory options* to address key issues. For example, a short list of infrastructure options for increasing water availability that warranted further investigation was identified. This not only included the most promising options for major storage but also the options of improved water use efficiency and managed aquifer recharge that had emerged during the strategy development phase.

The Strategic Framework provided a desired outcome: "To enable present and future generations to gain the greatest social, economic, recreational and cultural benefits from our water resources within an environmentally sustainable framework". In order to measure progress toward this outcome targets were defined. The targets comprised a set of goals applying from 2010 that reflect the fundamental principles of the CWMS and then the targets to be achieved for 2015, 2020 and 2040.

Targets which represented the final *decision areas* were defined for:

- Ecosystem health/biodiversity
- Natural character of braided rivers
- Kaitiakitanga
- Drinking water
- Recreational and amenity opportunities
- Water use efficiency
- Irrigated land area
- Energy security and efficiency
- Regional and national economies, and
- Environmental limits[22]

The target areas were given statutory backing through a revised Regional Policy Statement (Environment Canterbury 2013a), provided the framework for the work of the region and zone committees in formulating the region and zone implementation programmes, and, provided the basis for measuring progress on the implementation of the CWMS (Environment Canterbury 2013b).

A notable feature of the Strategic Framework is the specification of the governance framework for making decisions about implementation of the strategy, i.e. in the language of *Strategic Choice*, the understandings about the ways in which any deferred choices should be addressed. The intent was the continuation of the collaborative approach, at the local level through 10 Zone Water Management Committees, and at the regional level through a Regional Water Management Committee. These committees have been established and developed non-statutory Zone and Regional Implementation Programmes addressing the ten target areas.

A regional plan, The Land and Water Plan (Environment Canterbury 2015a), has been developed under RMA processes to give statutory backing to the components of the implementation programmes needing legal status. The Plan operates at two spatial scales. One is at the regional scale which defines objectives, policies and rules that apply across the entire Canterbury region. The second is at the subregional

[22] "Environmental limits" was created as a separate category. It brought together the requirements from the other nine target areas to recognise the role of RMA instruments to establish environmental limits for water bodies.

scale. The subregional sections contain policies and rules specific to the catchments and groundwater zones in the subregion. Plan changes to the Land and Water Plan are being progressively developed to give statutory backing to the Zone Implementation Programmes of the Zone Committees (Environment Canterbury 2013c, 2015b, c, d, 2016b).[23]

12.3 Sustainability Appraisal[24]

12.3.1 Introduction

New Zealand was an early leader in giving legal expression to sustainable management of natural resources through the enactment of the Resource Management Act 1991 (RMA).[25] More recently (2002), the Local Government Act (LGA) stipulates that sustainable development must be considered in all aspects of planning and specifies processes and outcomes that address economic, social, cultural and environmental well-beings.[26] However, there is no guidance on how to appraise policy, programmes or proposals in relation to achieving sustainable outcomes.

Sustainability appraisal (SA) is an emerging assessment methodology internationally. At the time of developing the approach (i.e. in 2008), literature examining progress on sustainability appraisal practices had highlighted the importance of taking an integrated, strategic approach that considered economic, environmental and social dimensions of public policy and planning (Gibson et al. 2005; Sadler 1999, 2006).

---

[23] Refer Sect. 3.3.2.

[24] The contents of this section are drawn from B.R. Jenkins, S. Russell, B. Sadler & M. Ward (2014) Application of sustainability appraisal to the Canterbury Water Management Strategy, Australasian Journal of Environmental Management, 21:1, 83–101. It is copyright © of Environment Institute of Australia and New Zealand, and reprinted by permission of Taylor & Francis Ltd., http://www.tandfonline.com on behalf of Environment Institute of Australia and New Zealand.

[25] RMA Section 5 states that the purpose of the Act "is to promote sustainable management of natural and physical resources" where sustainable management means "managing the use, development, and protection of natural and physical resources in a way, or at a rate, which enables people and communities to provide for their social, economic, and cultural well-being and for their health and safety while—(a) sustaining the potential of natural and physical resources (excluding minerals) to meet the reasonably foreseeable needs of future generations; and (b) safeguarding the life-supporting capacity of air, water, soil, and ecosystems; and (c) avoiding, remedying, or mitigating any adverse effects of activities on the environment.

[26] At the time of the sustainability appraisal the Act specifies the purpose of local government is "(a) to enable democratic local decision-making and action by, and behalf of, communities; and (b) to promote the social, economic, environmental wellbeing of communities, in the present and for the future" (section 10). The Act also highlights principles relating to local authorities including "in taking a sustainable development approach, a local authority should take into account – (i) the social, economic, and cultural wellbeing of people and communities; and (ii) the need to maintain and enhance the quality of the environment; and (iii) the reasonably foreseeable needs of future generations" (section 14(h)).

In the categorisation of sustainability-directed forms of assessment of Hacking and Guthrie (2008), the SA approach incorporates comprehensiveness (all sustainability themes covered), integratedness (all themes are aligned and compared) and strategicness (the focus is broad and forward looking). The SA approach also addresses some of the constraints to incorporating sustainability into assessment processes identified by Audouin and de Wet (2012). Their first constraint was a lack of engagement with underlying values of participants in the process. A key input to the SA approach was the value perspectives of the Canterbury community based on extensive consultation (Jenkins and Henley 2013) (refer Sect. 12.1.3). Their second constraint was the division of social-ecological systems into social, ecological and economic aspects to the point of dissociation. This was addressed in the strategy process by considering water management in the region as a nested adaptive system (Jenkins 2007) (refer Sect. 5.3.1). Their third constraint was the difficulties in engaging with a range of different types of knowledge in the assessment. The use of "capitals" across the four dimensions of sustainability and the development of compatible scales addressed this constraint (Russell and Ward 2010).

This section describes the place of sustainability appraisal approaches in the evolution of impact assessment methodologies. It then provides a description of the sustainability framework used in the case study. The application of the sustainability framework to the evaluation of the strategic options is described followed by the outcomes of the sustainability appraisal. The section closes with a discussion of the contribution and limitations of the process.

12.3.2 Sustainability Appraisal in the Evolution of Impact Assessment

A New Zealand model of Sustainability Appraisal has been proposed by Sadler et al. (2008). Their model was developed from the convergence of two main lines of approach: a 'top-down' strand stemming from developments in strategic planning driven by the importance of taking a 'holistic and inter-sectoral approach' to implementing sustainable development; and, a 'bottom-up' approach reflecting major trends in environmental impact assessment that are represented as paradigm shifts in Table 12.4.

Table 12.4 Evolving paradigm of impact assessment (Sadler 2002)

Paradigm	Key characteristics
First generation – Project level EIA	Includes social, health and other impacts, cumulative effects and biodiversity
Second generation – SEA	Applies to policy, plans, programmes and legislation
Third generation – Toward environmental sustainability assurance – ESA	Use of EIA and SEA to safeguard critical resource and ecological functions and offset residual damage
Next generation – Toward sustainability appraisal – SA	Integrated or full cost assessment of economic, environmental and social impacts of all types of proposals

Environmental impact assessment (EIA) of projects has been in place in New Zealand (Morgan 1999) for many years. The second generation of impact assessment has been the development of strategic environmental assessment (SEA) which applies to policies, plans and programmes (Dovers and Marsden 2002), i.e. assessment at a much earlier stage in the development process compared to EIA. Concern with the effectiveness of implementation of EIA recommendations has led to a third generation approach of environmental sustainability assurance (Sadler 1996). The next development is the concept of sustainability appraisal. This is an integrated assessment of economic, environmental and social effects, applied to optimise gains while avoiding undue risks and potentially significant adverse effects (Sadler 1999, 2002).

Generically, sustainability appraisal (SA) covers a broad family of approaches and tools for analysing and evaluating progress toward sustainable development (Dalal-Clayton and Sadler 2014; Bond et al. 2012). It can be applied ex-ante or ex-post to all types and levels of decision-making, but is considered to be particularly valuable when used proactively to assess the options for proposed courses of development that can deliver the best practicable sustainability outcome. As a decision tool, sustainability appraisal provides a means of informing specific choices and framing policy and public discourse on issues of sustainable development. Any sustainability appraisal will have two essential characteristics: (i) an integrated analysis of economic, environmental, and social effects of development proposals or actions[27]; and, (ii) an evaluation of their significance against identified principles or criteria for sustainable development (Sadler 2006; Dalal-Clayton and Sadler 2014).

12.3.3 Sustainability Appraisal Framework Used in the CWMS

The framework comprises steps and measures for an appraisal of planning options or strategies. It is derived from an internationally recognised three-pillar-model of sustainability (economic, natural, and social) but with the addition of the cultural dimension reflecting the four "well beings" of NZ legislation.

Impact assessment methods often focus on one dimension of sustainability (e.g. economic or social or environmental). Environmental assessment approaches, and especially the "assessment of environmental effects" process under the RMA in New Zealand, focus on the identification of adverse effects of an applicant's proposal compared to the baseline of the current situation and the mitigation of those effects (Fietje 2001). In contrast, SA offers a broader evaluation of alternative strategies (rather than an applicant's proposal) with respect to economic, social, cultural and environmental sustainability criteria.

This means that assessment occurs at a much earlier stage in the development process, i.e. at the strategy stage rather than the proposal stage (similar to strategic

[27] For effective integration, there should be: *substantive integration* of the different types of impact; *procedural integration* of analytical and consultative measures at key stages of the process; and *policy integration* of findings in decision-making and implementation (Dalal-Clayton and Sadler 2014).

environmental assessment). This facilitates a much broader consideration of alternatives. It also means that SA facilitates *integrated* decision-making to address key economic, environmental and social factors simultaneously (which is different to EIA and SEA where environmental factors are addressed after proposal definition or strategy development respectively).

Furthermore, SA involves evaluation within a *sustainability framework* derived from relevant international or national policies as well as local circumstances. The use of a sustainability framework has two important implications. First, sustainability criteria become the benchmark for the assessment of the significance of adverse effects rather than using the current situation as the baseline. This is particularly relevant where the current situation is unsustainable. Second, the sustainability framework identifies objectives to be achieved. Thus, SA includes the evaluation of options against both a *top line* of economic, environmental and social objectives and targets or norms to aim for, and a *bottom line* of key thresholds (baseline minima) or warning signs to avoid.

The SA framework comprises three core elements (Sadler et al. 2008):

- A ***'compass' of sustainability aims and principles*** for guiding policy options and choices against which overall progress and potential effects of proposals can be evaluated;
- A ***systematic procedure for assessing*** economic, environmental and social impacts and considerations of proposed actions;
- A ***set of 'rules of the game'*** for integrating and weighing different objectives in appraisal and decision making in support of sustainable development.

12.3.3.1 Compass of Sustainability Aims and Principles

The *compass* starts from the 'Brundtland' definition of sustainable development[28] and uses the concept of capital stocks as a proxy representation of the *opportunities* that are available to meet present and future human needs (World Bank 2005). Development at the aggregate level is considered to be non-sustainable if net capital wealth is being depleted or eroded, but sustainable if it is being maintained or increasing while also reducing intra-generational inequity. Sustainability as a non-declining stock of capital also requires consideration of the mix of different forms of capital or asset categories to be passed on to the next generation. The crux of this issue depends on the extent to which economic, natural and social capital are considered to be substitutes or complements to each other in determining future opportunities. This interpretation yields a reference level of sustainability against which development trends or actions will be evaluated.

[28] 'meets the needs of the present without compromising the ability of future generations to meet their needs'.

Following World Bank delimitations of the substitutability of natural capital, three levels of sustainability can be identified that offer a choice of frameworks for evaluating development (World Bank 2005):

- ***Weak sustainability*** involves maintaining total capital without regard to its composition and allows natural capital to be freely converted into economic capital and output (governed only by existing environmental policies, regulations and guidelines);
- ***Moderate sustainability*** requires attention being given to the mix of capital stocks with natural capital considered substitutable only up to certain critical limits or thresholds (which if not yet known can be formulated using the precautionary principle);
- ***Strong sustainability*** means maintaining natural capital more or less at current levels (no net loss) so that losses and damages from development must be offset in kind (which represent a stringent interpretation of the precautionary and polluter-pays principles).

Key objectives of sustainable development that provide direction for policy making are expressed by economic, environmental and social policies or sustainability strategies. The overarching policy aim may be expressed as: *to increase real, per-capita human welfare and wellbeing* through development that creates net wealth (or genuine savings) on a continuing basis, consistent with the economic, environmental and equity objectives.

12.3.3.2 Systematic Procedure for Assessment

Under the SA, the *systematic procedure for integrated assessment* can be initiated through three avenues (Dalal-Clayton and Sadler 2014):

1. Use an established process like environmental impact assessment (EIA) or strategic environmental assessment (SEA) as the assessment mainframe, and integrate specialised tools for economic and social analysis;
2. Conduct parallel streams of economic, environmental and social assessment bringing together findings at key stages (preliminary integration in scoping, and full integration in final decision-making); and
3. Rely on an integrative and interdisciplinary methodology such as multi-criteria analysis.

These approaches are not mutually exclusive and can be combined or modified to suit the circumstances. The main steps and activities that characterize impact assessment (screening, scoping, impact analysis, decision-making and monitoring) can be followed to identify potentially significant adverse social, economic, environmental and cultural impacts using a checklist of questions to gain preliminary insight on their sustainability implications.

Whatever path is taken, it is necessary to test the policy or strategy against objectives-led and effects-based sustainability criteria which assist with the determination of significance as the basis for sustainability assurance, i.e. making a policy judgement that the effects of proposals either, at a minimum, 'do no harm' or better still 'achieve much good' (World Bank 2005). Both 'objectives-led' (top line) and 'effects-based' (bottom line) significance criteria are critical to any assessment consistent with integrated decision-making.

In any operational form, applying the sustainability test and determining the eligibility of a proposal will be a subjective and often much qualified exercise. It will depend, in part, on the level of sustainability that is elected as a reference standard (i.e. weak, moderate or strong). The safe minima that identify bottom lines and objective targets that identify top lines may be found in national and regional strategies, and, regional and local plans. For strong sustainability, a stringent version of the precautionary approach should be applied to assess major proposals with potentially significant impacts.

12.3.3.3 Rules of the Game

One of the key concerns in impact assessment by proponents is the trade-off of adverse environmental effects for economic benefits (Sadler 1996). Gibson et al. have put forward suggested decision rules to deal with sustainability trade-offs (Gibson et al. 2005).

The SA approach requires a number of criteria and *rules of the game* for trade-off and decision-making to be followed. For example, during all stages of decision-making priority should be given to options and actions that do the most good, then to those that do no harm, and finally to those that have some adverse effects (but which still fall within acceptable levels). In order, first seek 'win-win-win' packages that will have lasting benefit, second, look for options that maximize net gains without any major adverse effects, and third accept options that have modest net gains but avoid potentially serious adverse effects.

In principle, all other configurations of choice would be unacceptable within a sustainability framework. In reality, to adhere strictly to this principle is not possible – politically and analytically. The process of identifying and tallying gains and losses, and undertaking the necessary trade-offs, is much messier and far more indeterminate than implied here. A 'best practicable sustainability option' is therefore sought to satisfy important objectives in all categories while avoiding critical thresholds or bottom lines. On some level, hard choices and trade-offs are an inevitable part of decision-making. A key to do so is to place the burden of proof on the proponent for all trade-offs that assume potentially significant adverse effects can be mitigated. This presumes that such effects are unacceptable unless their remedy can be substantiated.

12.3.3.4 Adapting the Sustainability Appraisal Framework Approach for New Zealand

Internationally sustainability endeavours are commonly based on a three-pillar model. In New Zealand there is a fourth, 'cultural', pillar – particularly to acknowledge Māori as a partner with the Crown in the Treaty of Waitangi.[29] The four pillars are expressed as economic, social, cultural and environmental in line with the 'wellbeing' terms of the LGA. While this fourth wellbeing is generally considered to provide for recognition of Māori values, importantly in this adaptation, the cultural dimension is not exclusively to address Māori values but where all elements of cultural wellbeing are considered.

To operationalise the SA framework approach in New Zealand a process was used of inventorying capital assets for the four pillars affected by the policy or activity under review, and using them to identify particular issues and their intergenerational and intra-generational equity dimensions. Sustainability criteria selected from those lists were then assigned 'top-lines' (objectives) and 'bottom-lines' (safe minima). The assets identified through this process may include a mix of process and outcome descriptors as well as assets.[30] Effective application of the assessment based on a four-pillar asset inventory in this way demands knowledge and information across each of the four pillars ensuring participation from all sectors and driving collaboration and integration.

12.3.4 Sustainability Appraisal Process

The purpose of sustainability appraisal was to assist the Steering Group and its officials to compare the sustainability implications of the four strategic options, and to identify an option or combination of options that best fit sustainable development objectives. The sustainability appraisal culminated in a two-day workshop to focus on the four strategic options (Russell and Ward 2010). The workshop followed 5 weeks of work by a facilitation team comprising SA specialists, members of the Officials group and other experts.

12.3.4.1 Sustainability Aims and Principles

The CWMS sets a vision that contributes to sustainability aims and principles for sustainability appraisal:

---

[29] The Treaty of Waitangi was signed in 1840 between the British Crown and many Māori chiefs. It provided the British Crown the right of governance and is generally considered the founding document of New Zealand as a nation. One of the key principles in relation to water management is that Māori were to retain *rangatiratanga* (management authority) over their resources.

[30] For example, 'Assets' related to natural capital including 'ecosystems'. Process descriptions include 'equity of water allocation'. Outcome indicators include 'regional value added'.

> To enable present and future generations to gain the greatest social, economic, recreational and cultural benefits from our water resources within an environmentally sustainable framework. (Canterbury Water 2009, p. 6)

In practical terms, sustainability aims and principles were set through tests for intergenerational equity and intra-generational equity and applied through an examination of the composition and rate of use of capital assets against an agreed level of sustainability. These were further tested against regional development objectives,[31] and the guiding principles (see Sect. 3.2.4). Despite general familiarity with the requirements of sustainable development in law, this approach was a significant departure from established assessment practice and required time to lay the groundwork in the workshop.

A related issue was whether the strategy was of sufficient scope and scale, or sufficiently bounded geographically, for capital stock substitution within the region to be meaningfully accommodated. For instance, future deployment of resources from outside the region to substitute for some that were lost or potentially lost in the implementation of the strategy might invalidate the application of an in-region sustainability test. However, it was agreed that the scale was sufficiently large and opportunities for inter-regional transfers of water were sufficiently remote, justifying the use of this approach.

12.3.4.2 Systematic Procedure for Assessment

For the CWMS application, a multi-criteria-analysis based approach was selected as the best 'fit-for-purpose' assessment methodology. The Officials Group selected criteria assisted by an expert group comprising economists, ecologists, and social scientists supported by the SA project team. Twenty-one criteria were initially selected with scale descriptors for five scoring points, two above and below a neutral position.

A key element in the SA work in the workshop was the identification of top and bottom lines and their reference to an agreed level of sustainability. The evaluation criteria were the foundation for this work. Setting bottom lines is familiar territory for New Zealand decision-makers in resource management as the RMA is effects-based and, in principle, allows activities that have no more than minor effect. Accordingly, national standards and regional plans, when they existed, gave guidance on these. Objectives-led planning (as incorporated in the LGA) is less familiar ground for many participants and process guidance was developed for the workshop application of this aspect of the work.

The focus on top and bottom lines sets this approach aside from 'classic' multi-criteria analysis which would involve weighting and scaling of different criteria. In

[31] as expressed in statutory and non-statutory planning documents such as Regional Policy Statements and Plans, Regional Environmental Report, the Community Outcomes Report and Long Term Council Community Plans (LTCCP).

this SA application, all criteria are ranked equally and the analysis sets out to identify aspects of alternatives where improvements are needed to ensure their overall sustainability.

12.3.4.3 Sustainability Appraisal Process Steps

Over 30 CWMS Steering Group members and officials attended the SA workshop. The workshop design sought to provide a consensus-based approach that was central to the operation of the Steering Group in all its work (Canterbury Water 2009, p. 40). Participants were allocated to four small groups, each incorporating a range of technical, regional and subject knowledge. They undertook a facilitated process with the following steps which are summarised in Table 12.5.

12.3.4.4 Rules of the Game

After each activity in small groups, plenary discussions were held to summarise and discuss findings. In addition, participants could identify any aspect of the process or the information that they were uncertain about or that was contested. In the event the time needed for consensus building was not always available. This meant that for some evaluation criteria the scoring was incomplete or there was still a range of views on the appropriate scoring. This required a further round of interaction with the Steering Group which could only be achieved by email rather than face-to-face discussion because of the time constraints for completing the CWMS document.

12.3.5 Sustainability Appraisal Outcomes

12.3.5.1 Sustainability Levels

After a presentation on capital assets and their substitutability, workshop participants discussed the three levels of sustainability. Given time pressures and a range of views, the moderate level of sustainability was selected as the reference point for analysis corresponding to the defining principles adopted by Statistics New Zealand for measuring progress toward sustainable development (Statistics New Zealand 2008).

12.3.5.2 Capital Assets

In the workshop, participants amended provisional lists of capital assets grouped under the four pillars of sustainability (Table 12.6). This had been prepared in advance by the facilitation team with reference to selected Steering Group members. Each group reviewed contributions made by the others. This activity generated

Table 12.5 Sustainability appraisal process steps

1. Sustainability level	Participants were introduced to key definitions and concepts of 'sustainability', specifically core pillars based on the four well-beings of the LGA, principles of inter- and intra-generational equity, sustainability levels and analytical approaches to be used for SA.
	Participants selected a sustainability level (weak, moderate, high) that could be used to evaluate each water management option and discussed the mix of capital assets to be maintained for current and future generations, and potential trade-offs among them, recognising that stocks of some assets (particularly natural assets) need to be maintained at safe minimum levels.
2. Capital assets	Participants compiled, annotated and prioritised the capital assets involved in water resources management in Canterbury using a preliminary set of capital assets produced by the SA project team with guidance from selected participants. Four small groups of participants added or deleted material and then noted those with particular significance for intergenerational and intra-generational equity.
3. Space-time dimensions	Participants considered the spatial and temporal scales of water management in relation to economic, environmental, social and cultural impacts of each option. Impacts were discussed and identified with reference to whether they were relevant to a sub-region and the short-term, to the region over a longer time, or to future generations regionally or nationally.
4. Reviewing evaluation criteria	Participants reviewed and revised the evaluation criteria under the four sustainability pillars previously developed by experts and officials. This activity challenged participants to identify inter-generational implications of use (or misuse) of regional resources and to consider if there were significant sub-regional differences in distribution. The activity provided for the evaluation criteria to be reviewed and amended and to consider inter-generational issues associated with water management.
5. Quadruple top and bottom lines	Participants identified on a five-point scale for each criterion the point that represents the safe base minimal position (Quadruple Bottom Line (QBL)) and the preferred objective position (Quadruple Top Line (QTL)).
6. Scoring options	Participants used the evaluation criteria to score the scale of the impact – From strong negative impact (−2) to strong positive impact (+2) – For each criterion in each option and compare the results with the QTL and QBL to draw a sustainability profile.
	Each group scored an option with regard to its regional implementation making compromises where necessary to achieve consensus. A half-point scale position was introduced in some cases. Due to time constraints within the workshop format, the four groups were unable to repeat the process for the other options and participants were asked to score options during the later extension process.
7. Sub-regional options	Participants considered sub-regional options to determine the best overall outcome.

a final list of assets associated with economic, environmental, social and cultural pillars. Individual participants were then asked to choose the six most important and relevant assets under each pillar. This ranking informed Activities 3 and 4 with regard to consideration of assets that are important for inter- and intra-generational

Table 12.6 Provisional 'asset' list for Canterbury Water Management

Social (human and social)	**Economic (produced and financial)**
Trust in institutions/processes	Schools, community halls, etc.
Sense of community/place	Roads, bridges
Whanaungatanga (kinship)	Dams and impoundments
Informal communication networks	Electricity generation plant & lines
Local knowledge	Irrigation infrastructure
Physical health of people	Water treatment & distribution infrastructure
Mental health of people	Farms (+ stock & machinery)
Skills in communities	Irrigated
Manaakitanga (sharing and caring for each other)	Irrigable
Arable farming knowledge/skill	Public finance
Dry stock farming knowledge/skill	Private finance
Dairy farming knowledge/skill	Ngāi Tahu finance
Communal decision-making	River-based tourism business
Environmental (natural)	**Cultural**
Air	Regional identity
Groundwater free from contaminants	Tastes (music, art, food, dress)
Surface water (at ecosystem sustaining flows)	Whakapapa (genealogy)
Mauri (natural state of being)	Sense of belonging
Reserve land	Attitudes and dispositions
Native bush in sustainable state	Customary rights
Native birds in sustainable populations	Sense of time
Native bird habitat	Culture and traditions
Native fish in sustainable habitat	Ahi kaa (land title through occupation)
Introduced fish	Language and linguistics/te reo
Coastal sediment budget	Tikanga and kawa (customs and ceremony)
Whenua (land)	Mana and rangatiratanga (authority and sovereignty over resources)
Soils	Monuments and significant historic sites

equity, and the review and amendment of evaluation criteria, respectively. Each group reached agreement on lists and ranking. Consensus was not sought between groups as this exercise was to inform the later stages of the process.

12.3.5.3 Space/Time

A space-and-time matrix was completed for each option one per small group which were then reviewed by all participants. This deceptively simple format (Table 12.7) was an effective canvass for introducing sub-regional issues as well as intergenerational ones. The independently generated matrices from the four groups had overlapping information but sufficient unique material to underline the importance of a

Table 12.7 Space-and-Time matrix

Scenario	Economic	Environmental	Social	Cultural
Sub-regionally & short-term				
Regionally & long-term				
Later, to safeguard future generations				

large and independently informed group for this process. There were no significant points of disagreement between groups.

12.3.5.4 Evaluation Criteria

A draft set of evaluation criteria were developed by the Expert Group in advance of the workshop. Workshop participants reviewed, and amended evaluation criteria and associated scale descriptors developed by the expert group and then developed other criteria and related scale descriptions with reference to the lists of capital assets – especially those assets that were highly ranked – generated under Activity 2. Six new criteria were incorporated into the list in Table 12.8: all were cultural and social criteria and are shown with an asterisk in Table 12.8. Facilitators annotated this list by creating descriptors for each new criterion. The list reflects the four capital asset categories of cultural, economic, environmental and social and incorporates five processes.

12.3.5.5 Scale Descriptors and Selecting Top and Bottom Lines

Informed by the preceding review and analysis the four small groups found selection of top and bottom line positions on the evaluation criteria scales not to be a demanding exercise for all but a few criteria. For example, in Table 12.9 a score of +1 was agreed as the bottom line because the current situation for aquatic and riparian biodiversity was considered unsatisfactory. The top line was agreed as +2. Newly adopted criteria proved harder to gain agreement on as little or no information had been gathered or supplied by the technical support group. Merging small group scores introduced a small number of additional divergences which were partly resolved in discussion. Process evaluation revealed that some individuals were not fully in agreement but without information to support their views did not disagree. Scale descriptors for new criteria were completed after the workshop.

Table 12.8 Evaluation criteria

Cultural	1	Opportunities for kaitiakitanga (stewardship)*
	2	Opportunities for rangatiratanga (self-management)*
	3	Sense of experience*
Economic	4	Employment impacts
	5	Household income
	6	Balance of total financial benefits to financial costs
	7	Regional value added
Environmental	8	Aquatic and riparian biodiversity
	9	Aquatic and riparian ecosystems
	10	Terrestrial biodiversity
	11	Water quality for ecosystem health
	12	Water quality for human health
	13	Water quality for recreation
Processes	14	Equity of water allocation – Access
	15	Equity of water allocation – Costs
	16	Feasibility – Alignment with policies and plans
	17	Feasibility – Public funds
	18	Resilience – Adaptability to long-term change
	19	Resilience – Flexibility of regulation and control
Social	20	Community cohesion
	21	Urban–rural cohesion
	22	Landscapes
	23	Recreation
	24	Trust and legitimacy – Institutions*
	25	Trust and legitimacy – Processes*
	26	Knowledge*

Asterisk indicates criteria added to list during workshop

12.3.5.6 Option Comparison

With top and bottom line positions identified the four strategic options were scored to determine whether their implementation met a sustainable (above bottom-line) or unsustainable (below bottom-line) criterion. The workshop allowed time only for scoring to be done by individual small groups without time for a plenary merging. Subsequently participants were invited to score the options for a group of criteria for which new or amended scale descriptors had been developed.

No score was recorded for some criteria. Reasons for no score comprised: there was a large spread of scores received from participants; there was insufficient information to score each criterion; and time constraints meant that not all groups completed the activity during the workshop. Also, for the new criteria there were no agreed scale descriptors. There were also gaps in data relating to metrics for some criteria, and finally, some groups did not want to score particular criteria for Canterbury as a whole, because of sub-regional differences. In sum, the SA process

Table 12.9 Example of evaluation criteria and scale descriptors

		Scale descriptors for impacts (vis-à-vis current state)				
		Strong negative impact	Moderate negative impact	Neutral impact	Moderate positive impact	Strong positive impact
Criterion	Brief description	−2	−1	0	+1	+2
Aquatic and riparian biodiversity	Aquatic and riparian indigenous biodiversity, including key species	Rapid or extensive reduction of biodiversity including loss of key species	Reduction of biodiversity in some areas and/or loss of key species	Biodiversity and key species maintained at current levels	Recovery of biodiversity in key areas and for key species	Extensive and sustained recovery of biodiversity and survival of all key species ensured

Table 12.10 Summary of scores for the four options

	Options	Below QBL	On or within QBL & QTL	Above QTL	n/s
A	Business as usual (the base case)	23	2	0	0
B	Advance environmental protection before developing significant infrastructure	4	16	0	5
C	Reconfigure consents and infrastructure for protection and repair of the environment, improved reliability of supply and for development	0	25	0	0
D	Advance infrastructure with strong requirements for environmental repair and protection	12	9	0	4

n/s not scored

provides only an initial, but still useful approximation of the CWMS 'best-practicable-sustainability option' (Table 12.10). To be "sustainable" the method requires a strategy to be above the QBL for all criteria.

Following the workshop and the completion of the scale descriptors for the new evaluation criteria, participants undertook a revised scoring activity as individuals or in small groups. Participation in this follow-up was offered to all CWMS Steering Group members and officials who attended the workshop. The final scores for each option were presented using line diagrams that incorporate the sustainability profile formed by the QBLs and QTLs (see Fig. 3.12). These show noticeable differences in the sustainability profile for each option, which reflects the varying emphasis of each option.

The scores under **Option A** fall below the QBL across 23 criteria, indicating that the current approach to water management is not adequate or aligned to sustainable development. **Option B** fares better in that 16 criteria were on or within the QBL and QTL parameters, especially environmental criteria (i.e. those numbered 8–13 in Table 12.8), reflecting the emphasis on environmental protection. In all criteria, **Option C** scored within the QBL and QTL parameters. In **Option D,** economic criteria (numbered 4–7 in Table 12.8) scored highly within the QBL and QTL parameters, reflecting the option's emphasis on economic development. The QBL is above the status quo illustrating the need for proactive programmes to address concerns.

As shown in Fig. 3.12 aspects of Options B, C and D have scored within QBL and QTL parameters. This contrasts with Option A, indicating that the current situation is not acceptable in terms of sustainable management of the region's water resources. Hence, for water management in the region to be sustainable, it may be possible to incorporate aspects of Options B, C, and D.

12.3.6 *Discussion*

12.3.6.1 Contribution of Sustainability Appraisal Framework

In undertaking the sustainability appraisal, a conceptual framework that had been adapted for New Zealand circumstances was a significant advantage, particularly in highlighting data gaps and analytical qualifications. Qualitative assessments based on subjective judgment were needed for some criteria and some issues were only partially resolved. However, there was sufficient information to reach relatively robust conclusions about the relative merits of strategic options. Firstly, the appraisal indicated that the status quo was not sustainable as reflected in the poor scores for Option A. Secondly, the appraisal indicated that to achieve sustainability there is a need to improve water use efficiency and land use practices (in relation to their effects on water quality) of existing users. (This was reflected in the positive scores for Option C.)

The intent of the SA approach is to develop a strategy where all criteria are at or above the QBL. This means that a sustainability strategy does not require trade-offs between criteria. What the SA identified was that for a strategy to be sustainable it requires integrated management of existing uses and new development to achieve acceptability on all criteria.

An important component was the multi-stakeholder engagement in the process and the support of a technical group. This enabled issues to be explored and new insights on potential resolution of issues to be developed. It also led participants to realize that their initial preferred solutions were not as robust as they first thought and that other issues needed to be addressed to achieve sustainability.

12.3.6.2 Limitations of the Approach

The application of the sustainability appraisal process to CWMS incorporated a range of activities that were challenging to incorporate into a two-day workshop. Despite these challenges, workshop participants identified the benefit of a structured framework for integrating complex and diverse information. The cooperative effort of participants and the open, honest, and frank discussion was remarked on. Also, there was general endorsement of the sustainability appraisal framework approach as an important stage in the overall process of evaluating options or considering a mix of options for the CWMS. The commitment and willingness to engage in the SA process was further exhibited by some participants during the extension exercise in October 2009.

12.4 Implications for Decision Making to Achieve Sustainability

12.4.1 The Value of Community Engagement as Decision Making

Using community engagement as the decision-making process led to the successful development of a regional water management strategy for Canterbury. This contrasts with planner-led technical approaches which were too narrow in scope and did not address all community concerns, and with process-led legal approaches which were unable to resolve conflicting perspectives. At the time of publication document, the approach set out in the *CWMS: Strategic Framework* document is still guiding water management in Canterbury (Jenkins 2013b).

A key element of the success of the collaborative approach for developing the strategic framework was the design of the community engagement process as the driver of the decision-making process. This included involvement in the design of the process, compared to having a predefined statutory process. The community involvement in defining the issues to be addressed, the options to be considered and the evaluation of the options differs from technical decision-making processes which are usually undertaken by technical experts and professional planners.

From a starting position where there was polarisation of community views about whether water storage and associated land use intensification should proceed, there developed widespread support for the strategic framework for integrated water management that delivers on multiple targets. The strategy development process shifted from a focus on water availability and storage to identification of community values and the wide range of uses and benefits associated with water. The acceptance of the strategy appeared to be related to the ability to be involved in and to influence the strategy development, as well as the outcomes of the process.

The use of a facilitated collaborative process resulted in a greater level of dialogue between different stakeholder interests compared to the adversarial style of statutory processes. This led to new concepts for increasing water availability being brought into the process such as different types of storage. Section 3.3.2 provided examples of off-river storage on the Rangitata River as an alternative to damming the Orari River, and tributary storage on the Waitohi River as an alternative to damming the Hurunui River. Concerns about the potential ability of one stakeholder group to dominate the process were countered: firstly, by having a diverse range of stakeholder backgrounds in the Steering Group that had oversight of the strategy development process; secondly, by having facilitated meetings at multiple locations so that it was not possible for one interest to dominate all meetings; and thirdly, by having an engagement process for the general public.

The emphasis on community engagement led to the introduction of innovative methods. *Open Strategies* enabled a stakeholder definition of uses and benefits for water. *Strategic Choice* could accommodate incomplete information and multiple interests with conflicting objectives. *Sustainability Appraisal* was based on the simultaneous achievement of multiple criteria rather than trade-offs between objectives.

The use of collective choice arrangements required innovative processes and methods compared to typical technical and statutory decision processes, and led to greater acceptance of the outcomes and improved likelihood of their implementation.

12.4.2 Decision Frameworks for Sustainability

The *Strategic Choice* framework provided an effective basis for dealing with the complex problem of water management at sustainability limits in the Canterbury region.

The earlier multi-stakeholder work in stage 3 of the strategy development had highlighted the need to consider multiple issues rather than just the initial focus on water availability. The use of a decision framework that considered many *decision areas* was needed. With the interconnected nature of multiple issues, a decision framework that explicitly dealt with *decision links* by mapping those links was invaluable.

Recognition of these interconnections significantly influenced the options being considered by different stakeholder participants in the workshop. There was a shift from the unidimensional and polarised debate of "storage versus no storage", to development of multidimensional strategies of "infrastructure incorporating environmental repair and protection" and "environmental protection first then infrastructure". Of even greater significance was the recognition of a "water use efficiency" strategy that could increase water availability and reduce the impacts of land use intensification.

Another key advantage of the *Strategic Choice* framework is the concept of a *commitment package.* It was not possible in the time available and the information available to provide a solution to all of the issues related to water management in Canterbury. However, it was possible to identify immediate actions to undertake, and, uncertainties that needed to be further explored, as well as establish the basis by which deferred decisions could be made.

The implementation component of the CWMS Strategic Framework document (Canterbury Water 2009) was designed as a *Commitment Package.* It contained the three key elements: (1) a set of proposed immediate actions, e.g. the Immediate Steps biodiversity programme and establishment of nutrient limits; (2) a set of investigations to deal with important areas of uncertainty, e.g. the identification of where environmental flows do not include flood peaks, flow variability, flood periodicity and channel forming flows, and setting of catchment load limits; and (3) definition of the way that deferred choices would made, i.e. the continuation of the collaborative approach, at the local level through 10 Zone Water Management Committees, and at the regional level through a Regional Water Management Committee, with the development of zone and regional implementation programmes.

While there are still issues to resolve, progress in implementation consistent with the strategic direction that was set in the CWMS Strategic Framework is being achieved (Sect. 3.3).

12.4.3 Evaluation Methods for Sustainability

Sustainability appraisal differs from other evaluation approaches that are used in selecting strategies or projects. Conceptual comparisons are made below with benefit-cost analysis, planning balance sheet and multi-criteria analysis. Also, the use of multi-criteria analysis by the proponents for the Hurunui Water project is compared with the collaborative decision-making process for storage proposals used by the Hurunui Waiau Zone Committee.

12.4.3.1 Comparison with Other Evaluation Methods

In benefit-cost analysis estimates are made of the economic benefits and economic costs in dollar terms. Future benefits and costs are discounted to present day equivalents through discounted cash flow analysis. The benefit-cost analysis of storage proposals for Canterbury was considered in Sect. 10.2.1. A strategy or project is viable if the benefit-cost ratio is greater than 1. The preferred alternative has the highest benefit-cost ratio. From a sustainability perspective, benefit-cost analysis has some significant shortcomings. One limitation is that it only considers the economic dimension. It does not directly consider environmental, social and cultural dimensions: consideration of these dimensions is only included where they can be calculated in dollar terms. A second key limitation is that benefit-cost analysis does not consider equity issues, neither intra-generational nor intergenerational equity. Benefit-cost analysis does not consider winners and losers but the net benefits and costs to whomsoever they accrue. Furthermore, with the use of discounting future benefits and costs, benefit-cost analysis discounts the effects on future generations.

The concept of the planning balance sheet was designed to address non-economic benefits and costs as well as equity considerations related to winners and losers (McAllister 1982). The planning balance sheet records the distribution of costs and benefits among different groups affected by a proposal. The planner preparing the balance sheet identified non-economic effects alongside the economic effects. While this is a way of addressing non-economic and equity issues it is a planner-led technical decision process. This leads to some key limitations. The non-economic effects are based on the planner's judgment. Also, there is no basis for reconciling distribution inequities. Furthermore, there is no basis for reconciling economic and non-economic effects.

The process of multi-criteria analysis involves: (1) the identification of options, (2) the determination of evaluation criteria to compare options, (3) the assessment of the effects of the options in relation to a numerical scale for the evaluation criteria, (4) in order to create an aggregate score to compare options relative weights are given to the evaluation criteria, (5) for each option the numerical scales for each evaluation criterion are weighted and then added to determine a total score for that option, and (6) the total scores for the various options are compared to identify a preferred option. Sustainability appraisal has components of multi-criteria analysis involving the first three steps described above. However, rather than generate an aggregates score for comparison, the objective is to develop an option which is above the sustainability bottom line and attempts to achieve a top line of desired economic, environmental, social and cultural objectives.

12.4.3.2 Evaluation of Storages in the Hurunui-Waiau Zone

A deliberative multi-criteria evaluation was undertaken for the Hurunui Water Project (Lennox et al. 2011). The proponent identified the list of options, determined the evaluation criteria and assessed the extent of effects for each option with respect to the evaluation criteria. Stakeholder involvement was introduced in relation to assigning the relative weights to the evaluation criteria. However, the list of options was limited to options with the lowest cost per unit of stored water: a dam of the south branch of the Hurunui River and control gates on Lake Sumner, either as individual projects or as a combined project.[32] In the stakeholder workshops to assign weights there was a divergence in emphasis between farming interests and environmental interests. Aggregate weighting favoured the combined project (i.e. a dam on the Hurunui south branch and control gates on Lake Sumner). Some participants considered that the resulting scores misrepresented their priorities. They also expressed concern about the accuracy of the impact scores for the options. Most of the environmental stakeholders withdrew from the process to pursue other decision making processes (Lennox et al. 2011).

The outcome of the multi-criteria analysis can be contrasted with the collaborative decision process as part of the Hurunui Waiau Zone Committee implementation of the CWMS (Canterbury Water 2011). A wider range of alternatives were canvassed by the Zone Committee. The alternatives were assessed against the ten target areas for the CWMS and criteria for economic viability, local community and multiple use. The Zone Committee also specified the desired characteristics for projects to deliver more water for the Hurunui-Waiau Zone.

From the deliberations of the Zone Committee, a tributary storage on the Waitohi River (with water diverted from the Hurunui River) was preferred to the dam of the Hurunui south branch and control gates on Lake Sumner. However as this was not the lowest cost per unit of stored water, the Zone Committee sought a review of the "affordability" of the Waitohi tributary storage alternative (Waitohi Selection Panel 2011). An independent panel reviewed three possible tributary storage schemes. The panel selected the Hurricane Valley scheme and determined that the scheme

[32] Refer Sect. 6.1.2 and in particular Table 6.3.

was financially viable. It also noted there are existing irrigation schemes that are more expensive than the preferred Waitohi scheme and are considered affordable in the current economic environment.

The Hurricane Valley scheme has been through the RMA consenting process with very few submissions. This contrasts with more than 1000 objections in the consenting process for the original Hurunui Water Project of a dam on the Hurunui south branch and control gates on Lake Sumner.

The comparison of decision making processes highlights the limitations of effects-based legislation like the RMA to generate sustainable development. With applicant-driven proposals under effects-based processes, the tendency is for the cheapest option to be proposed. The cheapest option can have significant adverse effects and can lead to opposition and adversarial decision processes.

With meaningful involvement of community interests in collaborative decision processes a wider range of options is considered. Desirable project characteristics can be identified rather than examining trade-offs between evaluation criteria. Rather than minimum cost options, all dimensions of sustainability can be considered. The economic dimension of financial viability and affordability must still be met but not at the expense of environmental, social and cultural dimensions.

In the CWMS application, it was found that sustainability appraisal can provide an effective framework for comparing alternative strategies, including the status quo, against sustainability objectives. In particular, the analysis highlighted that the status quo was not an appropriate baseline for assessment purposes because it was unlikely to be sustainable, and, that preferred strategies of different stakeholders were not sustainable across all criteria. This finding is considered robust by the stakeholders involved in the assessment even though it is based on qualitative rather than quantitative data for many parameters. Based on this work, it is concluded that a sustainable strategy not only needs to consider future water storage and its sustainability but also needs to take an integrated water management approach that addresses existing use particularly in relation to improvements in water use efficiency and changes in land use practices by existing users in order to reduce the effects of intensification based on irrigated agriculture. Furthermore, a sustainable strategy needs proactive programmes for environmental restoration, recreational use, cultural use by Māori and other improvement programmes for resource management.

12.4.4 Problems with Effects-Based Approaches

The experience in Canterbury highlights some of the key limitations of current effects-based approaches to water resource decision making when resource availability and cumulative effects are at sustainability limits. Any incremental increase in adverse effects will exceed the environmental bottom line which means no new development even with "less-than-minor" adverse effects should be approved.

Where there has been overallocation of water quantity or water quality at unacceptable levels, the existing situation is not an appropriate baseline for considering the effects of new development. The premise of effects-based legislation to protect the environment through a focus on limiting new development to minor impacts is no longer tenable. There is also a need to address existing allocations and uses.

Approaching sustainability limits also has implications for resource allocation. Reliance on "first-come first-served" allocation of resources or environmental capacity to accept discharges is problematic. Merit-based allocation is more appropriate to ensure resource productivity, impact minimization and equity in allocation. Equity has been identified as an issue in water allocation for tangata whenua, and, as an issue between existing and potential new development where nutrient limits are being set (refer Sect. 3.3.11 and Box 3.1). Furthermore, existing users have been given legal rights to resource allocation and levels of discharge through consents under effects-based legislation. This can be an impediment to requiring increased efficiency (and reduced allocations) or improved management (and reduced discharges).

Rather than effects-based approaches focused on reducing adverse effects of new development, there is a need for outcome-based approaches to keep overall resource extraction and impacts of use within sustainability limits. Furthermore, allocation of constrained resources needs to be merit-based rather than based on who asks first. For resources at sustainability limits to be managed to achieve sustainable development then there is a need for legislative change.

Two Canterbury examples of the failure of effects-based approaches to provide an adequate basis for sustainable development are provided in Box 12.1. One is in relation to water availability limits – the extraction of groundwater from the Rakaia-Selwyn groundwater zone. The other in relation to the cumulative effects of land use intensification on water quality – nitrate levels in the Selwyn River catchment. However alternative approaches have been introduced in other jurisdictions to address resource allocation in times of scarcity and to address cumulative effects of water quality. Two examples of approaches to address these types of issues are set out in Box 12.2. One is the approach taken in the South African Water Act (Republic of South Africa 1998) in areas where water is under stress, e.g. where demands exceed available supply or where water quality is under threat, in other words, at sustainability limits. The other is the management of salinity in the Murray-Darling Basin where sustainability targets have been set requiring reduction in salinity impacts associated with existing use but allowance has been made for new entrants through offsets (Murray-Darling Basin Ministerial Council 2015).

Box 12.1: Failure of Effects-Based Approaches to Achieve Sustainable Development

Rakaia-Selwyn Groundwater Extraction

The Rakaia-Selwyn groundwater zone is part of the Canterbury Plains unconfined aquifer system. There has been a significant increase in groundwater extraction primarily associated with the expansion of dairying. Groundwater allocation limits had been defined based on protecting flows of groundwater-fed streams (Aitchison-Earl et al. 2004). When the effective allocation for use of current consents exceeded the groundwater allocation limits, the regional council recommended that further consent applications for extraction from a groundwater zone be declined.

The first resource consent application affected by this recommendation was by Lynton Dairy. It was for a large volume of water in terms of a single consent application and represented about 2% of the total consented water volume in the Rakaia-Selwyn groundwater zone. The application was declined by hearing commissioners consistent with the council recommendation. The applicant appealed the recommendation to the Environment Court. The Court determined that because field measurements of flow have a measurement uncertainty of ±5%, there was no "probative evidence" of an adverse effect (i.e. you couldn't measure a 2% change). The Court granted 70% of the volume sought by the applicant (Environment Court 2005).

Subsequent decisions on further groundwater extraction applications by hearing commissioners and the Environment Court now mean that the effective allocation associated with groundwater consents is now 134% of the groundwater allocation limit (Environment Canterbury 2016a).

Central Plains Water Quality

The water quality example is the Central Plains Water Irrigation Scheme where nitrate leaching into groundwater from land use intensification was a significant concern. Groundwater is used for drinking water supplies and feeds lowland streams that discharge into a coastal lake. At the time of the hearings for the consent application, 3% of the monitoring wells exceeded the nitrate standard for drinking water (11.3 mg/L). In relation to nitrate toxicity, nitrate concentrations in the lower reaches of the Selwyn River exceeded the threshold for chronic toxicity of highly disturbed systems in environments that are considered measurably degraded (3.6 mg/L nitrate nitrogen median value). For algae in lowland streams the maximum limit for chlorophyll a is 200 mg/m^2. This is exceeded 95% of the time in the Selwyn River. The coastal lake had a Trophic Level Index of 7.0 while the objective was to achieve a TLI of 6.0. The catchment already exceeded the sustainability limits for water quality.

(continued)

Box 12.1 (continued)

The hearing commissioners acknowledged that the Scheme would increase nitrate concentrations in the aquifer, lowland streams and coastal lake. They also acknowledged that nitrate levels would be further increased from recent intensification because of the time lag in groundwater transport. They noted the conflict of the Scheme with water quality objectives and policies but considered the likely adverse effects would be minor. The consent was granted subject to the adoption of best management practices through Farm Environmental Plans to mitigate the impacts of land use intensification (Milne et al. 2010).

Subsequent cumulative effects analysis estimated that the current nitrogen load to the lake is 2650 tN/year. The equilibrium load (i.e. allowing for the time lag in groundwater transport) for the 2011 land use is estimated to be 4100 tN/year. With the addition of Central Plains Water Irrigation Scheme and from further gradual intensification the load is estimated to be 5600 tN/year (Canterbury Water 2013). This is more than double the nitrogen load where the sustainability limits of water quality have already been exceeded.

Box 12.2: Examples of Approaches to Manage Resources at Sustainability Limits

South African Reallocation Provisions

The South African Water Act has a provision for the responsible water authority to undertake compulsory licensing of any aspect of water use for existing and new users. The process can be undertaken to (a) achieve a fair allocation of water which is under water stress or to achieve equity in allocations; (b) to promote beneficial use of water in the public interest; (c) to facilitate efficient management of the water resource, or (d) to protect water quality. In the reallocation process the responsible water authority can consider a wide range of factors including existing lawful uses, investments already made, redress of past discrimination, socio-economic impacts, catchment strategies, effects on the resource and other users, water quality objectives, strategic importance of use, reserves for future use and international obligations, and duration of use.

New Zealand's effects-based legislation, the Resource Management Act (RMA), is far more limited with consent reviews limited to adverse effects better dealt with at a later stage, or, to water and discharge consents when operative regional plans introduce rules for flow rates, rates of water use or water quality standards. The factors that can be considered in resource allocation in the South African legislation are far more comprehensive in relation to sustainability outcomes compared to the first-come first-served allocation principles under the RMA.

(continued)

Box 12.2 (continued)

Salinity Management in the Murray-Darling Basin

Salinity of the River Murray has been a major concern in the Murray-Darling Basin, a very large catchment (1,061,469 km^2) involving four states in Australia (Queensland, New South Wales, Victoria and South Australia). A strategy has evolved over the last 30 years to reduce the salinity to achieve the target of 800 EC units[33] at Morgan (the offtake for Adelaide's drinking water supply) for 95% of the time (Murray-Darling Basin Ministerial Council 2015).

The focus of the Murray-Darling Basin Salinity Strategy is ensuring that for every new action that puts salt in the Murray River and for the delayed salinity impacts of past actions there is another action that reduces the salinity impacts of new actions and delayed effects of past actions by the same amount. A key element of the Strategy is the establishment of two salinity registers: Register A for new actions since the signing of the Murray-Darling Basin Agreement, and, Register B for the delayed salinity impacts of actions prior to the signing of the Agreement.

The Salinity Registers are a credit and debit based salinity accounting system which tracks all actions that are assessed to have a significant effect on river salinity. A significant effect is defined as a change in average daily salinity at Morgan that will be at least ±0.1 EC by 2100. The salinity registers provide the primary record of accountability for actions that affect river salinity.

Salinity credits (reductions in river salinity) can be achieved by investing in salt interception schemes, improving irrigation management to reduce saline drainage, ceasing irrigation, and increasing environmental flows in rivers. Salinity debits (increases in river salinity) primarily occur through new irrigation development. Salinity impact assessments estimate the average annual salinity debit or credit by modelling the effects of actions over a benchmark period (1975–2000).

Each State is required to prepare annual accounts to demonstrate that there are salinity credits to offset salinity debits. There is also a requirement to meet "end-of-valley salinity targets" for major tributaries, e.g. in Victoria this is a delegated responsibility of Catchment Management Authorities. Salinity credits can be earnt through joint works where all States contribute to the cost of salinity reduction measures (primarily salt interception schemes) or through measures undertaken within the State. While States are responsible for the costs, the cost of the credits is passed on to the beneficiaries of the credits through salinity levies. The financial cost per EC unit is determined annually.

[33] EC is a unit of measurement for electrical conductivity (1EC = 1 μS/cm) measured at 25 °C used as an indicator of water salinity (salt concentration).

Box 12.2 (continued)

For example, in Victoria charges are imposed on new water use licences in salinity impact zones to fund measures that will offset the salinity impact on the river as a result of increased water use. Detailed hydrogeological assessments underpin the salinity impact zones and the capital charges reflect the estimated salinity impact caused by irrigation in the zone (Victorian Department of Environment Land Water & Planning 2015).

The implementation of the Salinity Management Strategy has led to a reduction in modelled 95 percentile salinity at Morgan over the benchmark period from 1050 EC in 1988 to 710 EC in 2015.

References

Aitchison-Earl, P., Scott, D., & Sanders, R. (2004). *Groundwater allocation limits: Guidelines for the Canterbury region*. Christchurch: Environment Canterbury.

Andouin, M., & de Wet, B. (2012). Sustainability thinking in environmental assessment. *Impact Assessment and Project Appraisal, 304*(4), 264–274.

Bidwell, V., Lilburne, L., Thorley, M., & Scott, D. (2009). *Nitrate discharge to groundwater from agricultural land use: An initial assessment for the Canterbury plains*. Prepared for Steering Group of the CWMS, Christchurch.

Bond, A., Morrison-Saunders, A., & Pope, J. (2012). Sustainability assessment: the state of the art. *Impact Assessment and Project Appraisal, 30*(1), 53–62.

Booher, D., & Innes, J. (2010). Governance for resilience: CALFED as a complex adaptive network for resource management. *Ecology and Society, 15*(3), 35.

Brooks, M. P. (2002). *Planning theory for practitioners*. Chicago: American Planning Association.

Canterbury Water. (2008). *Milestone 4: Stakeholder & Public Consultation guide on uses and benefits of water in Canterbury*. Christchurch: Environment Canterbury.

Canterbury Water. (2009). *Canterbury Water Management Strategy Strategic Framework*. Christchurch: Environment Canterbury.

Canterbury Water. (2011). *Hurunui-Waiau Zone Implementation Programme*. Christchurch and Amberley: Environment Canterbury and Hurunui District Council.

Canterbury Water. (2013). *Selwyn Waihora ZIP Addendum*. Christchurch: Environment Canterbury.

Canterbury Water. (undated). *Zone and regional committees terms of reference*. http://ecan.govt.nz/publications/Council/cwms-tor-oct.pdf

Dalal-Clayton, B., & Sadler, B. (2014). *Sustainability appraisal: A sourcebook and reference guide to international experience*. Milton Park: Routledge.

Dark, A., Bright, J., & Sigle, S. (2008). *Canterbury Strategic Water Study (Stage 2)*. Christchurch: Aqualinc.

Department for Communities and Local Government, UK Government. (2014). *Planning portal*. http://www.planningportal.gov.uk/planning/applications/decisionmaking/. Accessed 22 Mar 2014.

Dovers, S., & Marsden, S. (2002). *Strategic environmental assessment in Australasia*. Leichardt: The Federation Press.

Driver, P. (2014). *Validating strategies: Linking projects and results to uses and benefits*. Farnham: Gower.

Dryzek, J. (2010). *Foundations and frontiers of deliberative governance*. Oxford: Oxford University Press.

Environment Canterbury. (2008). *Canterbury Regional Environment Report 2008*. Christchurch: Environment Canterbury.

Environment Canterbury. (2009a). *Creating a sustainable future for Canterbury water*. Christchurch: Environment Canterbury.

Environment Canterbury. (2009b). *Living streams: A guide to restoring rural waterways*. Christchurch: Environment Canterbury.

Environment Canterbury. (2011). *Natural Resources Regional Plan*. Christchurch: Environment Canterbury.

Environment Canterbury. (2013a). *Canterbury Regional Policy Statement 2013*. Christchurch: Environment Canterbury.

Environment Canterbury. (2013b). *Canterbury Water Management Strategy targets: Progress report 2013*. Christchurch: Environment Canterbury.

Environment Canterbury. (2013c). *Hurunui and Waiau River Regional Plan operative December 2013*. Christchurch: Environment Canterbury.

Environment Canterbury. (2015a). *Canterbury Land and Water Regional Plan December 2015*. Christchurch: Environment Canterbury.

Environment Canterbury. (2015b). Proposed Canterbury Land and Water Regional Plan Variation 1 Selwyn Te Waihora: Report and recommendations of Hearing Commissioners Adopted by council as its decision on 23 April 2015. Christchurch.

Environment Canterbury. (2015c) *Proposed Plan Change 6 to the Canterbury Land and Water Regional Plan – Section 10 – Banks Peninsula*. Christchurch: Environment Canterbury

Environment Canterbury. (2015d). *Proposed Variation 3 to the Proposed Canterbury Land and Water Regional Plan – Section 15 – Waitaki and South Coastal Canterbury*. Christchurch: Environment Canterbury

Environment Canterbury. (2015e). *Regional Committee*. http://ecan.govt.nz/get-involved/canterburywater/committees/regional-committee/Pages/default.aspx

Environment Canterbury. (2016a). *The groundwater allocation zone summary*. http://ecan.govt.nz/services/on-line-services/monitoring/groundwater-allocation/Pages/groundwater-allocation-summary.aspx

Environment Canterbury. (2016b). *Proposed Plan Change 2 to the Canterbury Land and Water Regional Plan – Section 13 – Ashburton*. Christchurch.

Environment Court. (2005). Lynton Dairy v Canterbury Regional Council NZ Resource Management Act decision C108/2005.

Fernando, A., & Werellagame, I. (2008). *Comparison of planning and consenting procedures for water resource projects in Sri Lanka and New Zealand*. Unitec Research Bank. http://unitec.researchbank.ac.nz/bitstream/handle/10652/1905/Fernando_Comparison%20of%20planning%20and%20consenting.pdf?sequence=1

Fietje, L. (2001). *Developing best practice in environmental impact assessment using risk management ideas, concepts and principles*. Christchurch: University of Canterbury.

Friend, J., & Hickling, A. (2005). *Planning under pressure: the strategic choice approach* (3rd ed.). Oxford: Elsevier.

Gibson, R., Hassan, S., Holtz, S., James, T., & Whitelaw, G. (2005). *Sustainability assessment: Criteria, processes and applications*. London: Earthscan.

Gunderson, L. H., & Holling, C. S. (2002). *Panarchy: Understanding transformations in human and natural systems*. Washington: Island Press.

Gunningham, N. (2008). *Innovative governance and regulatory design: Managing water resources*. Canberra: Australian National University.

Hacking, T., & Guthrie, P. (2008). A framework for clarifying the meaning of triple bottom-line, integrated and sustainability assessment. *Environmental Impact Assessment Review, 28*, 73–89.

Hemmati, M. (2002). *Multi-stakeholder processes for governance and sustainability*. London: Earthscan.

Jenkins, B. R. (2007, March 27–30). *Water allocation in Canterbury*. Paper presented at the NZ planning institute conference 2007, Palmerston North.

Jenkins, B. R. (2009a, March 19–21). *Best practice partnerships at the local government level for the environment*. Paper presented at the community boards' conference, Christchurch.

Jenkins, B. R. (2009b, August 18–20). *Canterbury Water Management Strategy: A region-wide approach to improve water management*. Paper presented at the National Water Conference 2009, Melbourne.

Jenkins, B. R. (2013a). *The development of sustainable alternatives to applicant's proposals using collaborative approaches*. Paper presented at the New Zealand Planning Institute 2013 annual conference, Hamilton, 30 April to 3 May 2013.

Jenkins, B. R. (2013b, August 28–30). *Progress of the Canterbury Water Management Strategy and some emerging issues*. Paper presented at the New Zealand Agricultural and Resource Economics Society conference Lincoln University.

Jenkins, B. R., & Henley, G. (2013, October 23–24). *Collaborative management: Community engagement process as the decision making process*. Paper presented at the Environment Institute of Australia and New Zealand, Melbourne.

Lennox, J., Proctor, W., & Russell, S. (2011). Structuring stakeholder participation in New Zealand's water resource governance. *Ecological Economics, 70*, 1381–1394.

Lomax, A., Memon, A., & Painter, B. (2010). *The Canterbury Water Management Strategy as a collaborative planning initiative: A preliminary assessment*. Lincoln ventures. https://researcharchive.lincoln.ac.nz/bitstream/10182/2463/1/CWMS_Appraisal_Report_20_May_2010.pdf

McAllister, D. (1982). *Evaluation in environmental planning: assessing environmental, social, economic and political trade-offs*. Cambridge, MA: MIT Press.

Midgley, G. (2000). *Systemic intervention: Philosophy, methodology, and practice*. New York: Kluwer Academic.

Milne, P., Fenemor, A., O'Callaghan, R., & Nixon, B. (2010). *Joint decision of the Hearing Commissioners in the matter of the various applications by Central Plains Water Trust*. Christchurch: Environment Canterbury.

Morgan, R. (1999). *Environmental impact assessment: a methodological perspective*. Dordrecht: Kluwer Academic.

Morgan, M., Bidwell, V., Bright, J., McIndoe, I., & Robb, C. (2002). *Canterbury Strategic Water Study*. Lincoln: Lincoln Environmental.

Murray-Darling Basin Ministerial Council. (2015). *Basin Salinity Management 2030 (BSM 2030)*. Canberra: Murray-Darling Basin Authority.

OECD. (2008). *Environmental performance of agriculture in OECD countries since 1990*. Berlin Centre: OECD.

Ostrom, E. (1990). *Governing the commons: The evolution of institutions for collective action*. New York: Cambridge University Press.

Republic of South Africa. (1998). *National Water Act*. Government Gazette 26 August 1998, Cape Town.

Russell, S., & Ward, M. (2010). *Sustainability appraisal: Application to the Canterbury Water Management Strategy*. Lincoln: Landcare Research.

Sadler, B. (1996). *Environmental assessment in a changing world*. Hull: Canadian Environmental Assessment Agency.

Sadler, B. (1999). A framework for environmental sustainability assessment and assurance. In J. Petts (Ed.), *Handbook of environmental impact assessment* (Vol. 1, pp. 12–32). Oxford: Blackwell.

Sadler, B. (2002). From environmental assessment to sustainability appraisal? In *Environmental assessment yearbook 2002. Institute of Environmental Management and Assessment* (pp. 145–152). Manchester: University of Manchester.

Sadler, B. (2006). Towards a framework for sustainable development assessment. In: *Annex to the report of the CBAC working party on biotechnology, sustainable development and Canada's future economy*. Industry Canada, Ottawa.

Sadler, B., Ward, M., & Frame, R. (2008). *A framework for sustainability appraisal in New Zealand*. Lincoln: Landcare Research.

Statistics New Zealand. (2008). *Statistics New Zealand's framework for measuring sustainable development*. Wellington: Statistics New Zealand.

Stevens, C. (2003). Waitaki catchment – Nonsense or national necessity. *Resource Management Journal, XI*(3), 9–11.

TV One News (online). (2009). *Hundreds protest Hurunui dam project*. http://tvnz.co.nz/national-news/hundreds-protest-hurunui-dam-project-3079784

Victorian Department of Environment Land Water & Planning. (2015). *Murray-Darling Basin Salinity Management Strategy: Victoria's 2013/2014 annual report*. Victorian department of environment, Land, water & planning, Melbourne.

Waitohi Selection Panel. (2011). *Report of the Waitohi Selection Panel*. Christchurch: Environment Canterbury.

Whitehouse, I., Pearce, A., & McFaddern, G. (2008). *Final report* – Canterbury Strategic Water Study Stage 3 Environment Canterbury, Christchurch.

World Bank. (2005). *Where is the Wealth of Nations measuring capital for the XXI century?* Washington, DC: World Bank.

Part IV
Implications for Water Management in the Canterbury Region

Chapter 13
Biophysical Limits and Sustainable Management

Abstract Adoption of collaborative governance has broadened strategic thinking in Canterbury from increased water availability through storage on alpine rivers, to sustainable management of water for multiple uses of importance to the community. Furthermore, improved water use efficiency was found to be a more effective way of increasing water availability. Water availability matters still to be resolved include: adaptation to climate change, institutional arrangements for infrastructure provision, and, measurement and management for enhanced water use efficiency.

Failure pathway analysis highlighted effects of water abstraction on river flows and groundwater levels, as well as effects of land use intensification on freshwater quality on nutrient, bacterial and sediment contamination. Nested adaptive system analysis found current levels of management interventions are insufficient to achieve sustainable outcomes. Also, greater attention is needed to the interactions between surface and groundwater for managing water quantity and quality issues.

Climate change projections indicate higher temperatures increasing potential evapotranspiration rates thereby increasing irrigation demand. Also, water availability in irrigation seasons is expected to decline from reduced winter rainfall to recharge aquifers and maintain lowland streamflow, lower foothill river flow, and changing flow patterns in alpine rivers from reduced snowmelt and increased winter rainfall. Higher winter flows in alpine rivers could be used to recharge aquifers.

Use of a nested approach for the region has demonstrated that at finer spatial scales there are differences in community priorities, differences in failure pathways, and differences in sustainability strategies. Introducing resilience assessments, sustainability strategies and managing cumulative effects places a greater reliance on modelling and monitoring. Management of extremes of droughts and floods requires managing the consequences of failure rather than for specific return-period events.

The RMA focuses on defining environmental bottom lines, however, experience with managing-to-limits indicates challenges with numerical uncertainties, model inaccuracies, natural variability, multiple variables, enforcement difficulties, contributions from legacy issues, lag times in effects, cause-effect attribution, and the range of possible management interventions. While limits are useful, managing based on nested adaptive cycles and integrating actions at individual, tributary and catchment scales are needed to achieve sustainable outcomes.

B.R. Jenkins, *Water Management in New Zealand's Canterbury Region*, Global Issues in Water Policy 19, https://doi.org/10.1007/978-94-024-1213-0_13

Keywords Sustainability limits • Water availability • Water quality management • Managing to limits

13.1 Biophysical Sustainability Limits

As an introduction to this section on biophysical sustainability limits the evolution of strategic thinking in Canterbury water management over the last 20 years is summarized. The Canterbury Water Management Strategy (CWMS) led to new insights into water availability. Progress on implementing these changes is outlined. However, there are water availability matters still to be resolved, in particular, climate change, the need for a water infrastructure and services entity, and, the management of water use efficiency. The key issues in relation to the impacts of water abstraction and use are then summarized and matters requiring greater recognition are discussed. Then the main implications of climate change for Canterbury are presented both in relation to adaptation to climate change and mitigation of greenhouse gases.

13.1.1 Evolution of Strategic Thinking in Canterbury

As discussed in Chap. 3, strategic thinking about water management in Canterbury commenced with concerns about water availability limits to meet future demand particularly during drought conditions (Morgan et al. 2002). Agriculture was the dominant consumptive use. Irrigation was the means of increasing agricultural production and irrigable land was available. Water availability was seen as the primary constraint on further economic development.

Reliability of supply for run-of-river irrigation schemes was dependent on instantaneous river flow. With Canterbury's climate variability, the low flow conditions made surface water irrigation schemes vulnerable. There was also a significant increase in groundwater abstraction. Sustainability limits in relation to maintaining environmental flows in rivers have been reached for run-of-river abstraction and in relation to groundwater recharge for groundwater abstraction.

The greatest pressure on rivers was on the groundwater-fed lowland streams from both direct abstraction and lowered groundwater levels, and on rain-fed foothill rivers from abstraction. The snow-fed alpine rivers carried the greatest proportion of runoff (88% of the region's annual flow) and were under less pressure from abstraction.

Under low flow conditions the peak demand for irrigation could not be met even in 2001. However, on an annual basis water was available to meet future demand but this would require storage. Storage on alpine rivers was seen as the strategic solution to water availability in the Canterbury region. Investigations of storage sites were undertaken.

A limited number of sites were found to be hydrologically feasible. However, evaluations of the potential storages brought out concerns about the impacts of dams and the effects of further intensification from irrigation using the water provided by storages. This led to significant community conflict between those advocating storage to facilitate further irrigation development and those opposed to storage and further land use intensification.

As well as the effects of water abstraction on reduced river flows and aquifer drawdown, there were concerns about the effects of storage particularly on the mainstems of alpine rivers, such as the reduction in braided river character, intrusion on areas of high naturalness, algal blooms downstream of storage, de-oxygenation in reservoirs, weed growth in reservoirs, reduced sediment transport and increased coastal erosion from decreased sediment supply. Principal concerns with land use intensification were water quality degradation in lakes, rivers and groundwater from increased nutrients (nitrogen and phosphorus), increased bacterial contamination, and increased sediment in the beds of rivers and lakes as well as suspended sediment.

The complexity of issues and the conflict in the community indicated the need for an integrated water management strategy for the Canterbury region. Also, the framework of the resource management legislation based on the management of effects within environmental bottom lines was inadequate for managing a resource at or beyond sustainability limits. Furthermore, the legalistic nature of RMA processes exacerbated conflict between different viewpoints.

Success of collaborative approaches at the tributary and catchment scale for generating innovative solutions between conflicting interests led to creating a collaborative approach at the regional level. As discussed in Chap. 12, extensive community engagement, innovative strategy development and sustainability appraisal led to a more comprehensive strategic framework for water management in the region compared to the initial focus on water availability. In addition to reliable water availability for increasing irrigated land area, the CWMS identified strategic outcomes for improving ecosystem health and biodiversity, maintaining the natural character of braided rivers, exercising kaitiakitanga, supplying drinking water of suitable quality and quantity, providing recreational and amenity opportunities, improving water use efficiency, and contributing to regional and national economies. Outcome targets and environmental limits were identified to guide the development of programmes for implementing the CWMS.

13.1.2 Insights from CWMS on Water Availability

As a result of the investigations and stakeholder engagement there was a broadening of strategies to address water availability from the initial approach of storage of alpine river water. A cheaper and more sustainable strategy was identified of increasing the efficiency of the use of water already allocated. This could be achieved by (a) shifting to more efficient irrigation technology, (b) matching irrigation application to times and degree of soil moisture deficit while leaving capacity for absorbing rainfall, (c) limiting application rates to prevent macropore flow, (d) using closed pipes for distribution rather than open canals, and (e) changing the spatial application of irrigation in a groundwater zone to using surface water in the upper part of the zone with increased leakage available for extraction in the lower part of the zone.

The improved water use efficiency would reduce but not eliminate the need for storage for increased water availability. To maintain the braided character of alpine rivers there were to be no new dams on the mainstems of major alpine braided rivers. However new approaches to storage were identified, not only to address water availability but also reliability of supply. Off-river storage of high river flows and diversions to storage on tributaries were alternative ways of accessing alpine water. Another option is groundwater recharge (i.e. managed aquifer recharge). Improved reliability could be achieved by on-farm storage and storage within irrigation schemes.

13.1.3 Implementation Progress on Water Availability

In relation to water efficiency, conversion from border dyke irrigation to centre pivot irrigation is progressing throughout the region.[1] Also, replacement of open canals by piped distribution is progressing in mid Canterbury with canal replacement by pipe occurring in the Ashburton Lyndhurst and Mayfield Hinds irrigation schemes. The new Central Plains Water irrigation development has piped distribution from an open head race. In addition one step in spatial reallocation has been identified with the Central Plains Water development: alpine water from the Rakaia and Waimakariri Rivers is to irrigate 30,000 ha of dryland (new irrigation) and replace groundwater takes on 30,000 ha of currently irrigated land in the upper groundwater zone (Canterbury Water 2013).

With respect to storage, the CWMS short-listed seven projects: (1) the use of Lake Coleridge for irrigation storage; (2) efficiency improvements in mid Canterbury; (3) groundwater storage in the Central Plains; (4) Hurunui integrated option; (5) Lees Valley storage; (6) Lake Tekapo water for South Canterbury; and (7) extension of Hunter Downs to the north. The use of Lake Coleridge has been consented and is being implemented. As noted above there has been progress in efficiency improvements in mid Canterbury. A trial of managed aquifer recharge is being advanced in the Hinds Catchment. Tributary storage on the Waitohi with diversions from the Hurunui has been consented. The Hunter Downs extension has also been consented. Investigations of the Lees valley storage and Lake Tekapo water for South Canterbury have indicated that these schemes are not financially viable.

[1] Between 2010 and 2015 the area under border dyke irrigation reduced from 53,400 ha to 27,800 ha (a 48% reduction) (Brown 2016).

Storage for irrigation schemes has also been implemented (e.g. the Carew storage at Mayfield Hinds Irrigation Scheme) or under consideration (e.g. the Wrights Road storage for the Waimakariri Irrigation Scheme). Numerous on-farm storages have also been constructed.

13.1.4 Water Availability Matters Still to Be Resolved

While progress with water availability issues is evident, there are still matters that need to be addressed, in particular, climate change, a water infrastructure and services entity, and water use efficiency measurement.

13.1.4.1 Climate Change

The CWMS identified climate change as a key challenge but did not include implementation outcomes. Chapter 7 indicated the significance of climate change projections for water management in Canterbury. Climate change is not addressed in the Regional Implementation Programme and is not included in the issues for fresh water in the Canterbury Regional Policy Statement. Some of the Zone Implementation Programmes (ZIPs) refer to climate change but do not incorporate adaptation or mitigation provisions in their recommendations. This issue is broader than water availability and is addressed further in Sect. 13.1.6 below.

13.1.4.2 Water Infrastructure and Services Entity

For integrated management, the CWMS saw the need for a Water Infrastructure and Services Entity. The Canterbury Strategic Water Study identified that there was no agency with the mandate for water resource development (Morgan et al. 2002). In a recent strategic review of national infrastructure, the water infrastructure sector ranked poorly (New Zealand Government 2011).

The Regional Committee has limited its consideration to developing a "big picture" for a regionally integrated approach to water supply and distribution infrastructure. Several Zone Committees have identified infrastructure components as part of their solutions packages for water quality management (e.g. flow augmentation of Wainono Lagoon, and managed aquifer recharge in the Hinds catchment) but did not address their implementation or funding.

While the private sector can be responsible for commercial investment in infrastructure there is also a need for public sector involvement in infrastructure for social, cultural and environmental well-being components of water infrastructure. Furthermore, there are legacy issues from past development and adaptation issues for future effects of climate change that need public sector involvement. This issue is considered further in Sect. 14.1.1 on infrastructure coordination and Sect. 14.1.4 on economics.

13.1.4.3 Measurement and Water Use Efficiency Indicators

Water use efficiency is a major theme of the CWMS. It was recognized that defining efficiency is not straightforward so development of benchmarks and reporting on them was part of the targets and was scheduled for completion in 2015. Water measurement, recording and reporting (as proposed in the RMA regulation on Measurement of Water Use) was seen in the CWMS as a key tool in the implementation of water use efficiency.[2] By July 2016, of the 5900 water take consents in Canterbury, 5400 have water meters installed and are compliant with their water take consents (Environment Canterbury 2016). Progress has been made with defining good management practice by industry and Irrigation NZ has developed a design Code of Practice (Irrigation New Zealand 2013a) and Standards (Irrigation New Zealand 2013b). However actual benchmarks of the efficient use of water are not yet available (Canterbury Water 2015). One of the 2015 targets for water efficiency was for 60% of water used for irrigation to be operating according to best practice. Using estimates of on-farm irrigation efficiency based on the type of irrigation system used (e.g. centre pivot 80–90% efficiency and border dyke 30–60% efficiency), it is estimated that approximately 55–60% of the area irrigated in Canterbury in 2015 achieved an application efficiency of 80% or higher (Brown 2016).

There are many definitions of irrigation and water use efficiency and there is a need for a clear definition of what is meant and how it is measured (Aqualinc 2012). A useful parameter that can be used to define water use efficiency and that is relevant to water quantity and quality considerations is the concept of "net use of irrigation water". The purpose of irrigation is to improve plant growth. The net use of irrigation is the difference between evapotranspiration from irrigated land and evapotranspiration from dryland:

$$\begin{aligned}\text{Net use} = {} & \text{evapotranspiration}\left(\text{irrigated}\right) \\ & -\text{evapotranspiration}\left(\text{dryland}\right)\end{aligned} \tag{13.1}$$

The concept is very useful in describing the water use efficiency of irrigation. The ratio of "net use" to "irrigation applied" is a direct measure of the efficiency of irrigation in converting water to plant growth.[3]

---

[2] The Resource Management (Measurement and Reporting of Water Takes) Regulations 2010 came into effect on 10 November 2010. There is staged implementation of metering requirements for consents granted before 10 November 2010 with takes of 20 L/s or more to be compliant by 10 November 2012, 10–20 L/s by 10 November 2014, and 5–10 L/s by 10 November 2016 (Ministry for the Environment 2016).

[3] This is comparable to the "field application efficiency" recommended by Land and Water Australia (Barrett Purcell & Associates 1999), and the "plant component" of "Farm Irrigation System

The significance of net use in water quantity management was identified in Stage 1 of the Canterbury Strategic Water Study (Morgan et al. 2002) highlighting that net use is reflected in increased evapotranspiration from increased plant growth while the balance of the applied irrigation (i.e. irrigation applied minus net use) returns as groundwater recharge or surface runoff.

Young and McColl consider the implications of water entitlement specifications and increased irrigation efficiency in the Murray-Darling Basin (Young and McColl 2009). Under gross water entitlements (i.e. the total volume of water received) water savings from efficiency gains can be used by the entitlement holder to increase the irrigated area. This reduces the groundwater leakage and/or surface water return flow thereby reducing water availability elsewhere in the catchment. Young argues that either entitlements need to be specified on a net use basis or there is an administrative regime that requires adjustment in allocations where water availability has been reduced (Young 2014).

This is an issue for groundwater management in the Canterbury plains where increases in irrigation efficiency have led to reduced groundwater leakage and declining groundwater levels (Sect. 3.3.4). Furthermore groundwater allocation limits include a component for leakage from irrigated areas (Scott 2004): as efficiency increases and leakage declines, groundwater allocation limits will need to be reduced. Thus, the measurement of net use is not only valuable for assessing water use efficiency at the farm level, it is also a valuable input to catchment water management.

The net use concept is also very useful for defining the additional groundwater or surface water runoff that is associated with intensification as a result of irrigation:

$$\begin{aligned}\text{Net use} = &\ \text{irrigation supplied} - \text{recharge / runoff}\left(\text{irrigated}\right)\\ &+ \text{recharge / runoff}\left(\text{dryland}\right)\end{aligned} \tag{13.2}$$

This means the additional recharge/runoff (with contaminants from land use intensification) is “irrigation supplied” less “net use”. Therefore, managing to improve net use improves water use efficiency and reduces contamination to surface water and groundwater.

From an overall irrigation system perspective of water use efficiency, it is important to identify different components of the system where losses can occur and where to target water use efficiency improvements. Losses can occur as water moves from the source e.g. reservoir, river or groundwater basin (source losses), conveyed and delivered at the farm gate (conveyance losses), applied to the farm (distribution losses) and finally consumed by the crop (plant losses) for crop production (Sharma et al. 2015). The water use efficiency improvements identified in Sect. 3.3.4 occur

Efficiency” recommended by Aqualinc (2012) for on-farm systems. Note the regional plan definition of “irrigation application efficiency” means “the volume of water stored in the plant root zone following irrigation, as a percentage of the total volume applied” (Environment Canterbury 2015a). However, Aqualinc (2012) recommend plant water use definitions rather than root-zone definitions because plant water use can be more readily estimated than water stored in the crop root zone.

at different points in the irrigation system. Improved irrigation methods and application rates reduce distribution losses. Piped distribution reduces conveyance losses while spatial application reduces source losses.

Net use can be measured or modelled at the farm scale (Aqualinc 2012). It is also possible to measure net use at the catchment scale. The regional council (eLEAF 2012) and GNS (Westerhoff and White 2013) have been experimenting with the use of remote sensing to measure evapotranspiration and soil moisture in the Canterbury Plains. The International Water Management Institute (Thenkabail et al. 2006) has developed a method of assessing irrigated area using remote sensing: this work has been further developed by eLEAF. NIWA has a virtual climate network for assessing rainfall across New Zealand. The regional council, NIWA and irrigation consultants have on-the-ground monitoring to calibrate and verify remote sensing analysis.

In situations of water scarcity not only is it important to improve water use efficiency, it is also important to enhance water productivity. Water productivity has two main dimensions. One is physical water productivity which relates to the amount of water needed to produce a product. The other is economic water productivity which involves the amount of water needed to earn a dollar from a product.

The relationship between output and water requirements (i.e. physical water productivity) can be expressed in two ways. One is the measure of the output of a given production system in relation to the water it consumes – measures of "crop per drop" (Cook et al. 2006). Typically the physical water productivity is expressed in kg/m^3 where crop production is measured in kg/ha and water use is estimated as depth of irrigation water applied or received as rainfall in mm (Sharma et al. 2015). It can be measured over different spatial scales such as farm, irrigation command area, or river basin (Molden 1997). These measures are useful in tracking improvements (or decline) in output for a limited availability of water.

The other is a measure of water requirements per unit of output – "water footprints" (Hoekstra et al. 2011). Water footprints, as discussed in Sect. 10.1.4, can be calculated for a variety of entities, e.g. a process, a product, a consumer, a group of consumers or producers, as well as countries. They are measured in water volumes consumed per unit of time for the entity under investigation. With components for rain-fed consumption (green footprint), for irrigation consumption (blue footprint), and for volume polluted by the entity (grey footprint), water footprints can be useful metrics for regions like Canterbury with a high dependence on irrigation, with significant climate variability (and therefore highly variable rainfall component available for production), and with concerns about diffuse pollution leading to water quality degradation.

In relation to water quantity management for a region or country water footprints can be aggregated for water used by producers in the economy and can be aggregated for water used by consumers in the economy. Water footprints facilitate the tracking of water embedded in products (i.e. "virtual water") throughout the economy. Water footprints are useful for tracking the reduction of water consumed by production, as well as the availability of water for an economy (like Canterbury and New Zealand) that is highly dependent on agricultural export.

Economic water productivity also has two approaches. One approach is to consider the economic output per unit of water volume in $/m^3. For managed water like irrigation (as distinct from rain water), productivity can also be expressed in terms of the economic value added per volume of water applied. McDonald and Paterson undertook analyses for several industries in Auckland, Northland and Waikato. However their analysis was constrained by the dearth of accurate information on actual water use by industry sectors (Ford et al. 2001). Furthermore it is difficult to isolate the contribution of water because added value is dependent upon many resources, e.g. fertilizer, labour and capital (Young 1996).

The second approach to economic water productivity is to consider the volume of water needed to generate a dollar of output. Analyses have usually been undertaken for an industry sector and referred to as the "water intensity" for that industry sector. There has been a detailed analysis of water intensity for all industry sectors in Australia as part of a triple bottom line reporting framework incorporating economic input-output modelling of the industry sectors (Foran et al. 2005). Water requirements are based on managed water (i.e. excluding rain water). Dairying was calculated to have a water intensity of 1450 litres per dollar, while rice production required about six times as much water at 8410 litres per dollar, and wheat production about one seventh the volume of water at 200 litres per dollar. In times of water scarcity reducing the water intensity of an industry or shifting to industries with low water intensity are two ways of improving economic resource productivity.

Because of the significance of water use efficiency for the implementation of the CWMS and the implications of water use efficiency for the Canterbury region's economy a comprehensive approach to measuring water use efficiency at the farm, irrigation scheme and catchment scales is needed.

13.1.5 Impacts of Use

The sustainability analysis has highlighted a range of impacts associated with water abstraction and water use. This has included:

- The effects on river flows with the need to retain adequate low flows to maintain freshwater habitat, adequate frequency of freshes to maintain habitat quality, sufficient floods to maintain braided river character, and flows at river mouths to manage the frequency and duration of river mouth closure.
- The effects on groundwater abstraction on maintaining flows in groundwater-fed streams (in particular during times of low rainfall recharge), of avoiding seawater intrusion, and the need for adaptive management of cumulative effects as well as managing interference effects on neighbouring bores.
- The effects on freshwater quality due to land use intensification on groundwater quality especially nitrate-nitrogen and bacterial contamination affecting drinking water quality; nutrient enrichment, algal blooms, faecal contamination, siltation, and nitrate toxicity in lowland streams in particular, but also the lower reaches of

foothill and alpine rivers; and, the increased trophic status from eutrophication for Waituna-type coastal lakes initiated by catchment clearing and land use change over the last 100–150 years, as well as more recent degradation of high country lakes from contemporary intensification.

While these effects were identified as separate effects, the analysis established there is a need for both greater integration of groundwater and surface water management, and, a greater integration of water quality and water quantity management.

There is increasing recognition of the need to manage groundwater and surface water in an integrated way. The inclusion of stream depletion considerations to incorporate the hydraulic connection between groundwater and surface water (refer Sect. 6.2.2), and the consideration of managed aquifer recharge to improve groundwater quality (Sect. 3.3.7) are two examples. The sustainability analysis identified other aspects of Canterbury water management that would benefit from greater integration of groundwater and surface water management, such as:

- The catchment scale interaction of river seepage to groundwater in upper reaches and groundwater inflow in lower reaches, with examples of the increased drying reach of the Selwyn River from groundwater extraction (Sect. 6.2.2) and the limitations of Christchurch's water supply (Sect. 5.1.2)
- A climate change adaptation strategy of using increased winter flows in alpine rivers for managed aquifer recharge to offset the decline in rainfall recharge to groundwater (Sect. 13.1.6)
- A water use efficiency strategy of using surface water for irrigation in the upper reaches of a groundwater zone and using groundwater in the lower reaches in order to increase recharge to the groundwater zone (Sect. 6.1.4)
- The improvements in water use efficiency from improved irrigation technology and using piped rather than canal distribution of run-of-river irrigation schemes resulting in less leakage and reduced recharge to groundwater (Sect. 6.1.4).

A broader concept of conjunctive use is needed to manage water availability rather than the historical focus on managing surface water and groundwater separately.

Some of the key sustainability issues to be addressed in Canterbury are multivariate issues. It is not sufficient to manage one critical variable, rather a number a number of critical variables need to be managed simultaneously in an integrated way:

- Algal blooms in rivers are influenced not only by nutrient concentrations but also the accrual period between flushing flows (Sect. 6.1.3 and 6.2.3)
- Algal blooms in lakes can be a combination of catchment inputs and re-mobilisation of in-lake contaminants (Sect. 11.2.1).
- River mouth closure is a combination of river flow, coastal processes and sea conditions (Sect. 6.2.1).

These issues indicate that to manage cumulative effects when resource availability and use are at sustainability limits needs a system-wide approach. Furthermore, with overallocation of water resources beyond sustainability limits and with cumulative effects of water resource use exceeding sustainability criteria, there is a need for proactive approaches to water resource improvement rather than effects-based approaches reliant on mitigation of adverse effects. This issue is discussed further in relation to the institutional arrangements for sustainable development (Sect. 14.1.3).

Collaborative approaches are generating new proactive approaches through solution packages for zones (Sect. 3.3.7) and delivering on-the-ground improvements through programmes like Living Streams (Sect. 12.1.1) and Whakaora Te Waihora (Sect. 5.3.2). However, the level of improvement achieved by on-the-ground measures is not yet sufficient to deliver sustainable water management. This was also evident in the analysis of six lakes in New Zealand (Sect. 11.2). While management interventions are reducing the rate of water quality degradation they are not sufficient to achieve the desired outcomes. Box 13.1 shows the example of Silverstream where farmers in collaboration with the regional council have made concerted efforts for a decade to reduce bacterial contamination from the Silverstream catchment to an important recreation site of Coes Ford on the Selwyn River. While noticeable improvement has been achieved, it has not been sufficient to achieve acceptable recreational water quality at Coes Ford.

Box 13.1: Water Quality Improvements in the Silverstream Catchment

The Silverstream catchment is representative of a groundwater-fed lowland stream catchment in the Selwyn-Waihora Zone. Concerns about the Silverstream catchment and its contribution to poor microbial water quality in the Selwyn River at Coes Ford were identified in a study conducted in 1994–5 by Environment Canterbury (Adamson and Main 1996). Direct dairy shed discharges were considered the most obvious sources of faecal coliform bacteria. However, the highest concentrations were recorded in McGraths Creek where there were no direct discharges. Run-off from agricultural land receiving animal effluent, particularly where the effluent loading was excessive and stock access to creeks and drains, were identified as other potential significant sources.

Since 2002, the Silverstream Water Improvement Group and the Environment Canterbury Living Streams programmes have been encouraging land owners to fence out reaches of tributaries and improve riparian vegetation. Comparisons between stream walks in 2006 and 2013 indicate that fencing to prevent stock access has increased – reducing unfenced areas from 20% of stream length to 9% (Glasgow 2013). In the ten years to 2012 water quality monitoring shows significant improvements in turbidity, ammonia-nitrogen, dissolved reactive phosphorus at Coes Ford (Robinson and Stevenson 2012). However the microbial contamination is still non-compliant for recreational use and has a suitability grading of "poor" for 2014/5 (Bolton-Ritchie and Robinson 2016).

Figure 13.1 shows *E. coli* monitoring results (95th percentile of 5 years' data over summer months November to March in MPN/100 mL) for Coes Ford and for Chamberlains Ford which is the site upstream of Coes Ford and the confluence with Silverstream. Chamberlains Ford is ranked as "good" as suitability for recreation grading and mostly in the microbiological assessment category (MAC) B – 131 to 260 MPN/100 mL. Coes Ford results show improvement from 2007/8 but are still well above the action guideline of 550 MPN/100 mL.

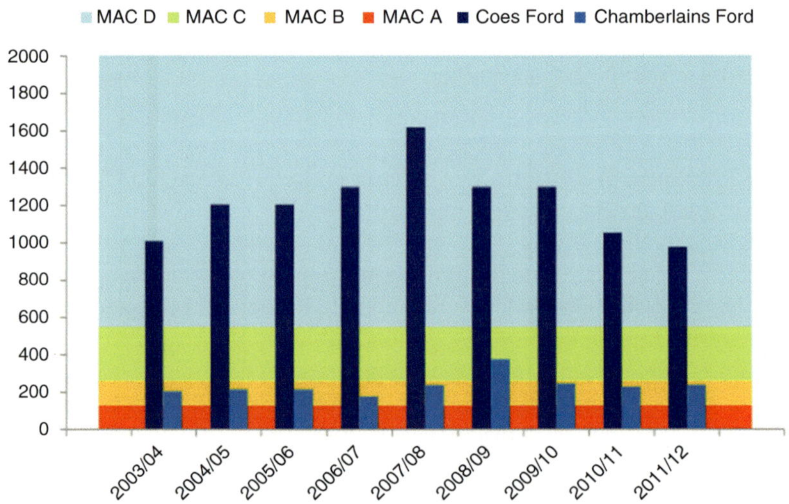

Fig. 13.1 Bacterial contamination at Chamberlains Ford and Coes Ford on the Selwyn River (Robinson and Stevenson 2012)
Note to Figure: Units are 95th percentile of *E. coli* (MPN/100 mL) over summer months (November to March) based on 5 years' data. Microbiological assessment categories (MAC) are: MAC A < 130 MPN/100 mL; MAC B 131 to 260 MPN/100 mL; MAC C 261 to 550 MPN/100 mL; and MAC D > 550 MPN/100 mL (Ministry for the Environment 2002)

13.1.6 Implications of Climate Change

New Zealand's greenhouse gas emission profile is different from other developed countries because agriculture is the largest contributing sector with 47.2% of the country's emissions. This results mainly from methane emissions from ruminant livestock and nitrous oxide from fertiliser use. Emissions from the sector are increasing because of the expansion of dairying particularly in Canterbury. Forest clearance for conversions to dairying is also reducing greenhouse sinks (Sect. 7.1.1).

Climate projections indicate an increase in temperature: about 1 °C in the next 50 years which is double the historical rate (1 °C in the last 100 years). For Canterbury on the dry east coast of the South Island, this will result in an increased potential evapotranspiration deficit of around 120–180 mm per year. Rainfall projections for Canterbury are for a small summer increase (2.5–5%) and a winter decrease (7.5–10%). Whereas the west coast is projected to have lower summer rain (0–2.5%) and increased winter rain (5–12.5%). With higher winter temperatures, it is projected that snow cover in the Southern Alps will decrease and snow lines rise (Sect. 7.1.2, 7.1.3, 7.1.4 and 7.1.5).

These climate change projections have significant implications for Canterbury where 89% of consumptive use is for irrigation of agriculture. An increase in potential evapotranspiration deficit would increase irrigation demand. A decrease in winter rainfall on the Canterbury Plains would reduce groundwater recharge while lowered groundwater levels would reduce flows in groundwater-fed lowland streams. Drier winters would also lead to lower flow in foothill rivers. For alpine rivers with their catchments extending into the Southern Alps, there would be an increase in annual flows from increased winter rain on the west coast. However, the reduced snow would lead to increased winter flows but reduced summer flows. More detailed analysis of monthly flows in the alpine river, the Waimakariri, indicate reduced reliability of supply for run-of-river irrigation schemes (Sect. 7.1.6 and 7.1.7).

These projections are significant for a region whose environmental and economic health is dependent on water. There is therefore a need for adaptation to the climate change projections but also an opportunity for greenhouse gas emission mitigation because of the significance of agricultural emissions.

13.1.6.1 Adaptation to Climate Change

While it is not possible to "future proof" the region, there are approaches that can increase the resilience of the ecological and economic systems that are dependent on water. There are a range of approaches for adapting to drought conditions of increasing the ability of farms to resist drought, increasing the flexibility of the farm system to deal with drought, minimizing the impact of drought, and, adjusting the farm structure to be more compatible with drier conditions (Sect. 7.3). For example, while there are concerns about storage on the main stems of alpine rivers, sustainable options exist for off-river storage: the increased winter flows of alpine rivers could be harvested and stored for summer use. This is occurring on the Rangitata at Arundel with off-river storage ponds (Sect. 3.3.3).

However, a more cost-effective solution for the Canterbury plains that would have both environmental and economic benefits is to use the increased winter flows for aquifer recharge. Managed aquifer recharge could maintain groundwater levels for abstraction and lowland stream flow as well as dilute groundwater contamination from land use intensification. It also avoids the evaporative losses and loss of land associated with surface storage. Analyses for the CWMS demonstrated that managed aquifer recharge on the Canterbury plains was only two thirds of the cost of equivalent surface water storage (SKM 2010). A pilot project is underway in the Hinds catchment (Sect. 3.3.7).

There can also be significant improvements in water use efficiency to reduce irrigation demand. This is beginning to occur with shifts to more efficient forms of irrigation and the use of piped rather than canal distribution of water in irrigation schemes (Sect. 3.3.4).

However more can be done through integrated surface and groundwater management. Integrated approaches would involve a predominant use of surface water when river flows allow, and a predominant use of groundwater when river flows are restricted. Also, the greater use of surface water for irrigation in the upper reaches of groundwater zones would enhance recharge and enable greater use of groundwater for irrigation in the lower reaches.

Increased water efficiency together with reduced groundwater leakage and reduced runoff can also be achieved by greater use of soil moisture demand irrigation. Irrigation is limited to times when soil moisture is below 80% (to avoid soil saturation leading to groundwater leakage) but kept above 50% (to ensure plant growth).

As well as considering water availability there is also the need to consider environmental flows in groundwater-fed lowland streams and foothill streams where diminished flows are anticipated. This is not just an issue for groundwater extraction for irrigation in rural areas but also for groundwater extraction for public water supply for Christchurch and the effect on the baseflow of urban groundwater-fed streams (Sect. 5.1.2). Maintaining groundwater levels through actions such as reduced abstraction or managed aquifer recharge (to offset projected rainfall decrease and evapotranspiration increase) or reduced leakage (because of increased water use efficiency) are needed.

There are ways that we can adapt to the projected effects of climate change but the adaptations involve changes in the way water is managed. Adaptive management approaches for water allocation to address climate variability (Sect. 6.3) can also be used for adapting to climate change.

One of the features of a successful strategy identified in the CWMS was that "the water management system will be better able to adapt to climate change in the future" (Canterbury Water 2009). The Canterbury Regional Policy Statement (CRPS) acknowledges regional climate change projections in Issue 7.1.4 relating to the abstraction and use of water for economic benefit. However there are no objectives or policies that directly address adaptation to climate change (Environment Canterbury 2013).

There is a need for a strategy of climate change adaptation that has statutory backing through the CRPS and Land and Water Regional Plan.

There are also implications for the institutions that are needed to govern its management: this is addressed further in Sect. 14.1.3 below.

13.1.6.2 Mitigation of Greenhouse Gas Emissions

Despite being the dominant source of New Zealand's greenhouse gas emissions, agricultural emissions are not part of the Emissions Trading Scheme (ETS), and greenhouse gas emissions from land use intensification, such as dairy conversions and forest clearance, are not subject to environmental impact assessment under the RMA. The agricultural sector is projected to provide 77% of the growth in emissions (Sustainability Council of New Zealand 2015). The Government's view was that the lack of mitigation options meant that agriculture should be kept out of the ETS (Editor 2011). However, there are mitigation measures and offsets available to address agricultural emissions (Sect. 7.2.3).

There is the potential for offsets of agricultural emissions through forestry plantings and incorporation of hydro electricity generation in association with irrigation schemes. The use of herd shelters can minimize the deposition of urine patches at high-risk times of the year, and, replacing nitrogen fertilizer inputs to boost pasture production with inputs of maize or cereal silage can reduce the amount of nitrogen ingested and excreted. Reductions in nitrous oxide emissions can be achieved by the application of nitrification inhibiters. However, the current legislative framework in New Zealand does not provide action forcing mechanisms to initiate appropriate changes to mitigate greenhouse gas emissions. The Western Australian experience demonstrates the value of including greenhouse gas emissions in the assessment of environmental effects process (Box 7.1). There is also a need to develop an international standard related to maximum residue levels in food products of dicyandiamide (DCD), the active ingredient of nitrification inhibitors.

13.2 Sustainable Management

This section on sustainable management first highlights the advantages of a nested approach in addressing the spatial variation in water management issues. It then considers aspects of sustainable management practices noting that innovations are occurring and that there is an increased role for modelling and monitoring in addressing the sustainable management of cumulative effects. The third section notes the importance of managing extremes. The final part looks at the concept of managing-to-limits showing that this is not sufficient to achieve sustainable management and that adaptive cycles are a more appropriate basis for management.

13.2.1 Spatial Variation of Issues

There has been an interesting shift from the CWMS and Natural Resources Regional Plan to ZIPs and Land and Water Regional Plan in relation to spatial scale and definition of issues. The spatial scale of the CWMS was predominantly regional in addressing the high-level issues associated with Canterbury water management, principally the sustainability limits of water availability and the cumulative effects of land use intensification. The spatial scale of the ZIPs is the water management zone addressing catchment-specific issues. While there are similarities of the types of issues because the ZIPs were designed to address the ten target areas of the CWMS[4], there are differences in priorities, differences in failure pathways, and differences in sustainability strategies for each of the zones (Sect. 3.3.7).

For the Hurunui-Waiau ZIP the major issue related to the implications of nitrate caps on the Hurunui River and the implications for further intensification and the equity of allocation of nitrate capacity. For the Selwyn Waihora Zone which is already over-allocated in terms of groundwater availability and exceeds nitrate water quality standards, the priority was on reducing dependence on groundwater and reducing the nitrate footprint of intensified farming. This led to the consideration of the affordability of improved management practices. For the Ashburton Zone where nitrate levels in groundwater in the Hinds catchment are high, the priority was balancing what combination of further intensification, nitrate level and managed aquifer recharge was sustainable. For the South Coastal Canterbury Zone where water quality of Wainono Lagoon was a key sustainability issue, the priority was the equity of allocation of nitrate caps, the affordability of improved management practices, and the dilution of lake nitrate concentrations with alpine river water. For the Banks Peninsula Zone, the priority issue was algal blooms in Wairewa/Lake Forsyth with a focus on the erosion of phosphorus-rich soils from the lake catchment due to the historical clearance of vegetation for agriculture and the remobilization of phosphorus from the deposition of eroded material as lake sediments. For the Christchurch West Melton zone, there were urban priorities relating to groundwater management for water quality and public water supply, and stormwater management to improve water quality in urban streams.

The ZIPs show a much more nuanced approach to water management issues at the catchment scale than was possible in a regional strategy. The increased sophistication is also reflected in the Canterbury Land and Water Regional Plan. The nested structure provides a more refined approach to addressing water management issues compared to the regional approach inherent in the earlier Natural Resources Regional Plan.

---

[4] The ten target areas were: ecosystem health/biodiversity; natural character of braided rivers; kaitiakitanga; drinking water; recreational and amenity opportunities, water-use efficiency, irrigated land area, energy security and efficiency, regional and national economies, and environmental limits.

13.2.2 *Sustainable Management Practices*

13.2.2.1 Innovation in Practice

There have been some significant innovations in management practices. Increased water availability is being achieved through increased efficiency and not just further abstraction. Progress is being made in irrigation efficiency through soil moisture management. Riparian margins are being fenced and planted to reduce water quality impacts. Constructed wetlands are being built. Lake beds are being treated to reduce remobilization of phosphorus. Flocculants are being added to remove phosphorus from streams.

New approaches to regulatory activities are being developed. Collaborative approaches are generating innovative solution packages to address issues. Audited self-management is giving farmers greater flexibility (and greater responsibility) in achieving environmental outcomes compared to command and control style oversight.

While the introduction of sustainable management practices is achieving improvements in water availability and water quality, the level of intervention and extent of adoption are frequently insufficient to achieve system sustainability.

13.2.2.2 Modelling and Monitoring of Cumulative Effects

Compared to effects-based management when the effects are within sustainability limits, the management of cumulative effects and the decision-making requirements of collaborative governance involve a much higher level of modelling capability and more comprehensive monitoring for effective implementation. Under effects-based management (within sustainability limits), the main emphasis of modelling is on predicting the adverse effects of individual projects (impact prediction), while the main focus of monitoring is the measurement of environmental condition (state-of-environment monitoring) and whether activities are meeting consent requirements (compliance monitoring). These elements are still required for cumulative effects management and collaborative governance but there are additional modelling and monitoring requirements.

With respect to modelling with resource availability or adverse effects at sustainability limits, the sustainability analyses in this book have involved:

- The modelling of cumulative effects at the catchment scale as well as activity level effects assessment (e.g. nitrate-nitrogen modelling of further intensification across the Canterbury region – Sect. 3.2.5)
- The modelling of alternative scenarios for future resource use to inform collaborative deliberations about economic, social and cultural outcomes (e.g. the Wainono Lagoon analysis of the options of on-farm mitigation and dilution with alpine water – Box 3.1)

- The modelling of management practices at the property scale in terms of environmental outcomes and financial costs (e.g. the Hinds catchment for assessing the cost of mitigation measures to achieve different nitrate outcomes – Sect. 14.1.4)
- The modelling of natural resource systems with respect to defining resilience thresholds, recovery processes and responses to management interventions (e.g. the analysis of lakes – Sect. 11.2)
- The modelling to distinguish natural changes (such as climate variability) from man-induced changes (such as water abstraction) (e.g. groundwater drawdown in the Selwyn catchment – Sect. 6.3)
- The modelling to estimate parameters in the absence of actual data (e.g. the modelling of catchments of high country lakes to estimate nitrogen and phosphorus loads – Sect. 6.2.3)

With respect to monitoring the management of cumulative effects and issues with multiple variables, sustainability analyses have required more than state-of-environment monitoring and compliance monitoring. In addition to condition monitoring, there have been examples of:

- Indicators of pressure and management response as well as environmental condition (e.g. for the resilience analysis of Te Waihora/Lake Ellesmere – Sect. 5.3.2);
- The spatial alignment of indicators to establish cause and effect with respect to environmental stressors and ecological values, and, ecological response to multiple variables such as changes in flow, habitat and biological interactions (e.g. the occurrence of periphyton – Sect. 6.2.3);
- The monitoring of catchment effects involving human-induced impacts (e.g. the upstream/downstream paired sites for periphyton – Sect. 6.2.3);
- The real-time monitoring of environmental changes and all human uses for the managing of cumulative effects and resource allocation for human use (e.g. monitoring of tides, groundwater levels and groundwater abstraction for Woolston/Heathcote groundwater zone – Sect. 6.2.2).

In addition to compliance monitoring, there is a need for:

- Monitoring of farmer performance against their farm environmental plans (e.g. audited self-management – Sect. 8.3.4);
- Monitoring of farmer collectives against their environmental management system (e.g. Hurunui Waiau farmer collectives – Sect. 14.1.1);
- Monitoring the effectiveness of management interventions (e.g. the management of lakes – Sect. 11.2.5);
- Not only monitoring the effectiveness of an intervention to address a particular issue but also the effects of the intervention on other issues (e.g. the effects on wader habitat of managing lake opening regimes for long-fin eel passage at Te Waihora/Lake Ellesmere – Sect. 5.3.2);
- Monitoring of the achievement of plan objectives and strategy targets (e.g. progress reporting of CWMS – Sect. 14.1.1).

13.2.3 Management of Extremes

The traditional approach of designing flood protection infrastructure to meet a design flow of a specified return period does not address the potential for catastrophic failure due to events greater than the design flow. A resilience approach that addresses the consequences of extreme events, such as the secondary stopbank design for the Waimakariri flood management scheme for Christchurch is a more sustainable approach (Sect. 7.4).

The experience of Hurricane Katrina has led the US Army Corps of Engineers to shift its approach for managing extreme events from a return period approach in the direction of a "nested adaptive systems" approach with consideration of failure pathways and sustainability strategies at multiple geographical scales (Box 7.2).

Sustainability approaches to drought response were also identified with strategies to increase the ability to resist drought (increase resilience), to minimize the impact of drought (reduce vulnerability), to be flexible in dealing with drought (enhancing adaptive capacity), and, to adjust farm structure for drought (enhance transformability) (Sect. 7.3).

13.2.4 Managing to Limits Is Not Enough

13.2.4.1 Managing to Limits and the RMA

The concept of managing within environmental limits underpins the original concept of the RMA. The intent of the Act was to provide a framework to establish objectives with a biophysical bottom line that must not be compromised (Upton 1995). The purpose of the Act was to enable people and communities to provide for their social, economic and cultural well-being through using and developing resources while meeting the following bottom lines: (a) sustaining resources to meet the needs of future generations, (b) safeguarding the life-supporting capacity of air, water, soil and ecosystems, and (c) avoiding, remedying or mitigating any adverse effects of activities on the environment (RMA Section 5(2)).

The Act also set out a clear and rigorous procedure for setting environmental standards (RMA Section 43) (Upton 1995). However the only water management National Environmental Standard since the RMA was passed in 1991 is for the Sources of Drinking Water established in 2007 (New Zealand Government 2007). There is also the ability for regional councils to set standards under a regional plan. While a number of catchments in Canterbury had regional plans, a regional plan for water issues in the Canterbury region as a whole was not notified by the regional council until 2004 (Environment Canterbury 2004) and it wasn't until after a lengthy hearing process by Hearing Commissioners that the plan became operative in 2011 (Environment Canterbury 2011).

The key decisions on limits primarily occurred during the consent process informed by criteria in the proposed regional plan, water quality guidelines, environmental flow studies and groundwater zone assessments with the limits being defined in consent conditions. It is also noteworthy that some of the key guidelines, such as the ANZEEC water quality guidelines (ANZECC 2000) and the New Zealand contact recreation guidelines (Ministry for the Environment 2002) are not expressed as specific limits but have "trigger values" that represent degrees of risk (Simpson Grierson 2010).

Furthermore, as argued by Skelton and Memon, it is a misconception that the purpose statement in the RMA (Sect. 5.2) is predominantly about prescribing biophysical environmental bottom lines. The purpose statement incorporates anthropocentric values in using and developing resources for social, economic and cultural well-being as well as environmental values such as the life-supporting capacity of water. Interpretation by the Courts has led to an "overall broad judgment" approach. It is a matter of the weight that is to be given to the various elements of sustainable management in the context of a particular case (Skelton and Memon 2002).

However, this interpretation of the purpose of the RMA has led to exceedances of water allocation limits and water quality criteria in Canterbury. Examples for groundwater extraction in the Rakaia-Selwyn Groundwater Zone and for the water quality effects of the Central Plains Scheme were provided in Box 12.1.

The introduction of the National Policy Statement in 2011 (Sect. 2.4) provided a stronger statutory basis for dealing with over-allocation by including provisions (a) that in relation to water quality "The overall quality of fresh water is maintained or improved while…improving the quality of fresh water in water bodies that have been degraded by human activities to the point of being over-allocated" (Objective A2(c)); and (b) in relation to water quantity "to avoid any further over-allocation of fresh water and phase out existing over-allocation" (Objective B2).

The addition of the National Objectives Framework in the National Policy Statement amendments of 2014 provided greater specification of water quality attribute states for setting limits. Attributes have been specified for human health for recreation and ecosystem health.[5] Attribute bands A, B, C and D of decreasing water quality have been defined with national bottom lines set at better than D. Where freshwater quality is below the national bottom line, the quality has to be improved over time. Regional councils are required to set freshwater objectives for each water body taking into account local and national values and aspirations and its existing condition.

[5] In relation to human health for recreation attributes have been defined for bacterial and cyanobacterial contamination; in relation to ecosystem health attributes have been defined for the trophic state of lakes (for total nitrogen, total phosphorus and phytoplankton), periphyton (as an indicator of algal blooms in rivers), nitrate toxicity in rivers, dissolved oxygen in rivers, and ammonia toxicity in lakes and rivers (New Zealand Government 2014).

13.2.4.2 Problems Experienced with Managing to Limits

However current experience with implementing the "managing to limits" concept in the National Policy Statement is showing some significant challenges in relation to this concept achieving sustainable outcomes. These challenges include: (a) uncertainty of the numerical value of the load; (b) inaccuracies and differences in models used for load estimation; (c) natural variability in the parameter used to define the limit; (d) multiple variables influencing the parameter used to define the limit; (e) enforceability of limits which lack clarity and certainty; (f) contributions from legacy issues as well as current activities; (g) lag time between the activity being managed and the effect in the environment; (h) ascertaining the cause of changes when monitoring for the limit; and, (i) the range of management interventions to address exceedances of the limit.

These challenges are considered below in the context of the current work in Canterbury to introduce nutrient load limits to manage eutrophication and algal blooms. The result of considering these challenges is that while managing to limits is valuable in identifying critical variables with respect to resilience thresholds, it is not enough to achieve sustainable outcomes. Rather there needs to be a transition from a regulatory approach of managing to bottom lines to a collaborative approach of the delivery of sustainable outcomes.

In the study of options for nutrient load limits in the Hurunui catchment, Norton and Kelly noted three sources of uncertainty: measurement error, incomplete data and models, and natural variability (Norton and Kelly 2010). Estimates of the current mean annual nutrient loads showed a range of values, e.g. the mean annual Dissolved Inorganic Nitrogen (DIN) load for the Hurunui River at State Highway 1 (SH1) was stated as 654–731 tN/year to reflect the estimated uncertainty.

Brown in his evidence to the consent hearing for Ngāi Tahu irrigation proposal for the Balmoral Forest identified four different valid methods for calculating the annual DIN load at SH1 (averaged over 5 years). The methods showed a range from 732 to 818 tN/year and Brown concluded that there was no perfect method (Brown 2014b).

In another example of modelling, the Cawthron Institute estimated the nutrient loading to high-country lakes in Canterbury (Kelly et al. 2014). Because there was limited environmental monitoring data a modelling approach was adopted using CLUES (Catchment Land Use for Environmental Sustainability) model. Vollenweider models (Vollenweider 1982) were also used to fit regression models to predict nutrient loading data. Analysis with available measurements indicated that the models explained 49% to 73% of the variation of in-lake Total Nitrogen (TN) and Total Phosphorus (TP) concentrations. It was concluded that it was highly likely that the modelled load estimates provide good estimates of actual nutrient loads. However, some lakes, like Lake Clearwater, were outliers. Modelled TN concentrations for Lake Clearwater were 219 mg/m^3 compared to measured TN concentrations of 562 mg/m^3.

In relation to natural variability, for the Hurunui nitrogen loadings Brown calculated the annual DIN load at SH1 for the seven-year period from 2005/6 to 2012/13.

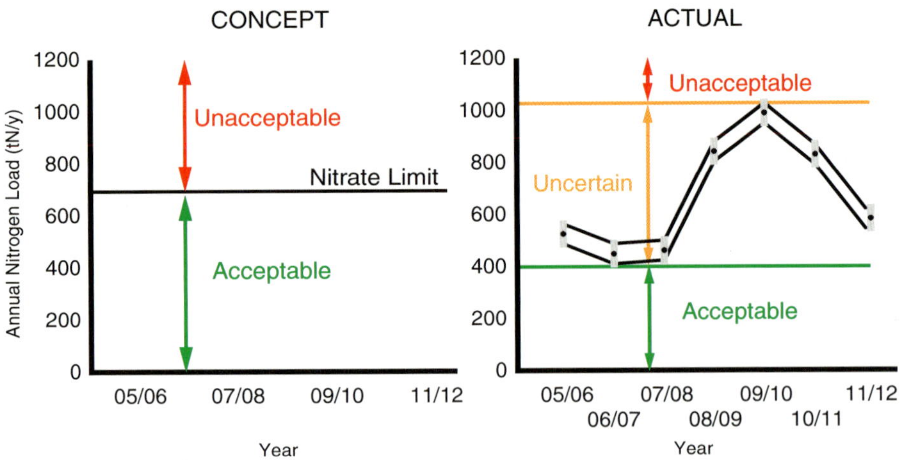

Fig. 13.2 Managing to limits allowing for natural variability and uncertainty

This showed an annual variation from 445 tN/year to 981 tN/year. Not only was there substantial variation in nitrogen loading from year to year, but also the greatest periphyton cover (49 days in excess of 20% cover) was in the year 2005/6 when the annual load was only 521 tN/year (Brown 2014a). Figure 13.2 shows the practical difficulty of managing to a limit. If the objective is maintaining the current annual average level (i.e. 693 tN/year) then the left-hand graph shows the intended concept of values below 693 tN/year being considered "acceptable" and values above 693 tN/year being "unacceptable". The right-hand graph shows the actual situation of year-to-year variations and an allowance for uncertainty. This indicates that there is uncertainty about whether the annual load is above or below the limit for results between 420 tN/year and 1019 tN/year.

It is not just natural variability that is highlighted in this example. It also indicates that there can be multiple variables involved in determining the sustainable outcome sought of managing periphyton cover to acceptable levels. The maximum abundance of periphyton is not only related to the nutrient concentrations. It is also related to the time available for biomass to accrue between flushing flows that dislodge periphyton from the streambed, to temperature, and, to the amount of light available (Snelder et al. 2013).

In developing the National Objectives Framework (NOF) for periphyton, consideration was given to natural variability by incorporating two factors in setting the objectives: one was the time period between freshes that can remove periphyton, and the other was the degree of natural nutrient-enrichment associated with the geology of the catchment. The periphyton objective is defined in terms of an abundance threshold measured as chlorophyll *a* and a maximum exceedance frequency. In general, the objective is for exceeding the threshold for no more than 8% of samples taken monthly over a three-year period. For dry climates (i.e. with longer periods between freshes) and for naturally nutrient-enriched catchments (i.e. catchments with soft sedimentary geology or volcanic soils) threshold exceedance is allowed for 17% of samples (Snelder et al. 2013). However, while this approach accommodates natural variability, threshold limit values that can only be assessed after 3 years of data. This represents a poor basis for a trigger mechanism for management intervention or regulatory action. Limits are only enforceable when they are timely, clear and certain.

Another influence on setting contaminant loads is discussed by Norton in his Section 42A Report for the proposed Hurunui and Waiau Regional Plan: the volume of streamflow is important in the dilution of nutrient loads (Norton 2012). As the total extraction of water from the river increases, the volume available for dilution decreases; this means that the instream concentrations of nutrients increase. Thus, the nutrient load that the environment can accommodate while still achieving the defined in-river outcome decreases with increasing allocation for extraction.

Algal growth also depends on the balance of nutrients. When the ratio of total nitrogen to total phosphorus (TN:TP) is near 7 then this is an indication that the supply of nitrogen and phosphorus is roughly balanced and co-limiting in relation to algal growth. Single nutrient limitation is more likely to occur when TN:TP is greater than 14 when algal growth will be limited by phosphorus, or when TN:TP is less than 3.5 when nitrogen levels will limit plant growth. For example, Lake Clearwater has a TN:TP of 36, indicating phosphorus limitation; however, in terms of contaminant thresholds TN is in NOF attribute band C (fair) while TP is in NOF attribute band B (Kelly et al. 2014). Nevertheless, phosphorus reductions would be a preferred nutrient management intervention to reduce plant and algae growth even though nitrogen is closer to the NOF limit.

Current catchment contaminant loads are not the only potential driver of algal blooms. There can also be contributions from legacy issues. In Te Roto o Wairewa/ Lake Forsyth algal blooms occurred when there were high levels of phosphorus from a combination of external catchment loading and phosphorus released from the sediment in the lake bed under conditions thought to be due to low redox conditions from lake stratification and photosynthetically induced high pH in the water column (Waters and Webster-Brown 2016).

Another limitation of managing to limits is when there is a time lag between the activity being managed and the effect on the environment. This can often occur with nitrate contamination of groundwater from land use intensification (Norton 2012). The ability to manage activities with a long lead time before adverse effects are evident is problematic.

Because of natural variability and multiple sources there are advantages in having independent means of identifying the likely sources of any increases in contaminant levels. Brown highlights in the case of the Hurunui catchment the need to monitor nutrient loss rates from farms, tributary loads and mainstem loads to identify likely sources of any increases (Brown 2014b). This means that in order to identify whether management and/or enforcement intervention is warranted managing to contaminant load limits is not sufficient. While managing to a specific load limits can be a useful component, it is also necessary to examine the overall system to develop management interventions to achieve sustainability. There is value in considering the adaptive cycles for nutrients to provide a basis for establishing appropriate management interventions. As an example of this approach the management of algal blooms in rivers is set out below.

13.2.4.3 Adaptive Cycles as the Basis for Management

For algal blooms in rivers the key phases of the adaptive cycle at the catchment scale are:

- Exploitation: the nutrient loss from land use intensification;
- Accumulation: the catchment load from accumulation of nutrient losses from properties in the catchment;
- Disturbance: the nutrient concentrations in surface runoff and seepage to groundwater;
- Reorganisation: any attenuation of nutrients prior to flow reaching the streambed.

At the scale of the streambed for periphyton growth the key phases of the adaptive cycle are:

- Exploitation: the nutrient levels, light availability and temperature for periphyton growth;
- Accumulation: the accrual period for biomass accumulation;
- Disturbance: the occurrence of algal blooms
- Reorganisation: either recovery through the occurrence of freshes to dislodge periphyton from the bed of the river or invertebrate grazing to reduce periphyton biomass, or, ongoing degradation with algal blooms.

The nesting of these adaptive cycles is depicted in Fig. 13.3.

The main thrust of introducing "managing to limits" is to reduce the impact of land use intensification on surface and groundwater quality, and, to reduce water extraction effects on streamflows and associated freshwater ecosystems or on groundwater levels and the associated spring-fed streams. In nested adaptive system terms with respect to managing for sustainability, this is a focus on reducing the pressure on the resource and is addressing the exploitation phase of the adaptive cycle at the catchment scale (Sect. 11.3.1). However, it is apparent from the limitations of "managing to limits" noted above and from examples of sustainability analysis elsewhere in this book that a more comprehensive approach to achieving sustainability is needed.

Other types of management interventions were identified in Sect. 11.3.1 that were directed at other phases of the adaptive cycle: addressing legacy issues at the accumulation phase, increasing resilience of the system at the disturbance phase, and rehabilitating adverse effects at the reorganization phase. These types of interventions are consistent with Chapin's four approaches to fostering sustainability: reduce vulnerability, enhance adaptive capacity, increase resilience, and enhance transformability (Sect. 4.2.8 and 11.3.1).

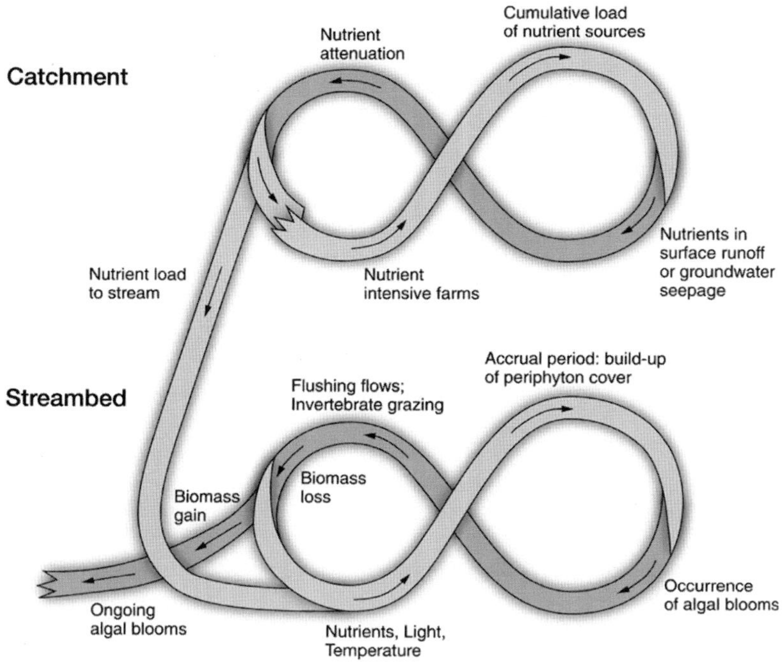

Fig. 13.3 Nested adaptive cycles for algal blooms in rivers

In the example of management of periphyton in rivers there are a suite of management interventions to achieve the delivery of sustainable outcomes at the catchment scale and the streambed scale. Table 13.1 sets out the phases of the adaptive cycle for the catchment scale and the streambed scale for periphyton in rivers. The table also indicates the critical variables and provides examples of management interventions for each phase. This includes managing to limits: firstly, nutrient loss rates relate to the catchment exploitation phase (nutrient intensive farming) with improved farm management practices to reduce loss rates as an example of a management intervention; secondly, catchment contaminant loads relate to catchment accumulation phase of the cumulative effect of diffuse and other nutrient sources with the specification of catchment nutrient limits as a management intervention; and thirdly, instream nutrient concentrations relate to the streambed exploitation phase with nutrient concentration limits as a management intervention.

The use of the nested adaptive systems framework also illustrates other avenues for management intervention. At the catchment disturbance phase (contamination of surface runoff and groundwater seepage) examples of management interventions are riparian planting for surface runoff and woodchip bioreactors for groundwater to reduce nutrient concentrations in runoff and groundwater respectively. At the catchment reorganization phase of nutrient attenuation, constructed wetlands are a possible management intervention to increase the critical variable of catchment attenuation factor.

Table 13.1 Adaptive Cycle Phases and Management Interventions for Periphyton in rivers

Adaptive cycle phase	Critical variable	Example of management intervention
Catchment exploitation: Nutrient intensive farming	Nutrient loss rates	Improved farm management practices to reduce loss rates
Catchment accumulation: Cumulative load of nutrient sources	Catchment contaminant load	Catchment limit on contaminant load
Catchment disturbance: Contamination of runoff and groundwater	Nutrient concentration in surface runoff and groundwater seepage	Riparian planting Woodchip bioreactors
Catchment reorganisation: Nutrient attenuation	Nutrient attenuation factors	Constructed wetlands
Streambed exploitation: Nutrient contamination of river	Nutrient concentrations, temperature, light	Concentration limits for nutrients Shading of streambed
Streambed accumulation: Build-up of periphyton	Accrual period between flushing flows	Maintenance of freshes in environmental flow requirements
Streambed disturbance: Potential for algal blooms	Periphyton cover Chlorophyll a level	Public health warnings
Streambed reorganisation: Recovery from algal blooms	Flushing flows Invertebrate grazing	Sediment removal to increase invertebrate habitat

At the streambed scale exploitation phase, in addition to nutrient concentration limits, light and temperature management through vegetation shading is also a potential management intervention. The accrual period (the time between flushing flows) is a critical variable for the streambed accumulation phase of periphyton build-up; this can be managed by environmental flow requirements to maintain freshes. The streambed disturbance phase is the occurrence of algal blooms with chlorophyll *a* levels or periphyton cover as the critical variables. Public health warnings are one form of management intervention. The final phase of streambed reorganization (the recovery from algal blooms) depends primarily on flushing flows and invertebrate grazing to reduce algal biomass. Management interventions include sediment removal to improve invertebrate habitat. Without adequate interventions, algal biomass will continue to grow leading to ongoing stream degradation.

13.2.4.4 Collaborative Rather than Regulatory Approach

The statutory instrument giving effect to the "managing to limits" approach in the National Policy Statement is a regional plan. While statutory backing is appropriate for aspects of the nested adaptive systems approach to management, a collaborative rather than a regulatory approach is more likely to facilitate sustainability outcomes.

Community engagement is implicit in the Objective CA1 for the National Objectives Framework that freshwater objectives recognize regional and local circumstances. As noted above with natural variability, multiple causes of stress, and scientific uncertainty influencing the measurement of contaminant loads, it is not always clear whether management intervention is warranted, and if warranted, what form of intervention is appropriate. Furthermore, with some limits based on percentage exceedances over multi-year records regulatory intervention is problematic.

However, this does not prevent agreement on setting limits and adopting management interventions. One example is the initial agreement among the Hurunui-Waiau Zone Committee members for nitrogen and phosphorus limits for the Hurunui catchment even when the uncertainties around the estimates were made explicit (Sect. 3.3.6). As demonstrated in the solutions package for South Coastal Canterbury, the proposed limits were not in question and agreement was reached on a combination of freshwater augmentation and nitrogen loss limits to manage the water quality in Wainono Lagoon (Sect. 3.3.7).

Variations in methods for determining limits, as well as different models (and even ongoing adjustments to the same model) make setting regulatory limits problematic. As noted by Brown above – there is not always a perfect method. Furthermore, methods and models get refined over time. Writing methods and models into statutory instruments means there is the potential for frequent changes in the statutory instrument as refinements occur.

Differences in approach and uncertainties in critical variables increase when the impact pathway is long. For periphyton management, there are relationships between farm nutrient loss rates and catchment loads, between catchment nutrient loads and river concentrations, and, between river nutrient concentrations and periphyton levels. The error margins can be large. Maintaining a link between individual discharges (i.e. nutrient losses from farms for periphyton management) and the sustainability outcome (i.e. algal blooms in rivers) is difficult to achieve with regulation based on individual consents but is more feasible with collaborative approaches as demonstrated by the solutions packages developed by Zone Committees.

There is also a socio-economic framework related to management intervention (Sect. 11.3.1) that can be expressed as an adaptive cycle of:

- Use of human, technical and economic resources to address sustainability issues (exploitation phase);
- Accumulation of knowledge, social, cultural and economic capital (accumulation phase);
- Formulation of new approaches that change existing practices (disturbance phase); and
- Development of new institutional arrangements (reorganization phase).

The output of the adaptive cycle is the adoption of management interventions.

In Canterbury, the Zone Committee process is an example where substantial resources have been applied, knowledge/social/cultural/economic capital has been accumulated, leading to new approaches and new institutional arrangements. The innovations of zone solutions packages, farmer collective environmental management systems, and, individual farm environment plans with audited self-management provide a potential means of linking the steps in the impact pathway from land use activity to the delivery of sustainability outcomes.

References

Adamson, B., & Main, M. (1996). *Potential sources of faecal coliform contamination at Coes Ford*. Christchurch: Environment Canterbury.

ANZECC. (2000). *Australian and New Zealand guidelines for fresh and marine water quality*. Canberra: Australian and New Zealand Environment and Conservation Council, and Agriculture and Resource Management Council of Australia and New Zealand.

Aqualinc. (2012). *What is irrigation efficiency?* Irrigation NZ Christchurch, Ministry for Agriculture and Forestry Wellington.

Barrett Purcell & Associates. (1999). *Determining a framework, terms and definitions for water use efficiency in irrigation*. Narrabri: Barrett Purcell & Associates.

Bolton-Ritchie, L., & Robinson, K. (2016). *Water quality monitoring for contact recreation: Summary of the 2014/2015 season*. Christchurch: Environment Canterbury.

Brown, P. (2014a). *Amuri Basin nitrogen load: Compliance with plan limits*. Christchurch: Aqualinc.

Brown, P. (2014b). *Statement of evidence of Peter Derek Brown for Amuri Irrigation Company Dated 20 May 2014*. Ngāi Tahu Forest Estates Ltd consent application CRC144606, Christchurch.

Brown, P. (2016). *Canterbury detailed irrigated area mapping*. Christchurch: Aqualinc.

Canterbury Water. (2009). *Canterbury Water Management Strategy Strategic Framework*. Christchurch: Environment Canterbury.

Canterbury Water. (2013). *Selwyn Waihora ZIP Addendum*. Christchurch: Environment Canterbury.

Canterbury Water. (2015). *Canterbury Water Management Strategy targets: Progress report June 2015*. Christchurch: Environment Canterbury.

Cook, S., Gichuki, F., & Turral, H. (2006). *Water productivity: Estimation at plot, farm and basin scale*. CGIAR Challenge Program on Water and Food.

Editor. (2011, September 20). *Farmers must share burden on emissions*. The Herald.

eLEAF. (2012). *Refined assessment of soil moisture and water consumption on the irrigated Canterbury Plains*. Christchurch: Environment Canterbury.

Environment Canterbury. (2004). *Variation 1 to the Proposed Canterbury Natural Resources Regional Plan*. Christchurch: Environment Canterbury.

Environment Canterbury. (2011). *Natural Resources Regional Plan*. Christchurch: Environment Canterbury.

Environment Canterbury. (2013). *Canterbury Regional Policy Statement 2013*. Christchurch: Environment Canterbury.

Environment Canterbury. (2015). *Canterbury Land and Water Regional Plan December 2015*. Christchurch: Environment Canterbury.

Environment Canterbury. (2016). *The implementation of water metering in Canterbury*. http://ecan.govt.nz/publications/General/LGOIMA-water-metering-072016.pdf. Accessed 27 Aug 2016.

Foran, B., Lenzen, M., & Dey, C. (2005). *Balancing act: A triple bottom line analysis of the Australian economy*. Sydney/Canberra: The University of Sydney and CSIRO.

Ford, S., Butcher, G., Edmonds, K., & Braggins, A. (2001). *Economic efficiency of water allocation*. Wellington: Ministry of Agriculture and Forestry.

Glasgow, K. (2013). *Silverstream catchment report August 2013*. Christchurch: Environment Canterbury.
Hoekstra, A., Chapagain, A., Aldaya, M., & Mekonnen, M. (2011). *The water footprint assessment manual: Setting the global standard*. London: Earthscan.
Irrigation New Zealand. (2013a). *The New Zealand piped irrigation systems design code of practice*. Christchurch: Irrigation New Zealand.
Irrigation New Zealand. (2013b). *The New Zealand piped irrigation systems design standards*. Christchurch: Irrigation New Zealand.
Kelly, D., Robertson, H., & Allen, C. (2014). *Nutrient loading to Canterbury high-country lakes for sustaining ecological values*. Nelson: Cawthron Institute.
Ministry for the Environment. (2002). *Microbiological water quality guidelines for marine and freshwater recreational areas*. Wellington: Ministry for the Environment.
Ministry for the Environment. (2016). *The Resource Management (Measurement and reporting of water takes) Regulations 2010*. http://www.mfe.govt.nz/fresh-water/regulations-measurement-and-reporting-water-takes. Accessed 27 Aug 2016.
Molden, D. (1997). *Accounting for water use and productivity*. Columbo: International Water Management Institute.
Morgan, M., Bidwell, V., Bright, J., McIndoe, I., & Robb, C. (2002). *Canterbury Strategic Water Study*. Lincoln: Lincoln Environmental.
New Zealand Government. (2007). *Resource Management (National Environmental Standards for sources of human drinking water) Regulations 2007*. Wellington: New Zealand Government.
New Zealand Government. (2011). *National infrastructure plan 2011*. Wellington: New Zealand Government.
New Zealand Government. (2014). *National Policy Statement for Freshwater Management 2014*. Wellington: New Zealand Government.
Norton, N. (2012). *Proposed Hurunui and Waiau River Regional Plan Section 42A report: Implications for water quality*. Christchurch: Environment Canterbury.
Norton, N., & Kelly, D. (2010). *Current nutrient loads and options for nutrient load limits for a case study catchment: Hurunui catchment*. Christchurch: Environment Canterbury.
Robinson, K., & Stevenson, M. (2012). *Microbial water quality of the lower Selwyn River/Waikirikiri and Silverstream tributary*. Christchurch: Environment Canterbury.
Scott, D. (2004). *Groundwater allocation limits: Land-based recharge estimates* (U04/97). Christchurch: Environment Canterbury.
Sharma, B., Molden, D., & Cook, S. (2015). Water use efficiency in agriculture: Measurement, current situation and trends. In P. Drechsel, P. Heffer, H. Magen, R. Mikkelsen, & D. Wichelns (Eds.), *Managing water and fertilizer for sustainable agricultural intensification*. Paris: International Fertilizer Industry Association, International Water Management Institute, International Plant Nutrition Institute and International Potash Institute.
Simpson Grierson. (2010). *Case law on limits for freshwater quality and environmental flow*. Wellington: Ministry for the Environment.
Skelton, P., & Memon, A. (2002). Adopting sustainability as an overarching environmental policy: A review of section 5 of the RMA. *Resource Management Law Journal, X*(1), 1–10.
SKM. (2010). *Managed aquifer recharge feasibility study*. Prepared for Environment Canterbury, Auckland.
Snelder, T., Biggs, B., Kilroy, C., & Booker, D. (2013). *National Objective Framework for periphyton*. Christchurch: NIWA.
Sustainability Council of New Zealand. (2015). *New Zealand's climate change: Targets, projections and liabilities*. http://www.sustainabilitynz.org/nzs-climate-change-targets-projections-and-liabilities/
Thenkabail, P., Biradar, C., Turrai, H., Noojipady, P., Li, Y., Vithanage, J., Dheeravath, V., Velpuri, M., Schull, M., Cai, X., & Dutta, R. (2006). *An irrigated area map of the world (1999) derived from remote sensing*. Columbo: IWMI.

Upton, S. (1995). Purpose and principle in the Resource Management Act. *Waikato Law Review, 3*, 17–55.

Vollenweider, R. A. (1982). *Eutrophication of waters: Monitoring, assessment and control*. Paris: Organisation for Economic Co-operation and Development.

Waters, S., & Webster-Brown, J. (2016). The use of a mass balance phosphorus budget for informing nutrient management in shallow coastal lakes. *Journal of Hydro-environmental Research, 10*, 32–49.

Westerhoff, R., & White, P. (2013). *Application of equilibrium water table estimates using satellite measurements to the Canterbury region, New Zealand*. Wellington: GNS.

Young, R. (1996). *Measuring economic benefits of water investments and policies*. Washington, DC: The World Bank.

Young, M. D. (2014). Designing water abstraction regimes for an ever-changing and ever-varying future. *Agricultural Water Management, 145*, 32–38.

Young, M. D., & McColl, J. C. (2009). Double trouble: The importance of accounting for and defining water entitlements consistent with hydrologocial realities. *The Australian Journal of Agricultural and Resource Economics, 53*, 19–35.

Chapter 14
Socio-economic Issues and Collaborative Governance

Abstract Collaborative governance has evolved through four stages in Canterbury: an experimental stage at the tributary and catchment scale, the regional strategy, the development of zone implementation programmes, and an operational stage that has just commenced. As water management moves from strategy to implementation programmes and then operations there is a trend of decreasing dominant spatial scale (from regional to subregional to catchment/irrigation district), increasing formality from (from non-statutory to statutory), and, decreasing scope of decision (from all issues, to ten target areas, to selection of management approach). There are a number of unresolved issues. While solution packages for water quality management devised by Zone Committees will improve water quality compared to current management, they will not achieve desired community water quality outcomes. There is a need for institutional arrangements and funding mechanisms to provide regional infrastructure for improving water quality and achieving infrastructure coordination.

While there has been significant progress against many strategic targets, some have not been achieved. This includes recreational and amenity opportunities, economics of externalities, the economic value-added contribution of water, costs of restoration, and habitat protection for freshwater species and river birds. The failure to address issues that are the prime concern of some stakeholders has undermined trust in the collaborative process and the accountability of the governance of the regional council leading to disengagement of some groups from the collaborative process.

Water management issues in Canterbury have changed since the RMA was designed in the 1980s. The concept of sustainable development has also evolved since then. Changes in legislation and institutional arrangements are needed to address these changes. This includes water framework legislation, regional management strategies, strategic assessments and project authorisations consistent with water strategies. In New Zealand, water charging is based on recovery of financial costs involved in water supply. There is a need for mechanisms to internalize environmental externalities and economic opportunity costs associated with issues such as water quality impairment and inefficient water use. Zone committees have limited recommendations for on-farm mitigation to what they consider "affordable" for farmers. The means of funding solution packages and infrastructure is yet to be addressed.

B.R. Jenkins, *Water Management in New Zealand's Canterbury Region*, Global Issues in Water Policy 19, https://doi.org/10.1007/978-94-024-1213-0_14

The reliance on Section 32 reports on whether regional plans meet the purpose of the RMA and the assessment of environmental effects for proposed activities are not sufficient to achieve sustainable development. There also needs to be strategic assessments involving sustainability appraisal for defining acceptable development.

Keywords Collaborative governance • Infrastructure development and coordination • Legislative framework • Funding of environmental improvements • Tiered assessment

14.1 Socio-economic Issues

This section considers some of the key socio-economic issues in relation to water management in Canterbury. It begins with a description and analysis of the evolution of collaborative governance in Canterbury through four stages: (1) experimental, (2) regional strategy formulation, (3) development of zone implementation programmes, and (4) the commencement of the operational stage. It then examines the deficiencies in infrastructure development and coordination, and after that considers the unresolved issue of funding infrastructure and solution packages. The initial outcomes of operational implementation are examined with a review of CWMS implementation progress and the first audits from the adoption of audited self-management.

The development of the CWMS and the operation of the zone committees has attracted some research interest. The findings of these research evaluations are summarized. An analysis of the Canterbury approach is then made using the "empowered participatory governance" framework of Fung and Wright (2003).

An analysis is provided of the legislation and organizational arrangements. The limitations of the RMA are discussed and a legislative framework described to address the deficiencies of the RMA and the changing role of government in sustainable development. Organisational arrangements to address infrastructure implementation and funding are put forward.

The final subsection considers the issue of economics and addresses the issue of externalities associated with water abstraction and land use intensification, and how these externalities could be addressed. The costs of water quality management at the property and catchment scale are discussed and economic ways of funding catchment interventions described.

14.1.1 *Evolution of Collaborative Governance in Canterbury*

14.1.1.1 The Four Stages of Collaborative Governance

The use of collaborative governance in Canterbury has evolved through several stages (Jenkins 2017). It began with an experimental stage at the tributary and catchment scale before proceeding to a second stage of regional strategy development leading to the Canterbury Water Management Strategy. The third stage was the development of implementation programmes through the Region and Zone Committees. It is now entering a fourth operational stage focused on farmer collectives. The characteristics of these different stages are described below and then compared in relation to their dominant spatial scale, their governance arrangements, communication with stakeholders and the community, approach to decision making, funding, and the objectives they are trying to achieve.

Stage 1: Experimental stage at tributary and catchment level.

With concerns about the inability of the RMA to provide the basis for managing water at sustainability limits and the increasing adversarial nature of court-based decisions, Environment Canterbury began to introduce the principles of Ostrom's "self-governing communities" to address water management issues. One programme was "Living Streams" (Sect. 12.1.1) commencing in 2003 that was targeted at tributary catchments with degraded water quality (Jenkins 2009). It was a council-led process of interaction with the community that had four phases (1) engaging the community and awareness raising, (2) achieving understanding within the community, (3) the community taking actions, and (4) monitoring and reviewing success. The work programme had three stages: firstly, an investigation stage involving data compilation and stream walks to produce a catchment report to identify water and land management issues; secondly, an involvement stage with landowners and community groups to develop an action plan for voluntary projects; and thirdly, an improvement stage of undertaking actions, securing funding, monitoring outcomes and reassessing the need for further action. For example, a 5-year programme in the Pahau catchment including on-farm projects, riparian management projects and irrigation management improvements led to a threefold reduction in bacterial contamination and a twofold reduction in phosphorus concentration in the Pahau River. Participation was voluntary and decisions on actions were made by landowners and community groups. Participants funded the projects, often with financial assistance from the Council's Environmental Enhancement Fund. Monitoring of outcomes was by the regional council.

There was also collaborative catchment management programmes focused on resolving community conflict around water management issues. The programme was often initiated by community concerns being brought to the council's attention. The programme involved the following steps (1) getting stakeholder engagement to define issues and request information, (2) compiling information for stakeholder evaluation, (3) option development in consultation with stakeholders, (4) responding to requests for analysis and means of resolving differences, (5) reaching agreement and negotiating compromises, and (6) where needed, giving statutory backing to the agreements. This was achieved through open public meetings and in some cases with a community steering group. Agreed actions were implemented through funding by key participants and in some cases with council assistance. For example, to address algal blooms downstream of Opuha Dam, an agreement was reached between irrigators, the dam operator, conservationists and fishermen to provide flushing flows from the dam while allowing reduced minimum flows. The flow management actions were undertaken by the dam operator.

Stage 2: Strategy development at the regional level

The success of the collaborative approaches at the tributary and catchment scale noted above and the recognition of the need for community engagement in the forming of a regional strategy led to the Canterbury Water Management Strategy being developed in a collaborative governance framework. (The CWMS was discussed in Sect. 3.2.4 in technical terms and in Sect. 12.1 in relation to community engagement in decision making.) There was recognition of the need for a nested approach with four spatial scales of: the region, subregions related to interconnected catchments and groundwater zones, tributary catchments, and individual properties. However, the focus was at the regional scale and the governance structure was at the regional scale with oversight by the Canterbury Mayoral Forum (the mayors of the city and district councils in the region, the chair of the regional council, and the chief executives of the councils) and a multi-stakeholder group with members selected from across the region.

Reliance on open meetings for community engagement was not logistically possible for a region about 400 km in length and 100 km in width. A programme of structured stakeholder engagement and region-wide community consultation was developed. Decision making was by the multi-stakeholder steering group informed by community input and then endorsement by the Mayoral Forum. Funding of the strategy development was by the regional council with some minor assistance by central government in the latter stages of strategy finalization.

Stage 3: Implementation programme development.

The CWMS defined the governance structure for the development of implementation programmes for the strategy. It was a nested (rather than hierarchical) system with a Regional Committee to recommend programmes relevant to regional issues (such as water storage and distribution across the region) and ten Zone Committees to recommend programmes relevant to subregional issues (such as changes in land use practices to improve water quality).

The Zone Committees are joint committees of the regional, district and city councils in the zone area and each council is represented on the Zone Committee. Rūnanga whose rohe is in the zone area are represented on the committee. Applications are sought for 4–7 community members. Applicants are assessed on skills, expertise and experience as well as their ability to work together to develop water management solutions that deliver economic, social, cultural and environmental values. The community members need to include people with a range of backgrounds and interests in the community. The purpose of the committee is to facilitate community involvement in the Zone Implementation Programme (ZIP) and monitor progress of the ZIP implementation (Canterbury Water undated). The objectives of the Zone Committee include developing the ZIP and overseeing its delivery, as well as engaging stakeholders and ensuring community input to the ZIP. Decisions are by consensus. If consensus cannot be reached, then the committee is to be replaced. There is a code of conduct which defines the operating philosophy for a collaborative, co-operative, participatory and solutions-focused approach by all members (Canterbury Water 2014b). The operations of the Zone Committees are funded by the regional council with contributions from the city and district councils.

The Regional Committee is a committee of the regional council with 2 regional council members, a member of Christchurch City Council, 3 district council members (one from a southern, central and northern district council), one representative from Ngāi Tahu, three rūnanga representatives (one from South, Mid and North Canterbury, 5 to 7 community representatives bringing expertise related to fisheries, energy, biodiversity, agriculture, recreation and regional development, with observers from central government and Canterbury District Health Board. The purpose of the Regional Committee is to monitor progress of CWMS implementation and provide advice on regional issues. It has a similar decision making and operating philosophy as the Zone Committees.

The dominant component of the implementation programme development stage has been at the zone level.

Stage 4: Operational management.

In relation to operational management the focus has been on water quality in rivers and lakes. The main operational elements are having farmers adopt good management practice, setting nutrient contaminant limits with respect to rivers and lakes, linking these river and lake limits to catchment nutrient loads, and, allocating the catchment loads among existing users while trying to create headroom for new users. The primary governance element is the establishment of farmer collectives based on irrigation districts, tributary catchments (or stream allocation zones), or farm enterprises. Collectives need an approved Environmental Management System (EMS) that defines water quality outcomes for the collective consistent with regional plan requirements. The EMS also requires an inventory of nutrient loss rates, identification of the nutrient risks and how those risks will be managed including a statement of best nutrient management practices. The EMS also defines the contractual arrangements with members including a Farm Environmental Plan (FEP) consistent with the EMS, and, how the FEPs will be audited and compliance achieved. The FEP has to address irrigation management, soils management, nutrient management, effluent management as well as wetland and riparian management. The compliance approach is based on audited self-management (Sect. 8.3.4). This includes an audit process of assessing performance against management actions and outcomes at the individual property level.[1] The EMS sets out the record keeping requirements, how audit results will be fed back to members and shared with the wider community[2], and how issues of poor performance are to be managed.[3]

14.1.1.2 Comparison of Collaborative Governance Stages

The four stages are compared in Table 14.1. In terms of scale, the evolution has been from the first experimental stage as small scale (tributary and catchment) addressing specifics issues to the second stage of region-wide strategy looking at multiple issues. For the third stage of implementation programme development, the dominant scale was at the sub-regional zone scale focused on achieving the ten target areas identified in the CWMS. The fourth stage of operational management involved farmer collectives at the tributary or irrigation scheme scale with an emphasis on management practices for water quality management.

The governance arrangements were relatively informal at the first experimental stage with Living Streams based on voluntary council-led informal meetings and slightly greater formality with catchment groups. The second stage strategy development was non-statutory but with increasing formality under the Mayoral Forum (a non-statutory body) and the multi-stakeholder steering group as well as structured stakeholder engagement and community consultation across the region. The CWMS, although non-statutory, has been influential over the last 7 years in framing water management in the region. The third stage of implementation programme development was more formal as the Region and Zone Committees were constituted under the Local Government Act. The RIP and ZIPs, although themselves only advisory documents, led to statutory backing under the Land and Water Regional Plan. Even greater formality characterizes the fourth operational stage associated with the farmer collective approach. This involves the formation of Collectives, setting water quality outcomes, and defining the contents of EMSs and FEPs as set out in the Canterbury Land and Water Regional Plan (Environment Canterbury 2015a).

With respect to decision making and funding, for the Living Streams programme in the experimental stage there was a need for sufficient landowner support for a programme to proceed. Decisions and funding of actions was a voluntary decision of landowners albeit with the possibility of a contribution from the regional council's Environmental Enhancement Fund. Decisions for catchment groups were by consensus among the stakeholders participating, with funding typically borne by the stakeholders with some funding of components by the regional council. For the

[1] Assessing performance is to address the "assurance problem" of voluntary approaches (Sect. 8.3.4).

[2] Sharing data with the wider community is to address the "credibility obstacle" of voluntary approaches (Sect. 8.3.4).

[3] Managing poor performance is to address the "collective action problem" of voluntary approaches (Sect. 8.3.4).

Table 14.1 Four stages of collaborative approaches in Canterbury water management (Jenkins 2017)

Collaborative programme	Dominant spatial scale	Governance	Communication	Decision making	Funding	Objective
Experimental stage						
Living streams	Tributary	Regional council led informal meetings	Voluntary participation	Participant decision on actions taken	Participant assisted by Environmental Enhancement Fund	Water quality improvement in degraded streams
Catchment groups	Catchment	Regional council led regular meetings	Open meetings	Community consensus	Participant and Council funding	Resolution of conflict or community concerns
Regional strategy development						
Canterbury Water Management Strategy	Region	Mayoral forum led. Appointed stakeholder Steering Group	Structured stakeholder engagement. Region-wide community consultation	Steering Group informed by community with Mayoral forum endorsement	Regional Council	Regional water management strategy
Implementation programme development						
Regional Implementation Programme	Region	Regional council appointed committee	Open public meetings	Committee consensus which is subject to RMA processes	Regional council	Regional Implementation Programme
Zone Implementation Programme	Subregional zones	Regional and district council appointed committee	Open public meetings and community engagement	Committee consensus which is subject to RMA processes	Regional and district councils	Zone Implementation Programmes
Operational management						
Farmer Collectives	Tributary or Irrigation District	Collective and farmer audited self-management	Collective	Collective strategy and farm plans to meet Regional Plan limits	Farmer/Collective	Achieve water quality outcomes

regional strategy, the decisions were made by agreement among the multi-stakeholder steering group influenced by community input and endorsed by the Mayoral Forum. Funding of the process and investigations was primarily by the regional council. Similarly funding and staffing for the Region and Zone Committee processes was primarily by the regional council. The recommendations of the Region and Zone Committees on ways to achieve the ten target areas are advisory. The statutory components have to be drafted by the regional council and are then subject to RMA hearing processes (Sect. 2.2). The funding implications of the implementation of the decisions were borne by water users (in relation to land and water management requirements) and by the regional council (in relation to biodiversity programmes and further investigations). For the farmer collectives, the members could define their own governance arrangements within the requirements of the regional plan and could choose the management approaches to deliver the outcomes specified in the regional plan.

There is a clear trend in the collaborative governance arrangements as water management moves from strategy to implementation programme to operational management of (1) decreasing dominant spatial scale (from region to subregional zone to catchment/irrigation district), (2) increasing formality (from non-statutory to statutory), and (3) decreasing scope of decisions (from all issues to ten target areas to selection of management approach).

Even with the decreasing dominant spatial scale moving from strategy to operations, multiple scales from the region to the individual land parcel are relevant to all stages. One of the unresolved issues at the operational scale is how implementation of tasks, in particular infrastructure, that are beyond the scale of the farmer collective will be managed and funded. The solutions packages from the Zone Committee addenda include major infrastructure components at the catchment scale (e.g. augmentation of Wainono Lagoon with high quality Waitaki River water, managed aquifer recharge in the Hinds catchment, a sedimentation basin in the Wairewa catchment, and, constructed wetlands for water quality improvement in the St Leonards catchment). There is not a funding mechanism identified or an implementation agency specified for this infrastructure.

Furthermore the Regional Committee recognized that while new water supply and distribution projects must be economically viable, these infrastructure elements need to be developed in a coordinated way to achieve an integrated regional approach (Canterbury Water 2012). This concept is given statutory support in Policy 4.8 of the Land and Water Regional Plan that "the harvest and storage of new irrigation or new hydro-electricity generation schemes contribute to or do not frustrate the attainment of the regional concept for water harvest, storage and distribution…" (Environment Canterbury 2015a). The Regional Committee also saw the potential for water quality improvements and other benefits through the development of 'environmental infrastructure' such as constructed wetlands and on-farm treatment swales that can be incorporated into water storage and supply networks (Canterbury Water 2012). However, there is not an operational programme for implementing and funding the regional concept.

Another unresolved issue is that the solutions packages being developed by the Zone Committees fall short of the targets defined in the CWMS. Furthermore, while the progress of the implementation of the CWMS has been significant, not all the milestones identified for completion by 2015 have been achieved (Canterbury Water 2015). These unresolved issues of infrastructure development, ability of solutions packages to meet desired outcomes, funding of infrastructure and solutions packages, and implementation progress are discussed further below.

14.1.1.3 Infrastructure Development and Coordination

The issue of infrastructure development and coordination in Canterbury in particular, but also in New Zealand in general, has not been adequately addressed since the abolition of the Ministry of Works and Development in 1988. The concept of the RMA was that people know best what it is that they are after in pursuing their well-being and that responsibility for defining proposals was left to proponents (Sect. 2.1).

However as early as 2002, the Canterbury Strategic Water Study (Morgan et al. 2002) concluded that the future development of Canterbury's water resources would require strategic integrated water resource management and identified there was no agency with the mandate to plan the long-term development of the region's water resources (Sect. 3.2.2).

The Canterbury Water Management Strategy (Canterbury Water 2009) identified the need for a "Water Infrastructure and Services Entity" to facilitate both private sector investment in infrastructure such as storage that demonstrate a reasonable rate of return, and public service obligations to maintain surface and groundwater flows and fund ecosystem restoration projects.

In the strategic review of national infrastructure, the water infrastructure sector ranked poorly in relation to the five guiding principles considered in the review. The sector was considered "ineffective" in terms of investment analysis, funding mechanisms and regulation. It was also considered that the sector "could be further developed" in relation to accountability and performance, and, coordination (New Zealand Government 2011).

The Land and Water Forum recommended that "regional planning on a collaborative basis must occur so that rural infrastructure can be developed in a way that provides a range of social, economic, cultural and environmental benefits" (recommendation 31 (Land and Water Forum 2010)). It also recommended the establishment of a National Land and Water Commission to develop and oversee the implementation of a National Land and Water Strategy (recommendations 37 and 38). One of the components of the Strategy is to identify opportunities for enhancing cultural, economic, environmental and social value in an integrated way from water resources, including water infrastructure development (recommendation 39).

As noted in Sect. 8.2, the implementation and funding of the public good component of water infrastructure is problematic. There is no central government agency to do this. Regional councils have a regulatory rather than a development role under the RMA. The scale and implications of the infrastructure are regional in nature and therefore beyond the boundaries of individual territorial authorities.

The implications for regional councils have been highlighted in the Hawkes Bay where the Hawkes Bay Regional Council (HBRC) was proposing the Ruathaniwha Dam on the Tukituki River in a catchment where the nitrate levels exceed the values set in the council's regional plan. In relation to the consent application for the dam, the council advised the central government Environment Protection Authority: "because the HBRC investment arm is the proponent of the RWSS (Ruathaniwha Water Storage Scheme) it is appropriate that the resource consents are processed by a body that is independent of the Regional Council" (Environmental Protection Authority 2013).

This is not just an issue for New Zealand. The difficulty faced by the Mekong River Commission in the conflict between its development role and basin management role is described in Box 14.1.

Organisation theory recommends that differentiated tasks (in this case the water resource development task and the basin management task) should be placed in

Box 14.1 Role Conflict for the Mekong River Commission

The 1995 Agreement on Cooperative Sustainable Development of the Mekong River basin established the Mekong River Commission (Mekong River Commission 2000). The agreement established a river basin organization with both development and basin management roles explicitly included. As discussed by Campbell the two roles are not compatible: "An organization promoting development is not accepted as an "honest broker" when providing advice on basin management, and, as an honest basin manager is liable to be criticized by the development lobby for insufficiently promoting development projects" (Campbell 2009).

Campbell indicates that the Mekong River Commission resolved this contradiction between the two roles by changing from one to the other under different chief executives. Under the first CEO, Yanasabu Matoba, the work programme was primarily focused on infrastructure projects for investment (Ha 2011). Joern Kristensen, who was CEO from 2000 to 2003, identified the basin management role as the mission for the Mekong River Commission. Under his leadership the organisation produced the first State of the Basin Report (Mekong River Commission 2003b), a social atlas (Hook et al. 2003), and a river awareness kit (Mekong River Commission 2003a). Then Oliver Cogels (CEO from 2004 to 2007) saw the role of the Mekong River Commission in terms of economic development and the emphasis was on projects in hydropower, navigation, fisheries, irrigated agriculture, environmental management and watershed management. Under the next CEO, Jeremy Bird, the MRC again shifted positions with greater consideration being given to the social and environmental impacts related to project implementation (Ha 2011).

separate organisations. However, provision must also be made for organizational arrangements to integrate these tasks with appropriate conflict resolution processes and coordination mechanisms (Lawrence and Lorsch 1969). The issue of changes in organisational arrangements for water resource development and basin management is discussed further below in Sect. 14.1.3.

14.1.1.4 Ability of Solution Packages to Achieve Desired Outcomes

The Selwyn-Waihora Zone Committee in its Addendum to the Zone Implementation Programme (ZIP) which set out a "solution package" for water quality management for the catchment of Te Waihora / Lake Ellesmere, stated: "While the Package is a significant first step it does not achieve all of the Selwyn Waihora ZIP outcomes and continual improvement is needed over time" (Canterbury Water 2013). The Ashburton Zone Committee in its solution package for the Hinds catchment considered: "this package of recommendations will not achieve[4] all of the desired outcomes for the Hinds Plains area but will be a significant first step" (Canterbury Water 2014a). In a report assessing the solution package in the Upper Waitaki ZIP Addendum in terms of its implications for the ecology of rivers and streams, it was stated: "While the ZCSP (Zone Committee Solutions Package) is considered to be an improvement over the current state for recreational water quality and biodiversity protection it is not predicted to fully meet Zone Implementation Plan outcomes" (Gray 2015).

The Banks Peninsula Zone Committee sought a target TLI for Te Roto o Wairewa / Lake Forsyth of 4. Technical experts commented that this target was "ambitious" but the Zone Committee retained the target to provide a "stretch goal". The Zone Committee also noted that: "Currently there are no easy solutions that will immediately remedy the poor water quality of the lake, restore mahinga kai, or ensure no more flooding in the catchment" (Canterbury Water 2014d).

In the South Coastal Canterbury ZIP Addendum, the solution package aimed over time to reduce the TLI of Wainono Lagoon from 6.5 to 6. However while this is an improvement it does not achieve the ecological health threshold for brackish lagoons (Schallenberg 2013). The solution package was dependent on flow augmentation of Wainono Lagoon with high quality water from the lower Waitaki River. The Zone Committee noted that: "If augmentation does not occur, then a plan change will be required along with either another significant intervention or considerably better than Good Management Practice" (for nitrate losses from land use) (Canterbury Water 2014c).

In all cases the solution package will be an improvement in water quality compared to the current management but will not achieve the desired community outcome.

[4] Underlining emphasis in the original.

14.1.1.5 Funding of Infrastructure and Solution Packages

Zone Committees have agreed solution packages to address significant water management issues within their respective zones. Many of these solution packages include infrastructure elements. The Zone Committees have identified that the infrastructure elements will need to be funded and have recommended that funding mechanisms be developed.

For the Wainono Lagoon augmentation, the South Coastal Canterbury Zone Committee makes two recommendations: (1) "The Zone Committee supports the seeking of a one-off national funding contribution towards the cost of augmentation infrastructure", and (2) "Environment Canterbury and Waimate District Council investigate a funding mechanism to support augmentation based on district and regional contributions" (Canterbury Water 2014c). In its solution package for Te Roto o Wairewa / Lake Forsyth, the Banks Peninsula Zone Committee recommended that: "The community, Environment Canterbury and Christchurch City Council work together to develop funding models to support the implementation of these recommendations that include, but not be limited to, council annual and long-term plans and budgets, and community, regional and national funding opportunities" (Canterbury Water 2014d).

In the Selwyn-Waihora ZIP Addendum the Zone Committee concludes its summary with the statement: "A programme of action, including funding, needs to be developed and then put in place to implement this package over the next 20 years" (Canterbury Water 2013). For the Hinds catchment solution package, the Ashburton Zone Committee states: "Funding will be required over the next 20 or more years for the programme of work to implement the solution package. The Zone Committee's focus has been to identify the most significant activity, not on how actions may be funded" (Canterbury Water 2014a).

In the Upper Waitaki ZIP Addendum, the Zone Committee used the Mackenzie Agreement to inform its work. One of the elements referred to is to provide incentives for development areas that are set aside for conservation but the Zone Committee considered that funding for external payments was outside the committee's scope.

The Regional Committee has recommended to the Environment Canterbury Commissioners: "While public funding should be a last resort, there could be a case for Environment Canterbury to provide public funding (through a rate) to contribute to only the public benefit elements of an infrastructure project, if the following criteria were satisfied. The project: (1) Delivers significant, demonstrable ecological, social and cultural benefits over and above the alternatives (including doing nothing); (2) Requires only a one-off capital investment (i.e. other funding mechanisms are appropriate for ongoing activities); (3) Is a cost-effective way to achieve goals; (4) Benefits a group wider than the immediate users (i.e. clear identification of beneficiaries is required); (5) Environment Canterbury should not help underwrite private gain; (6) Contributes to the achievement of other public policies or strategies (if relevant); and (7) Has obtained resource use consents that may be required (including any obligation to avoid, remedy or mitigate effects).

Good investment principles should be followed when assessing the project (including a risk assessment and cost-benefit analysis). In addition, an assessment of the scale of the benefits and the affordability of the project, including the ability of a local community to meet the costs, would help to determine the mix of funding and how to rate (i.e. targeted or regional)." These recommendations appear to be focused on private investment and do not address the environmental infrastructure identified in solution packages. Recent agendas of the Regional Committee have at least included environmental infrastructure as a matter to be addressed.

Environment Canterbury's Long Term Plan 2015–2025 (Environment Canterbury 2015b) includes a regional water infrastructure component with Programme 5 to "partner on actions for environmental restoration, improved water quality, improved water use efficiency and land management practices". The Long Term Plan includes a 30-year investment programme for regional water infrastructure. However, the Long Term Plan states that investment in increased levels of service is mostly related to improving the level of flood protection in the Waimakariri system over the next 5 years. It also states that "this strategy highlights that most of the regional infrastructure is already in place and the level of new infrastructure required over the next 30 years is minimal. A separate issue is keeping of these infrastructure assets in as new condition. The cost of maintenance required to do this more significant." These quotes indicate that no allowance has been made in the regional council infrastructure budget to fund the infrastructure in the solutions packages.

Possible approaches to funding arrangements for solutions packages are considered further in Sec. 14.1.4 below.

14.1.1.6 Collaborative Governance and Accountability

Collaborative governance involves different accountabilities for the achievement of community outcomes compared to regulatory governance approaches (hierarchical governance). Hierarchical governance has clear lines of accountability: the consent holder to the regulatory agency; the regulatory agency to its governing board; and the governing board to the community through public reporting and election processes: this is referred to as "vertical accountability". In collaborative governance arrangements accountability is more diffuse. There are multiple stakeholders with greater involvement in decision making and implementation. While vertical accountabilities may remain, for collaborative governance "horizontal accountability" also exists between consent holders to each other, between consent holders and the community, and between the regulatory agency and the community (Jenkins 2014).

Additional (horizontal) accountabilities are introduced with collaborative governance approaches. Partnership arrangements require mutual accountability to community outcomes at a higher level than the agency's mandate. They also require agency accountability for agreed contributions to community outcomes. Collective water quality management requires agreed actions and accountabilities. Management of constrained water availability needs data on individual and collective takes to compare with river flows. However, the horizontal accountabilities are inherent in managing partnerships, constrained resources and cumulative effects rather than the collaborative governance arrangements themselves.

The CWMS identified goals to be applied from 2010 and set targets for 2015, 2020 and 2040. The regional council reported against the 2015 targets (Canterbury Water 2015). While there was significant progress against many targets, there were also targets which were not achieved. Two target areas were noted where progress was significantly behind the 2015 milestones: recreational and amenity opportunities, and, regional and national economies. There were also other target areas where some of the targets were not being achieved. These are discussed below.

In relation to recreational water quality, a target had been set of 80% of river bathing sites graded as suitable for contact recreation: an increase from 74% graded as suitable in 2009/10. However, the gradings for the 2013/14 monitoring period indicated 67% as suitable. A positive trend in availability and/or quality of fresh water angling opportunities was the target for 2015. This could not be assessed. Another target was for a positive trend in each zone in the availability and/or quality of recreational opportunities following a baseline survey of existing opportunities. Only a baseline survey of kayaking had been completed. A fourth target was to have identified where environmental flows are not met or require change to meet recreational outcomes and implemented actions to rectify. While environmental flows have been set, angling interests are not satisfied with the outcome (Shutt 2013).

With respect to targets for Regional and National Economics, there were to be assessments of regional economic value factors in externalities (e.g. water quality treatment costs, climate change emissions, changed recreational values) and the cost of repair and restoration.[5] No assessments have been made. Also, there was a target to increase the "value added" and employment per unit of water. No assessments have been made.[6]

In relation to the Ecosystem Health and Biodiversity targets there was reasonable progress through the Immediate Steps Biodiversity Programmes. However, there was limited progress in other target areas. For example, the programme for freshwater species and their habitat was only assessed as "started'; protecting and enhancing ecological health of lowland streams showed limited progress; maintaining the upper catchments of Canterbury's alpine braided rivers as largely natural ecosystems and landscapes had not started; and, enhancing and protecting breeding populations of indigenous braided river birds was only assessed as started.

The failure to address issues that are the prime concern of certain groups of stakeholders has certainly undermined the trust in the collaborative process and the accountability of the governance of the regional council. This is discussed in the analysis of the process as empowered participatory governance (Sect. 14.1.2).

---

[5] Note that earlier assessments indicate externality costs equivalent to regional economic benefits (Sect. 10.2.2)

[6] Note the discussion of economic water productivity measures on matters still to be addressed in Sect. 13.1.4.

14.1.1.7 Operational Implementation

Operational implementation is at an early stage. The plan requirements are in place for the Hurunui Waiau Zone (Environment Canterbury 2013b) while the plan change for the rest of Canterbury has been notified (Environment Canterbury 2016). The first round of auditing has been completed for the Amuri Irrigators Environmental Collective (Rutherford 2016), one of the first farmer collectives established in the Hurunui Waiau Zone. The audit looks at each of the six management areas (Amuri Irrigation Company 2015) and gives a high, medium, or low confidence that objectives are being met.[7] These confidence rankings are then used to give an overall audit grade A, B, C or D for the farms. For the collective, 11% were graded A (high confidence that all objectives are being met), 62% were graded B (at least one medium, but the farm is on track to meet Good Management Practice for each management area), 20% were graded as C (at least one medium, but the farm is not on track to meet Good Management Practice for each management area), and 7% were graded as D (at least one low confidence ranking). These results are encouraging that the new compliance process has the potential to be effective; however, it is too early to be convincing that the management of cumulative effects will be achieved.

14.1.2 Reviews of Collaborative Governance Arrangements in Canterbury

14.1.2.1 Evaluations of CWMS and Zone Committees

There have been several evaluations of the approach to collaborative arrangements adopted in Canterbury. The key conclusions from these evaluations are summarized below.

On the basis of interviewing 21 key informants, Lomax and her colleagues provided an assessment about whether the collaborative governance process for developing the Canterbury Water Management Strategy (CWMS) was successful (Lomax et al. 2010). Drawing on the framework developed by Innes for process conditions for producing high quality outcomes during deliberative planning processes (Innes et al. 2007), the conclusions of Lomax and her colleagues are set out in Table 14.2.

Their conclusion was that "the process of crafting the CWMS has been very successful as an exercise in building consensus around a strategic framework to manage a highly contested resource" (Lomax et al. 2010). They also identified the following unresolved tensions which could potentially derail the consensus: (1) the long-term sustainability of intensive, irrigation-based agriculture in Canterbury; (2)

[7] The six management areas are: (1) irrigation management, (2) nutrient management, (3) collected animal effluent management, (4) wetland and riparian management, (5) soils management, and (6) management of 'hotspots'. These are derived from Schedule 2 of the Hurunui Waiau Regional Plan (Environment Canterbury 2013b).

Table 14.2 Extent to which process conditions for deliberation are satisfied by CWMS development process (Lomax et al. 2010)

Process condition	Extent to which condition satisfied
There is a practical shared task	General agreement that not enough water in the right place at the right time
	Frustration with adversarial process, legal battles – must be a better way
All interests are included	People were handpicked to make sure about this
	But very limited representation of women in Stage 3
	Breadth of recreational interests under-represented in some processes and energy sector declined to be involved at first
	Ngāi Tahu – limited involvement at hapū level
	But flexible process allowed groups to join later in process, e.g. energy
	Open invitations to media to several public meetings
	All had opportunities to comment on options, draft, etc.
	Numerous additional informal meetings
The process is self-organising rather than externally controlled.	Initiated by small groups of people
	Became more self-organising to the extent that it passed to the steering group which was set up for that task.
There is high quality, agreed upon information	High degree of satisfaction with scientific and technical knowledge
	Some agreement that uncertainty exists with respect to issues such as groundwater
There is productive dialogue	Produced concrete outputs – sustainability assessment process, fundamental principles, strategy itself, draft targets and now proposals of strategy are being operationalised as zonal committees set up, staff appointed to water executive
There is creative thinking	Novel suggestions for ways to identify community preferences and incorporate into existing statutory processes or develop new statutory processes – details not yet clear
The status quo is challenged	Yes

conflict within the local government sector about institutional arrangements to implement the Strategy; and (3) conflict with central government's short term macro-economic objectives (Lomax et al. 2010).

Salmon looked at three questions in relation to the role of the CWMS multi-stakeholder steering group process: (1) was it democratic? (2) did the process achieve integration of all issues? and (3) did the strategy change institutional incentives and risks? (Salmon 2012).

In relation to the first question, he concluded that: "the overall democratic legitimacy of the CWMS Steering Group process rates highly", but he noted two qualifying factors: the exclusion of economic issues being raised by the public about greater sharing of the economic benefits of the commercial use of water for irrigation, and, the detrimental effects of the enactment of the ECan Act (Sect. 2.4). He also noted that the ongoing willingness of sector representatives to be involved is likely to depend on the development of a convention that elected decision-makers do not substantially change the consensus outcomes of collaborative processes.

Regarding the second question, Salmon stated: "The CWMS Steering Group Process was relatively successful at achieving integration of issues in the policy process. He did identify two unresolved issues: (1) whether the extent of proposed land use intensification across Canterbury was consistent with the restoration of healthy ecosystems in lowland streams and coastal lagoons, and (2) whether the allocation of public water resources to private landowners could result in a fair sharing of economic benefits.

With respect to the third question, Salmon expressed concern that, while acknowledging the Immediate Steps Biodiversity Fund, the strategy did not address economic fairness issues in relation to a charge on water users for community purposes. The issue of externalities was discussed in Sect. 10.1.2 while the issue of internalizing externalities is considered further in Sect. 14.1.4.

Thomas undertook a review of the Hurunui-Waiau Zone Committee process from a feminist political ecology perspective through 42 semi-structured interviews (Thomas 2014). She argued that multiple processes worked to channel particular understandings of nature and facilitated the enclosure of freshwater for economic advantage. This channeling occurred in three ways. (1) the reregulation in Canterbury (i.e. the replacement of the elected councilors by appointed commissioners) removed many democratic rights thereby limiting opportunities for participation in water politics; (2) the devolved collaborative and consensus based water committee was constrained by targets and discourses that determined that more water needed to be enclosed to serve a neoliberal growth agenda; and (3) the community was privileged as a scale democracy with a sense of community belonging inhibiting public debate. Thomas argues that: "consideration of power must be at the forefront of democratic design and uneven power relations need to be engaged with in such a way that multiple understandings of nature and society can be articulated and seen to be legitimate" (Thomas 2014).

Eppel, in a more recent evaluation of 5 years of the implementation of the CWMS through the Regional and Zone Committee processes, identified several issues which have implications for water governance at the national and regional levels (Eppel 2015). She contends: "First under conditions of changing economic development, failure to specify environmental bottom lines can very quickly lead to rapid decline in the quality of fresh water…Second once damage to the environment has been done, through diffuse means as occurred in Canterbury, it is a very complex, costly, multi-actor and long-term project to bring about remediation, if it can be done at all…For other regions, the lesson is that a region-wide view of water resources is needed, one that takes a dynamic view of ecosystem health and its resilience and sustainability, and which recognises that there are bound to be episodic shifts in how water is used and how intensively."

The issues raised in these evaluations are discussed below. A common theme in the evaluations relates to the long-term sustainability of irrigated agriculture (raised by Lomax and her colleagues), whether the extent of proposed land use intensification is consistent with the restoration of healthy ecosystems in lowland streams and coastal lagoons (raised by Salmon), and the effort required to bring about remediation (raised by Eppel). This has been the focus of the "solution packages" being developed in the Zone Implementation Programme addenda (Sect. 3.3.7). As identified above (Sect. 14.1.1), the solution packages will deliver an improvement in water quality but they will not be sufficient to achieve the water quality and ecosystem outcomes sought by the strategy.

The tension between the economic growth agenda of central government and possible constraints on development to meet environmental limits (as raised by Lomax) is related to this theme. The work of the Selwyn Waihora Zone Committee in limiting the degree of estimated cost to farmers in having to adopt more comprehensive packages for nitrate production is a reflection of this tension (Sect. 3.3.6).

This is also related to the unresolved issue of whether the allocation of public water resources to private landowners would result in a fair sharing of economic benefits (as raised by Salmon). This issue is still unresolved. Even though the contributions to economic wealth and assessment of economic value factors in externalities are included as a target area in the CWMS, there has been little work undertaken to address this issue since the adoption of the strategy (Sect. 14.1.1). Resolution of economic issues is discussed further in Sect. 14.1.4 below.

With respect to Eppel's issue about specifying environmental bottom lines, as discussed in Sect. 13.2.4, this is not sufficient to achieve sustainability. Firstly, the court interpretation of the RMA purpose is a weighing of resource use and environmental effects rather than adherence to environmental bottom lines. Secondly, implementing the "managing to limits" approach has significant challenges in achieving sustainable outcomes.

The unresolved tension concerning the conflict within the local government sector about institutional arrangements to implement the strategy (raised by Lomax) appears to have been addressed through the Zone Committees being joint committees of the regional council and the territorial authorities within the zone.

The relevance of power relationships in the design of the collaborative arrangements (as raised by Thomas) was identified in Sec. 10.3.1 as an issue when discussing Israel's water committees with the Palestinian Authority and with Jordan (Box 10.2). It is discussed further in the next section.

14.1.2.2 Analysis as Empowered Participatory Governance

From the practical applications of what they term "empowered participatory governance", Fung and Wright have abstracted three general principles and three institutional design features for effective implementation of empowered participatory governance (Fung and Wright 2003). The collaborative governance process for the Zone Committees to implement the ten strategic targets of the Canterbury Water Management Strategy is reviewed with respect to these principles and design features.

Fung and Wright also recognize ways in which inequalities of power can subvert collaborative governance processes and consider the need for a "countervailing power" to address issues relating to managing imbalances of power. This aspect is also considered for the Canterbury zone committee process (and addresses the issue raised by Thomas).

Empowered participatory democracy is a concept of coordinated decentralisation incorporating three design principles: (1) devolution to local units (like zone committees); (2) centralized supervision and coordination (with zone facilitators and links to regional council resources); and (3) direct links to government decision making (through the zone implementation programmes). The institutional design features are: (1) there needs to be a focus on specific tangible problems (the ten target areas of the CWMS); (2) there needs to be involvement of ordinary people affected by those problems (community members on the zone committees with the role of facilitating community involvement in implementation programme development); and (3) there is genuine deliberation in the development of solutions to these problems (zone committee members are to work in a collaborative and co-operative manner using best endeavours to reach solutions that take account of the interests of all sectors of the community).

According to Fung and Wright (2003), the concept has the potential to produce effective outcomes because the process:

- Involves and empowers individuals close to the points of action who possess intimate knowledge about relevant situations: these individuals know best how to improve the situation;
- Is likely to generate superior solutions compared to hierarchically imposed solutions or decision processes with less deliberative approaches (e.g. voting) where participants have less opportunity to consider alternative solutions more deeply;
- Shortens the distance and time between decisions, actions, effect, observation and reconsideration;
- With multiple groups exploring different approaches and with centralised coordination these approaches can be made available to all groups, thereby enhancing the learning capacity as a whole.

In relation to Canterbury water management, the development of the implementation programmes, in particular the solutions packages, illustrate the involvement of communities within the zone. However, the empowerment of the solution packages is dependent on, firstly, statutory backing through RMA hearing processes where changes can occur (as is the case of catchment nitrogen loads in the Hurunui), and secondly, on funding arrangements that have yet to be devised.

The solution packages reflect a detailed knowledge of the local situation. They are also a more sophisticated approach than that contained in the Natural Resources Regional Plan which was prepared following Schedule 1 of the RMA. While the solution packages achieve significant improvements compared to the current situation, they will not achieve the outcomes sought by the CWMS.

The time frame for moving from zone implementation programme to statutory backing is still in progress but appears faster than the traditional process following Schedule 1 of the RMA.

There is also evidence of learning from processes in other zones. The allocation arrangements for the nitrate contaminant load in the South Coastal Canterbury Zone considered the arrangements developed in the Selwyn-Waihora Zone as a starting point. However, the Selwyn-Waihora arrangements were considered inappropriate for the circumstances facing South Coastal Canterbury so a different approach was adopted (Box 3.1).

Fung and Wright also identify six critical concerns about empowered participatory governance: (1) vulnerability to problems of power and domination in deliberations; (2) limitations of scope of deliberative decisions and actions; (3) capture by well-informed or interested parties; (4) potential for fragmentation of the region into competing zones; (5) participation involves unrealistically high level of commitment; and (6) difficult to sustain over the long term.

In relation to these concerns, there have been issues raised by environmental interests about farmer domination of rural zone committees (item 1), development interests being better resourced (item 3), and, increased irrigated area and infrastructure being given priority while environmental and recreational targets have not been addressed (item 2) (Coalition for Clean Water 2013). Fragmentation into competing zones does not appear to be evident (item 4). A review of the CWMS process highlighted the personal toll of the workloads being experienced by zone committee members (item 5) (Henley 2014a). However, the process has been sustained beyond the initial zone implementation programmes with the development of the solution packages in ZIP Addenda and strong applications for zone committee membership after the initial 3-year appointments (item 6).

In relation to power imbalances, Fung and Wright discuss the need for a "countervailing power – a variety of mechanisms that reduce, and perhaps even neutralise, the power advantages of ordinarily powerful actors". They distinguish collaborative countervailing power from adversarial countervailing power. Adversarial countervailing power relates to interest groups seeking to advance their interest in government processes. An example is the submitters to Schedule 1 and consenting processes of the RMA (Table 2.1) which enable interest groups to oppose or support plan provisions or development proposals in hearing and court processes. In contrast, collaborative countervailing power seeks to solve problems across multiple interests trying to identify commonalities rather than differences.

Fung and Wright believe that collaborative governance without an appropriate form of countervailing power is likely to fail for the following reasons: (1) groups well organized for adversarial processes can oppose collaborative processes which are seen as risky, costly and demobilizing; (2) collaborative processes can favour entrenched, previously organized or concentrated interests by limiting the range of issues or reducing the influence of collaboration to mere advice that can be heeded or ignored; or (3) concentrated or entrenched interests will more ably advance their interests over others unless countervailing forms of power mitigate these general advantages (Fung and Wright 2003).

Fung and Wright suggest possible pathways for establishing collaborative countervailing power. One is the redeployment of adversarial groups into collaborative groups. This could be through local adversarial organisations or through national adversarial organisations. This does require the ability to operate at local scales (rather than larger political scales), develop competence in problem solving and implementation (rather than advocacy and mobilizing support), and, a cognitive frame of openness to exploring complex causes of problems (rather than solidarity for a particular cause). Another pathway can be through political leaders who view participatory collaboration as good politics as well as good policy. They support policies that create venues for participatory collaboration to reap the democratic and technical benefits.

In relation to water management in Canterbury there are some adversarial groups that became active as collaborative participants in the development of the Canterbury Water Management Strategy and were initially supportive of the zone committee process. However, some, such as the Water Rights Trust and Fish & Game, have now withdrawn from participation in zone committees and reverted to adversarial approaches.[8] An important event in this change was in the Hurunui Waiau Zone when agreement reached by the Zone Committee on nutrient limits was circumvented by the appointed Environment Canterbury Commissioners. In the proposed regional plan change to give statutory backing to Zone Committee agreements, the Commissioners decided to increase the nitrogen limits after lobbying from the dairy industry and without consulting the Zone Committee (Sect. 3.3.6). The situation has been exacerbated by the unwillingness of the Environment Canterbury Commissioners to meet with some of these interest groups (in effect closing the pathway for collaborative countervailing power through political leaders). Also, the limited progress on targets relating to the concerns of these groups has led to their disenchantment with the process.

As pointed out by Henley: "the CWMS is built on a social contract. Arguably the most important part of the CWMS is the targets. It was the commitment to the targets which was the single most powerful force in bringing disparate groups together to support the Strategy at its inception" (Henley 2014b).

The lack of engagement with some environmental interest groups can be contrasted with the extensive engagement with farming industry groups through the Matrix of Good Management project (MGM). The MGM is a collaborative project between Environment Canterbury, primary industries (DairyNZ, Deer Industry New Zealand, New Zealand Pork, Beef + Lamb New Zealand, Horticulture New Zealand and Foundation for Arable Research) and Crown Research Institutes (AgResearch, Plant and Food Research, and Landcare Research) (FAR et al. 2015). MGM is overseen by a cross-sectoral governance stakeholder group. MGM is designed to establish industry-agreed Good Management Practices and to estimate expected nitrogen and phosphorus losses across a range of farming systems, soils and climates operating at Good Management Practice across the Canterbury region. Phase 1 of MGM

[8] In April 2016, North Canterbury Fish & Game produced a report analysing Environment Canterbury's Enforcement of Waterway Protection Rules. The report focussed on (1) slow response times and failure to respond to many complaints, (2) a reluctance to use formal enforcement tools, (3) a lack of consistency in rule enforcement, and (4) poor follow up where breaches had been identified (North Canterbury Fish & Game 2016).

involved the characterisation of current management practices, the development of software to process nutrient loss models, and, the definition and modelling of good management practices. It has delivered a matrix of nitrogen and phosphorus losses for agreed Good Management Practices for various sectors (dairy, horticulture, arable, sheep/beef/deer, and outdoor pigs) across a range of soils and climates (Robson 2015). This work was an important component for the development of Farm Environment Plans and audited self-management for Plan Change 5 to the Land and Water Regional Plan (Sect. 3.3.8). Further work will be undertaken to build the MGM as a tool taking into account changes in farm management practices and improvements in science.

Distrust of collaborative processes has also been compounded by the failure of central government to adopt the outcomes of collaborative processes. One was the Mackenzie Basin Agreement that resulted from a collaborative process initiated by central government to address highly contentious land management issues in the Mackenzie Basin (Upper Waitaki Shared Vision Forum 2013). The process involved 22 groups including farmers, dairy industry, tourism interests, conservation interests, environmental groups and mountain climbing clubs. A shared vision, strategy, and an approach to its implementation were agreed. However this agreement uniting opposing sides to protect the Mackenzie landscape faltered after the Government refused to support some of its key recommendations (Mitchell 2015). Many parties exited the process.

The second collaborative process is the Land and Water Forum. While this forum has been very influential in informing government policy (Sect. 2.4.1), the great bulk of their recommendations remain unimplemented (Land and Water Forum 2015). The first recommendation of its fourth report was that: "The government should complete implementing the Forum's recommendations from its previous reports as soon as possible". Fish & Game, Forest & Bird and ECO (Environment and Conservation Organisations of Aotearoa New Zealand) have now left the Land and Water Forum process.

Considering the zone committee process in light of the issues identified by Fung and Wright, there is evidence of:

- Effective involvement of local interest in developing innovative solutions to zone problems;
- A degree of empowerment but with limitations relating to commissioner override, RMA hearing processes and funding;
- Differential progress in the implementation of targets leading to disenchantment of interest groups related to the targets not being achieved;
- Concerns about power and dominance and the absence of a collaborative countervailing power;
- Disengagement from the process by environmental and recreational interests.

For the long-term viability of the collaborative process there is a need to address the issues of empowerment, differential progress in implementation of targets, the absence of a countervailing power, and re-engagement of environmental and recreational interests.

14.1.3 Legislation and Organisational Arrangements

Institutional arrangements in the form of government authorities and their enabling legislation are designed to address the issues facing society. The Resource Management Act 1991 (RMA) was developed in the late 1980s to address water management issues in New Zealand that are significantly different from the water management issues facing New Zealand today, particularly in the Canterbury region. Furthermore the role and structure of government authorities with the reorganisation that occurred with the introduction of the RMA reflected new philosophies and approaches to support a shift from the welfare state to a society less reliant on government (i.e. neoliberalism) and emerging concepts of sustainable development[9] (Upton 1995).

Since the late 1980s there has been a significant expansion of irrigation; water availability is now an issue for the dry east coast of New Zealand. Water availability is now a significant constraint on further agricultural development. Agriculture which was considered a 'sunset' industry in the 1980s is now the major export earner for New Zealand. With the focus of the RMA of managing the adverse effects of development, point source discharges are now being better managed. The main water management issues are currently associated with water scarcity and the cumulative effects of diffuse sources of land use intensification which were not recognized as significant issues in the 1980s.

14.1.3.1 Limitations of the RMA

The issues facing New Zealand have changed and as noted throughout the book there are limitations of the RMA in addressing these new issues. Management of diffuse sources requires a focus on catchments rather than a focus on new developments.[10] The RMA has separated the management of land (which is the responsibility of territorial authorities) from the management of water (which is the responsibility of regional councils).[11] With water now the limiting resource on further agricultural development[12] there is a need for merit-based allocation that can address issues such as resource productivity, water use efficiency, equity of allocation and nutrient management[13], rather than allocation based on first-come first-served which is suited to an abundant resource.

Management at sustainability limits of resource availability and for cumulative effects needs consideration of system resilience as well as assessment of adverse

---

[9] The World Commission on Environment and Development emphasized the concept of "sustainable development", i.e. development that "meets the needs of the present without compromising the ability of future generations to meet their own needs" (World Commission on Environment and Development 1987).

[10] Refer Sect. 3.3.7 for discussion of solution packages to address significant water quality issues in the various Canterbury zones.

[11] Refer Sect. 2.1 on New Zealand institutional arrangements.

[12] Refer Sect. 3.2.2 on water availability

[13] Refer Box 12.2 for the South African reallocation provisions

effects.[14] An "overall balancing approach" can lead to over-allocation and exceedances of water quality thresholds.[15] This may meet an interpretation of Sect. 5 of the RMA but it is not sustainable management. As a common pool resource, the activities of a new applicant influence the well-being of others and not just the well-being of the applicant. Economic externalities of further development are estimated to be of similar magnitude to the economic returns of that development.[16] For new development there is a need for a demonstration of sustainability to achieve community outcomes rather than a demonstration of effects less than minor: an outcome-based approach rather than an effects-based approach is needed.

Because of scientific uncertainty about acceptable thresholds, prudent resource management requires a precautionary rather than a balancing approach to setting limits. Furthermore, where there is evidence of unacceptable outcomes, the ability to change management requirements (i.e. adaptive management) is needed.[17]

Limiting assessment to environmental effects is constraining in developing a sustainability strategy. It is relevant to consider economic, social, cultural and environmental issues as an integrated package. A development pattern based on applicant-driven proposals is unduely limiting. Infrastructure coordination for both water development and environmental management at the catchment scale can be more cost-effective.[18] This involves a more active role for government than a reactive regulatory role as envisaged in the RMA. It also implies a more collaborative involvement of other water users rather than the adversarial approach in plan making and resource consents that is contained in the RMA.

14.1.3.2 Evolution of Sustainable Development

The concept of sustainable development was in its infancy in the late 1980s when the RMA was being developed. However, since that time the concept has evolved. This book has used the approach of nested adaptive systems which has been part of that evolution. The management of socio-ecological systems (rather than new developments), assessment of system resilience (rather than adverse environmental effects of projects), consideration of multiple scales (rather than project scale), and developing sustainability strategies to address failure pathways (rather than developing project-level mitigation measures to reduce adverse environmental effects) provides a more suitable framework for managing resources at their sustainability limits.

At the international level there has also been an evolution in thinking (Paul 2008). The primary emphasis at the 1972 Stockholm Conference on the Human Environment was on environmental issues. At the 1992 Rio de Janeiro Earth Summit there was a shared focus on environmental, social and economic development. The Millennium Summit in New York in 2000 raised the importance of poverty allevia-

[14] Refer Sect. 13.2.4 on Managing to limits is not enough.

[15] Refer Box 12.1 on Failure of Effects-based Approaches to Achieve Sustainable Development

[16] Refer Sect. 10.1.2 on Externalities of Dairying Conversions

[17] Refer Sect. 13.2.4 on Managing to limits is not enough.

[18] Refer Sect. 14.1.4 on Costs of Water Quality Management

tion. It also led to the multi-dimensional Millennium Development Goals and the Millennium Ecosystem Assessment using the concept of ecosystem services (Millennium Ecosystem Assessment 2005). Most recently has been the adoption in 2015 by the United Nations of Sustainable Development Goals for the year 2030. One set of goals relates to water and covers (a) equitable access to safe drinking water, (b) equitable access to adequate sanitation, (c) improving water quality, (d) increasing water use efficiency and ensuring sustainable withdrawals, (e) implementing integrated water management at all levels, (f) protecting and restoring water-related ecosystems, (g) expanding international cooperation, and (h) supporting and strengthening the participation of local communities in water management (United Nations 2015). These goals are similar to the targets of the Canterbury Water Management Strategy and go beyond the concepts underpinning the RMA.

There has also been a shift in the instruments and action forcing mechanisms for implementing the approach to development. Under the RMA the key mechanisms for managing water resource development are: (1) regional plans which set objectives, policies and rules in relation to the environmental effects of development and (2) resource consents requiring applicants for development to undertake an assessment of environmental effects of proposed developments. There has been the progression into analyzing the environmental effects of policies, plans and programmes with strategic environmental assessment (SEA). For example, SEA has been introduced in Western Australia (Malcolm 2002), European Union (European Commission 2001) and Canada (Government of Canada 2010). California has introduced tiered assessment with the first tier at the programme level and the second tier at the level of the project within a programme. Projects where effects have been adequately assessed at the programme level require no further assessment (Bass and Herson 1994). Arizona in its Groundwater Code (Arizona Department of Water Resources 1980) has identified groundwater basins where water extraction is at sustainability limits as "active management areas' (AMAs). In these areas developers have to demonstrate that water of sufficient quantity and quality is available to sustain the proposed development for 100 years including demonstrating consistency with the AMA's Groundwater Management Plan.

Another significant change has been the European Union Water Framework Directive (European Commission 2000). The Directive commits EU member states to achieve good qualitative and quantitative status for all water bodies. It is based on River Basin Districts defined by hydrological boundaries rather than administrative or political boundaries. For each river basin district, some of which traverse national frontiers, a new "river basin management plan" must be established and updated every 6 years. An EU Directive river basin management plan is a detailed account of how objectives set for the river basin (ecological status, quantitative status, chemical status and protected area objectives) are to be reached within specified timescales. An economic analysis of water use within the river basin must be carried out. Another major element is the public participation requirements because the measures to achieve objectives involve balancing the interests of various groups. These are proactive plans for actions by member states (rather than RMA regulatory style regional plans defining objectives policies and rules for the environmental effects of development) (European Commission 2015).

14.1.3.3 Changing Role of Government in Sustainable Development

As noted in Sect. 2.1, the RMA was designed for the role of government as a regulator of the adverse environmental effects of development activities. As noted above, the issues facing New Zealand have changed since the RMA was introduced and there are significant limitations of the RMA in addressing these issues. Furthermore, the concept of sustainable development has evolved with more appropriate tools for managing resources at or beyond their sustainability limits. While government still has a regulatory role, it also has a role as a facilitator of sustainability strategies. In addition, there are public good infrastructure elements for water quality management and ecological restoration. The public good infrastructure needs to be coordinated and integrated with private sector infrastructure. Funding mechanisms are needed for this infrastructure and an agency responsible for its implementation.

There is a proactive component to government's role in sustainable development. It would be appropriate for the legislative and institutional framework to reflect this changing role. Figure 14.1 sets out a revised planning and authorisation framework to address the changing role of government and the evolving concepts of sustainable development. This comprises:

- Water framework legislation that establishes the range of sustainable development goals for water resources in New Zealand (similar to the EU Water Framework Directive but also providing the basis for addressing the UN Sustainable Development Goals for water);
- The authority through the water framework legislation to develop a national water strategy (as recommended by the Land and Water Forum) and regional strategies (so that the Canterbury Water Management Strategy has a statutory basis);
- The provision for strategic assessments to evaluate the environmental, social, cultural and economic outcomes of programmes associated with a regional strategy (i.e. broader than a strategic environmental assessment and similar to the sustainability appraisal approach of the CWMS);
- Regulatory legislation, an updated RMA, would be retained as the means of managing applicants' proposals for development;
- The regulatory instruments under the RMA (i.e. the National Policy Statements, National Environmental Standards and Regulations at the national level; and, Regional Policy Statements, Regional Plans and Resource Consents at the regional level) would be retained;
- Proactive legislation, an updated Local Government Act (LGA), would be needed as a means of enabling local authority contribution to sustainable water management;
- Regional and zone implementation programmes would have a statutory basis under the updated LGA: the regulatory elements would be incorporated into regional plans while the government actions would be channeled to a project authorization process for funding and implementation under the updated LGA.

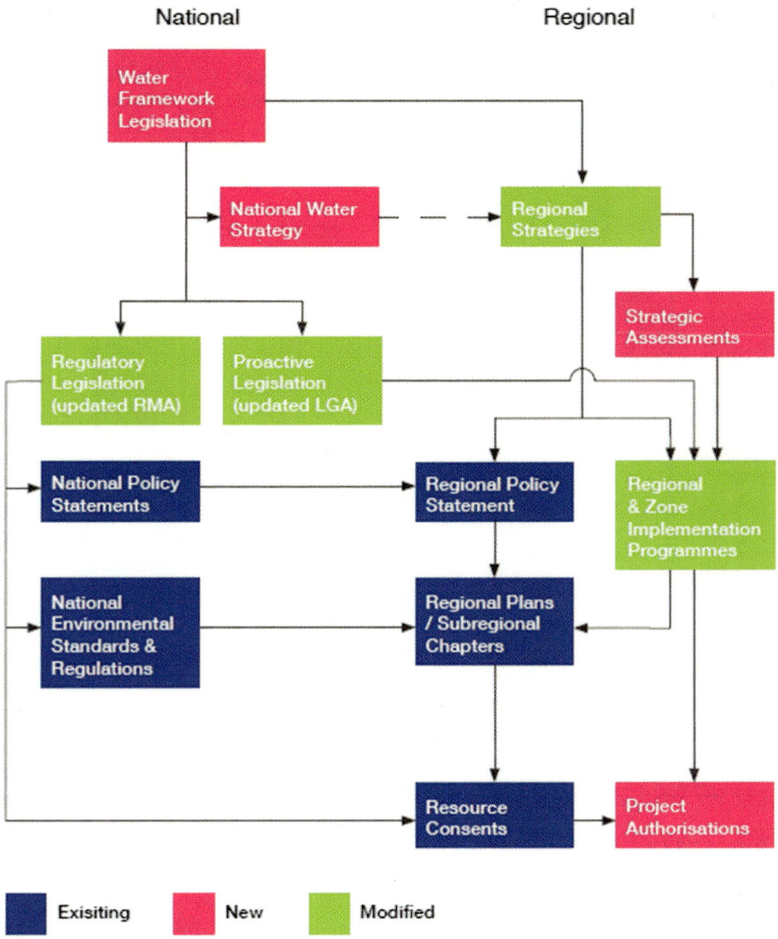

Fig. 14.1 Planning and project authorisation framework

The provision for strategic assessments can also be made available to applicants' proposals. The resource consent process for evaluating proposals creates uncertainty and delay for applicants at the time of high investment risk exposure and limited flexibility to incorporate changes to meet environmental and other sustainability requirements. It is preferable to define environmental and other sustainability design requirements prior to the formulation of applicants' proposals so that they can be incorporated into the proposal design concepts.

Figure 14.2 shows a proposed process where regional strategies are in place and an applicant's feasibility study can be subject to a strategic assessment in relation to the desired results of the regional strategy. The outputs of the strategic assessment are the environmental and other sustainability requirements that the development proposal is expected to achieve. Rather than a full assessment of environmental

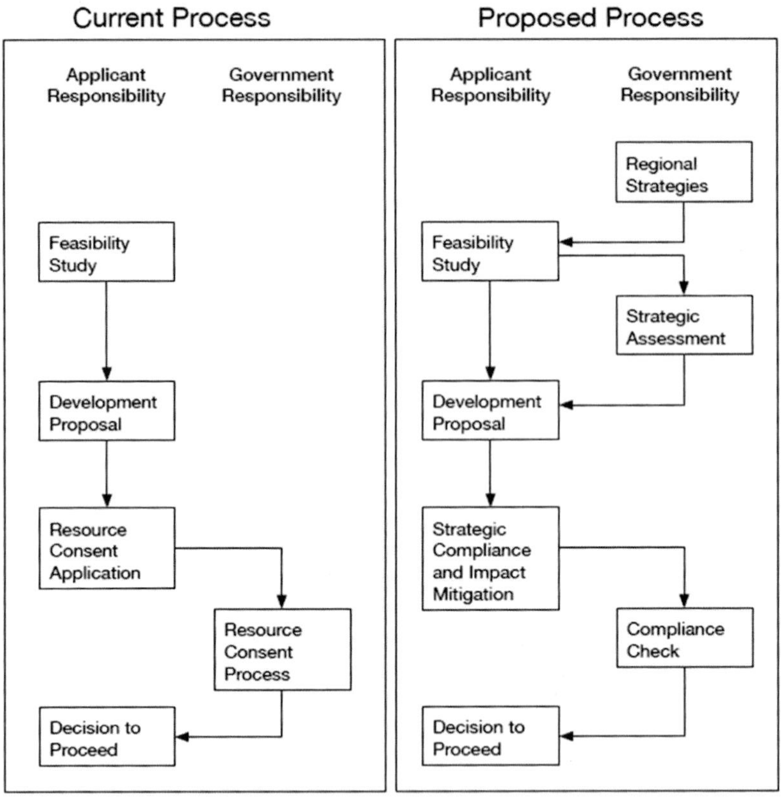

Fig. 14.2 Comparison of current process and proposed process incorporating strategic assessment

effects there is a report on strategic compliance and impact mitigation. There is then a compliance check to determine if the environmental and other sustainability requirements are met.[19] Where these are met a decision to proceed can be made.

14.1.3.4 Infrastructure Implementation and Funding

In the late 1980s the New Zealand Government shifted from a development role in water infrastructure to a regulatory role (Sects. 2.1 and 8.2.1). The also meant there was no agency with the mandate to plan the long-term development and management of the region's water resources (as identified in the Stage 1 strategic study of water availability in Canterbury – Sect. 3.2.2). Furthermore, in the strategic review of national infrastructure, the water infrastructure sector ranked poorly (Sect. 8.2.2).

[19]This is similar to the Californian two-tiered assessment and Arizona requirements for demonstrating sustainability and consistency with the groundwater management plan.

There is a diversity of agencies, some public and some private, with a role in water infrastructure with a policy, design, operation or regulatory role in water infrastructure (Table 8.4).

However, for lake restoration projects there is growing evidence of collaborative organizational arrangements with one-off intergovernmental partnerships and financial agreements (Sect. 11.2.5). In addition, following recommendations of the Land and Water Forum, Central Government has established a Fresh Water Clean-Up Fund to restore waterways affected by historic pollution (Sect. 2.4.2). In relation to irrigation development Central Government has developed an Irrigation Acceleration Fund to support the development of irrigation proposals to the "investment ready" stage and the Crown Irrigation Investments Ltd. for direct capital investment in regional-scale irrigation schemes (Sect. 2.4.2).

In the sustainability analyses throughout this book, a number of infrastructure issues have been identified: (1) major storage; (2) water use efficiency, (3) water quality management, and, (4) climate change adaptation. Minimum cost storage proposals from applicants under RMA processes can have environmental sustainability issues (e.g. the Orari Dam proposal and the Hurunui Water Project – Sect. 3.3.3); or social sustainability issues (e.g. the Waianiwaniwa storage – Sect. 10.3.1 – and the Wrights Road storage – Sect. 3.3.3). However, where collaborative processes have been adopted affordable alternatives have been identified.

The CWMS identified water use efficiency proposals as a more cost-effective means to increase water availability (Sect. 3.3.4). There has been private sector investment in storage at farm scale and irrigation scheme scale, replacement of distribution canals with pipes, and changing from border dyke to spray irrigation. However there has been less attention to soil moisture demand management and spatial reallocation of surface and groundwater to enhance recharge (Sect. 3.3.4).

The solutions packages in the ZIP Addenda identified catchment-based infrastructure components for addressing water quality management issues, e.g. managed aquifer recharge in the Hinds catchment, and augmentation with high quality alpine water for Wainono Lagoon. However, the responsibility for implementation and funding is still to be determined (Sect. 3.3.7).

While identified in the CWMS as a matter needing attention, climate change adaptation has not received detailed consideration in the regional and zone implementation programmes. There are significant implications for water management in the Canterbury region associated with climate change projections (Sect. 7.1.6).

Issues associated with organizational arrangements for infrastructure provision have also been identified in the sustainability analyses. One is the potential conflict between development roles and basin management roles, which is appropriately undertaken by separate organisations but also integrated (Sect. 14.1.1). The issues of appropriate scale and affordability of management interventions for public water supply infrastructure was considered in Sect. 9.3.5. The research indicates there are economies of scale but there is a critical level of output after which scale economies are exhausted. Surveys comparing private versus public ownership are inconclusive. However, in relation to decisions concerning affordability, public control has advantages over privatization. To achieve equity in water supply and municipal wastewater treatment the advantages of larger scale were evident with the improvement in Akaroa water supply (Box 9.1) after the small Banks Peninsula District Council was amalgamated with the larger Christchurch City Council, while the disadvantages of small scale were evident with the affordability issues facing Hurunui District Council with improving drinking water supplies because of the small rating base (Sect. 9.3.1).

New Zealand experience indicates that the public sector hasn't always been successful in providing water resource development that is commercially viable (Sect. 8.2.1). Furthermore, integration of private sector irrigation can be achieved as indicated by the Barrhill Chertsey scheme (Box 8.2). Provision of water supply infrastructure by adequately sized local government appears to be effective. This does not appear to require political amalgamation, only technical integration as evidenced by the Wellington Water example (Sect. 9.3.5).

Catchment-wide public good infrastructure for sustainable flood management appears achievable by the regional council (Sect. 7.4). There are also provisions for community acceptance and control of funding for flood control works: firstly there is the requirement to form a river rating district which requires a 60% majority of people in the proposed district[20]; secondly a decision on the level of expenditure is needed from a river rating district liaison committee (of elected landowner representatives); and thirdly funding decisions are included in the council's annual plan that is subject to a public submission and hearing process. This means the beneficiaries of the flood control works determine whether works are undertaken. It also means the beneficiaries determine the balance between the level of protection provided and the cost of protection.

Funding agreements between all levels of government, Ngāi Tahu and industry have been achieved for Canterbury lake restoration projects, Whakaora Te Waihora and Wainono Lagoon. However, the level of funding committed is well below the solutions packages that have been identified for both lakes, e.g. the current commitments to Whakaora Te Waihora are $12million whereas the solutions package is estimated at $190million. Furthermore, the solutions packages are not sufficient on their own to achieve the desired water quality outcomes for these and other lakes (Sect. 14.1.1).

For climate change management, no infrastructure or funding provision has been made. An example of a strategy implementation where this issue has been addressed is the Long Term Sustainability Plan for the Great Barrier Reef on the north Queensland coast of Australia (Australian Government and Queensland Government 2015). The development of the plan – Reef 2050 – has followed a collaborative approach very similar to the CWMS (Box 14.2). Reef 2050 Long Term Sustainability Plan considers actions to address the combined cumulative effects of climate change, land use change, land-based runoff and direct impacts. An important component on the Reef 2050 was an investment framework to assess the current funding commitments and additional funding needs to deliver the actions over the next 5 years to meet

---

[20]Typically, river rating districts involve a targeted rate on direct beneficiaries dependent of the level of flood risk reduction to their properties and a uniform rate to indirect beneficiaries related to damage reduction to community infrastructure and reduced economic and social disruption.

Box 14.2 Reef 2050 Long-Term Sustainability Plan for the Great Barrier Reef

The Great Barrier Reef Outlook Report 2014 concluded that in the southern two thirds of the reef system that cumulative impacts (especially from climate change, land-based runoff, coastal land-use change and some direct uses) have resulted in deterioration of its ecosystem health and world heritage values (Great Barrier Reef Marine Park Authority 2014a). The Reef 2050 Long-Term Sustainability Plan provides an overarching strategy for the management of the Great Barrier Reef to address these impacts (Australian Government and Queensland Government 2015). The Plan was developed by a Partnership Group that was jointly chaired by the Queensland and Commonwealth Environment Ministers and brought together representatives of government, Traditional Owners, key industry organisations, scientists and interest groups. It was guided by two complementary strategic assessments: one for the marine component (Great Barrier Reef Marine Park Authority 2014b) and one for the coastal component (Department of State Development Infrastructure and Planning 2014).

The Reef 2050 Plan establishes a vision for the Reef and "Outcomes" which are to be achieved by 2050 to deliver the vision. "Actions" have been identified to achieve "Targets" for 2020 (a 5-year horizon) with five-yearly reviews to achieve "Objectives" by 2035 (the medium term) linking to the 2050 Outcomes.

The Reef 2050 Plan states that implementation will be underpinned by a robust investment framework that (1) establishes current investments in reef protection, (2) determines investment priorities for the future, and, (3) sets out a strategy for boosting investment and diversifying its sources. The Investment Framework prioritises the actions in the Plan based on expert advice and stakeholder input, identifies the priority areas where additional funding is most needed, and develops strategies for mobilizing private sector and philanthropic investment. It provides a comprehensive view of the additional investment required, where it is needed and strategies for how to get there. In the next 5 years the current investments of $1.3 billion have been committed and further funding needs ranging from $143 million to $408 million have been estimated to meet the targets for 2020 (Australian Government and Queensland Government 2016).

There is an Intergovernmental Agreement between the Australian and Queensland Governments and a Great Barrier Reef Ministerial Forum that oversees the Plan's implementation with annual reports on progress. The Reef Trust has been established to provide innovative targeted investment focused on improving water quality, restoring coastal ecosystem health and enhancing species protection in the Great Barrier Reef region. There is a Reef Trust Joint Steering Committee which comprises senior government representatives

(continued)

Box 14.2 (continued)
including the Great Barrier Reef Marine Park Authority. There is a Reef 2050 Advisory Committee to facilitate engagement with the broader community and industry, and consists of senior representatives from key industry and community bodies to provide strategic advice on implementation. There is also a Reef 2050 Plan Independent Expert Panel to advise on funding priorities, with eminent experts from a number of scientific fields.

the targets for that period on the first 5-year step in the journey to achieve the outcome sought in 2050 (Australian Government and Queensland Government 2016).

There is a need in the CWMS implementation for an investment framework to deliver the actions from the solutions packages and climate change adaptation in order to achieve the next steps in the journey for sustainable management of water in the Canterbury region.

There is also a need for an infrastructure coordination entity which brings together central, regional and local government, irrigation companies and hydro generators to integrate water infrastructure related to major storage, regional water use efficiency, water quality management, and climate change adaptation. The role would be at least that of the Great Barrier Reef Partnership Group for strategic coordination. It could also facilitate co-governance, co-management, and funding agreements for specific issues which so far have been limited to isolated examples, e.g. Te Waihora / Lake Ellesmere with the Te Waihora Co-Governance Agreement between the regional council and Ngāi Tahu, and the funding agreement for Whakaora Te Waihora between central government, Ngāi Tahu, regional council, Selwyn District Council, Fonterra, Lincoln University and the local community.

There would also be value in the technical integration of the city and district councils water supply and wastewater infrastructure to achieve the benefit of economies of scale.

The regional council could undertake basin management infrastructure for water quality management without creating a conflict of interest with commercial water resource development or its regulatory role. Basin management projects would need to be part of a regional strategy produced on a collaborative basis, subject to strategic assessment and incorporated in a regional or zone implementation programme (refer Figs. 14.1 and 14.2). In terms of funding of basin management infrastructure, a mechanism like the Murray Darling Basin Salinity Strategy (Box 12.2) is needed rather than the flood control works approach. Water quality management infrastructure needs to achieve the water quality outcome defined by the collaborative strategy which determines the required infrastructure and associated costs. The distribution of costs is appropriately based on a polluter pays basis (rather than the flood management approach where the beneficiaries can reasonably determine the level of protection and associated costs). The funding mechanism could be achieved by the financial contribution provisions of the RMA with the approach specified in a regional plan.

14.1.4 Economics

14.1.4.1 Internalising Externalities

In Sect. 10.1.2 it was identified that the externalities associated with the expansion of dairying were comparable with the net economic returns (i.e. benefits less costs) that the expansion was estimated to achieve. It was also identified from the CWMS investigations that increasing efficiency from existing water allocations was more cost effective than storage (Sect. 3.2.5). The risks with establishing markets based on setting caps and trading allocations were identified in the Lake Taupo nitrogen management approach (Sect. 11.2.4). The nitrogen cap appears to have been set too high so that the water quality objectives will not be met. Furthermore, the nitrogen reductions were mostly achieved by government purchase of farming properties rather than trading in nitrogen discharge allowances resulting from reductions achieved by better farming practices. The limited volume of water trading through Hydrotrader in Canterbury is an indication that while trading assists in water allocation it is unlikely to achieve a major shift in allocation to more productive uses (Jenkins 2015).

From an economic perspective, the full cost of water should be borne by the user.[21] However in New Zealand, a government charge should be set at no more than the amount necessary to recover costs, unless the government authority is expressly authorized to do otherwise. Setting a charge that recovers more than the costs of providing the goods or services could be viewed as a tax. Unless expressly authorised by statute this could breach the constitutional principle that Parliament's explicit approval is needed to impose a tax (Office of the Auditor-General 2008).

Thus, for the commercial use of water from public water supply, charges are limited to the financial costs associated with the construction, operation and maintenance of the infrastructure associated with water supply. For private water supplies, such as most irrigation schemes, these financial costs are borne by the scheme owners. A resource consent is required but under the RMA (section 36) a consent applicant can't be charged more for a water resource consent than the cost of administering it (Guerin 2006).

However, there are provisions for territorial authorities (but not regional councils) to charge a development contribution under the Local Government Act (section 106) and for local authorities (including regional councils) to charge financial contributions under the RMA (section 108) (Office of the Auditor-General 2013). Development contributions are primarily a funding mechanism for the portion of new infrastructure that is related to growth. Development contributions can be charged for network infrastructure (including water, wastewater, and stormwater), reserves, and community infrastructure.

[21] The EU Water Framework Directive in Article 9 states that Member States shall "take account of the principle of recovery of the costs of water services, including environmental and resource costs, having regard to the economic analysis according to Annex III and in accordance with the polluter pays principle".

The intention of financial contributions under the RMA is to promote sustainable management of natural resources in terms of the purpose of the RMA. For financial contributions to be imposed, the purpose and level of contribution have to be specified in a Plan or proposed Plan. Financial contributions can be taken to provide off-site mitigation to offset adverse effects of development. They are not widely used by regional councils.[22] Financial contributions tend to focus on the direct marginal impact of the effects of particular developments without considering the wider cumulative impact of multiple developments (The RMA Quality Planning Resource 2012).

Furthermore, the Resource Legislation Amendment Bill 2015 provides for only one charging regime for infrastructure by phasing out financial contributions under the RMA and instead using development contributions under the Local Government Act. It would still be possible to offset environmental effects but only if volunteered by the applicant (Ministry for the Environment 2015).

This can be contrasted with the approach in Europe where full cost pricing is part of the EU Water Framework Directive (Article 9) (European Commission 2000). Full cost pricing not only includes the capital charges and operating and maintenance costs but also includes opportunity costs, economic externalities and environmental externalities (Agarwal et al. 2000). Economic analysis is to take into account the principle of recovery of costs of water allowing for long term forecasts of supply and demand and infrastructure investments. Costs are to include environmental and resource costs in accordance with the polluter pays principle. This includes funding of preventative or remedial measures to the objectives of the EU Water Directive.

14.1.4.2 Costs of Water Quality Management

As noted in Sect. 14.1.1, the ZIP Addenda for the various zones in Canterbury have been identifying solutions packages as the initial steps in addressing deteriorating water quality in the respective zones.

The economic analysis and water quality modelling as inputs to the Zone Committee decision making have provided useful information on the costs associated with achieving water quality outcomes in the Canterbury region. The example of the Hinds catchment in relation to the costs of on-farm mitigation for reducing water quality impacts and the use of managed aquifer recharge to offset nitrate concentrations are considered below.

The current load for the Hinds catchment is calculated to be 4500 tN/y. It is estimated that land use change associated with increased intensive dairying activities

[22] Environment Canterbury has no financial contribution provisions. Bay of Plenty Regional Council has financial contributions for land use change in the catchments of Rotorua Lakes for the purpose of reducing nutrients in the lakes, e.g. for properties with septic tanks that are not connected to a reticulated sewerage system or have not upgraded to an Aerated Wastewater Treatment system with Nutrient Reducing capabilities require a resource consent and may be required to pay a financial contribution to remedy or mitigate the effects of the septic tank discharge (Bay of Plenty Regional Council 2014).

could further increase the catchment load to 5600 tN/y. This would lead to a nitrate concentration in shallow groundwater of about 14 mgN/L. This is well over the chronic toxicity levels for most aquatic species and exceeds the New Zealand Drinking Water Standard of 11.3 mgN/L (Canterbury Water 2014a). The national bottom line for ecosystem health for rivers for nitrate is 6.9 mgN/L (annual median) (New Zealand Government 2014) and this was set by the Ashburton Zone Committee as the water quality objective for shallow groundwater which is the source for groundwater-fed streams in the Hinds catchment.

To achieve this nitrate concentration target, options for on-farm mitigation of nitrate leaching and for dilution through managed aquifer recharge were investigated. Four levels of on-farm mitigation were analysed: (1) GMP – good management practices, (2) AM1 – Advanced Mitigation level 1 (e.g. soil moisture monitoring to manage irrigation), (3) AM2 – Advanced Mitigation level 2 (e.g. covered feed pads), and (4) AM3 – Advanced Mitigation level 3 (e.g. reducing stocking rates and fertilizer application rates). Eleven different farm systems were analysed for different soil and rainfall conditions (Everest 2013). Three levels of managed aquifer recharge were analysed: 0 m^3/s, 2.5 m^3/s and 5 m^3/s. Different levels of irrigation expansion were compared to the current baseline (48,000 ha irrigated): an increase of 15,000 ha and an increase of 30,000 ha (Scott 2013).

To achieve the combination of economic and environmental objectives for the Zone, the Zone Committee's favored option was for 30,000 ha of irrigation expansion, for the dairy and dairy support farming systems (i.e. the major contributors to nitrate discharges) to implement at least AM1, and for 5 m^3/s of managed aquifer recharge (Canterbury Water 2014a). The water quality modelling indicated that the water quality target could be achieved with AM3 without the need for aquifer recharge. However, the Zone Committee did not consider AM3 as a viable option based on the economic modelling. Net Profit After Tax (NPAT) was the key indicator: NPAT incorporates both operational profitability and capital investment in mitigation measures. While operational profitability can be maintained with advanced mitigation measures, increased farm infrastructure means that NPAT reduces as mitigation levels increase (Everest 2013).

Drawing upon the data from economic and water quality modelling, Fig. 14.3 shows an example of the loss in net profit after tax (in $/ha) for one farm system (dairy farm system 2: 3.4 cows/ha with a mixture of irrigation types) associated with different levels of mitigation. The results are plotted against the modelled nitrate levels in shallow groundwater without managed aquifer recharge and with 30,000 ha of irrigation expansion.

As shown in the figure, AM3 mitigations achieve a modelled nitrate level in shallow groundwater of 5.2 mgN/L. This is below the 6.9 mgN/L water quality target set by the Zone Committee consistent with the national bottom line for nitrate toxicity in streams. However, for Dairy Farm System 2, AM3 mitigation comes at an estimated loss in NPAT of $776/ha (compared to a current estimate of $835/ha, i.e. a 93% reduction).

The Zone Committee considered that the threshold of affordability for most dairy farmers was AM1 mitigation. As shown in the figure AM1 mitigation would

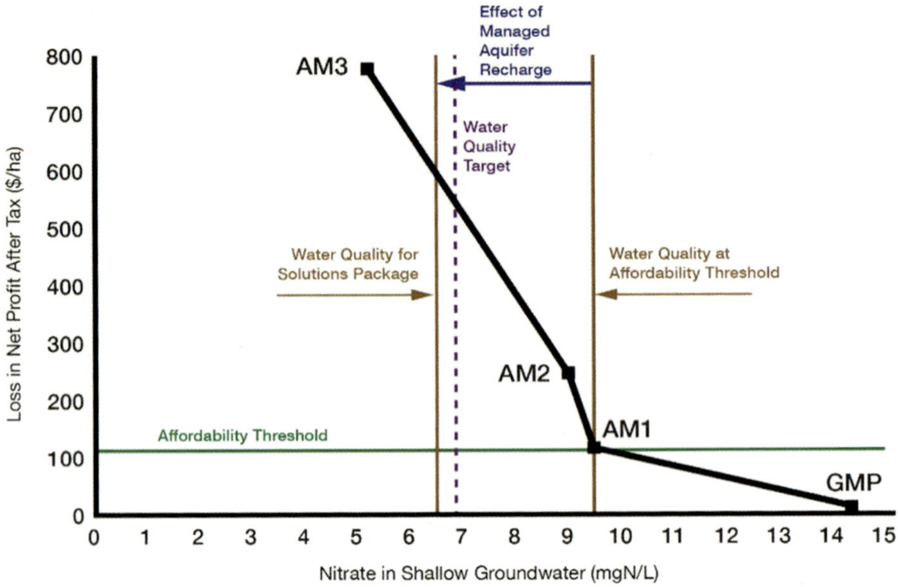

Fig. 14.3 Cost of mitigation measures to achieve nitrate reductions

result in an estimated loss in NPAT of \$116/ha, or 14% of the current practice NPAT of \$835/ha. However, this only achieves a modelled nitrate level in shallow groundwater of 9.5 mgN/L. The figure shows the further reduction in nitrate levels achieved by incorporating 5m^3/s of managed aquifer recharge – modelled to be 6.5 mgN/L.

The capital cost of managed aquifer recharge has been estimated to be \$1.2 m (Environment Canterbury 2013a). For an irrigated area of 75,000 ha, this represents an average cost of \$16/ha. In economic terms MAR is a cost-effective way of achieving the water quality target. However, the Zone Committees have not addressed the issue of how the solution packages will be funded.

14.1.4.3 Funding of Environmental Infrastructure

In the subsection above on internalizing externalities, it was noted that there is only limited provision in New Zealand legislation to fund environmental infrastructure related to the cumulative effects of multiple developments. Furthermore, there are proposed amendments to the RMA to limit the scope of financial contributions to those volunteered by applicants.

However, there are international examples of successfully managing water quality issues associated with diffuse sources with funding mechanisms to support catchment level interventions and encouraging on-farm mitigation. Box 12.2 summarises the approach to salinity management in the Murray-Darling Basin in Australia. In terms of environmental economics, the Murray-Darling Basin salinity strategy uses a "mitigation cost method" (Pascual and Muradin 2010). Dischargers of the contaminant are charged the marginal cost of the next cheapest catchment level intervention to achieve or maintain the environmental target. This encourages on-farm mitigation where the cost of on-farm mitigation is less than the marginal cost of the catchment level intervention.

Maintaining the contaminant accounts (i.e. the salinity registers in the Murray-Darling Basin example) is the responsibility of the States. The costs of catchment mitigation are the responsibility of the States and recouped through contaminant charges (i.e. salinity levies) imposed on water use licences and based on the cost per EC unit [23].

In the New Zealand context, the cost of catchment interventions, such as managed aquifer recharge in the Hinds catchment, could be the responsibility of a water infrastructure agency with the cost recouped through environmental contributions as a condition of the resource consent for water use.

14.2 Evaluation

14.2.1 Multiple Bottom Lines and Multiple Actors

At present the main evaluation tools for water resource management are the regional plans, especially the Section 32 reports designed to examine the extent to which plan objectives and provisions meet the purpose of the RMA; and, resource consents, especially the assessment of environmental effects (AEEs) for activities subject to a resource consent application.

For sustainable development, there is a need for considering multiple bottom lines and the functions and activities of multiple stakeholders and agencies in any evaluation. AEEs are limited to the consideration of environmental effects of proposed activities. Section 32 reports involve a more comprehensive coverage of issues involving the benefits and costs of environmental, economic, social and cultural effects of the provisions of a regional plan. However regional plans relate to the functions of the regional council. In relation to water this is primarily associated with the taking and use of water and the discharge of contaminants. As demonstrated by the issues addressed in this book, a broader range of considerations is needed for sustainable development.

The CWMS sustainability appraisal not only considered multiple criteria relating to environmental, economic, social and cultural capital, but also sought the achievement of sustainable bottom lines for all evaluation criteria. It identified desired objectives for all evaluation criteria as well. It was not a trade-off between benefits and costs for different evaluation criteria. Furthermore, it considered the activities of multiple stakeholders and not just the functions of the regional council.

[23] EC is a unit of measurement for Electrical Conductivity. 1 EC = 1 μS/cm, measures at 25 °C used as an indicator of water salinity (salt concentration).

The concept of strategic assessment of regional strategies as set out in Sect. 14.1.3 and Fig. 14.1 is needed to achieve sustainable development.

14.2.2 Sustainability Appraisal

Section 12.3 described the concept of sustainability appraisal tailored to New Zealand requirements. This concept is seen as the next generation of approaches to evaluation that has evolved from project level environmental impact assessment (Jenkins 2016). Some of the key points of difference are:

- The focus on maintaining and improving environmental, economic, social and cultural capital rather than assessing the acceptability of adverse environmental effects;
- The aim of achieving sustainable outcomes for all evaluation criteria rather than trading off benefits and costs;
- The involvement of stakeholders as well as technical experts in the evaluation process rather than just assessments by technical experts;
- The consideration of socio-ecological systems and the integrated assessment of different evaluation criteria rather than independent environmental, economic, social and cultural assessments;
- The evaluation occurs as part of the selection of a sustainable strategy rather than the evaluation of a specific proposal;
- Sustainability outcomes are the benchmark for the assessment of changes, rather than assessing against the current situation as the benchmark: this is particularly relevant where the current situation is not sustainable
- It is not only a development proposal that is evaluated but also the development proposal along with existing and proposed developments so that cumulative effects are addressed.

14.2.3 Tiered Assessment for Sustainability

One of the significant challenges when resources are at sustainability limits is how to evaluate further development without exacerbating the extent of adverse effects. When the cumulative effects of existing development are at or exceeding the acceptable limit then any additional adverse effect cannot be acceptable. The case of Central Plains Water irrigation scheme (Box 12.1) is an example where allowing further development makes an already unsatisfactory situation in terms of water quality even worse.

The concept of tiered assessment as set out in Sect. 14.1.3 and Fig. 14.2 has an initial strategic assessment which indicates the types of activities that can proceed or the conditions under which activities would be acceptable. This facilitates a more streamlined process for approvals of development proposals consistent with the strategic assessment. The evaluation becomes a compliance check against the requirements defined by the strategic assessment. This means that proponents of activities know in advance the requirements that new proposals need to meet and can determine in advance of the assessment of their proposal whether an activity is viable. It would still be possible to evaluate development proposals that are inconsistent with the strategic assessment through the traditional means of impact assessment but with an extremely low expectation of approval.

References

Agarwal, A., de los Angeles, M., Bhatia, R., Cheret, I., Davila-Poblete, S., Falkenmark, M., Gonzalez, V., Jonch-Clausen, T., Ait Kadi, M., Kindler, J., Rees, J., Roberts, P., Rogers, P., Solanes, M., & Wright, A. (2000). *Integrated water resources management*. Stockholm: Global Water Partnership and Technical Advisory Committee.

Amuri Irrigation Company. (2015). *AIC irrigation scheme management plan*. Culverden: Amuri Irrigation Company.

Arizona Department of Water Resources. (1980). *Groundwater management code 45-576*. http://apps.azsos.gov/public_services/Title_12/12-15.pdf

Australian Government and Queensland Government. (2015). *Reef 2050 long-term sustainability plan*. Canberra: Commonwealth of Australia.

Australian Government and Queensland Government. (2016). *Reef 2050 plan investment framework*. Canberra: Commonwealth of Australia.

Bass, R. E., & Herson, A. I. (1994). *Successful CEQA compliance: A step-by-step approach*. Point Arena: Solano Press Books.

Bay of Plenty Regional Council. (2014). *On-site effluent treatment regional plan newsletter No 9 December 2014*. Whakatane: Bay of Plenty Regional Council.

Campbell, I. C. (2009). *The Mekong: Biophysical environment of an international river basin*. New York: Elsevier.

Canterbury Water. (2009). *Canterbury Water Management Strategy Strategic Framework*. Christchurch: Environment Canterbury.

Canterbury Water. (2012). *Canterbury Water Management Strategy regional implementation programme including annex*. Christchurch: Environment Canterbury.

Canterbury Water. (2013). *Selwyn Waihora ZIP Addendum*. Christchurch: Environment Canterbury.

Canterbury Water. (2014a). *Ashburton ZIP Addendum: Hinds Plains area. Environment Canterbury, Christchurch*. Ashburton: Ashburton District Council.

Canterbury Water. (2014b). *Canterbury Water Management Strategy Zone Committees: Code of conduct*. http://ecan.govt.nz/publications/General/code-of-conduct-all-committees-2014.pdf

Canterbury Water. (2014c). *South Coastal Canterbury ZIP Addendum*. Christchurch: Environment Canterbury.

Canterbury Water. (2014d). *Wairewa ZIP Addendum*. Christchurch: Environment Canterbury, Christchurch City Council.

Canterbury Water. (2015). *Canterbury Water Management Strategy targets: Progress report June 2015*. Christchurch: Environment Canterbury.

Canterbury Water. (undated). *Zone and regional committees terms of reference*. http://ecan.govt.nz/publications/Council/cwms-tor-oct.pdf

Coalition for Clean Water. (2013, August 28). *Letter to Environment Canterbury Commissioners*.

Department of State Development Infrastructure and Planning. (2014). *Great Barrier Reef Coastal Zone Strategic Assessment: Strategic assessment report*. Brisbane: Department of State Development, Infrastructure and Planning.

Environment Canterbury. (2013a). *Groundwater replenishment using Managed Aquifer Recharge (MAR): Catchment-scale potential options for water quality and quantity – Hinds Plains Zone*. Christchurch: Environment Canterbury.

Environment Canterbury. (2013b). *Hurunui and Waiau River Regional Plan operative December 2013*. Christchurch: Environment Canterbury.

Environment Canterbury. (2015a). *Canterbury Land and Water Regional Plan December 2015*. Christchurch: Environment Canterbury.

Environment Canterbury. (2015b). *Long-term Plan 2015–2025*. Christchurch: Environment Canterbury.

Environment Canterbury. (2016). *Plan Change 5 to the Canterbury Land and Water Regional Plan*. Christchurch: Environment Canterbury.

Environmental Protection Authority. (2013). *Recommendation of the Environmental Protection Authority: Tukituki catchment proposal*. Wellington: EPA.

Eppel, E. (2015). Canterbury Water Management Strategy – 'a better way'. *Policy Quarterly, 11*(4), 49–57.

European Commission. (2000). Directive 2000/60/EC of the European Parliament and of the Council establishing a framework of community action in the field of water policy. *Official Journal of the European Communities, 43*(L327), 1–73.

European Commission (2001) *Directive 2001/42/EC of the European Parliament and of the council of 27 June 2001 on the assessment of the effects of certain plans and programmes on the environment.* Official Journal L 197, 21/07/2001 P. 0030 – 0037.

European Commission. (2015). *Report on the progress in implementation of the Water Framework Directive programme of measures*. Brussels: European Commission.

Everest, M. (2013). *Hinds catchment nutrient and on-farm economic modelling*. Christchurch: Environment Canterbury.

Foundation for Arable Research, New Zealand Pork, DairyNZ, Beef+Lamb New Zealand, Horticulture New Zealand, Deer Industry New Zealand. (2015). *Industry-agreed good management practices relating to water quality*. Christchurch: Environment Canterbury.

Fung, A., & Wright, E. (2003). *Institutional innovations in empowered participatory governance*. London: Verso.

Government of Canada. (2010). *The cabinet directive on the environmental assessment of policy, plan and program proposals*. Ottawa: Privy Council Office and Canadian Environmental Assessment Agency.

Gray, D. (2015). *Waitaki limit setting process, predicting the consequences of future scenarios in the Upper Waitaki rivers and streams: Water quality and ecology*. Christchurch: Environment Canterbury.

Great Barrier Reef Marine Park Authority. (2014a). *Great Barrier Reef outlook report 2014*. Townsville: Great Barrier Reef Marine Park Authority.

Great Barrier Reef Marine Park Authority. (2014b). *Great Barrier Reef strategic assessment: Strategic assessment report*. Townsville: Great Barrier Reef Marine Park Authority.

Guerin, K. (2006). *Principles for Royalties on Non-Mineral Natural Resources in New Zealand*. Wellington: NZ Treasury.

Ha, M.-L. (2011). The role of regional institutions in sustainable development: A review of the Mekong River Commission's first 15 years. *Consilience: The Journal of Sustainable Development, 5*(1), 125–140.

Henley, G. (2014a). *Review of limit setting process in four zones – Hurunui, Selwyn/Waihora, Hinds, and Southern Coastal and Streams*. Wellington: Network.

Henley, G. (2014b). *ZIP delivery programme review*. Wellington: HenleyHutchings.

Hook, J., Novak, S., & Johnston, R. (2003). *Social Atlas of the lower Mekong basin*. Phnom Penh: Mekong River Commission.

Innes, J., Connick, S., & Booher, D. (2007). Informality as a planning strategy. *Journal of American Planning Association, 73*(2), 195–210.

Jenkins, B. R. (2009, March 19–21). *Best practice partnerships at the local government level for the environment*. Paper presented at the Community Boards' Conference, Christchurch.

Jenkins, B. R. (2014, October 30–31). *Collaborative governance and accountability*. Paper presented at the Environment Institute of Australia and New Zealand 2014 Annual Conference, Hobart.

Jenkins, B. R. (2015). New Zealand water pricing. In: A. Dinar, V. Pochat & J. Albiac-Murillo (Eds.), *Water pricing experiences and innovations* (pp. 263–288). Springer: New York.

Jenkins, B. R. (2016). From mitigation to sustainability: The evolution of incorporating environmental factors into development decisions in Australasia. *International Journal of Sustainable Development and Planning, 11*(6), 920–929. https://doi.org/10.2495/SDP-V11-N6-920-929.

Jenkins, B. R. (2017, July 10–14). *Evolution of collaborative governance in Canterbury water management*. Paper presented at the XVI Biennial IASC Conference Practicing the Commons: Self-Governance, Cooperation and Institutional Change, Utrecht.

Land and Water Forum. (2010). *Report of the Land and Water Forum: A fresh start to fresh water*.

Land and Water Forum. (2015). *Fourth report of the Land and Water Forum*. Wellington: Land and Water Trust.

Lawrence, P. R., & Lorsch, J. W. (1969). *Organization and environment: Managing differentiation and integration*. Homewood: Irwin.

Lomax, A., Memon, A., & Painter, B. (2010). *The Canterbury Water Management Strategy as a collaborative planning initiative: A preliminary assessment*. Lincoln Ventures. https://researcharchive.lincoln.ac.nz/bitstream/10182/2463/1/CWMS_Appraisal_Report_20_May_2010.pdf

Malcolm, J. (2002). Strategic environmental assessment: Legislative developments in Western Australia. In S. Dovers & S. Marsden (Eds.), *Strategic environmental assessment in Australasia*. Sydney: The federation Press.

Mekong River Commission. (2000). *Agreement on cooperation for sustainable development of the Mekong River Basin*. 5 April 1995. Phnom Penh: MRC Secretariat.

Mekong River Commission. (2003a). *Mekong River awareness kit: Interactive self-study CD-ROM*. Phnom Penh: Mekong River Commission.

Mekong River Commission. (2003b). *State of the Basin report 2003*. Phnom Penh: Mekong River Commission.

Millennium Ecosystem Assessment. (2005). *Ecosystems and human well being: Synthesis*. Washington, DC: Island Press.

Ministry for the Environment. (2015). *The second phase of Resource Management Act Reform*. Wellington: New Zealand Government.

Mitchell, C. (2015). *Historic agreement to protect Mackenzie Basin falters*. http://www.stuff.co.nz/the-press/news/73581719/Historic-agreement-to-protect-Mackenzie-Basin-falters

Morgan, M., Bidwell, V., Bright, J., McIndoe, I., & Robb, C. (2002). *Canterbury Strategic Water Study*. Lincoln Environmental: Lincoln.

New Zealand Government. (2011). *National infrastructure plan 2011*. Wellington: New Zealand Government.

New Zealand Government. (2014). *National Policy Statement for Freshwater Management 2014*. Wellington: New Zealand Government.

North Canterbury Fish & Game. (2016). *An analysis of Environment Canterbury's enforcement of waterways protection rules*. Rangiora: North Canterbury Fish & Game.

Office of the Auditor-General. (2008). *Charging fees for public sector goods and services*. Wellington: Office of the Auditor-General.

Office of the Auditor-General. (2013). *Using development contributions and financial contributions to fund local authorities' growth-related assets*. Wellington: Office of the Auditor-General.

Pascual, U., & Muradin, R. (2010). The economics of valuing ecosystem services and biodiversity. In P. Kumar (Ed.), *The economics of ecosystems and biodiversity: The ecological and economic foundations*. London and Washington, DC: Earthscan.

Paul, B. D. (2008). A history of the concept of sustainable development: Literature review. *Annals of the University of Oradea Economic Science Series, 17*(2), 576–580.
Robson, M. (2015). *Matrix of Good Management: Overview report.* Christchurch: Environment Canterbury.
Rutherford, A. (2016). *Progress report and update on farm environment plan auditing.* Culverden: Amuri Irrigators Environmental Collective.
Salmon, G. (2012). *Canterbury Water Management Strategy – a case study in collaborative governance.* Ecologic, Nelson.
Schallenberg, M. (2013). *Nutrient loading thresholds relevant to Wainono Lagoon (Canterbury).* Dunedin: Hydrosphere Research.
Scott, L. (2013). *Hinds Plains water quality modelling for the limit setting process.* Christchurch: Environment Canterbury.
Shutt, P. (2013). *Update on areas of concern tabled at AGM.* http://www.stuff.co.nz/timaru-herald/sport/opinion/peter-shutt/9543049/Update-on-areas-of-concern-tabled-at-AGM
The RMA Quality Planning Resource. (2012). *Related laws: Relationship between the Local Government Act 2002 and the Resource Management Act 1991.* Wellington: Ministry for the Environment.
Thomas, A. (2014). *Accessing nature: The battle of the Hurunui River.* Wellington: Victoria University.
United Nations. (2015). *Sustainable Development Goals.* http://www.un.org/sustainabledevelopment/water-and-sanitation/. Accessed 30 Dec 2016.
Upper Waitaki Shared Vision Forum. (2013). The Mackenzie Country: A shared vision and strategy, and a proposal for a Mackenzie Country Trust www.mackenziecountry.org.nz
Upton, S. (1995). Purpose and principle in the Resource Management Act. *Waikato Law Review, 3*, 17–55.
World Commission on Environment and Development. (1987). *Our common future.* Geneva: Report of the World Commission on Environment and Development.

Chapter 15
Concluding Comments

Abstract The RMA is based on the assessment of the effects of proposed development to meet environmental bottom lines. However, when environmental bottom lines have been exceeded proactive measures to address the causes of these exceedances are needed to ensure sustainable development. Managing to limits is not enough to achieve environmental outcomes. Where there are cumulative effects from multiple sources there is a need for a systems-based approach, like nested adaptive systems, to devise management interventions to achieve sustainable resource management.

For the Canterbury Water Management Strategy, improved water use efficiency is a key element. However, a more comprehensive approach to measuring and managing water use efficiency is needed at the farm and catchment scale. While solutions packages have been developed for water quality improvement, they only represent a first step towards sustainable water quality outcomes. Further interventions are needed.

Implementation of actions towards some targets has been significant and there has been considerable collaboration with industry in defining good management practices. However other targets have received insufficient attention, leading to the disengagement of some stakeholders from the collaborative process. Adaptation to projected effects of climate change on reduced water availability and reducing the increasing emissions from agricultural intensification have not been adequately addressed in strategy implementation.

Shifting from proponent-led development to devising alternatives and interventions for improved sustainability outcomes highlights the issue of affordability and funding of the actions to be taken. There is a need for an agency with a mandate for environmental infrastructure and infrastructure coordination and an investment framework for funding proactive measures. There is also the need for water framework legislation which addresses sustainable development of water resources.

Keywords Management based on adaptive cycles • Water use efficiency • Water quality improvements • Affordability • Water framework legislation

B.R. Jenkins, *Water Management in New Zealand's Canterbury Region*, Global Issues in Water Policy 19, https://doi.org/10.1007/978-94-024-1213-0_15

15.1 Implications for Canterbury Water Management

This book has developed and applied a framework for sustainability analysis to water management in Canterbury. The framework was based on nested adaptive systems and collaborative governance. This framework was the basis of the development of the Canterbury Water Management Strategy (CWMS) as well as the basis for the analysis of its implementation and the institutional arrangements for water management in Canterbury. The operational implementation of the CWMS is in its early days. It has generated a paradigm shift in Canterbury water management but further change is needed to achieve sustainability at the operational level. This concluding section looks at some of the key issues from the sustainability analysis that need to be considered if sustainability of water management is to be achieved. These comprise:

- Proactive measures
- Management based on adaptive cycles
- Water use efficiency
- Water quality improvements
- Delivery of the social contract of the CWMS
- Climate change adaptation and greenhouse gas mitigation
- Affordability as a constraint on adaptive capacity
- Implementation and funding of proactive measures
- Introduction of water framework legislation and strategic assessment

15.1.1 Need for Proactive Measures

With water resource abstraction and the effects of intensification from water use at or beyond sustainability limits, reactive measures, such as limiting resource availability or the effects of resource use, are insufficient to achieve sustainable management. Rather there is a need for proactive measures (i.e. management interventions) to achieve sustainable outcomes.

However, many of the legislative and management tools are not designed to facilitate management interventions that address the cause of the problem. The principal instrument for natural resource management in New Zealand, the Resource Management Act (RMA), is designed for government to have a regulatory role. Furthermore, the application of the RMA involves an "overall broad judgement" between further resource use and additional effects when sustainability limits have been reached (Sect. 2.1). Decisions favouring resource use in Canterbury have led to overallocation of water resources and exceedances of water quality criteria (Sect. 12.3.4 and Box 12.1).

The public health approaches for management of waterborne disease are varied in relation to facilitating proactive measures (Sect. 9.3). Water safety plans for drinking water require improvement plans for water supplies for greater than 500 people where there is risk of contamination. Water quality management for contact recreation includes an assessment of risk of contamination from upstream land use and pollution sources. However, when monitored levels exceed public health criteria the guidelines do not specify that the cause of failure be rectified only that the public is informed. Management of cyanobacteria at health threatening levels only requires public warnings. Commercial shellfish management involves a comprehensive hazard analysis but management interventions are limited to closure of harvesting areas and corrective actions for contaminated product rather than addressing the source of contamination.

For sustainability, legislation and management approaches need to facilitate proactive measures that include addressing the cause of the problems.

15.1.2 Management Based on Adaptive Cycles

One of the key elements of the sustainability framework set out in this book is the development of sustainability strategies to address potential failure pathways (Sect. 11.3). As examined in Sect. 13.2.4 for periphyton in rivers the managing to limits for individual pollutant loads to achieve sustainable outcomes is not enough. This is because of load uncertainties, inaccuracies in load estimation, natural variability, multiple variables affecting outcomes, contributions from legacy issues as well as current activities, lag times, unresolved cause-effect relationships, and difficulties in enforcing limits that lack certainty. Furthermore, with multiple geographical scales, many potential points of intervention, and multiple actors, a collaborative approach is needed with agreed accountabilities for delivering a sustainability strategy to achieve sustainable outcomes.

Cumulative effects management does require controls on all pollutant load contributions but catchment level interventions may be more cost-effective than advanced mitigation measures at individual properties. Management approaches that integrate property level, sub-catchment level and river basin level interventions, such as the Murray-Darling Basin salinity strategy, are more appropriate for the achievement of sustainable outcomes.

A systems-based approach like nested adaptive cycles is also needed for the management of the interaction between surface water and groundwater (e.g. the public water supply for Christchurch in Sect. 5.1.2) and the interaction between water quantity and water quality (e.g. the management of algal blooms in rivers in Sect. 13.2.4).

15.1.3 Water Use Efficiency

One of the outcomes of the sustainability appraisal was that the “efficiency-led” option scored above the sustainability bottom line on nearly all criteria. Water use efficiency was defined as one of the ten target areas for the CWMS (Sect. 3.2.5).

Improved water use efficiency of existing users increases water availability without requiring further abstraction and reduces surface runoff and groundwater leakage contaminated by land use intensification.

However, measurement of water use efficiency and setting benchmarks for best practice water use is yet to be achieved.[1] One of the 2015 targets for water use efficiency was for 60% of water used for irrigation to be at best practice. In the absence of estimates of farms at best practice or efficiency based on recorded water use, water use efficiency estimates have been based on irrigation system type (Brown 2016). Section 13.1.4 set out a basis for measurement of water use efficiency and indicators of water use efficiency at the farm and catchment scale.

As water scarcity is a major constraint on agricultural production, there is also value in indicators for physical water productivity (e.g. crop production improvement associated with the volume of water applied) and water footprints (i.e. a measure of water requirements per unit of output); and, for economic water productivity either as economic output per unit of water volume, or, water intensity – the volume of water required to generate a dollar of output.

15.1.4 Water Quality Improvements

The decline in aquatic ecological health and increase in contaminant concentrations in lakes, rivers and groundwater associated with land use intensification in Canterbury was one of the key drivers for developing the CWMS (Sect. 3.1.6). It has also been the focus of attention in the development of the ZIPs and their Addenda (Sect. 3.3.7). The identification of solutions packages for significant water quality issues in the various zones represent important strategies for water quality improvement. However as acknowledged in the ZIPs, the solution packages represent the first step towards achieving the desired water quality outcomes and that further interventions are needed (Sect. 14.1.1).

It was also found from the sustainability analysis of six New Zealand lakes (Sect. 11.2) and for the Silverstream catchment (Box 13.1) that current management interventions led to water quality improvements but not sufficient to achieve water quality outcomes.

Sustainable water quality management in Canterbury needs a greater level of intervention than is currently proposed in the zone implementation programmes.

[1] Good management practices relating to water quality have been agreed (Foundation for Arable Research et al. 2015) but not water quantity.

15.1.5 Delivery of the Social Contract of the CWMS

A key element of the acceptance of the CWMS was the commitment to targets that reflected the breadth of uses and benefits that the community sought from water management in Canterbury (Sect. 12.1.4). However there has been differential progress in the implementation of these targets. In particular, the 2015 targets for recreational and amenity opportunities, ecosystem health and biodiversity, and economic externalities have not been met (Sect. 14.1.1 Collaborative Governance and Accountability).

In addition, the overriding of the Hurunui-Waiau Zone Committee decision in relation to nitrogen caps by the Environment Canterbury Commissioners (Sect. 3.3.6), and the Commissioners' unwillingness to meet with some of the environmental and recreational interests (Sect. 14.1.2) has led to disengagement of these interests from the collaborative process. They consider that the Commissioners are focusing on increased irrigated land and infrastructure, and have effectively changed the priority of the targets as set in the CWMS.

As noted in Sect. 14.1.2, addressing the issues of differential progress of target implementation, empowerment of all water stakeholders, and re-engagement of environmental and recreational interests is needed to ensure the long-term viability of the collaborative process.

15.1.6 Climate Change Adaptation and Greenhouse Gas Emission Reduction

The main implications of climate change projections for Canterbury were summarized in Sect. 7.1.6 of (1) an increase in potential evapotranspiration deficit leading to increased irrigation demand, (2) a decrease in winter rainfall on the Canterbury Plains reducing aquifer recharge and groundwater levels thereby reducing flows in groundwater-fed lowland streams, (3) a drier east coast leading to lower flows in foothill rivers, and (4) a wetter west coast and warmer winters leading to reduced snow and increased winter flows but reduced summer flows in alpine rivers. These projected changes have significant consequences in increasing irrigation requirements and increasing water scarcity in the irrigation season.

New Zealand's response to climate change was summarized in Sect. 7.2. This indicated increasing greenhouse gas emissions and the preclusion of consideration of greenhouse gas emissions by regional councils. Effects of climate change were added as one of the other matters that decision makers "shall have particular regard to" rather than a matter of national significance that decision makers "shall recognize and provide for". Agricultural land use intensification is forecast to represent 77% of the growth in emissions but is not subject to any mitigation requirements despite offsets and mitigation measures being available.

Sustainability strategies for drought management were considered in Sect. 7.3. Adaptation to climate change projections were considered in Sect. 13.1.6. The need to incorporate climate change strategies in regional policies and regional plans as well as zone implementation programmes was identified in Sect. 14.1.3.

With the projected consequences of climate change there is a need for a regional strategy for climate change adaptation and greenhouse gas mitigation.

15.1.7 Affordability as a Constraint on Adaptive Capacity

An important issue identified in the sustainability analysis in the adoption of proactive approaches to addressing failure pathways is the issue of affordability. One instance was the provision of storage to address water availability in the Hurunui catchment (Sects. 6.1.2 and 12.4.3). The adverse effects associated with the lowest cost water availability strategy of a storage on the Hurunui south branch and control gates on Lake Sumner were considered unacceptable. The option of a tributary storage on the Waitohi river was preferred for environmental, social and cultural reasons. The Waitohi tributary storage was then subject to an affordability analysis which placed it in the cost range of current irrigation schemes albeit at the upper end (Sect. 3.3.3).

Another instance is the affordability of the treatment costs for water supply in Cheviot (Sect. 9.3.1). The water safety plan identified failure pathways for contaminants however the financial costs to the community to implement the treatment improvements were considered to be too great. The issue of affordability in relation to the security of water supply was demonstrated by Vorosmarty and his colleagues as a global issue (Vorosmarty et al. 2010). They demonstrated how high-income countries were able to address issues related to the security of water supply by investments in water infrastructure whereas less wealthy nations remain vulnerable because they didn't have the financial resources to afford infrastructure investment.

The issue of affordability was also evident in relation to the solutions packages for water quality improvement in the Addenda to ZIPs (Sect. 14.1.4). While the solutions packages are designed to achieve improvements in water quality they are not sufficient to achieve sustainable outcomes. One of the key considerations was the affordability of improved management practices. Furthermore, the question of funding catchment level interventions is unresolved.[2] This is addressed in the next section.

[2] It is noteworthy in the European Commission report on progress in implementation of the EU Water Framework Directive Programme of Measures by Member States that lack of finance was the most frequently cited obstacle to the delivery of the Programme of Measures (European Commission 2015).

15.1.8 Implementation and Funding of Proactive Measures

While there is a need for proactive measures to achieve sustainable water management, there is no agency with a mandate to undertake such measures. The institutional changes introduced in New Zealand in the late 1980s envisaged that the role of government was to regulate the effects of activities by others (Sect. 2.1). However, the CWMS recognized the need for a "water infrastructure and services entity" (Sect. 8.2.3). Furthermore, the national water infrastructure sector ranked poorly in a strategic review based on investment analysis, funding mechanisms, regulation that facilitates investment, resilience to disruption and changing circumstances, and, accountability and performance (Sect. 8.2.2).

While the private sector can address commercial water resources development it is not well placed to address lake or river restoration, climate change strategies, managed aquifer recharge, biodiversity projects and catchment-wide public good infrastructure. There is a need for a basin management role such as that performed by the Murray-Darling Basin Commission (MDBC) in managing the salinity strategy for the Murray-Darling Basin.

As discussed in Sect. 14.1.3 (Infrastructure Implementation and Funding), basin management projects would need to be part of a regional strategy produced by a collaborative process subject to strategic assessment and with statutory backing. The legislative framework to achieve this was identified in Sect. 14.1.3 to reflect the changing role of government in sustainable development.

The deficiencies in New Zealand's water infrastructure could be addressed in the following way:

- Regional strategies including water infrastructure projects need to be supported by an investment framework such as that produced for the Great Barrier Reef Long-term Sustainability Plan (Box 14.2).
- Funding mechanisms such as the mitigation cost approach of the Murray-Darling Basin Salinity Strategy can be introduced using existing environmental contributions of the RMA (Sect. 14.1.4).
- The water framework legislation provides a regulatory basis for facilitating investment consistent with a regional sustainability strategy (Fig. 14.2).
- A regional sustainability strategy based on nested adaptive systems can address resilience to disruption and changing circumstances (such as climate change).
- There is a need for a more comprehensive accountability and performance framework that incorporates auditing of farm environmental plans and farmer collective environment management systems (Sects. 3.3.8 and 8.3.5), water quality contaminant accounts (similar to the MDBC salinity accounts – Box 12.2), managing to adaptive cycles and not just limits (Sect. 13.2.4) as part of a collaborative governance and accountability framework (Sect. 14.1.1).

15.1.9 Water Framework Legislation and Regional Sustainability Strategies

While the RMA provides a framework for regulation of activities it does not provide a framework for identifying failure pathways and developing sustainability strategies for water resource management. As identified in Sect. 14.1.3, the RMA is not well suited to managing water scarcity and the cumulative effects of diffuse sources from land use intensification.

This can be achieved by putting in place Water Framework Legislation and the requirement for regional sustainability strategies (Fig. 14.2). This approach can also define a streamlined pathway for development that implements an agreed regional strategy (Fig. 14.3). Evaluations of proposed regional strategies can be designed to incorporate multiple bottom lines relating to environmental, economic, social and cultural criteria in the form of sustainability appraisals (Sect. 14.2).

The concepts of sustainable development have evolved since the framing of the RMA. The role of government has also changed. It is appropriate to change the legislative and institutional framework to reflect these evolving concepts of sustainability and the changing role of government.

References

Brown, P. (2016). *Canterbury detailed irrigated area mapping*. Christchurch: Aqualinc.

European Commission. (2015). *Report on the progress in implementation of the Water Framework Directive Programme of Measures*. Brussels: European Commission.

Foundation for Arable Research, New Zealand Pork, DairyNZ, Beef+Lamb New Zealand, Horticulture New Zealand, & Deer Industry New Zealand. (2015). *Industry-agreed good management practices relating to water quality*. Christchurch: Environment Canterbury.

Vorosmarty, C., McIntyre, P., Gessner, M., Dudgeon, D., Prusevich, A., Green, P., Glidden, S., Bunn, S., Sullivan, C., Reidy Liermann, C., & Davies, P. (2010). Global threats to human security and river biodiversity. *Nature, 467*, 555–561.

Glossary

1:1 Flow sharing as river flow increases 50% of the additional flow can be extracted while 50% is specified to remain in the river.

A Block the volume of water that can be withdrawn from a river that meets a specified level of reliability with no material effect on instream values. It is defined by an Allocation Limit (the limit on the aggregate volume of permits that can be granted) and a Minimum Flow (the flow at which taking or diverting water authorized by resource consent must cease).

A1B emission scenario A1B scenario is one of the Special Report on Emission Scenarios (Nakicenovic and Swart 2000). It assumes rapid economic growth and global population that peaks mid-century and declines thereafter, rapid introduction of new and more efficient technologies and a balance of fossil and non-fossil energy sources. This assumes a doubling of global emissions from 1990 to 2050 and declining thereafter.

A1FI emission scenario A1FI scenario is one of the Special Report on Emission Scenarios (Nakicenovic and Swart 2000). It assumes rapid economic growth and global population that peaks mid-century and declines thereafter, rapid introduction of new and more efficient technologies with a technological emphasis on fossil-intensive sources.

A2 emission scenario A2 scenario assumes a heterogeneous world, increasing global population, regionally oriented economic development and slower technological change compared to other scenarios. This assumes a doubling of emissions from 1990 to 2040 and ongoing increases to 2100 (Nakicenovic and Swart 2000).

Adaptive Capacity the capacity to respond to change in the system.

Adaptive Cycle a description of biophysical and socio-economic systems as four phases. The first phase is the "exploitation" phase, which is the use of resources from a biophysical or socioeconomic system. This leads to a second phase of "accumulation" in which a build-up of energy or material results from the exploitation of resources. The accumulation phase can be disrupted by a "disturbance" phase that leads to the release of accumulated energy or material and

B.R. Jenkins, *Water Management in New Zealand's Canterbury Region*, Global Issues in Water Policy 19, https://doi.org/10.1007/978-94-024-1213-0

can potentially change the structure and function of the system. Following the disturbance phase is a "reorganization" phase involving the restructuring of the system. System response can be a recovery of the original system or a shift to an alternative system.

Agronomic drought a deficit of water in the landscape either in the groundwater reserves or in the surface hydrological system such as rivers, streams and lakes.

Allocation block the amount of water either set as a flow rate or set as a volume that is set aside for abstraction where all users allocated a proportion of that water will be subject to the same management controls.

Alpine rivers rivers in Canterbury with their upper reaches in the Southern Alps so that they are snow-fed with summer peak flows.

Anabaena a genus of filamentous cyanobacteria that exist as plankton and are known for nitrogen-fixing abilities.

Anoxia in waters refers to an absence of dissolved oxygen.

Audited Self-Management a system of compliance which involves the company developing their own policies, procedures and plans to achieve environmental outcomes set with the regulator, requires third party certification of their environmental systems and auditing of environmental performance, and, public reporting of performance.

B Block the volume of water above the A Block that can be withdrawn from a river without interfering with the allocation to A block water permits and taking into account environmental flow requirements. It is defined by an Allocation Limit (the limit on the aggregate volume of water that can be granted), a Minimum Flow which is greater than the sum of the A Block minimum flow and allocation limit, and may include other environmental flow requirements to protect flushing flows.

B1 scenario B1 scenario is one of the Special Report on Emission Scenarios (Nakicenovic and Swart 2000). It is based on a convergent world with the global population, that peaks in mid-century and declines thereafter, as in the A1 scenarios, but with rapid change in economic structures toward a service and information economy, with reductions in material intensity and the introduction of clean and resource-efficient technologies. The emphasis is on global solutions to economic, social and environmental sustainability, including improved equity, but without additional climate initiatives.

Common Pool Resource a resource that is readily accessible and difficult to exclude access to, and, is in limited supply so that resource use by one user diminishes the availability for others.

Commitment Package an assemblage of immediate actions (action schemes), exploratory options, deferred choices, and a set of understandings about the way in which any deferred choices should be addressed.

Connectedness the nature and degree of links between processes.

Critical Variables the main issues that are determining the viability of a system.

Depuration immersion of shellfish in tanks of clean seawater to allow contaminants to be purged.

Effective allocation public and commercial/industrial users were assumed to use 100% of their consented daily volume over 365 days. Irrigation users were assumed to require an average of 60% of their consented volume over a 150-day period.

Exponentially-weighted moving average (EWMA) a smoothing statistic giving decreasing weight to data more distant in time.

Failure Pathways processes that can cause system failure and shift the system to an alternative degraded state.

Flocculation the process by which clay-sized particles aggregate into clot-like masses or precipitate as lumps.

Foothill rivers rivers in Canterbury with rain-fed catchments and winter peak flows.

Freshes flows that are sufficient to move fine particles and algal growth and maintain habitat quality and are typically three to six times the median flow.

Grandparenting allocation directly related to historical discharges.

Hapua coast-parallel bodies of predominantly freshwater impounded by a long narrow spit of coarse sediments formed by longshore drift resulting in the river outlet being offset from the main channel alignment.

Headroom the capacity within a catchment to tolerate further nutrient discharges. Headroom is available when the current nutrient load is lower than the load limit. Headroom is equal to the difference between the load limit and current load.

Hydrological drought a deficit of water in the landscape either in the groundwater reserves or in the surface hydrological system such as rivers, streams and lakes.

Irrigation Reliability means, in relation to irrigation, the ability of the water supply to meet demand from one or more abstractors, when operating within the flow and allocation regime or the allocation limits.

Iwi means Māori community or people.

Kaitiakitanga means the exercise of guardianship by the tangata whenua of an area in accordance with tikanga Māori (Māori custom) in relation to natural and physical resources; and includes the ethic of stewardship.

Kondratiev cycles cycles in the world economy with a cycle period of forty to sixty years characterised by four phases of prosperity, recession, depression and recovery which can be related respectively to the exploitation-accumulation, accumulation-disturbance, disturbance-reorganisation, and reorganisation-exploitation sequences of the adaptive cycle.

Land and Water Forum A forum of key stakeholders in water management established to engage directly with each other to find a way forward for water management in New Zealand. The Forum was formed initially with 58 organisations as plenary members with a "small group" of 21 major stakeholders with six active government observers as the main developers of the Forum's recommendations.

Lake Submerged Plant Index (LakeSPI) LakeSPI Index is a synthesis of components from both the native condition and invasive condition of a lake and provides an overall measure of the lake's ecological condition. The Native Condition Index captures the native character of vegetation in a lake based on the diversity and quality of native plant communities. The Invasive Impact Index captures the invasive character of vegetation in a lake based on the degree of impact by invasive weed species.

Lissajous figure a curve formed by combining two perpendicular simple harmonic motions commonly exhibited by an oscilloscope.

Lowland streams rivers in Canterbury that are fed from groundwater.

Mahinga kai means maintaining healthy populations of food species and their habitats.

Managed Aquifer Recharge supplementing natural recharge to an aquifer system under controlled conditions by diversion of water into recharge wells or infiltration of water through the floor of infiltration basins, galleries or river beds.

Mauri means the life-giving essence of a resource.

Meteorological drought The state of the climate system that creates abnormally dry weather, prolonged enough for the lack of rainfall to cause serious hydrological imbalances.

Nested Adaptive Systems adaptive cycles operating at different spatial and/or time scales which are linked.

Ngāi Tahu the Māori tribe whose rohe (tribal territory) includes the Canterbury region.

Nodularia a genus of nitrogen-fixing cyanobacteria that occur mainly in brackish or saline waters.

Picocyanobacteria cyanobacteria having dimensions less than 2 micrometres.

Profile Available Water (PAW) the amount of water potentially available to plant growth that can be stored in the soil to 100 cm depth.

Red zone (groundwater) a groundwater basin where the effective allocation of water exceeds the allocation limit and is considered fully allocated.

Relaying transferring shellfish to another growing area for contaminant reduction.

Resilience the capacity of a system to absorb disturbance and still retain its basic function and structure.

Restorative Justice This is based on a theory of justice that crime is an offence against an individual or community rather than the state. The victims of crime are active in the judicial process. Offenders are encouraged to take responsibility for their actions by apologising, making reparations or community service.

Rūnanga Māori groupings centred on the whanau (family) and hāpu (sub-tribe) of marae (tribal meeting place) based communities.

Socio-ecological systems linked socio-economic and biophysical systems at multiple spatial and temporal scales.

Southern Oscillation Index index based on the pressure difference between Tahiti and Darwin. Negative values (El Nino) are associated with sustained warming of the central and eastern Pacific Ocean.

Strategic Choice a methodology to deal with multi-criteria, multi-stakeholder decision-making characterized by urgency and uncertainty within a collaborative approach.

Stream Depletion Effect the effect of groundwater pumping on reducing the flow in surface waterways.

Sustainability Appraisal a decision tool which provides a means of evaluating alternative approaches in terms of sustainable development. Any sustainability appraisal will have two essential characteristics: (i) an integrated analysis of economic, environmental, and social effects of development proposals or actions, and (ii) an evaluation of their significance against identified principles or criteria for sustainable development.

Tangata Whenua means 'people of the land' in Māori and refers to the roles and interests of indigenous people.

Taonga means a treasure, considered to be of value including socially and culturally valuable resources.

Tenure review the process of reviewing leasehold tenure of some high country land. It involves individual lessees selling their leasehold interest to the Crown and negotiating to buy back freehold title to productive land while the Crown retained land of conservation and recreational value.

Tidal prism the volume of seawater that enters an inlet or estuary during a tidal cycle.

Tikanga Māori means rights, customs, accepted protocol, rule, Māori traditions, lore or law, the correct Māori way.

Tino rangatiratanga political control by Māori people over Māori affairs.

Thermal stratification the process of heating of the upper layer of a lake during spring and summer leading to increased temperature and decreased density in the upper layer. The density difference causes stratification between the upper and lower layers of the lake.

Treaty of Waitangi the treaty first signed on 6 February 1840 by representatives of the British Crown and various Māori chiefs. It resulted in British sovereignty over New Zealand and is generally considered the founding document of the nation. There is a preamble and three articles. The first article addresses Crown sovereignty. The second article addresses Māori rights in land and other resources. The third article guarantees Māori the same rights as other British subjects.

Virtual water the virtual water content of a product can be defined as the volume of freshwater used to produce the product, measured at the place where the product was produced.

Vulnerability assessment the analysis of potential failure pathways for a natural resource system that could threaten its sustainability.

Wāhi taonga means sites of significance.

Waituna-type lagoons shallow lakes that develop landward of barrier beaches with intermittent connection to the sea.

Water footprint the water volumes consumed (evaporated or otherwise not returned) or polluted per unit of time.

Weighted useable area an index of habitat availability based on the wetted area of a stream weighted by its suitability for use by aquatic organisms.

Zoonotic a disease that normally exists in animals but that can affect humans.

Index

B.R. Jenkins, *Water Management in New Zealand's Canterbury Region*, Global Issues in Water Policy 19, https://doi.org/10.1007/978-94-024-1213-0

Printed by Printforce, the Netherlands